Salvia Biotechnology

Vasil Georgiev · Atanas Pavlov
Editors

Salvia Biotechnology

Springer

Editors
Vasil Georgiev
Laboratory of Applied Biotechnology
The Stephan Angeloff Institute
of Microbiology-BAS
Plovdiv
Bulgaria

and

Department of Organic Chemistry
University of Food Technologies
Plovdiv
Bulgaria

Atanas Pavlov
Department of Analytical Chemistry
University of Food Technologies
Plovdiv
Bulgaria

and

Laboratory of Applied Biotechnology
The Stephan Angeloff Institute
of Microbiology-BAS
Plovdiv
Bulgaria

ISBN 978-3-030-08881-1 ISBN 978-3-319-73900-7 (eBook)
https://doi.org/10.1007/978-3-319-73900-7

Printed on acid-free paper

This Springer imprint is published by the registered company Springer International Publishing AG part of Springer Nature
The registered company address is: Gewerbestrasse 11, 6330 Cham, Switzerland

Preface

The genus *Salvia* represents nearly 1000 species that are widely distributed around the world. It is the largest in the *Lamiaceae* family. Traditionally, infusions of *Salvia* species have been widely used to treat oral inflammation, throat and headaches, and digestive disturbances in various folklore- and ethnomedicine practices worldwide. The antispasmodic, antiseptic, and hypoglycemia effects of their extracts have been recognized by the ancient healers long times ago before the development of modern medicine. With the advance in phytochemistry and pharmacology, terpenes, polyphenols, and volatile compounds have been recognized as the source of bioactivity in *Salvia* extracts. Nowadays, because of their valuable pharmaceutical and nutraceutical properties, many *Salvia* species have been widely used as ingredients in food, pharmacy, and cosmetic industries. The economic importance of *Salvia* plants continues to increase, following closely the growing interest to the concept for modern healthy lifestyle, based on prevention by consuming quality foods and nutraceutical supplements of natural origin.

The growing demand has led to overexploitation of natural habitats, and in the last few years, many wild growing *Salvia* species have shrunk or fallen under threat. Obviously, to deal with that problem and to prevent ecological crisis, there is an urgent need for alternative, renewable source of *Salvia* biomass. Plant biotechnology can solve this problem. Plant in vitro culture technology is a powerful method for continuous production of plant secondary metabolites under controlled conditions, recently adapted to various *Salvia* species. *Salvia* in vitro systems are harmless to natural plant populations and can be grown independently of environmental factors, geographical latitude, climatic change, and seasonal variations. Several bioactive metabolites from rare and endangered *Salvia* plants can be produced by employing different plants in vitro systems. However, the achievements in development of large-scale biotechnology, based on *Salvia* in vitro systems, are still in early stage and many points still have to be addressed before the commercialization to take place.

In the book, leading scientists presented recent achievements on distribution and chemodiversity of *Salvia* species (Chapter “European Species of Genus *Salvia*: Distribution, Chemodiversity and Biological Activity”); phytochemistry (Chapters

"Bioactive Constituents of Anatolian Salvia Species and Composition of Aroma Compounds in *Salvia* spp. from Bulgaria and Their Potential Practical Applications"); biological activity of metabolites, ethnobotanical and ethnopharmacological aspects of genus *Salvia* (Chapters "European Species of Genus *Salvia*: Distribution, Chemodiversity and Biological Activity–Anticancer Activity of *Salvia miltiorrhiza* and Its Secondary Metabolites"). The information about *Salvia* in vitro systems as biological matrixes for the production of pharmaceutically important secondary metabolites is summarized as well. Deep reviews are made on production by different types of plant in vitro systems of rosmarinic acid (Chapters "Strategies for Optimization of the Production of Rosmarinic Acid in *Salvia officinalis* L. and *Salvia dolomitica* Codd Biomass with Several Biotechnological Approaches and Biosynthesis of *Salvia* Specialized Metabolites and Biotechnological Approaches to Increase Their Production"), as well as polyphenols and terpenoids (Chapter "Biosynthesis of *Salvia* Specialized Metabolites and Biotechnological Approaches to Increase Their Production"). A special attention is paid on hairy roots of *Salvia* as perspective platform for production of bioactive substances (Chapter "Hairy Roots of *Salvia* Species for Bioactive Substances Production"). Chapters "Molecular Markers in *Salvia* L.: Past, Present and Future and Genetic Engineering and Manipulation of Metabolite Pathways in *Salvia* spp." summarize the recent achievements in molecular biology techniques, applied in *Salvia* research. Finally, in Chapter "Microbial Diversity of Lactic Acid Bacteria in Genus *Salvia*," *Salvia* species are described as perspective biological systems that could be used as source for isolation of lactic acid bacteria with a special application as starter cultures for tailor-made foods with improved health benefits.

In the end, we would like to acknowledge all contributing authors for their professionalism and immense efforts in realization of the book. We hope this book will be a base for further development in biotechnology of *Salvia* species.

Plovdiv, Bulgaria

Vasil Georgiev
Atanas Pavlov

Contents

Editors and Contributors

About the Editors

Dr. Vasil Georgiev obtained his Ph.D. at The Stephan Angeloff Institute of Microbiology, Bulgarian Academy of Sciences in 2009. He is an expert on bioprocess engineering and secondary metabolites production by plant cells, tissue and organ cultures, functional genomics, metabolite profiling, development and improvement of plant in vitro production platforms. In 2011, he was awarded by the Bulgarian Academy of Sciences with the "Prof. Marin Drinov" Prize for the outstanding research in biological sciences. Dr. Georgiev did postdoctoral research at Florida A&M University, USA (2012–2016). Since 2016, he is Associate Professor in the Laboratory of Applied Biotechnology, The Stephan Angeloff Institute of Microbiology, Bulgarian Academy of Sciences, and also joined the Department of Organic Chemistry at the University of Food Technologies, Plovdiv, Bulgaria.

Prof. Atanas Pavlov completed his Ph.D. in biotechnology at The Stephan Angeloff Institute of Microbiology, Bulgarian Academy of Sciences, in 1998. From 2003 to 2005, he did postdoctoral research at the Institute of Food Technologies and Bioprocess Engineering, University of Technologies, Dresden, Germany. In 2005, he joined The Stephan Angeloff Institute of Microbiology, Bulgarian Academy of Sciences, as an Associate Professor. In 2009, he received his habilitation (DSci) and is since 2010 a Professor at the Laboratory of Applied Biotechnology. Since 2011 he also holds a second Professor position at the University of Food Technologies, Plovdiv, Bulgaria. His research is mainly focused on the production of specific bioactive metabolites by plant cell, tissue, and organ cultures.

Contributors

Ina Aneva Institute of Biodiversity and Ecosystem Research, Bulgarian Academy of Sciences, Sofia, Bulgaria

Sara Barberini CREA Research Centre for Vegetable and Ornamental Crops, Sanremo, Italy

Laura Bassolino CREA Research, Centre for Genomics and Bioinformatics, Montanaso Lombardo, Italy

Dora Beshkova Laboratory of Applied Biotechnologies, The Stephan Angeloff Institute of Microbiology, Bulgarian Academy of Sciences, Plovdiv, Bulgaria

Teresa Docimo Institute of Biosciences and Bioresources, Research (CNR-IBBR), National Research Council of Italy, Portici, NA, Italy

Vincenzo D'Amelia Department of Agricultural Sciences, University of Naples "Federico II", Portici, NA, Italy

Thomas Efferth Department of Pharmaceutical Biology, Institute of Pharmacy and Biochemistry, Johannes Gutenberg University, Mainz, Germany

M. Emilia Carretero Department of Pharmacology, Faculty of Pharmacy, Complutense University of Madrid, Madrid, Spain

Vasil Georgiev Laboratory of Applied Biotechnologies, Institute of Microbiology, Bulgarian Academy of Sciences, Plovdiv, Bulgaria; University of Food Technologies, Plovdiv, Bulgaria

Tanya Girova Department of Biochemistry and Microbiology, "Paisii Hilendarski" University of Plovdiv, Plovdiv, Bulgaria

Velizar Gochev Department of Biochemistry and Microbiology, "Paisii Hilendarski" University of Plovdiv, Plovdiv, Bulgaria

Elena González-Burgos Department of Pharmacology, Faculty of Pharmacy, Complutense University of Madrid, Madrid, Spain

Christiane Haas Technische Universität Dresden, Institute of Natural Materials Technology, Dresden, Germany

Yana Hristova Food Research and Development Institute at Agricultutal Academy, Plovdiv, Bulgaria

Ayse Gul Ince Vocational School of Technical Sciences, Akdeniz University, Antalya, Turkey

Mehmet Karaca Department of Field Crops, Akdeniz University, Antalya, Turkey

Sibylle Kümmritz Technische Universität Dresden, Institute of Natural Materials Technology, Dresden, Germany

Antonietta Leone Department of Pharmacy, University of Salerno, Fisciano, SA, Italy

Eric Mozzanini CREA Research Centre for Vegetable and Ornamental Crops, Sanremo, Italy

Milena Nikolova Institute of Biodiversity and Ecosystem Research, Bulgarian Academy of Sciences, Sofia, Bulgaria

Atanas Pavlov Laboratory of Applied Biotechnologies, Institute of Microbiology, Bulgarian Academy of Sciences, Plovdiv, Bulgaria; University of Food Technologies, Plovdiv, Bulgaria

M. Pilar Gómez-Serranillos Department of Pharmacology, Faculty of Pharmacy, Complutense University of Madrid, Madrid, Spain

Laura Pistelli Department of Agrarian, Food and Agro-environmental Sciences, University of Pisa, Pisa, Italy

Luisa Pistelli Department of Pharmacy, University of Pisa, Pisa, Italy

María Porres-Martínez Department of Pharmacology, Faculty of Pharmacy, Complutense University of Madrid, Madrid, Spain

Barbara Ruffoni CREA Research Centre for Vegetable and Ornamental Crops, Sanremo, Italy

Alessandra Ruggiero Institute of Biosciences and Bioresources, Research (CNR-IBBR), National Research Council of Italy, Portici, NA, Italy

Marco Savona CREA Research Centre for Vegetable and Ornamental Crops, Sanremo, Italy

Halil Şenol Department of Pharmaceutical Chemistry, Faculty of Pharmacy, Bezmialem Vakif University, Istanbul, Turkey

Albena Stoyanova Department of Tobacco, Sugar, Vegetable and Essential Oils, University of Food Technologies, Plovdiv, Bulgaria

Tsvetanka Teneva-Angelova Laboratory of Applied Biotechnologies, The Stephan Angeloff Institute of Microbiology, Bulgarian Academy of Sciences, Plovdiv, Bulgaria

Gülaçtı Topçu Department of Pharmacognosy and Phytochemistry, Faculty of Pharmacy, Bezmialem Vakif University, Istanbul, Turkey

Valentina Tranchida-Lombardo Institute of Biosciences and Bioresources, Research (CNR-IBBR), National Research Council of Italy, Portici, NA, Italy

Marina Tucci Institute of Biosciences and Bioresources, Research (CNR-IBBR), National Research Council of Italy, Portici, NA, Italy

Katja Winkler Technische Universität Dresden, Institute of Natural Materials Technology, Dresden, Germany

Ching-Fen Wu Department of Pharmaceutical Biology, Institute of Pharmacy and Biochemistry, Johannes Gutenberg University, Mainz, Germany

Rümeysa Yücer Department of Pharmacognosy and Phytochemistry, Faculty of Pharmacy, Bezmialem Vakif University, Istanbul, Turkey; Department of Pharmacognosy, Faculty of Pharmacy, University of Health Sciences, Istanbul, Turkey

European Species of Genus *Salvia*: Distribution, Chemodiversity and Biological Activity

Milena Nikolova and Ina Aneva

Abstract The present review summarizes the information published in the scientific literature during the last two decades on the species of genus *Salvia* occurring naturally in Europe. The reviewed species were selected following the Flora Europea, vol 3, where 36 numbered species of the genus are listed, together with some provisory taxa, hybrids, and species of uncertain status.

Keywords Acetylcholinesterase · Antimicrobial · Antioxidant
Butyrylcholinesterase · Cytotoxic · Essential oils · Flavonoids

Abbreviations

AchE	Acetylcholinesterase
BchE	Butyrylcholinesterase
MIC	Minimal inhibitory concentration
BHA	Butylated hydroxyanisole
BHT	Butylated hydroxytoluene
DPPH	2,2-Diphenyl-1-picrylhydrazyl
ABTS	2,2′-Azino-bis(3-ethylbenzothiazoline-6-sulphonic acid)

1 Introduction

Salvia, the largest genus of the family Lamiaceae, represents an enormous and cosmopolitan assemblage of nearly 1000 species displaying a remarkable range of variation only 36 of which occur in Europe [1]. Iberian, and particularly Balkan

M. Nikolova (✉) · I. Aneva
Institute of Biodiversity and Ecosystem Research, Bulgarian Academy of Sciences, Sofia, Bulgaria
e-mail: mtihomirova@gmail.com

V. Georgiev and A. Pavlov (eds.), *Salvia Biotechnology*,
https://doi.org/10.1007/978-3-319-73900-7_1

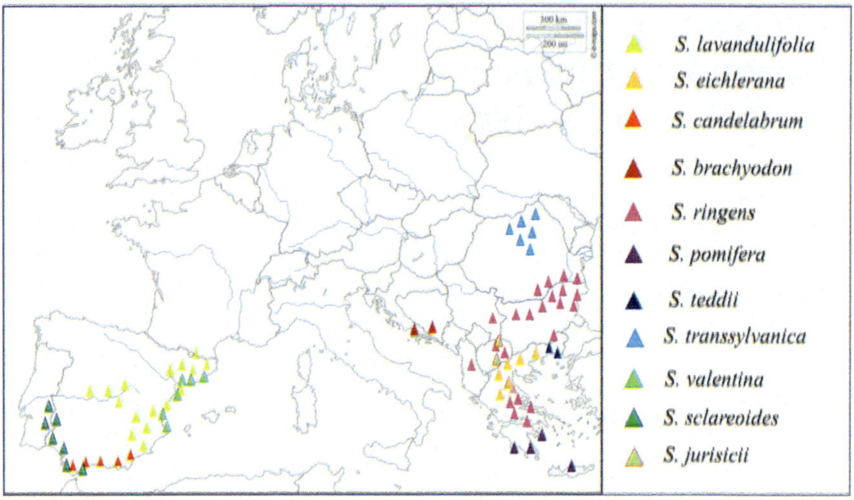

Fig. 1 Schematic distribution of the endemic species of genus *Salvia* in Europe

Peninsula are rich in endemic species. Altogether 11 *Salvia* species can be considered endemics to a restricted geographic area (Fig. 1).

Most species prefer warmer and sunny areas, usually irrespective of the soil quality and often grow in stony and rocky habitats (Fig. 2).

Essential oils, fatty acids, flavonoid compounds and to a lesser extent—di- and triterpenoids are predominant biologically active compounds studied of Salvia species. Antimicrobial, antioxidant, cytotoxic activity, the acetylcholinesterase (AchE) and butyrylcholinesterase (BchE) inhibitory activity were among the most frequently investigated pharmacological actions of crude extracts and individual compounds of *Salvia* species.

The species *Salvia officinalis*, *S. lavandulifolia* and *S. fruticosa* were the most extensively studied ones. Research on rare and endemic species are also of interest with a view to establish new promising biological substances and actions.

2 European Species of Genus *Salvia*

2.1 Salvia aethiopis *L.*

The species occurs in xeric habitats in Southern and Eatern Europe from France and Portugal to the west through Central Europe to Ukraine, Turkey and Iran to the east. Its natural range of distribution covers also Iberian, Italian and Balkan Peninsulas. *S. aethiopis* grows on steppes, igneous and limestone slopes, fallow fields, and roadside banks [1–3]. The plant has been used in Iranian herbal medicine as a carminative and tonic agent [4]. Other studies reported cytotoxic, antioxidant,

Fig. 2 Some *Salvia* species with contrasting habitats. **a** *Salvia viridis* in open dry grasslands. **b** *Salvia sclarea* in dry stony slopes. **c** *Salvia glutinosa* at the edge of a beech forest. **d** *Salvia scabiosifolia* in its typical habitat in Bulgaria. **e** *Salvia ringens* in the zone with Mediterranean influence. **f** *Salvia nemorosa* near agricultural fields

antimicrobial, inhibition of 5-lipoxygenase activity, antiinflammatory, antibacterial, analgesic, antipyretic, haematological effects and acetylcholinesterase activity in *S. aethiopis* [5].

Variability in essential oil composition of the species from different geographical regions has been detected but germacrene D, β-caryophyllene, and α-copaene were found to be among the major constituents of essential oils [4]. Hatipoglu et al. [5] in their extensive study of 45 Turkish species of the genus *Salvia* for volatile organic compounds placed *S. aethiopis* in the group of species rich of caryophyllene oxide

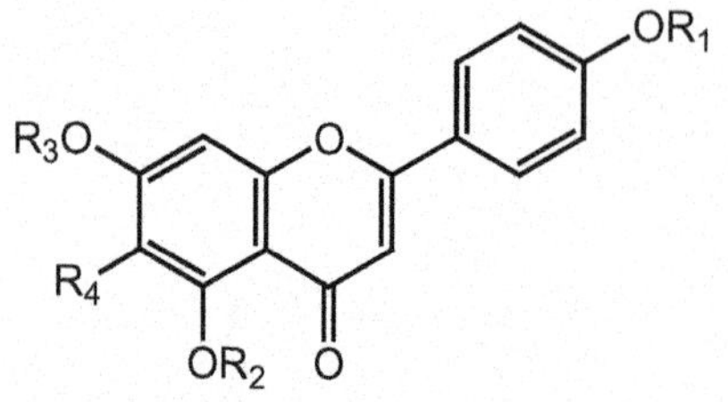

apigenin $R_1=R_2=R_3=R_4=H$
apigenin 7-methyl ether $R_1=R_2=R_4=H$ $R_3=CH_3$
apigenin 4'-methyl ether $R_2=R_3=R_4=H$ $R_1=CH_3$
apigenin 7,4'-dimethyl ether $R_2=R_4=H$ $R_1=R_3=CH_3$
scutellarein $R_1=R_2=R_3=H$ $R_4=OH$
scutellarein 6-methyl ether $R_1=R_2=R_3=H$ $R_4=OCH_3$
scutellarein 6, 4'-dimethyl ether $R_2=R_3=H$ $R_1= CH_3$ $R_4=OCH_3$
scutellarein 6,7-dimethyl ether $R_1=R_2=H$ $R_3= CH_3$ $R_4=OCH_3$
scutellarein 7,4'-dimethyl ether $R_2= R_4=H$ $R_1=R_3=CH_3$
scutellarein 6,7,4'-trimethyl ether $R_2=H$ $R_1=R_3= CH_3$ $R_4=OCH_3$
scutellarein 5,6,7,4'-tetramethyl ether $R_1=R_2=R_3= CH_3R_4=OCH_3$

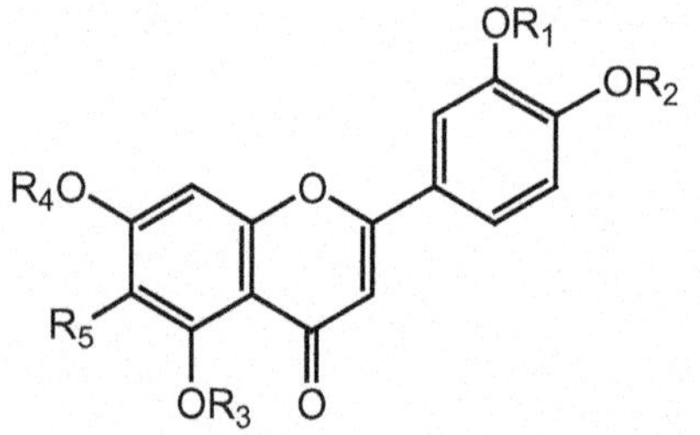

luteolin $R_1=R_2=R_3=R_4=R_5=H$
luteolin 7-methyl ether $R_1=R_2=R_3=R_5=H$ $R_4=CH_3$
6-hydroxyluteolin 6-methyl ether $R_1=R_2=R_3=R_4= H$ $R_5=OCH_3$
6-hydroxyluteolin 6,7-dimethyl ether $R_1=R_2=R_3=H$ $R_4= CH_3$ $R_5=OCH_3$
6-hydroxyluteolin 6,7,3'-trimethyl ether $R_2=R_3= H$ $R_1= R_4= CH_3$ $R_5=OCH_3$

Fig. 3 Structure of main flavonoid aglycones detected of *Salvia* species

(caryophyllene oxide chemotype), based on cluster analysis. Carvacrol, betulin, oleanolic acid, palmitic acid have been also reported for *S. aethiopsis* as predominant volatile compounds.

Ethanol extract of *S. aethiopis* of wild-growing plants in Serbia was studied for nonpolar compounds, and μ-thujone, 1,8-cineole and μ-humulene have been reported as major components. The same extract expressed activity on Gram-positive (*Staphylococus aureus*) and a Gram-negative (*Escherichia coli*) bacteria with 2.2 mg/mL MIC values [6].

Aethiopinone (*o*-naphthoquinone diterpene) isolated from the roots of *S. aethiopsis* were showed to possess strong anti-inflammatory and antinocicetive activity, similar to that of non steroidal-anti-inflammatory drugs. Also aethiopinone has been shown to have analgesic effect [7].

Kilic et al. [8] in their study on the fatty acid composition of seeds of *S. aethiopis* determined 9,12-octadecadienoic acid (linoleic acid, 18:2) as dominant component of seed oil, followed by hexadecanoic acid (palmitic acid, 16:0) and 9-octadecenoic acid (oleic acid, 18:1). Seeds extract of *S. aethiopis* showed activity against *Staphylococus aureus, Escherichia coli, and Mycobacterium smegmatis*, as well as against a fungus *Candida albicans.*

The methanol and dichloromethane extracts of *S. aethiopis* were proven to exhibit cytotoxic activity against 3 human cancer cell lines: HL60 (human acute promyelocytic leukemia cells), K562 (human chronic myelogenous leukemia cells) and MCF-7 (human breast adenocarcinoma cells). The extracts demonstrated also antimicrobial activity against *Klebsiella pneumoniae* and *Salmonella typhi* [9].

Flavonoid compounds reported for *S. aethiopis* are summarized and presented in Table 1.

2.2 Salvia argentea *L.*

The species is a perennial herb distributed in Mediterranean region and Northwestern Africa. It occurs primarily on stony hillside meadows, basalt, volcanic soils and rocky bluffs. (http://ww2.bgbm.org/euroPlusMed/). The young leaves of *S. argentea* were topically used as haemostatic whereas the basal leaves, peeled and stewed, were consumed as food in Spain ("gordolobo") [1, 2, 10].

Data concerning chemical profile of essential oil of *S. argentea* populations from different regions were summarized by Riccobono et al. [10]. From the review it is clear that there is great variability of essential oil profiles. Caryophyllene oxide was reported as the main component of the oil from Macedonian populations, while viridiflorol from the Serbian ones. Moroccan oil was characterised by the presence of camphor. The profile of essential oils of two Tunisian populations was quite similar to the Serbian ones. 14-Hydroxy-a-humulene (40.1%) has been recognised as the main constituents of the essential oil of Sicilian *S. argentea.*

Sarac and Ugur [11] investigated the antimicrobial activities of the ethanolic extracts of *S. argentea* against several pathogenic bacteria and observed that there was antimicrobial activity expressed to Gram-positive bacteria, especially *Staphylococci*. Stagos et al. [12] observed high protective activity of *S. argentea* collected from the Mt. Parnassos (C. Greece) against free radical-induced DNA damage. Acetylcholinesterase (AChE) and butyrylcholinesterase (BChE) inhibitory activity as well as larvicidal activity against the mosquito *Culex pipiens* were

Table 1 Flavoniod compounds identified of Salvia species

Salvia species	Flavonoid compounds[a]	References
S. aethiopis	Scutellarein 6,7,4′-trimethyl ether, luteolin 7,3′,4′-trimethyl ether	[120]
S. argentea	Apigenin, apigenin 7-methyl ether, apigenin 7,4′-dimethyl ether, scutellarein 7,4′-dimethyl ether; scutellarein 6,4′-dimethyl ether; scutellarein 6,7-dimethyl ether; scutellarein 6,7,4′-trimethyl ether, luteolin; the 6,7-dimethyl and 6,7,3′-trimethyl ethers of 6-hydroxyluteolin	[121]
S. austriaca	Apigenin, apigenin 7-methyl ether, scutellarein 6,7-dimethyl ether, scutellarein 6,4′-dimethyl ether, luteolin	[122]
S. barrelieri (*syn. S. bicolor*)	Apigenin Luteolin 7-*O*-glucoside	[22]
S. candelabrum	Apigenin, apigenin 7-methyl ether, scutellarein 6,7,4′-trimethyl ether, luteolin	[123]
S. candidissima	Scutellarein 6,7,4′-trimethyl ether, luteolin 3′-methyl ether, luteolin 4′-methyl ether, luteolin 7,3′-dimethyl ether	[121]
S. glutinosa	Apigenin, apigenin 7-methyl ether, the 6,7-dimethyl and 6,7,3′-trimethyl ethers of 6-hydroxyluteolin, kaempferol 3-methyl ether, kaempferol 3,7-dimethyl ether, quercetin 3-methyl ether, quercetin 3,7,4′-trimethyl ether, quercetin 3,7,3′,4′-tetramethyl ether, quercetin 3,7,3′-trimethyl ether	[120, 121]
S. fruticulosa (*syn. S. triloba*)	Scutellarein 6-methyl ether, scutellarein 6,7-dimethyl ether, scutellarein 6,7,4′-trimethyl ether, 6-hydroxyluteolin 6,3′-dimethyl ether Apigenin 7-*O*-glucoside, apigenin 7-*O*-glucoronide, scutellarein 7-*O*-glucoside, apigenin-6-C-glucoside-7-*O*-glucoside, apigenin 6,8-di-C-glucoside, luteolin 7-*O*-glucoside, luteolin 7-*O*-glucuronide, luteolin 7-*O*-rutinoside, luteolin 3′-methyl ether 7-*O*-glucoronide, luteolin 7-*O*-cellobioside, 6-hydroxyluteolin 7-glucoside, 7 glucoronide, luteolin diglucuronide	[49, 120, 124]
S. jurisicii	Apigenin, apigenin 7-methyl ether, apigenin 7,4′-dimethyl ether, scutellarein 6,7,4′-trimethyl ether, quercetin Luteolin 5-*O*-(6″-*O*-malonilglucoside), genkwanin 5-*O*-glucoside, genkwanin 5-*O*-(6″-*O*-malonylglucoside), genkwanin 4′-*O*-glucoside, luteolin 5-*O*-glucoside, kaempferol 3-*O*-(6″-*O*-malonylglucoside)-7-*O* glucoside, kaempferol 3-*O*-(6″-*O*-acetilglucoside)-7-*O*-rhamnoside, kaempferol 3-*O* glucoside, kaempferol 7-*O*-rhamnoside, quercetin 3,7-*O*-diglucoside, hyperoside	[51, 122]
S. lavandulifolia	Apigenin, apigenin 7-methyl ether, apigenin 7,4′-dimethyl ether, scutellarein 6-methyl ether, scutellarein 6,7-dimethyl ether, scutellarein 6,7,4′-trimethyl ether, the 6-methyl and 6,7,4′-trimethyl ethers of 6-hydroxyluteolin Luteolin 7-*O*-glucoside, luleolin 3′-*O*-glucuronide, luteolin 7-*O*-rutinoside, luteolin 5-*O*-rutinoside, quercetin 3-*O*-glucoside	[120]

(continued)

Table 1 (continued)

Salvia species	Flavonoid compounds[a]	References
S. nemorosa	Apigenin, apigenin 7-methyl ether, apigenin 7,4′-dimethyl ether, scutellarein 7,4′-dimethyl ether, scutellarein 6,7-dimethyl ether, scutellarein 6,7,4′-trimethyl ether, luteolin, the 6-methyl 6,7-dimethyl 6,3′,4′-trimethyl and 6,7,3′-trimethyl ethers of 6-hydroxyluteolin, 6-methyllutcolin (2, 5,7,3′,4′-tetrahydroxy-6-methyl-flavone), quercetin	[60, 61, 121, 122]
S. nutans	Luteolin, luteolin 4′-methyl ether Luteolin 7-*O*-glucoside	[120, 122, 125]
S. officinalis	Apigenin, apigenin 7-methyl ether, apigenin 7,4′-dimethyl ether, luteolin, luteolin 7-methyl ether, scutellarein, scutellarein 6-methyl ether, scutellarein 6,7-dimethyl ether, scutellarein 7,4′-dimethyl ether, scutellarein 6,7,4′-trimethyl ether, scutellarein 5,6,7,4′-tetramethyl ether, the 6-methyl and 6,7-dimethyl ethers of 6-hydroxyluteolin, 8-hydroxyapigenin (isoscutellarein), 5,7,3′-hydroxy4′-methoxyflavanones (hesperetin) Apigenin 7-*O*-glucoside,scutellarein 7-*O*-glucoside, apigenin 6-C-glucoside-7-*O*-glucoside, apigenin 6,8-di-C-glucoside luteolin 7-*O*-glucoside, luteolin 7-*O*-glucuronide, luleolin 3′-*O*-glucoronide, 6-hydroxyluteolin 7-*O*-glucoside, 6-hydroxyluteolin 7-*O*-glucuronide, luteolin 7-*O*-rutinoside, luteolin-diglucuronide	[120, 121, 124]
S. pinnata	Apigenin, luteolin, 6-hydroxyluteolin-6-methyl ether Apigenin 7-*O*-glucoside, luteolin 7-*O*-glucoside, quercetin 3-*O*-glycoside	[120]
S. pratensis	Apigenin, luteolin, luteolin 4′-methyl ether Apigenin 7-*O*-glucoside, luteolin 7-*O*-glucoside luteolin 7-*O*-glucuronide	[120, 125]
S. ringens	Apigenin, apigenin 7-methyl ether, apigenin 7,4′-dimethyl ether, scutellarein 6,7-dimethyl ether, scutellarein 7,4′-dimethyl ether, scutellarein 6,7,4′-trimethyl ether, luteolin, the 6,7-dimethyl and 6,7,3′-trimethyl ethers of 6-hydroxyluteolin, quercetin-3-methyl ether, kaempferol-3-methyl ether	[121]
S. scabiosifolia	Apigenin, apigenin 7-methyl ether, apigenin 7,4′-dimethyl ether, scutellarein 7,4′-dimethyl ether, scutellarein 6,7-dimethyl ether, luteolin, luteolin 7,3′-dimethyl ether, the 6,7-dimethyl 6,7,3′-trimethyl and 6,7,4′-trimethyl ethers of 6-hydroxyluteolin	[121]
S. sclarea	Apigenin, apigenin 7-methyl ether, apigenin 7,4′-dimethyl ether, scutellarein 6,7-dimethyl ether, scutellarein 6,7,4′-trimethyl ether, luteolin, luteolin 7,3′-dimethyl ether, the 6,7-dimethyl 6, 3′,4′-trimethyl and 7,3′,4-trimethyl ethers of 6-hydroxyluteolin, 5,7-dihydroxy-6,8,4′-trimethoxyflavone (nevadensin), naringenin, taxifolin (dihydroflavonols), pseudobaptigenin (isoflavones) Kaempferol 3-*O*-glucoside, quercetin 3-*O*-rutinoside, fisetin 3-*O*-glucoside	[121,124, 126–128]

(continued)

Table 1 (continued)

Salvia species	Flavonoid compounds[a]	References
S. sclareoides	Quercetin, catechin Luteolin 4′-*O*-glucoside, luteolin 3′,7-di-*O*-glucoside, luteolin 7-*O*-(6″-*O*-acetylglucoside, kaempferol 3-*O*-glucoside	[93, 94]
S. tomentosa	Apigenin, apigenin 7-methyl ether, scutellarein 6,7-dimethyl ether, scutellarein 6,4′-dimethyl ether, scutellarein 7,4′-dimethyl ether, luteolin, luteolin 4′-methyl ether, the 6-methyl 6,3′-dimethyl 6,7-dimethyl and 6,7,3′-trimethyl 6,3′,4-trimethyl and 6,7,3′,4′-tetramethyl ethers of 6-hydroxyluteolin Luteolin 7-*O*-glucoside, 6-hydroxyluteolin 7-glucoside, 6-hydroxyluteolin 6-methyl ether 7-glucoside, 6-methyl ether luteolin 5-*O*-glucoside	[120, 121]
S. transsylvanica	Quercetrin, narcissin (flavonol 3-*O*-rutinose)	[100]
S. verbenaca	Apigenin, apigenin 7,4′-dimethyl ether, scutellarein 6,7,4′-trimethyl ether, luteolin, 6-hydroxyluteolin 6,7-dimethyl ether, quercetin 3,7,3′,4′-tetramethyl ether (retusin), 5-hydroxy 3,4′,7-trimethoxyflavone, 5-hydroxy 7, 4′-dimethoxyflavone Luteolin 7-*O*-glucoside, luteolin7-cellobioside, 6,7dimethyl ether of 6-hydroxyluteolin 5-glucoside, 5, 4′-dihydroxyflavone-5-*O*-β-d-glucopyranoside, verbenacoside	[107, 120]
S. verticillata	Scutellarein 6,7-dimethyl ether, myricetin, quercetin, kaempferol, naringenin, naringin Luteolin 7-*O*-glucoside, luteolin 7-*O*-glucuronide, 6-hydroxyluteolin 5-*O*-glycoside, apigenin 4′-methyl ether 7-*O*-glucuronide, salvigenin 5-*O*-glucoside, quercetin 3-*O*-rutinoside	[112, 120, 127]
S. viridis	Apigenin 7-*O*-glucoside, luteolun 7-*O*-glucoside, luteolin 7-*O*-galactoside, luteolin 7-*O*-rutinoside, luteolin 7-*O*-rhamnoside, luteolin 7-*O*-rhamnogalactoside	[111]
S. virgata	Luteolin, luteolin 3′-methyl ether, luteolin 7,3′,4′-trimethyl ether, scutellarein 6,7,4′-trimethyl ether Luteolin 7-*O*-glucoside, salvigenin 5-glucoside	[118, 120]

[a]Structure of the main flavonoid aglycones detected of *Salvia* species are presented at Fig. 3

reported for different extracts of *S. argentea* [10, 13]. Phytochemical studies of the plant indicated the presence of abietane diterpenoids, oleanane and ursane derivatives, and ursane triterpenoids [10].

Lakhal et al. [14] showed that triterpenoids that have been isolated from the exudate of *S. argentea* var. *aurasiaca* (Pomel) Batt. & Trab. have chemotaxonomic importance.

The flavonoid compounds detected of *S. argentea* are presented in Table 1.

2.3 Salvia amplexicaulis *Lam.*

S. amplexicaulis is a perennial plant distributed on the Balkan Peninsula, Southern Romania and in Turkey (Asia Minor) and is reported also as an invasive plant in Hungary. It inhabits dry rocky pastures, wood margins and bushes, usually in lowlands on limestone [1, 2, 15].

Essential oil of *S. amplexicaulis* was characterized by a high amount of sesquiterpenes with germacrene D, viridiflorol, caryophyllene oxide and β-caryophyllene as main components [16, 17]. The oil showed microbial growth inhibitory effect against gram-positive bacteria *Micrococcus luteus* followed by the effects against *Staphylococcus epidermidis*, *S. aureus* and a yeast *Candida albicans* [15].

Methanolic and ethanolic extracts of *S. amplexicaulis* demonstrated very strong in vitro antioxidative activity against the DPPH free radical, which is similar to that of the reference antioxidants (BHA and BHT) [18, 19]. Also, Orhan et al. [18] found that *S. amplexicaulis* alcoholic extracts possess a moderate inhibitory activity against the enzymes linked to neurodegeneration—acetylcholinesterase, butyrylcholinesterase, lipoxygenase and tyrosinase.

Diterpenoids, steroids, triterpenoids and one flavonoid salvigenin have been isolated from the roots of *S. amplexicaulis*. The crude extract of the species' roots and the pure compounds (diterpenoids) were shown to possess a vasodepressor effect [20].

Data concerning flavonoid conpounds of *S. amplexicaulis* are summarized in Table 1.

2.4 *Salvia austriaca Jacq.*

Austrian sage is a medicinal herbaceous plant native to high altitudes across Russia, Central and Eastern Europe [1, 2].

The analysis of chemical composition of the essential oils of *S. austriaca* revealed that sesquiterpene fractions were the main part of volatiles. Spathulenol, hexahydrofarnezyl acetone, isobornyl acetate, palmitic acid, *trans*-phytol and caryophyllene oxide were identified as the main components [17].

The roots of *S. austriaca* accumulate abietane-type diterpenoids. It was demonstrated that root extract of the species exhibited the acetylcholinesterase (AchE) and butyrylcholinesterase (BchE) inhibitory activity with IC_{50} = 139.5 μg/mL and IC_{50} = 23.6 μg/mL, respectively. Moreover, the authors showed that taxodione, an abietane diterpenoid isolated from *S. austriaca* transformed roots grown in vitro, expressed activity towards human acetylcholinesterase IC_{50} = 54.84 μg/ml. Also, the root extract and taxodione was shown to have cytotoxicity against A549 cell line with IC_{50} = 75.7 μg/mL and IC_{50} = 9.1 μg/ml, respectively [21].

Flavonoid compounds identified of *S. austriaca* are summarized in Table 1.

2.5 Salvia barrelieri *Etl.* (syn. S. bicolor *Desf.*) [2]

This species occurs in Northwestern Africa and only Southwestern Spain in Europe. It inhabits dry and sunny places in the mountains from 500 to 1200 m a.s.l. [1, 2].

Methanolic and petroleum ether extracts of *S. bicolor* were shown to exhibit significant antioxidant, anti-inflammatory, antimicrobial and analgesic properties. Aamong fatty acids, β-sitosterol and β-amyrin, as well as linolenic and erucic acids have been determined as the main constituents of petroleum ether extract. In the methanolic extract protocatchuic acid was detected as predominant (75.22 mg/g dry sample), followed by p-coumaric, gallic and synergic acids, luteolin 7-*O*-glucoside and apigenin among flavonoid compounds [22].

Monoterpenes: α-pinene, α-thujone, β-caryophyllene and 1,8-cineole were reported as major components of the essential oil of *S. bicolor*. The essential oil demonstrated a remarkable growth inhibitory effect on *Staphylococcus aureus*, *Bacillus subtilis*, *Escherichia coli* and *Candida albicans* at the concentration range 300–400 μg/mL [23].

2.6 Salvia brachyodon *Vandas*

S. brachyodon is a species endemic to the northwest part of Balkan Peninsula [1, 2].

The essential oils of *S. brachyodon* collected from Montenegro, were analysed by GC–MS. The oil was characterized by a high content of oxygenated monoterpenes. The most abundant compounds were 1,8-cineole, bornyl acetate, α- and β-pinene, and camphene [24, 25].

Diffrences in the essential oil composition was observed between the corolla, calyx and leaves. The leaf oil composition was the closest to that previously reported [24, 25]. The corolla oil was the most complex, because of the presence of higher concentration of sesquiterpenes and presence of diterpenes, which were not recorded in the other samples investigated. The oil isolated from the calyx contained beta-pinene (27.6%), alpha-pinene (16.2%), camphene (13.7%), borneol (13.2%) and camphor (12.4%) as major components and this oil possessed the best antifungal activity against all the fungal species tested (*Aspergillus niger*, *Trichophyton mentagrophytes* and *Candida albicans*) [26].

2.7 Salvia bracteata *Banks & Sol.*

The species is distributed in Western Asia, including Mediterranean part of the Near East and its occurrence in Turkey in Europe is the only one on the continent [1, 2].

The essential oils of *S. bracteata* collected in Turkey, Iran and Lebanon were reported to have different profiles. Although caryophyllene oxide was determined as the main component of the oil of Turkish and Lebanon populations, differences in other components have been found. For example, α-pinene and β-pinene were determined as major components of essential oils of Iranian plants. These changes in the essential oil compositions might arise from different environmental (climatic, seasonal, and geographical) and genetic differences [27].

Essential oils of aerial parts of *S. bracteata* were proven to have a good antibacterial activity (MIC = 50 μg/ml) against Gram-positive bacteria. Also the essential oils exhibited an inhibitory effect on on M14 human melanoma cell. Anticancer activity may be related to active sesquiterpenes associated to the synergism of other natural present in the essential oil composition [27].

The seed extract of *S. bracteata* have been studied for fatty acid composition. It is noteworthy that 9,12-octadecenoic acid (linoleic acid, 18:2) amounted to 68.5% of total fatty acid content [28].

2.8 Salvia blancoana *Webb & Heldr.*

In Europe this species occurs in Eastern and Southeastern Spain and spreads to the south in Northwest Africa—Algeria and Morocco [1, 2].

The species has been used as spice, digestive, emmenagogue, antitussive, sedative, hypotensive tea, in perfumes, and as ornamental plant. Essential oil composition of flowering aerial parts of *S. blancoana* subsp. *mariolensis* have been analyzed using gas chromatography-flame ionization detection (GCFID) and gas chromatography-mass spectrometry (GC-MS). 1,8-Cineole and camphor have been found as the main constituents of essential oils [29].

2.9 Salvia candidissima *Vahl.*

S. candidissima is a herbaceous perennial native throughout western Greece, and parts of Turkey, Iraq, and Iran [1, 2].

Different extracts (petroleum ether, chloroform, ethyl acetate, and methanol) of *S. candidissima* ssp. *occidentalis* have been shown to have weak acetylcholinesterase (AChE) and butyrylcholinesterase (BChE) inhibitory activity at 0.2 and 1 mg/mL [30]. Şenol et al. [31] reported that dichloromethane and ethyl acetate extracts of the species at concentration under 0.1 mg/mL did not display inhibitory activity against AchE and BchE enzymes. Also, the exctacts showed low (dichloromethane extact) to strong (ethyl acetate, and methanol extracts) activity against DPPH radicals. Özler [32] reported high antioxidant activity of ethanol and hexane extracts of the plant.

Seed extract of *S. candidissima* ssp. *candidissima* was analysed for fatty acid composition. Hexadecanoic acid (16:0, palmitic acid) 20.8%, 9,12-Octadecadienoic acid (18:2, linoleic acid) 20.8%, 9-Octadecenoic acid (18:1, oleic acid) 20.4%

9,12,15-Octadecatrienoic acid (18:3, linolenic acid) 18.5% were the main components. The seed extract showed activity against Gram-negative bacterium *Mycobacterium smegmatis* with a MIC value of 0.25 mg/mL [8]. Also the seed fatty acid extract of *S. candidissima* was found to have high antimicrobial activity to dermatophyte species *Epidermophyton* sp. [33].

Hydroalcholic and chloroform extracts of *S. candidissima* dose dependently reduce edema and hyperalgesia during CFA-induced inflammation [34].

The species has been classified as borneol chemotype from data of cluster analysis of Hatipoglu et al. [5] Camphene, verbenone, spathulenol have been determined as other dominant volatile compounds [5].

Flavonoid compounds determined of *S. candidissima* are presented in Table 1.

2.10 Salvia candelabrum *Boiss.*

S. candelabrum is a perennial plant to small shrub endemic to Southern Spain [1, 2].

Máthé et al. [35] reported 1,8-cineole (20.5%), camphor (9.0%), α-pinene (6.5%), and β-pinene (7.4%) as the main component of essential oils of *S. candelabrum* collected in the experimental populations of Hungaria.

Seven abietane and seco-abietane diterpenes have been isolated from methanolic extracts of aerial parts of *S. candelabrum* under antioxidant assay guidance. The isolated compounds exhibited antioxidant activity in enzyme dependent and enzyme-independent systems of lipid peroxidation [36].

Flavonoid compounds reported for *S. candelabrum* are presented in Table 1.

2.11 Salvia cremenecensis *Bess.*

S. cremenecensis is a rare taxon, endemic to Western Ukraine (Podolsk Hills) [37]. However, its species status was questioned by Hedge [1], who considered both *S. cremenecensis* and *S. cernua* Czern. ex Schost. as taxa of hybrid origin with one parent being *Salvia nutans* L. Therefore, in later treatment [38] it is considered as a synonym of *S. nutans*.

2.12 Salvia dumetorum Andrz. ex Besser

The species occurs in Europe only in Romania, while its main range of distribution covers Ukraine, Kazakhstan and Russia, eastward to Siberia [1, 2].

Janicsak et al. [39, 40] investigated the oleanolic and ursolic acid contents as well as antioxidant activity of some species of the Lamiaceae, one of which is *S. dumetorum*. Among the studied species *S. dumetorum* demonstrates one of the highest antioxidant effects. The studies confirmed the close correlation between the antioxidant activities and the total phenol contents of the extracts.

2.13 Salvia eichleriana *Heldr. ex Halacsy*

It is endemic to Northern and Central Greece [1, 2].

2.14 Salvia forskaohlei *L.*

It occurs in Southeastern part of Balkan Peninsula and Northern Anatolia in Asia Minor [1].

The studies performed on *S. forskaohlei* focused to anticholinesterase and antioxidant effects of extracts. Moderate inhibition effect only against butyrylcholinesterase (BchE) at 0.2 mg/ml was found for petrol ether and chloroform extracts. Methanolic and ethyl acetate extracts showed antioxidant activity for xanthine oxidase (XO) inhibition and DPPH assay [30].

It is one of the garden species found at gardening market in Japan and was, therefore, one of the species used for determining of salvinorin A and salvinorin B in *S. divinorum*-related products [41].

Ciesla et al. [42] found low quantities of the methanol-soluble constituents in three species, including *S. forskaohlei*, *S. azurea*, and *S. nemorosa*.

Sajewicz et al. [43] compared binary fingerprints (chromatographic profiles) of six different sage species, one of them being *S. forskaohlei*, using HPLC coupled with two different detectors. Protocatechuic acid was detected in *S. azurea*, *S. forskaohlei* and *S. officinalis*.

Forskalinone (abietane diterpene isolated from roots of *S. forskahlei* was proven to possess slight antimicrobial activity against *Enterococcus faecalis* (168 μg/ml) [7].

2.15 Salvia fruticosa *Mill.*

There are 19 synonyms of *S. fruticosa* according to the Kew checklist, most frequently used being *S. triloba*, *S. libanotica*, *S. lobryana*, and *S. cypria* [2].

The species is native to Eastern Mediterranean including Southern Italy, southern parts of the Balkan Peninsula to West Syria. It is found on dry rocky limestone soils or the edges of pine forests, riverbeds, and roadsides, at altitude from 100 to 800 m a.s.l. [44, 45]. Due to the value of its essential oil it has been naturalized in Western Mediterranean region in Malta, Spain and Portugal. The leaves of the plant are boiled as an herbal tea for the relief of headaches, stomachaches, abdominal pain and many other disorders. The aqueous and oil extracts of the species have been shown to possess antioxidant, anti-inflammatory, anticancer and antimicrobial activities. *Salvia fruticosa* has been used to enhance memory, as a sedative and to treat headaches. It has also been accepted as a medicinal *Salvia* species by the European Pharmacopeia and British Pharmacopeia [45, 46].

Topçu et al. [47] summarized data related to essential oil composition of populations of *S. fruticosa* from different geographical regions. In most cases monoterpene 1,8-cineol (eucalyptol) has been found to be a major constituent of the oil, and rarely α-thujone has been also reported. It was shown that essential oil rich of eucalyptol exhibited inhibition against AChE (73.52%) at 100 μg/mL. Also Gali-Muhtasib [46] reported that essential oil has potent chemopreventive effects against skin tumor promotion in mice. Many substances were isolated from the aerial part of *S. fruticosa*, among them abietane diterpenoids (carnosol, carnosic acid, carnosic acid 12-methyl ether, rosmadial, isorosmanol and ferruginol), labdane diterpenoid manool, the triterpenoids (α-amyryltetracosanoate, ursolic acid, oleanolic acid, and erythrodiol), the steroid 3-acetylsitosterol, and the favonoid salvigenin. Strong antioxidant activity of methanol extract of *S. fruticosa* was reported. Also, the extract exhibited high inhibition values (approximately 90%) against AChE and BChE at 100 μg/mL [47]. Şenol et al. [31] reported that dichloromethane extract of *S. fruticosa* caused inhibition (51.07%) of AChE at 100 μg/mL.

Orhan and Aslan [48] revealed that hydroalcoholic extract of *Salvia triloba* (one of synonyms of *S. fruticosa*) exhibited mild inhibition as compared to galanthamine.

Ursolic acid, carnosol, oleanolic acid, salvigenin, rosmanol, cirsimaritin and hispidulin isolated of *S. fruticosa* have been assayed for cognition enhancing properties. Majority of compounds demonstrated some inhibitory activity at α1, β2, γ2L GABAA receptors, which supports the concept that the herb may have cognition enhancing properties [49].

When studied for antiproliferative activity, crude ethanol extract of *S. fruticosa* showed IC_{50} values lower than 30 μg/mL against the four cell lines—human tumor of breast cancer; MCF-7, T47D, ZR-75-1, and BT 474. Besides, *S. fruticosa* extract demonstrated minimal toxicity against normal human periodontal fibroblasts. This selectivity in action indicates a potential source for novel anticancer therapy [50].

S. fruticosa displayed the best activity on Gram-negative bacteria with MIC 640 µg/mL against *Salmonella typhimurium* and *Enterobacter aerogenes. Carvacrol rosmarinic and caffeic acids* were determined as major components defining this action [51].

Flavonoid compounds detected of *S. fruticosa* are presented at Table 1.

2.16 Salvia jurisicii *Kosanin*

It is a hardy herbaceous perennial native to the high mountains in the southern regions of Serbia and Macedonia [1, 2].

The main class of compounds of the essential oils of in *S. jurisicii* were identified as sesquiterpenes, consisting of spathulenol, ß-bourbonen, and n-nonanal [52].

Hydroethanolic extracts of *S. jurisicii* showed in vitro antioxidant activity against DPPH and ABTS free radicals [53].

2.17 Salvia glutinosa *L.*

S. glutinosa is a herbaceous perennial plant native to Central and East Europe, and West Asia. It is found in forested areas in deciduous and mixed woods especially in the shade and partial shade (Fig. 2c) and in calcareous soils, at an altitude of 100–1600 m a.s.l. [1, 2].

Antioxidant activity of *S. glutinosa* aerial parts have been investigated employing different in vitro assay systems. The extracts exhibited weak nitric oxide and H_2O_2 scavenging activity, but good reducing power that is comparable with Vit C and significant DPPH radical scavenging activity [54].

Veličković et al. [55] reported that *S. glutinosa* extracts have better antioxidant capacity than *S. officinalis* estimated by DPPH assay. Further Loizzo et al. [54] found that *S. glutinosa* extract exhibit higher DPPH radical-scavenging activity than ascorbinic acid (IC_{50} = 5.0 µg/mL) with an IC_{50} value of 3.2 µg/mL.

Hydroalcoholic extracts of *S. glutinosa* expressed great activity against the yeasts: *Staphylococcus aureus* and *Candida albicans* [55].

Hatipoglu et al. [5] in their extensive study of Turkish *Salvia* species for volatile compounds found that *S. glutinosa* is rich in borneol and is a part of borneol chemotype according to the cluster analysys.

S. glutinosa exhibited IC_{50} values of 30.4, 29.5 and 29.1 µg/mL against lung carcinoma large cell ACHN, human renal cell adenocarcinoma, cell line and human breast cancer, respectively [56].

Flavonoid compounds reported for *S. glutinisa* are presented in Table 1.

2.18 Salvia grandiflora *Etl. Is a Synonym of* Salvia tomentosa *Mill. [2]*

2.19 Salvia lavandulifolia *Vahl., Is a Synonym of* Salvia officinalis *Subsp.* Lavandulifolia *(Vahl.) Gams [2]*

Natural distribution of *S. lavandulifolia* is restricted to Iberian Peninsula, its main localities being in Central, Southern and Eastern Spain, but it also penetrates in Southern France [1, 2].

Porres-Martínez et al. [57] summarized the available data concerning the phytochemistry and pharmacology of *S. lavandulifolia.* The plant has been traditionally used for its spasmolytic, antiseptic, analgesic, sedative and anesthesic activities. Pharmacological studies confirmed that the plant can act as memory enhancing and anti-dementia drug, due to the inhibition of cholinesterase enzyme. The species also possesses a wide range of pharmacological uses including: antioxidant, anti-inflammatory, estrogenic and central nervous system depressant properties, relevant to the treatment of Alzheimer's disease.

Porres-Martínez et al. [57] reported that essential oil of *S. lavandulifolia* have a selective acetylcholinesterase-inhibiting effect with an IC_{50} value of 0.03 μg/ml. The main reason for this activity is believed to be the presence of monoterpenes 1,8-cineole and α-pinene, which have IC_{50} values of 0.67 and 0.63 mM, respectively.

Flavonoid compounds identified in *S. lavandulifolia* are presented in Table 1.

2.20 Salvia napifolia *Jacq.*

Its distribution is restricted to East Aegean Islands and Western Turkey [1, 2].

Şenol et al. [31] reported that dichloromethane and ethyl acetate extracts of the species under concentration of 0.1 mg/mL did not display inhibitory activity against AchE and BchE enzymes. Also, the exctacts showed low (dichloromethane extact) to strong (ethyl acetate, and methanol extracts) activity against DPPH radicals.

Acetone extract of the roots of *S. napifolia* contained diterpenoids and diterpenes [58].

2.21 Salvia nemorosa *L.*

S. nemorosa L. (*syn. S. sylvestris*) is a species growing in southeastern Europe, southern Siberia, and Asia Minor [1, 2]. It grows both in deciduous forests and in open areas including near agricultural fields (Fig. 2f).

In the Bulgarian traditional medicine *S. nemorosa* is used mainly for treatment of stomach ache, diarrhea, hemorrhages, and furuncles [59]. Leaves of the species have been used in Turkish medicine to stop bleeding when applied externally. Aqueous and ethanolic extracts from the aerial parts showed anti-nociceptive activity with an efficacy similar to classical analgesic drugs. Its aerial parts contained diterpenoid compounds, megastigmane glycosides, triterpenes, and sterols [60]. Data about flavonoid components of *S. nemorosa* are presented in Table 1.

The studies on essential oils of Iranian *S. nemorosa* lead to isolation of *β*-caryohpyllene, germacrene B, caryophyllene oxide, *cis*-*β*-farnesene and germacrene D, which were its main components [5].

Şenol et al. [31] reported that dichloromethane and ethyl acetate extracts of the species did not display inhibitory activity against AchE and BchE enzymes at concentration of 0.1 mg/mL, and the exctracts showed low (dichloromethane extract) to strong (ethyl acetate, and methanol extracts) activity against DPPH radicals. The extracts displayed moderate to high aldehyde oxidase, xanthine oxidase, acetylcholinesterase, and a-glucosidase inhibitory activity and moderate cytotoxicity. The oxygenated sesquiterpenes were the major components. Methanolic extract showed the high DPPH radical scavenging activity (IC_{50} = 82 μg/mL) [61].

Aqueous methanolic extract of *S. nemorosa* l *albiflora* was demonstrated to have lipid peroxidation-inhibiting activities more than that of ascorbic acids [39].

2.22 Salvia nutans L.

S. nutans occurs in Central and Eastern Europe to Asia Minor and Caucasus. It is found in most cases in plainy and steppe habitats, occasionally in the mountains [1, 2].

Zlatanov and Antova [62] examined the content of glyceride oil, phospholipid, sterol and tocopherol of seeds of *S. nutans*. The content of phospholipids in the oils was found to be within the range 1.0–1.6%. Phosphatidylcholine phosphatidylinositol and phophatidic acids were the main components in the phospholipid fraction. In the sterol fraction β-sitosterol predominated, followed by campesterol and stigmasterol. Mainly α- and γ-tocopherol were found in this class of compounds.

Flavonoid compounds detected of *S. nutans* are presented in Table 1.

2.23 Salvia officinalis *L.*

S. officinalis is native to Mediterranean region but is cultivated in numerous countries worldwide with Mediterranean and temperate continental climate. Therefore, a clear distinction between its populations of natural and cultivated origin is not always possible [1, 2].

Grdiša et al. and Hamidpour et al. [46, 63] did a literature survey concerning pharmacological actions and chemical composition of *S. offcinalis*.

Flavonoid compounds identified of *S. officinalis* are presented in Table 1.

2.24 Salvia pinnata *L.*

This species occurs in Europe only in the southeastern corner of Balkan Peninsula, in Bulgaria and Turkey. Eastwards its range spreads through Asia Minor and Cyprus to Lebanon and Palestina [1, 2].

The main components of the essential oils were identified as bornyl acetate, camphor, camphene, bornyl formate, a-pinene and borneol. The oils showed antimicrobial activity against standard bacteria and yeast [64].

Flavonoid compounds detected of *S. pinnata* are presented in Table 1.

2.25 Salvia phlomoides *Asso*

In Europe this species occurs in Spain and in Northwest Africa (Algeria, Morocco, and Tunisia) [1, 2].

Abietane diterpenoids isolated from acetone extract of the root of *S. phlomoides* were reported [65].

2.26 Salvia pomifera *L.*

This species is endemic to South Aegean region, occurring in Southern Greece, East Aegean islands and Crete. It inhabits *Pinus brutia* woodland, calcareous scrub, garigue and rocky places at altitude 200–1200 m [1, 2].

Essential oils of the aerial parts of *S. pomifera* subsp. *calycina* were analyzed from different geographical regions by different authors. Monoterpene fraction was the most abundant. The main oil components were in all cases α- and/or β-thujone [66–69]. The oil and oxygenated monoterpenes α- and β-thujone exhibited fungistatic effect to soil borne pathogens—*Rhizoctonia solani* and *Sclerotinia sclerotiorum* [68].

2.27 Salvia pratensis *L.*

This is a widespread species native to Europe, western Asia and northern Africa. It grows in meadows, fields, banks [1, 2].

S. pratensis is poor in essential oil content. Caryophyllene was determined as major compound in the essential oil [6, 70–72].

Antimicrobial activity of ethanolic extracts of the plant has been reported by Veličković et al. [6].

Flavonoid compounds detected of *S. pratensis* are presented in Table 1.

2.28 Salvia ringens *Sibth. et Sm.*

S. ringens is a hardy herbaceous perennial plant, distributed in South and Eastern parts of Balkan Peninsula, mostly in areas with Mediterranean (Fig. 2e) or transitional climate conditions [1, 2].

Main components of the essential oil of *S. ringens* originating from Macedonia were identified as monoterpenes 1.8-cineole and α-pinene [73–75], whereas the main constituents of essential oils, obtained of *S. ringens* growing wild in Bulgaria, were camphor and borneol [76].

Essential oil, ethanol, and water extracts showed antimicrobial activity [73–75].

Ethanol and water extracts demonstrated cytotoxic activity against colon carcinoma HCT-116 cell line [73]. Acetone extract of the roots of *S. ringens* and seven abietane diterpenes isolated by bio-assay guided fraction showed marked concentration-dependent cytotoxic activity against HeLa cells [77]. Alcohol extracts *S. ringens* exhibited strong antioxidant activity [78, 79].

Flavonoid compounds reported for *S. ringens* are presented in Table 1.

2.29 Salvia scabiosifolia *Lam.*

Salvia scabiosifolia is a rare plant with very restricted habitats in the world. It can be found only on the Crimean Peninsula and in one isolated locality in Bulgaria. Its habitat is typically in dry and rocky places (Fig. 2d) [1, 2].

The basic components identified in the essential oils of *S. scabiosifolia* were camphor, borneol and camphene [80]. Methanolic extract demonstrated higher antioxidant effect (EC_{50} 5.4 µg/mL) (lipid peroxidation-inhibiting activitie) than that of ascorbic acid (EC50 123.8 µ/mL) evaluated in an enzyme-independent biological system [39].

Flavonoid compounds detected in *S. scabiosifolia* are presented in Table 1.

2.30 Salvia sclarea *L.*

S. sclarea is native to Southern Europe, particularly Mediterranean region, Northern Africa and Central Asia (Fig. 2b). It is cultivated worldwide [1, 2].

Variability of essential oil profiles of *S. scalrea* has been extensively studied. Generally linalyl acetate, linalool, germacrene had been determined as major components [81–85]. Monoterpenes linalool and linalyl acetate were characteristic substances of flowers, whereas the sesquiterpenes occurred in higher proportions in leaves [86].

Essential oil and alcohol extracts of *S. sclarea* were shown to have antioxidant, antibacterial, antifungal, antiinflammatory, analgesic, antinociceptive and cytotoxic activities [83, 85, 87–91].

Abietane diterpenoids have been shown to be bacteriostatic as well as bacteriocidal for the cultures of *S. aureus* and *S. epidermidis* strains, regardless of their antibiotic susceptibility profile [88].

Flavonoid compounds detected in *S. sclarea* are presented in Table 1.

2.31 Salvia sclareoides *Brot.*

S. sclareoides is an aromatic herb native to the western part of Iberian Peninsula [1, 2].

New compound lupene triterpenetriol and other, already known compounds, like nepetidin, nepeticin, lupendiol (1beta,11alpha)-dihydroxy-lup-20(29)-en-3-one, ursolic acid, sumaresinolic acid and hederagenin have been isolated from the acetone extracts of the aerial parts of *S. sclareoides*. Acetone, ethanol, *n*-butanol and water extracts of *S. sclareoides* have been shown to have inhibitory activity of acetylcholinesterase (AChE) enzyme at 10 μg/mL [92, 93]. Rosmarinic acid was determined as the only explicit binder for AchE [94]. The n-butanol extract showed antioxidant activity (IC_{50} = 7.9 μg/mL) higher than that of the synthetic antioxidant BHT (IC_{50} = 29.7 μg/mL) in DPPH radical scavenging method. Methanol, ethanol and water extracts had IC_{50} values of 114, 134 and > 500 μg/mL, respectively. In b-carotene/linoleic acid bleaching assay, the acetone extract presented the highest activity (IC_{50} = 38.4 μg/mL), similar to that of ascorbic acid (IC_{50} = 26.1 μg/mL). Similar values were obtained for the n-butanol, methanol, ethanol and water extracts. Extracts were not cytotoxic nor genotoxic, making them suitable for nutraceutical and medicinal purposes. Several phenolics—vanillic, gallic, gentisic, caffeic, syringic, coumaric and ferulic acids have been identified in the extracts [92, 93].

Flavonoid compounds detected in *S. scleroides* are presented in Table 1.

2.32 Salvia sonklarii *Pant.*

It is distributed in the Northwestern part of the Balkan Peninsula [1, 2].

2.33 Salvia teddii *Turrill*

This species is endemic to Greece [1, 2].

2.34 Salvia tomentosa *Mill.*

S. tomentosa is distributed in Southeastern Europe (mostly Balkan Peninsula and Crimea) and part of Western Asia (Anatolia and Near East) [1, 2].

The species is traditionally used for preparation of high quality sage tea. In folk medicine, the plant is popular for treatment of stomachache, asthma, cold and inflamed wounds. *S. tomentosa* is used in the Turkish folk medicine, and it has been shown that its oils have antibacterial and antioxidant activities [51, 95].

A considerable variation between the populations with respect to the oil composition was reported. Most often α- and β-pinene, and borneol have been identified as the main constituents, as well as 1,8-cineole, camphor and camphene [96]. Handlidou et al. [97] reported for the first time that the oils contained cis-thujone.

The essential oil showed considerable acetylcholinesterase inhibitory activity ($IC_{50} = 0.28 \pm 0.06$ μg/mL), comparable with that of galanthamine. Also the high antioxidant capacity and moderate antifungal and antibacterial activities against *Candida albicans* and Gram-positive bacteria of the essential oil were reported [98].

Flavonoid compounds detected in *S. tomentosa* are presented in Table 1.

2.35 Salvia transsylvanica *(Schur ex Griseb. & Schenk) Schur*

S. transsylvanica is endemic to Northern and Central Romania [1, 2].

Ethanol extract of the species induced significant analgesic and antipyretic effect which is similar to that of acetylsalicylic acid. Also the extract have been shown to exhibited antiinflammatory, anticonvulsan and antiulcerogenict activity. The plant proved to be relatively nontoxic [99].

The following compounds were isolated and identified from the aerial parts of *S. transsylvanica*: lupeol acetate, β-amyrin, β-sitosterol, caffeic acid, quercetrin, narcissin and chlorogenic acid [100].

Strong antioxidant activity of methanolic extract of plant assayed in an enzyme-independent biological system was reported by Janicsák et al. [39].

Flavonoid compounds detected in *S. transsylvanica* are presented in Table 1.

2.36 Salvia valentina *Vahl.*

S. valentina is endemic species from the Mediterranean coastal region of Spain [1, 2].

The plant was successfully regenerated in vitro from adult plants [101].

2.37 Salvia verbenaca *L.*

S. verbenaca has wide natural distribution ranging from Macaronesia throught the Western Europe and Mediterranean region to Caucasus [1, 2].

In the Sicilian traditional medicine the aerial parts are used to resolve cases of kidney stones, chewing the fresh leaves or in decoction. The plant is also known as bactericide against respiratory ailments, as healing in wounds and ulcers, and above all as eyedrops, because fruits or seeds when applied on the eyes remove impurities or dust particles [102].

Strong antioxidant activity has been reported for *S. verbenaca*. The studies showed that the rosmarinic acid, its derivatives and the high amount of phenolic compounds are more likely to be responsible for the observed antioxidant activities [103–105].

The profile of essential oil of *S. verbenaca* from different geographical regions showed noticeably qualitative and quantitative differences [102, 106].

The oil exhibited a good activity as inhibitor of growth of Gram-positive bacteria [102].

Anthihypertensive acticvity of ethanolic extracts of *S. verbenaca* as well as of individual flavonoid compounds isolated of the extract was reported. Crude extract and 5-hydroxy 3,4′,7-trimethyl flavones displayed significant decrease of blood pressure and heart rate with 36.0 and 30.0 (mmHg) decrease for BP and 18.8% and 28.5% for HR, respectively [107].

S. verbenaca exhibited cytotoxic effects on human colon cancer cells (HT-29) and human breast cancer cells (MCF-7) with IC_{50} values under 57.0 μg/mL [108].

The flavonoid compound identified in *S. verbenaca* are presented in Table 1.

2.38 Salvia verticillata *L.*

This species is a herbaceous perennial native to a wide area ranging from central Europe to western Asia, and naturalized in northern Europe and North America.

S. verticillata was introduced as an antioxidant plant, which contains polyphenols and diterpenoids. It has radical scavenging effect which is the skill to protect cells from different kinds of oxidative tensions. There are some reports about biological activity of this plant such as antibacterial and anti-diabetic effects. *S. verticillata* may have the potentiality of being used in lost or declining cognitive functions [109].

Variable essential oil profile of *S. verticillata* among populations from different geographical regions has been observed. Generally the oil consists mainly of caryophyllene, but also of β-phellandrene, humulene, germacrene and D-spathulenol [71, 110, 111].

Essential oil of the plant showed cytotoxic effect on colorectal adenocarcinoma (Caco-2) cell line [111].

Antibacterial and antioxidant properties were reportd for *Salvia verticillata* L. var. *amasiaca* extracts [112].

Using activity-guided fractionation chrysoeriol has been isolated as active antioxidant compound [113]. The flavonoid compounds identified in *S. verticillata* are presented in Table 1.

Hydro-alcoholic extract from *S. verticillata* produced significant anticonvulsant activity and significant antidepressant effect [109].

Different extracts (petroleum ether, chloroform, ethyl acetate, and methanol) of *S. verticillata* were shown to have weak or not detectable at 0.2 and 1 mg/mL acetylcholinesterase (AChE) and butyrylcholinesterase (BChE) inhibitory activity [30].

2.39 Salvia viridis *L.*

S. viridis (synonym *Salvia horminum* L.) is distributed in Europe, particularly the Mediterranean region (Fig. 2a) and has widely been cultivated in Britain as an ornamental plant [1, 2]. It has been used in traditional medicine, for example, an infusion of leaves as a gargle for sore gums, and has also been employed to increase the quality of liquor by putting leaves and seeds into the fermentation tank.

The presences of triterpenoids and flavonoids in the aerial part of *S. viridis* and diterpenoids in the root part have been reported [114–116]. It has been shown that tri- and diterpenoids of the plant exhibited markedly antibacterial activity [115, 116].

Verbascoside (acteoside), which is a major component in *Salvia viridis*, showed a significant protective effect against UVA induced damage in a human skin fibroblast model in vitro [114].

Significant free radical-scavenging capacity of the methanolic extract of *S viridis* has been established [117].

The flavonoid compound identified of *S. viridis* are presented in Table 1.

2.40 Salvia virgata *Jacq.*

This species is distributed in Southeastern Europe, mostly on Balkan Peninsula, and occupies also parts of Western and Central Asia [1, 2].

In Turkey it is used for treatment of skin diseases, wounds and against blood cancer [118]. The ethanol extract of *S. virgata* showed a dose-dependent inhibitory effect on the α-amylase activity. Flavonoid aglycone chrysoeriol has been determined as antidiabetic compounds that inhibited α-amylase activity in a dose-dependent manner [119]. The flavonoid compounds identified of *S. virgata* are presented in Table 1.

Akkol et al. [118] have been established that methanol extract from the aerial parts of *S. virgata* significantly inhibited carrageenan-induced paw edema and p-benzoquinone-induced abdominal constriction at 100 mg/kg dose. Also anti-inflammatory and antinociceptive activities of the extract have been reported.

3 Conclusions and Future Prospects

Salvia is taxonomically difficult genus requiring extensive further experimental investigations. The taxonomic complexity is illustrated by the numerous synonyms of the species. This is a prerequisite for the difficulties in the interpretation of results obtained in phytochemical analyzes into the general context and taxonomic scheme of the genus. The increased interest to the chemistry and bioactivity of the species of genus *Salvia* is reflected in the substantial increase of number of published papers dealing with these topics. The present review is focused on the last two decades and covers mainly studies on the type of polyphenols and essential oils content, on the one hand, and on the antioxidant and antimicrobial activities on the other. Little information concerning the main terpenoids was included in the review, because this class of substances was subjected to extensive studies over the last century. The trend of the next investigations should be targeted first, to the genetic differentiation within and among the species in relation to their chemical properties, and second, to the qualitative composition of the main biologically active substances, like poliphenols and essential oils. The isolation of these compounds and testing their biological activity, combined with the studies on the mechanism of their action, would be the key for new drug discoveries.

Acknowledgements We would like to express our sincere thanks to Dr. Peter Zhelev for his support, for the critical reading and valuable suggestions, which improved the earlier draft of the manuscript.

References

1. Hedge IC (1972) Salvia L. In: Tutin TG, Heywood VH, Burges NA et al (eds) Flora Europaea, vol 3. Cambridge University Press, Cambridge, pp 188–192
2. World Checklist of Selected Plant Families (WCSP) (2016) Facilitated by the Royal Botanic Gardens, Kew. Published on the Internet. http://apps.kew.org/wcsp/
3. Güllüce M, Özer H, Barış Ö et al (2006) Chemical composition of the essential oil of *Salvia aethiopis* L. Tur J Biol 30:231–233
4. Morteza-Semnani K, Goodarzi A, Azadbakht M (2005) The essential oil of *Salvia aethiopis* L. J Essent Oil Res 17:274–275
5. Hatipoglu SD, Zorlu N, Dirmenci T et al (2016) Determination of volatile organic compounds in fourty five *Salvia* species by thermal desorption-GC-MS technique. Rec Nat Prod 10:659–700
6. Veličković DT, Ranđelović NV, Ristić MS et al (2002) Chemical composition and antimicrobial action of the ethanol extracts of *Salvia pratensis* L., *Salvia glutinosa* L. and *Salvia aethiopis* L. J Serb Chem Soc 67:639–646
7. Baricevic D, Bartol T (2000) The biological/pharmacological activity of the Salvia genus. In: Kintzios SE (ed) Sage the Genus Salvia. Harwood Academic Publishers, Amsterdam, pp 143–184
8. Kilic T, Dirmenci T, Satil F et al (2005) Fatty acid compositions of seed oils of three Turkish *Salvia* species and biological activities. Chem Nat Compd 41:276
9. Firuzi O, Miri R, Asadollahi M et al (2013) Cytotoxic, antioxidant and antimicrobial activities and phenolic contents of eleven *Salvia* species from Iran. Iran J Pharm Res 12:801–810
10. Riccobono L, Maggio A, Rosselli S et al (2016) Chemical composition of volatile and fixed oils from of *Salvia argentea* L. (Lamiaceae) growing wild in Sicily. Nat Prod Res 30:25–34
11. Sarac N, Ugur A (2007) Antimicrobial activities and usage in folkloric medicine of some Lamiaceae species growing in Mugla, Turkey. Eurasia J Biosci 4:28–37
12. Stagos D, Portesis N, Spanou C et al (2012) Correlation of total polyphenolic content with antioxidant and antibacterial activity of 24 extracts from Greek domestic Lamiaceae species. Food Chem Toxicol 50:4115–4124
13. Orhan IE, Senol FS, Ercetin T et al (2013) Assessment of anticholinesterase and antioxidant properties of selected sage (*Salvia*) species with their total phenol and flavonoid contents. Ind Crops Prod 41:21–30
14. Lakhal H, Kabouche A, Alabdul Magid A et al (2014) Triterpenoids from *Salvia argentea var. aurasiaca* (Pomel) Batt. & Trab. and their chemotaxonomic significance. Phytochemistry 102:145–151
15. Petrović S, Pavlović M, Tzakou O et al (2009) Composition and antimicrobial activity of *Salvia amplexicaulis* Lam. essential oil J Essent Oil Res 21:563–566
16. Alimpic A, PljevlakušicD Šavikin K et al (2015) *Salvia jurisicii* Košanin and *Salvia amplexicaulis* Lam. chemical composition of the essential oils. Food Anal 26:64–67
17. Veličković TD, Ristić SM, Milosavljević PN et al (2012) Chemical composition of the essential oils of *Salvia austriaca* Jacq. and *Salvia amplexicaulis Lam.* from Serbia. Agro Food Ind Hi Tech 23:7–9
18. Orhan IE, Senol FS, Ozturk N et al (2012) Profiling of in vitro neurobiological effects and phenolic acids of selected endemic *Salvia* species. Food Chem 132:1360–1367

19. Alimpić A, Oaldje M, Matevski V et al (2014) Antioxidant activity and total phenolic and flavonoid contents of *Salvia amplexicaulis* Lam. extracts. Arch Biol Sci 66:307–316
20. Kolak U, Ari S, Birman H et al (2001) Cardioactive diterpenoids from the roots of *Salvia amplexicaulis*. Planta Med 67:761–763
21. Kuźma Ł, Wysokińska H, Sikora J et al (2016) Taxodione and extracts from *Salvia austriaca* roots as human cholinesterase inhibitors. Phytother Res 30:234–42
22. Ibrahim TA (2012) Chemical composition and biological activity of extracts from *Salvia bicolor* Desf. growing in Egypt. Molecules 17:11315–34
23. Ibrahim TA (2014) Chemical composition and antimicrobial activity of essential oil of *Salvia bicolor* Desf. growing in Egypt. J Essent OilBear Pl 17:104–111
24. Savikin-Fodulovic KP, Slavoljub RT, Menkovic NR (2002) The essential oil of *Salvia brachyodon* Vandas. Lamiaceae. J Essent Oil Res 14:342–343
25. Tzakou O, Couladis M, Slavkovska V et al (2003) The essential oil composition of *Salvia brachyodon* Vandas. Flavour Fragr J 18:2–4
26. Sokovic M, Grubisic D, Ristic M (2005) Chemical composition and antifungal activities of essential oils from leaves, calyx and corolla of *Salvia brachyodon* Vandas. J Essent Oil Res 17:227–229
27. Cardile V, Russo A, Formisano C et al (2009) Essential oils of *Salvia bracteata* and *Salvia rubifolia* from Lebanon: chemical composition, antimicrobial activity and inhibitory effect on human melanoma cells. J Ethnopharmacol 126:265–72
28. Bagci E, Vural M, Dirmenci T et al (2004) Fatty acid and tocochromanol patterns of some *Salvia* L. species. Z Naturforsch B 59c:305–309
29. Martinez-Frannces V, Hahn E, Juan-Vicedo J et al (2012) Ethnobotanical study of the sages used in traditional Valencian medicine and as essential oil: characterization of an endemic *Salvia* and its contribution to local development. Contrib Sci 8:77–84
30. Orhan I, Kartal M, Naz Q et al (2007) Antioxidant and anticholinesterase evaluation of selected Turkish *Salvia* species. Food Chem 103:1247–12254
31. Şenol FS, Orhan I, Celep F et al (2010) Survey of 55 Turkish *Salvia* taxa for their acetylcholinesterase inhibitory and antioxidant activities. Food Chem 120:34–43
32. Özler MA, Duru ME, Diri HA et al (2009) *Antioxidant activity* and *chemical composition* of the *essential oil* of *Salvia candidissima Vahl. growing wild* in *Turkey*. Acta Hortic 826:363–370
33. Kursat M, Erecevit P, Sari A, Emre I, Kirbag S, Civelek S (2012) The antimicrobial activities of seed fatty acid extracts from some *Salvia* L. species. Turk J Sci Technol 7:31–36
34. Zadeh AMK, Ghalam JZ, Sonboli A et al (2014) Effects of hydroalcholic and chloroformic extracts of *Salvia candidissima* on hyperalgesia and Edema during adjuvant-induced arthritis. Koomesh—J Semnan Univer Med Sci 16:239–245
35. Máthé I, Máthé Á, Hohmann J et al (2010) Volatile and some non-volatile chemical constituents of Mediterranean *Salvia* species. Isr J Plant Sci 58:273–277
36. Janicsák G, Hohmann J, Zupkó I et al (2003) Diterpenes from the aerial parts of *Salvia candelabrum* and their protective effects against lipid peroxidation. Planta Med 69:1156–1159
37. Akimov IA (ed) (2009) Red data book of Ukraine. Plant Kingdom. Globalkonsalting, Kiyv, p 912
38. Euro+Med (2006): Euro + Med PlantBase—the information resource for Euro-Mediterranean plant diversity. Published on the Internet http://ww2.bgbm.org/EuroPlusMed/. Accessed 15 Jul 2008
39. Janicsák G, Zupkó I, Máthé I et al (2010) Comparative study of the antioxidant activities of eleven Salvia species. Nat Prod Commun 5:227–230
40. Janicsák G, VeresK Kakasyc AZ et al (2006) Study of the oleanolic and ursolic acid contents of some species of the Lamiaceae. Biochem Syst Ecol 34:392–396
41. Tsujikawa K, KuwayamaK Miyaguchi H et al (2008) Determination of salvinorin A and salvinorin B in *Salvia divinorum*-related products circulated in Japan. Forensic Sci Int 180:105–109

42. Ciesla L, Hajnos M, Staszek D et al (2010) Validated binary high-performance thin-layer chromatographic fingerprints of polyphenolics for distinguishing different Salvia species. J Chromatogr Sci 48:421–427
43. Sajewicz M, Staszek D, Wróbel MS et al (2012) The HPLC/DAD fingerprints and chemometric analysis of flavonoid extracts from the selected sage (*Salvia*) species chromatography. Res Int, Article ID 230903. https://doi.org/10.1155/2012/230903
44. Radosavljević I, Satovic Z, Jakse J et al (2012) Development of new microsatellite markers for *Salvia officinalis* L. and its potential use in conservation-genetic studies of narrow endemic *Salvia brachyodon* Vandas. Int J Mol Sci 13:12082–12093
45. Grdiša M, Jug-Dujaković M, Lončarić M et al (2015) Dalmatian sage (*Salvia officinalis* L.): a review of biochemical contents, medical properties and genetic diversity. Agric Conspec Sci 80:69–78
46. Gali-Muhtasib H (2006) Anticancer and medicinal properties of essential oil and extracts of East Mediterranean sage (*Salvia triloba*). In: Khan M Ather A (eds) Advances in phytomedicine, vol 2: Lead molecules from natural products: discovery and new trends. Elsevier Science, Netherlands, pp 169–180
47. Topçu G, Ozturk M, Kusman T et al (2013) Terpenoids, essential oil composition, fatty acid profile, and biological activities of Anatolian *Salvia fruticosa* Mill. Turk J Chem 37:619–632
48. Orhan I, Aslan M (2009) Appraisal of scopolamine-induced antiamnesic effect in mice and in vitro antiacetylcholinesterase and antioxidant activities of some traditionally used Lamiaceae plants. J Ethnopharmacol 122:327–332
49. Abdelhalim A, Chebib M, Aburjai T et al (2014) GABAA Receptor modulation by compounds isolated from *Salvia triloba* L. Adv Biol Chem 4:148–159
50. Abu-Dahab R, Afifi F, Kasabri V et al (2012) Comparison of the antiproliferative activity of crude ethanol extracts of nine *Salvia* species grown in Jordan against breast cancer cell line models. Pharmacog Mag 8:319–324
51. Askun T, Tumen G, Satil F et al (2009) Characterization of the phenolic composition and antimicrobial activities of Turkish medicinal plants. Pharm Biol 47:563–571
52. Alimpić A, Knežević A, Šavikin K et al (2016) Composition and biological activities of different extracts of *Salvia jurisicii*, a rare and endemic Macedonian species. Plant Biosyst. https://doi.org/10.1080/11263504.2016.1219414
53. Alimpić A, Duletić-Laušević S, Matevski V et al (2015) Antioxidant activity of *Salvia jurisicii* Košanin ethanol extracts. Bot Serb 39:53–58
54. Esmaeili A, Tavassoli A, Ebrahimzadeh MA (2009) Antioxidant activity and free radical scavenging activity of *Salvia glutiosa* growing in Iran. Pharmacology 2:109–116
55. Veličković DT, Karabegović IT, Stojičević SS et al (2011) Comparison of antioxidant and antimicrobial activities of extracts obtained from *Salvia glutinosa* L. and *Salvia officinalis* L. Hem Ind 65:599–605
56. Loizzo MR, Abouali M, Salehi P et al (2014) In vitro antioxidant and antiproliferative activities of nine *Salvia* species. Nat Prod Res 28:2278–2285
57. Porres-Martínez M, Carretero ME, Gomez-Serranillos MP (2013) Pharmacological activity of *Salvia lavandulifolia*Vahl. and chemical components of its essential oil. Lazaroa 34:237–254
58. Ulubelen A, Topcu G, Sönmez U et al (1995) Abietane diterpenes from *Salvia napifolia*. Phytochemistry 40:861–864
59. Tz Daskalova (2004) On some specificities of seed formation in *Salvia nemorosa* (*Lamiaceae*). Phytol Balcan 10(1):79–84
60. Skala E, Wysokinska H (2004) In vitro regeneration of *Salvia nemorosa* L. from shoot tips and leaf explants. In Vitro Cell Dev Biol Plant 40:596–602
61. Bahadori B, Asghari B, Dinparast L et al (2017) *Salvia nemorosa* L.: a novel source of bioactive agents with functional connections. Food Sci Technol 75:42–50
62. Zlatanov MD, Antova GA (2004) Composition of biologically active lipids of *Lamiaceae* seed oils. Grasas Aceites 55:143–147

63. Hamidpour M, Hamidpour R, Hamidpour S et al (2014) Chemistry, pharmacology, and medicinal property of sage (*Salvia*) to prevent and cure illnesses such as obesity, diabetes, depression, dementia, lupus, autism, heart disease, and cancer. J Tradit Complement Med 4:82–88
64. Somer NU, Sarikaya BB, Erac B et al (2015) Chemical composition and antimicrobial activity of essential oils from the aerial parts of *Salvia pinnata* L. Rec Nat Prod 9:614–618
65. Rodríguez B (2003) A methoxyabietane diterpenoid from the root of *Salvia phlomoides* and structural correction of another diterpene from *Cryptomeria japonica*. Z Naturforsch—B 58b, 324–327
66. Karousou R, Vokou D, Kokkini S (1998) Distribution and essential oils of *Salvia pomifera* subsp. *pomifera* (Labiatae) on the island of Crete (S Greece). Biochem Syst Ecol 26:889–897
67. Pitarokili D, Tzakou O, Couladis M et al (1999) Composition and antifungal activity of the essential oil of *Salvia pomifera* subsp. *calycina* growing wild in Greece. J Essent Oil Rese 11:655–659
68. Glamočlija J, Soković M, Vukojević J et al (2006) Chemical composition and antifungal activities of essential oils of *Satureja thymbra* L. and *Salvia pomifera* ssp. *calycina* (Sm.) Hayek. J Essent Oil Res 18:115–117
69. Pitarokili D, Tzakou O, Kalamarakis A (2002) Activity of the essential oil of *Salvia pomifera* L. ssp. *calyclna* (Sm.) Hayek against soil borne pathogens. *J Essent Oil Res* 14(1):72–75
70. Velickovic D, Ristic M, Velickovic A (2001) GC-MS analysis of the essential oils from several species of salvia from the Southeast Serbia. Lekovite sirovine 21:51–60
71. Smekalova K, Dusek K, Duskova E (2010) *Salvia verticillata* L. and *Salvia pratensis* L.—the variability of essential oil content in the Czech Republic. Acta Hortic 860:51–60
72. Anačkov G, Božin B, Zorić L et al (2009) Chemical composition of essential oil and leaf anatomy of *Salvia bertolonii* Vis. and *Salvia pratensis* L. (Sect. *Plethiosphace*, Lamiaceae). Molecules 14:1–9
73. Tzakou O, Pitarokili D, Chinou IB et al (2001) Composition and antimicrobial activity of the essential oil of *Salvia ringens*. Planta Med 67:81–83
74. Šavikin KP, Ristić MS, Zdunić GM et al (2008) Chemical composition and antimicrobial activity of essential oil of *Salvia ringens* Sibth. et Sm. var. *baldacciana* Briq. J Essent Oil Res 20:363–365
75. Alimpić A, Pljevljakušić D, Šavikin K et al (2015) Composition and biological effects of Salvia ringens (Lamiaceae) essential oil and extracts. Ind Crops Prod 76:702–709
76. Georgiev V, Marchev A, Nikolova M et al (2013) Chemical compositions of essential oils from leaves and flowers of *Salvia ringens* Sibth. et Sm. growing wild in Bulgaria. J Essent Oil Bear Pl 16:624–629
77. Janicsák G, Zupkó I, Nikolova M et al (2011) Bioactivity-guided study of antiproliferative activities of *Salvia* extracts. Nat Prod Commun 6:575–579
78. Couladis M, Tzakou O, Stojanovic D et al (2001) The essential oil composition of *Salvia argentea* L. Flavour Fragr J 16:227–229
79. Tusevski O, Kostovska A, Iloska A et al (2014) Phenolic production and antioxidant properties of some Macedonian medicinal plants. Cent Eur J Biol 9:888–900
80. Marchev A, Georgiev V, Nikolova M et al (2012) Chemical composition of essential oil of *Salvia scabiosifolia* Lam. from Bulgaria. J Essent Oil Bear Pl 15:908–914
81. Sharopov SF, Setzer WN (2012) The essential oil of *Salvia sclarea* L. from Tajikistan. Rec Nat Prod 6:75–79
82. Pešić PŽ, Banković VM (2003) Investigation on the essential oil of cultivated *Salvia sclarea* L. Flavour Fragr J 18:228–230
83. Džamić A, Soković M, Ristić M et al (2008) Chemical composition and antifungal activity of *Salvia sclarea* (Lamiaceae) essential oil. Arch Biol Sci 60:233–237
84. Verma RS (2010) Chemical investigation of decanted and hydrophilic fractions of *Salvia sclarea* essential oil. Asian J Tradit Med 5:102–108

85. Taârit MB, Msaada K, Hosni K et al (2014) GC analyses of *Salvia* seeds as valuable essential oil source. Adv Chem Article ID 838162, 6 pp
86. Schmiderer C, Grassi P, Novak J et al (2008) Diversity of essential oil glands of clary sage (*Salvia sclarea* L., Lamiaceae). Plant Biol 10:433–40
87. Gulcin I, Oguz MT, Beydemir S et al (2004) Evaluation of the antioxidant and antimicrobial activities of clary sage (*Salvia sclarea* L.). Turk J Agric For 28:25–33
88. Kuźma Ł, Rózalski M, Walencka E et al (2007) Antimicrobial activity of diterpenoids from hairy roots of *Salvia sclarea* L.: Salvipisone as a potential anti-biofilm agent active against antibiotic resistant Staphylococci. Phytomedicine 14:31–35
89. Yuce E, Yildirim N, Yildirim NC et al (2014) Essential oil composition, antioxidant and antifungal activities of *Salvia sclarea* L. from Munzur Valley in Tunceli, Turkey. Cell Mol Biol 60:1–5
90. Kuźma Ł, Kalemba D, Różalski M et al (2009) Chemical composition and biological activities of essential oil from *Salvia sclarea* plants regenerated in vitro. Molecules 14:1438–1447
91. Eidi A, Eidi M, Mazooji A et al (2011) *Antinociceptive* effects of ethanolic extracts of *Salvia sclarea* L. aerial parts in mice. J Sci (Islamic Azad University) 20:61–70
92. Rauter AP, Branco I, Lopes RG et al (2007) A new lupene triterpenetriol and anticholinesterase activity of *Salvia sclareoides*. Fitoterapia 78:474–81
93. Rauter AP, Dias C, Martins A et al (2012) Non-toxic *Salvia sclareoides* Brot. extracts as a source of functional food ingredients: phenolic profile, antioxidant activity and prion binding properties. Food Chem 132:1930–1935
94. Marcelo F, Dias C, Martins A et al (2013) Molecular recognition of rosmarinic acid from *Salvia sclareoides* extracts by acetylcholinesterase: a new binding site detected by NMR spectroscopy. Chemistry 19:6641–6649
95. Georgiev V, Marchev A, Haas C et al (2011) Production of oleanolic and ursolic acids by callus cultures of *Salvia tomentosa* Mill. Biotechnol Biotechnol Equip Suppl 34–38
96. Avci AB (2013) Essential oil content and composition of *Salvia tomentosa* Mill. from Gölcük, Isparta Suleyman Demirel University. J Nat Appl Sci 17:1–4
97. Hanlidou E, Karousou R, Lazari D (2014) Essential-oil diversity of *Salvia tomentosa* Mill. in Greece. Chem Biodiv 11:1205–1215
98. Marchev A, Ivanov I, Denev P et al (2015) Acetylcholinesterase inhibitory, antioxidant, and antimicrobial activities of *Salvia tomentosa* Mill. essential oil. J BioSci Biotech 4:219–229
99. Maklad YA, Aboutabl EA, El-Sherei MM et al (1999) Bioactivity studies of *Salvia transsylvanica* (Schur ex Griseb) grown in Egypt. Phytother Res 13:147–150
100. Aboutabl EA, El-Sherei MM, Meselhy KM (1999) *Investigation of lipoidal* and *phenolic contents* of *Salvia transsylvanica* (*Schur ex Griseb) cultivated* in *Egypt*. Bull Fac Pharm Cairo Univ 40:55–62
101. Cuenca S, Amo-Marco JB (2000) In vitro propagation of two Spanish endemic species of *Salvia* through bud proliferation. In Vitro Cell Dev Biol Plant 36:225–229
102. Canzoneri M, Bruno M, Rosselli S et al (2011) Chemical composition and biological activity of *Salvia verbenaca* essential oil. Nat Prod Commun 6:1023–1026
103. Khlifi S, El Hachimi Y, Khalil A et al (2006) In vitro antioxidant properties of *Salvia verbenaca* L. hydromethanolic extract. Ind J Pharm 38:276–80
104. Tepe B (2008) Antioxidant potentials and rosmarinic acid levels of the methanolic extracts of *Salvia virgata* (Jacq), *Salvia staminea* (Montbret & Aucher ex Bentham) and *Salvia verbenaca* (L.) from Turkey. Bioresour Technol 99:1584–1588
105. Farhat MB, Chaouch-Hamada R, Sotomayor JA et al (2015) Antioxidant properties and evaluation of phytochemical composition of *Salvia verbenaca* L. extracts at different developmental dtages. Plant Foods Hum Nutr 70:15–20
106. Taârit MB, Msaada K, Hosni K et al (2010) Essential oil composition of *Salvia verbenaca* L. growing wild in Tunisia. J Food Biochem 34:142–151

107. Ahmed B, Al-Howiriny TA, Mossa JS et al (2005) Isolation, antihypertensive activity and structure activity relationship of flavonoids from three medicinal plants. Ind J Chem 44B:400–404
108. Jiang YY, Zhang L, Rupasinghe HPV (2016) The anticancer properties of phytochemical extracts from *Salvia* plants. Botanics 6:25–44
109. Naderi N, Akhavan N, Aziz Ahari F et al (2011) Effects of hydroalcoholic extract from *Salvia verticillata* on pharmacological models of seizure, anxiety and depression in mice. Iran J Pharm Res 10:535–545
110. Krstic L, Malencic D Anackov G (2006) Structural investigations of trichomes and essential oil composition of *Salvia verticillata*. Bot Helv 116:159–168
111. Khosravi Dehaghi N, Ostad SN et al (2014) Cytotoxic activity of the essential oil of *Salvia verticillata* L. Res J Phcog 1:27–33
112. Erbil N, Digrak M (2015) Total phenolic and flavonoid contents, antimicrobial and antioxidant properties of *Salvia verticillata* L. var. *amasiaca* and *Salvia microstegia* Boiss & Bal from Turkish Flora. J Microbiol Antimicrob Agents 1:23–29
113. Nickavar B, Rezaee J, Nickavar A (2016) Effect-directed analysis for the antioxidant compound in *Salvia verticillata*. Iran J Pharm Res 15(1):241–246
114. Rungsimakan S (2011) Phytochemical and biological activity studies on *Salvia viridis* L. thesis submitted for Ph.D. University of Bath
115. Rungsimakan S, Rowan MG (2014) Terpenoids, flavonoids and caffeic acid derivatives from *Salvia viridis* L. cvar. Blue Jeans. Phytochemistry 108:177–188
116. Ulubelen A, Oksüz S, Kolak U et al (2000) Antibacterial diterpenes from the roots of *Salvia viridis*. Planta Med 66:458–462
117. Yazdinezhad AR, Malekzadeh M (2015) Evaluation of antioxidant effect, total phenols, anthocyanins and flavonoids contents of methanolic extract of *Salvia viridis* L. collected from Zanjan. J Zanjan Univ Med Sci Health Serv 23:100–108
118. Akkol EK, Goger F, Kosar M et al (2008) Phenolic composition and biological activities of *Salvia halophila* and *Salvia virgata* from Turkey. Food Chem 108:942–949
119. Nickavar B, Abolhasani L (2013) Bioactivity-guided separation of an α-amylase inhibitor flavonoid from *Salvia virgata*. Iran J Pharm Res 12(1):57–61
120. Lu Y, Foo YL (2002) Polyphenolics of Salvia—a review. Phytochemistry 59:117–140
121. Nikolova M, Renée JG, Genova E et al (2006) Exudate flavonoids from Bulgarian species of *Salvia*. Biochem Syst Ecol 34:360–364
122. Nikolova M, Janicsák G, Genova E et al (2006) Comparative analysis on external flavonoids of Bulgarian and Hungarian samples of *Salvia* species. Acta Bot Hung 48:361–367
123. Valant-Vetschera KM, Roitman JN, Wollenweber E (2003) Chemodiversity of exudate flavonoids in some members of the Lamiaceae. Biochem Syst Ecol 31:1279–1289
124. Cvetkovikj I, Stefkov G, Acevska J et al (2013) Polyphenolic characterization and chromatographic methods for fast assessment of culinary Salvia species from South East Europe. J Chromatogr A 1282:38–45
125. Coisin M, Necula R, Grigoras V (2012) Phytochemical evaluation of some Salvia species from Romanian flora. Analele Şti Univ "Al I Cuza" Iaşi, Ser. Nouă, 2.a. Biol Veget 58:35–44
126. Gavrilin MB, Popova OI, Gubanova EA (2010) *Phenolic compounds of the aboveground part of Salvia sclarea* L., cultivated in Stavropol region. Chem Plant Raw Mat 4:99–104 (in Russian)
127. Demirezer OL, Gurbuz P, Kuruuzum-Uz A et al (2012) Chemical constituents of two sages with free radical scavenging activity. Nat Prod Commun 7:187–90
128. Kharazian N (2013) Identification of flavonoids in leaves of seven wild growing Salvia L. (Lamiaceae) species from Iran. Prog Biol Sci 3:81–98

Bioactive Constituents of Anatolian *Salvia* Species

Gülaçtı Topçu, Rümeysa Yücer and Halil Şenol

Abstract *Salvia* (sage) species are grown world-wide, mainly in the three regions; South-East Asia, South Europe to Central and South America with approximately 1000 species. The genus *Salvia* is represented by over 100 species in Turkey. Since ancient times, they have been used in folk medicine for many purposes in the world, for more than sixty different ailments ranging from aches to epilepsy, and to treat colds, bronchitis, tuberculosis, haemorrhage, some cardiovascular and menstrual disorders. In Anatolia, *Salvia* species are used against cold and flu, and in the treatment of stomach, liver and rheumatism problems and as wound healing agents. A literature survey and our experiences indicated that terpenoids and flavonoids are the main constituents of them. 45 Anatolian *Salvia* species have been investigated for their chemical constituents with some biological activities. Di- and triterpenoids were found to be the main non-volatile constituents besides flavonoids and phenolics. The aerial parts, particularly flowers and leaves of Anatolian *Salvia* species contain flavonoids, triterpenoids, and monoterpenes while diterpenoids are found mostly in the roots. Totally 317 compounds were isolated, major part belonging to diterpenoids, among isolated 158 diterpenoids from Anatolian *Salvia* species, almost 130 of which have abietane skeleton identified by spectroscopic methods, particularly based on extensive NMR and Mass techniques. In addition to diterpenoids, 60 triterpenoids, 15 steroids, 5 sesterterpenes and 9 sesquiterpenes and 41 flavonoids and 29 other phenolics were isolated. In this chapter, Anatolian *Salvia* plant extracts and/or its constituents with various biological activities were presented, including antioxidative, antimicrobial, anti-inflammatory, cytotoxic/

G. Topçu (✉) · R. Yücer
Department of Pharmacognosy and Phytochemistry, Faculty of Pharmacy,
Bezmialem Vakif University, 34093 Istanbul, Turkey
e-mail: gtopcu@bezmialem.edu.tr; gulacti_topcu@yahoo.com

R. Yücer
Department of Pharmacognosy, Faculty of Pharmacy, University of Health Sciences,
34668 Istanbul, Turkey

H. Şenol
Department of Pharmaceutical Chemistry, Faculty of Pharmacy,
Bezmialem Vakif University, 34093 Istanbul, Turkey

V. Georgiev and A. Pavlov (eds.), *Salvia Biotechnology*,
https://doi.org/10.1007/978-3-319-73900-7_2

antitumor, cardiovascular, cholinesterase and some other enzymes inhibitory activities. In addition, some investigations on *Salvia* species consist of biological properties which carried out on the extracts rather than pure isolates were presented.

Keywords *Salvia* · Lamiaceae · Secondary metabolite · Sesquiterpenoid Diterpenoid · Sesterterpenoid · Triterpenoid · Steroid · Flavonoid Phenolic · Bioactivity · Anatolia

Abbreviations

UA	Ursolic acid
OA	Oleanolic acid
BHA	Butylated hydroxyanisole
BHT	Butylated hydroxytoluene
DPPH	1,1-diphenyl-2-picrylhydrazyl
AAE	Ascorbic acid equivalent
SFW	*Salvia* fruticosa Wildly grown samples
SFC	*Salvia* fruticosa Cultivated sample
FRAP	Ferric-reducing antioxidant power
CUPRAC	Cupric reducing antioxidant capacity
TE	Trolox equivalent
AD	Alzheimer's disease
NMDA	*N*-methyl D-aspartate
AChE	Acetylcholinesterase
BChE	Butyrylcholinesterase
BC 1	Human breast cancer
LU 2	Human lung cancer
COL 2	Human colon cancer
KB	Human epidermoidal carcinoma in mouth
KB-VI	Multidrug resistant-KB
LNCaP	Hormone-dependent human prostate cancer
KOPr	Kappa Opioid Peptide Receptor

1 Introduction

One of the largest and most well-known Lamiaceae (Labiatae) family plants is *Salvia* (sage) which has wide distribution with over 900 species in the three regions of the world: Central and South America (>500 spp.), Central Asia/Mediterranean (280 spp.), and Eastern Asia (120 spp.) [1]. Sage derives from the Latin verb salvare that means to heal/to save, and the Latin proverb "*Cur moriatur homo, ciu calvia in horto*" (Why should a man die who has sage in his garden?) confirms its use for the centuries. Sage became very popular in China in the eighteenth century where the merchants would exchange two crates of best tea for a crate of sage [2]. Since antiquity, *Salvia* species have been used in the treatment of cold, flu, bronchitis and tuberculosis, menstrual disorders and menoposal syndromes [2–4], hemorrhage and some cardiovascular problems, such as angina pectoris and myocardial infarction [5]. Salvia extracts, essential oils and/or their constituents have shown antimicrobial, antioxidant, anti-inflammatory, and anti-cholinesterase properties which may help for cognitive brain functions as neuroprotective agents in various neurodegenerative diseases as well as in inflammatory and infectional diseases [6–8].

Some sage products have been used as protective agents in food industry, especially in Europe and Asia [9]. Recently, some sage oral solutions have been prepared and used in the treatment of some mouth wards and dental plaques and other problems due to their antioxidant, antiseptic and antimicrobial properties [10, 11].

In Anatolia (the Asian part of Turkey), *Salvia* species have been reportedly used in folk medicine for wound healing, and in alleviating stomach, liver, and rheumatism pains and for treating the common cold in the form of infusion and decoction [12]. They have also used to cure tuberculosis, diabetes as well as skin diseases such as psoriasis and eczema and some species are consumed as tea, especially in rural areas.

There are over 100 species wildly growing in Turkey, half of them being endemic. *Salvia* species are 30–100 cm tall, herbaceous, suffruticose or shrubby perennial, rarely biennial or annual, often strongly aromatic plants [13] (Fig. 1).

Fig. 1 *Salvia* species growing in Turkey

Salvia species growing in Turkey have been studied since 1965 for chemically and in subsequent years for their biological activity, and most of Anatolian *Salvia* species have been investigated particularly for chemical constituents by Ulubelen and her former Ph.D. student Topcu until now [14–17].

In the last two decades, Anatolian *Salvia* extracts and their main constituents; di-and triterpenoids were intensively investigated for their biological activities, especially antibacterial, cytotoxic, antioxidant/antiradical and anticholinesterase properties. Ulubelen and Topcu have initiated to investigate antibacterial [18], antituberculous [19] and cytotoxic effects [20] of Anatolian *Salvia* terpenoids before 1990s. Their antioxidant, anticholinesterase and some other properties have been searched since 2000 by the Topcu and her group and some other groups in Turkey [21, 22].

In the present chapter, non-volatile constituents of *Salvia* extracts rather than volatile constituents are reviewed. Because, essential oils and some other volatile constituents of Turkish *Salvia* species [23] have been deeply investigated by Prof. Baser and his group [24] and some other research groups [25] and published in many research articles, review articles and in the books.

Until today, several chapters and review articles were published on the genus *Salvia* plants growing throughout the world. One of the reviews was published in 2001 on Polyphenolics of *Salvia* which covers 160 polyphenols with bioactivity, some of which are unique to the genus [26]. In 2006, a first review article on "Bioactive Triterpenoids of *Salvia* Species" was published by Topcu G. in J. Nat. Prod. [27]. A recent review article, presented all chemical constituents of *Salvia* species with their bioactivity reported until 2011 including Anatolian *Salvia* species was published in 2012 in the journal "Chemical Review" by Wu et al. [28].

A more recent article covers studied triterpenoid constituents of the *Salvia* plants to date consisting about 214 triterpenoids of different classes, isolated and characterized from 113 species, as the second review article on the triterpenoids of the genus [1]. In the present chapter, chemical constituents of the *Salvia* species growing in Turkey with their bioactivity or/and their extracts activities will be given, published till 2017 September.

2 Constituents

Salvia species contain namely terpenoids and flavonoids and other phenolics which are classified as seen below, in general. The aerial parts of *Salvia* species usually rich in triterpenic compounds and flavonoids, and other phenolics while the roots contain much more diterpenoids.

2.1 Terpenoids

- *Mono- and Sesquiterpenoids*
- *Diterpenoids and Sesterterpenoids*
- *Triterpenoids and Steroids.*

2.2 Flavonoids and Other Phenolics

- *Flavone and Flavonol aglycones and glycosides*
- *Anthocyanins and Proanthocyanidins*
- *Caffeic acid metabolites as phenolics.*

3 Anatolian *Salvia* Species Constituents

Turkish *Salvia* species are rich in di- and triterpenoids which have mainly abietane and rearranged abietane, rarely pimarane or labdane type skeletons.

Turkish *Salvia* species were found to be also rich in flavone and flavonol aglycones and partially their glycosides. Some simple coumarins, especially esculetin was also isolated from several *Salvia* species.

So far, approximately 50 Anatolian *Salvia* species have been studied chemically yielding 158 diterpenoids consist of over 130 abietanes, some labdanes and pimaranes, 60 triterpenoids, 15 steroids, 5 sesterterpenoids, 9 sesquiterpenoids, and 70 flavonoids and other phenolics were isolated, and their chemical structures were elucidated through NMR, mass and other spectroscopic techniques [17, 29].

In the main text of this chapter, the constituents of each investigated Anatolian *Salvia* species will be given based on classification of their secondary metabolites. And all the isolated compounds from the each investigated species are further listed in Table 8.

Ulubelen and Topçu have investigated about half of the *Salvia* species growing in Turkey, particularly chemically, when available for their biological activities [27, 29–31]. In the beginning of 21th century, bioactivity studies on Anatolian *Salvia* plants are increased and varied including antioxidant, antimicrobial, cytotoxic/antitumoral, cardioactive and enzyme inhibition test assays carried out by different groups [32–37].

In Turkish *Salvia* species, the roots were found to be rich in diterpenoids rather than triterpenoids and other constituents. Triterpenes and other constituents were isolated from the aerial parts of the *Salvia* plants, in general.

The common triterpenoids found in all *Salvia* species have ursane or oleanane skeletons, the most common ones ursolic acid (UA) and oleanolic acid (OA), and most of the isolated diterpenoids have abietane skeleton while sesquiterpenes and sesterterpenes are present fairly rare in Anatolian *Salvia* species.

3.1 Terpenoids

3.1.1 Mono- and Sesquiterpenoids

Since mono- and sesquiterpenoids are the main constituents of essential oils, they were not subjected in this review. However, several sesquiterpenes, which were obtained during isolation from organic solvent extracts of Anatolian *Salvia* species, are presented herein [38].

Although *Salvia officinalis* L. is a medicinal plant of Europe, *S. fruticosa* Miller (syn = *S. triloba*) is the medicinal plant of Turkey, because *S. officinalis* can only be cultivated in Turkey. It should be emphasized that *S. fruticosa* essential oil is rich monoterpenoids, particularly in eucalyptol (1,8-cineole) in a range of 40–60%, and the other major monoterpenoids were found to be camphor, borneol, terpineol while *S. officinalis* is rich in thujone which has toxic effects, therefore its usage is limited.

Sesquiterpenes are rarely found in *Salvia* species either growing in Turkey on in the other countries. As seen from Table 1, spathulenol was the most common sesquiterpene, isolated from six Anatolian species. A linear sesquiterpene salvinine was only

Table 1 Isolated sesquiterpenes from Turkish *Salvia* species

No	Name	Plant Name	Ref
1	spathulenol	*S. candidissima* Vahl. subsp. *occidentalis* Hedge	[41]
		S. cyanescens Boiss. & Balansa	[42]
		S. limbata C.A. Meyer	[43]
		S. sclarea L.	[44]
		S. staminea Montbret et Aucher ex Benth.	[45]
		S. yosgadensis Freyn. et Bornm.	[40]
2	caryophyllene oxide	*S. cyanescens* Boiss. & Balansa	[42]
		S. macrochlamys Boiss. et Kotschy	[46]
		S. sclarea L.	[44]
		S. staminea Montbret et Aucher ex Benth.	[45]
3	salvinine	*S. divaricata* Montbret et Aucher ex Bentham	[39]
4	Istanbulin D	*S. yosgadensis* Freyn. et Bornm.	[40]
5	(+)-ledol	*S. tchihatcheffii* (Fisch. et Mey.) Boiss.	[47]
6	β-eudesmol	*S. microphylla* Kunth.	[38]
7	8α-hydroxy-β-eudesmol	*S. microphylla* Kunth.	[38]
8	1-acetoxyeudesm-4(15),7(11)-dien-8,12-olide	*S. yosgadensis* Freyn. et Bornm.	[40]
9	1-acetoxy-8-hydroxyeudesm-4(15),7(11)-dien-8,12-olide	*S. potentillifolia* (Boiss. et Heldr.) ex Bentham	[48]

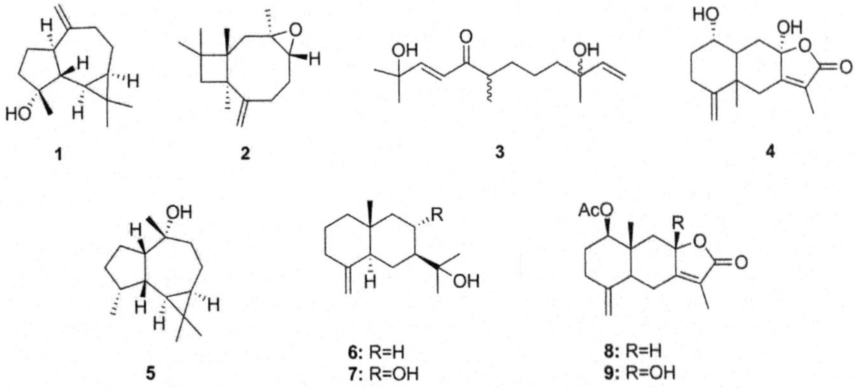

Fig. 2 Chemical structures of the isolated Anatolian *Salvia* sesquiterpenes from Turkish *Salvia* species, in agreement with Table 1

isolated from *S. divaricata* (endemic) [39] while three sesquiterpene lactones (1-acetoxy-8-hydroxyeudesm-4(15),7(11)-dien-8,12-olide, 1-acetoxyeudesm-4(15), 7-dien-8,12-olide, Istanbulin D) were isolated from *S. yosgadensis* [40]. *S. microphylla* was one of the Anatolian *Salvia* species cultivated which afforded two sesquiterpenes β-eudesmol and 8α-hydroxy-β-eudesmol [38] (Fig. 2).

3.1.2 Diterpenoids and Sesterterpenoids

Salvia species have shown a rich diversity for their diterpenic skeletons. In European and Asian *Salvia* species, abietane diterpenoids have been found as the most common terpenoids while *Salvia* species growing in South America region contain neo-clerodane diterpenoids rather than abietanes.

Some of the species may also bear labdane and/or pimarane diterpenoids. However, *Salvia* species didn't show such rich diversity for their triterpenoids which most have oleanane or ursane skeleton, and some of them lupane skeleton.

Anatolian *Salvia* species are highly rich in abietane diterpenoids with high biodiversity. A total number of the isolated diterpenoids from Anatolian *Salvia* species is about 200, almost 160 having abietane skeleton, and a huge number of them ($\sim$120) are new. Some of them, such as ferruginol, horminone, 7-*O*-acetylhorminone, royleanone were isolated from most of the studied Anatolian *Salvia* species (Figs. 3 and 4).

The first studied Anatolian specie is *S. virgata* by Ulubelen et al. [104]. However, in this first investigation no diterpenoid was isolated from the plant extract. Virgatol, a new abietane diterpene was isolated from the roots of *Salvia virgata* [105] in a later study.

Salvia tomentosa Mill., widespread in the Mediterranean and Aegean regions of Turkey, has a folkloric use to reduce abdominal pains and to heal wounds. One of

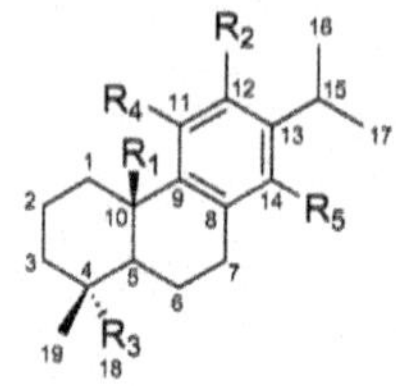

12: R_1=OAc, R_2=OH, R_3=Me, R_4=R_5=H
14: R_1=R_3=Me, R_2=OMe, R_4=OH, R_5=H
19: R_1=COOH, R_2=OMe, R_3=Me, R_4=R_5=H
22: R_1=R_3=Me, R_2=OMe, R_4=R_5=H
23: R_1=COOH, R_2=OMe, R_3=Me, R_4=R_5=H
24: R_1=R_3=Me, R_2= R_5=OH, R_4=H
26: R_1=H, R_3=CH_2OH, R_2=R_4=R_5=H, 7-oxo
27: R_1=Me, R_3=COOH, R_2=R_4=R_5=H, 1-oxo, 5α-OH
29: R_1=R_3=Me, R_2=OH, R_4=R_5=H, 1-oxo
38: R_1=R_3=Me, R_2=OMe, R_4=H, R_5=OH, 3-oxo
39: R_1=COOH, R_2=R_3=R_4=R_5=H, 3-oxo
40: R_1=R_3=Me, R_2=R_4=R_5=H, 3-oxo
42: R_1= R_2=R_4=R_5=H, R_3=COOH, $\Delta^{15(16)}$
46: R_1=R_3=Me, R_2=R_4=OH, R_5=H, 5α-OH
49: R_1=R_3=Me, R_2=R_5=OH, R_4=H, 6-OH, Δ^6
50: R_1=R_3=Me, R_2=R_4=H, 6-OH, 7-oxo, Δ^5
52: R_1=R_3=Me, R_2=OOH, R_4=R_5=H, 6-oxo
53: R_1=R_3=Me, R_2=OH, R_4=R_5=H, 6-oxo
59: R_1=R_3=Me, R_2=R_4=OH, R_5=H, 6-αOH, 7-oxo, Δ^5
69: R_1=COOMe, R_2=R_4=R_5=OH, R_3=Me, 7-oxo, Δ^5
71: R_1=R_3=Me, R_2=R_4=R_5=H, 7-oxo
72: R_1=Me, R_2=OH, R_3=CHO, R_4=R_5=H, 7-oxo
77: R_1=R_3=Me, R_2=R_4=R_5=H
83: R_1=Me, R_3=COOH, R_2=R_4=R_5=H, Δ^{15}, 3β-OH
84: R_1=COOH, R_2=OMe, R_3=Me, R_4=R_5=H, Δ^5
87: R_1=COOH, R_2=R_4=OH, R_3=Me, R_5=H
88: R_1=COOH, R_2=OMe, R_3=Me, R_4=OH, R_5=H
90: R_1=R_3=Me, R_2=OH, R_4=R_5=OH, Δ^6
91: R_1=R_3=Me, R_2=OMe, R_4=OH, R_5=H, 7-oxo
92: R_1=Me, R_2=R_4=R_5=H, R_3=COOH
93: R_1=R_3=Me, R_2=R_4=H, 7-oxo
96: R_1=R_3=Me, R_2=OH, R_4=R_5=H
97: R_1=Me, R_2=OH, R_3=CHO, R_4=R_5=H
98: R_1=R_3=Me, R_2=OMe, R_4=R_5=H
100: R_1=CH_2OH, R_2=OMe, R_3=Me, R_4=R_5=OH, 6,7=dioxo
102: R_1=R_3=Me, R_2=OH, R_4=R_5=H, 2-oxo
105: R_1=R_4=R_5=H, R_3=Me, R_2=OH, 6β-OH, 1,7-dioxo
106: R_1=R_3=Me, R_2=OH, R_4=R_5=H, 15-OH, 7-oxo
108: R_1=R_3=Me, R_2=OH, R_4=R_5=H, 3-oxo, Δ^6
109: R_1=R_3=Me, R_2=R_4=H, R_5=OH, 1-oxo
120: R_1=R_3=Me, R_2=R_4=OH, R_5=H, 6-OH, 7-oxo, Δ^5
121: R_1=R_3=Me, R_2=OMe, R_4=OH, 6-OH, R_5=H, 7-oxo, Δ^5
124: R_1=Me, R_2=R_4=H, R_3=COOH, R_5=OH, 2α-OH
126: R_1=COOMe, R_2=OMe, R_3=Me, R_4=R_5=H
127: R_1=R_3=Me, R_2=R_4=R_5=H, 6-oxo, Δ^5
128: R_1=CHO, R_2=OH, R_3=Me, R_4=R_5=H
129: R_1=COOH, R_2=OH, R_3=Me, R_4=R_5=H
130: R_1=Me, R_3=CH_2OH, R_2=R_4=R_5=H
131: R_1=Me, R_3=CH_2OAc, R_2=OH, R_4=R_5=H
132: R_1=Me, R_2=OMe, R_3=COOH, R_4=OH, R_5=H
133: R_1=R_3=Me, R_2=R_4=R_5=H, 2α-OH
134: R_1=R_3=Me, R_2=R_4=R_5=H, 2α-OH, 7-oxo
135: R_1=R_3=Me, R_2=R_4=R_5=H, 2β-OH, 7-oxo, 15-OH
136: R_1=Me, R_2=R_4=R_5=H, R_3=CH_2OAc
144: R_1=R_3=Me, R_2=R_4=OH, R_5=H, 7-oxo, Δ^5
145: R_1=R_3=Me, R_2=OMe, R_4=OH, R_5=H, 7-oxo, Δ^5
151: R_1=R_3=Me, R_2=R_5=H, R_4=OH, Δ^5
152: R_1=R_3=Me, R_2=OMe R_4=OH, R_5=H, 6α-OH, 7β-OH
158: R_1=R_3=Me, R_2=OH, R_4=R_5=H, 7-oxo
160: R_1=R_3=Me, R_4=OH, R_2=R_5=H, Δ^{15}
161: R_1=R_3=Me, R_2=OH, R_4=R_5=H, 7-OH, Δ^6
162: R_1=R_3=Me, R_2=OH, R_4=R_5=H, 6-oxo, 7-OH, Δ^{15}

Fig. 3 Chemical structures of the isolated Anatolian *Salvia* diterpenoids, in agreement with Table 2

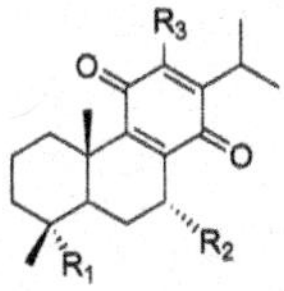

20: R_1= CH_3, R_2=OAc, R_3=OMe, Δ^5
21: R_1= CH_3, R_2=OH, R_3=OMe, Δ^5
23: R_1= CH_3, R_2=R_3=OMe
36: R_1= CH_3, R_2=H, R_3=OH, 2β-OH
48: R_1= CH_3, R_2=H, R_3=OH, Δ^6
51: R_1= COOH, R_2=H, R_3=OMe, 6-oxo
54: R_1= COOH, R_2=H, R_3=OH, 6-oxo
63: R_1= CH_3, R_2=H, R_3=OH, 6β-OH
66: R_1= CH_3, R_2=OAc, R_3=OH
68: R_1= CH_3, R_2=Ac, R_3=OH
73: R_1= CH_3, R_2=H, R_3=OH, 7-oxo
74: R_1= CH_3, R_2=Ac, R_3=OH, 16=17=OH
78: R_1= CH_3, R_2=R_3=OAc
85: R_1= CH_3, R_2=H, R_3=OH, 4α-OMe, Δ^{15}
103: R_1= CH_3, R_2=R_3=OH
104: R_1= COOH, R_2=R_3=OH
110: R_1= CH_3, R_2=H, R_3=OH, 5-αOH, Δ^6
138: R_1= CH_3, R_2=H, R_3=OH, 16=17=OH
163: R_1= CH_3, R_2=H, R_3=OH

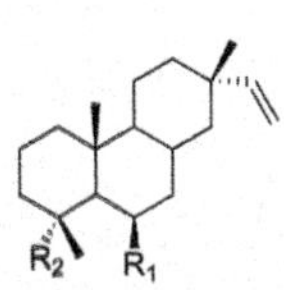

25: R_1=H, R_2=COOH, 14-oxo
61: R_1=OH, R_2=CH_3, $\Delta^{8(9)}$, 1-oxo
62: R_1=OH, R_2=COOH, Δ^7
75: R_1=H, R_2=COOH, 7β-OH, $\Delta^{8(14)}$
111: R_1=H, R_2=CHO, Δ^7
112: R_1=H, R_2=COOH, Δ^7
146: R_1=H, R_2=Me, 3,11-dioxo, 7β,9β-diOH, $\Delta^{8(14)}$
153: R_1=H, R_2=COOH, $\Delta^{8(14)}$

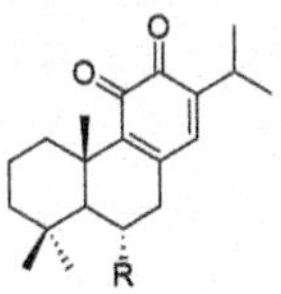

11: R=H
56: R=OH

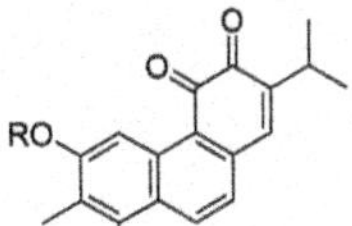

34: R=H
123: R=Me

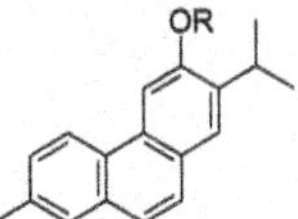

16: R=H
122: R=Me

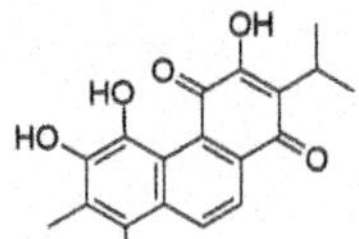

114

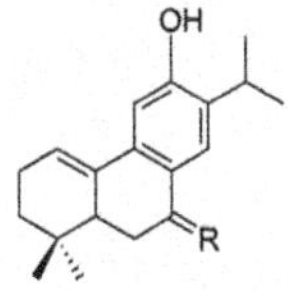

148: R=H_2
149: R=O

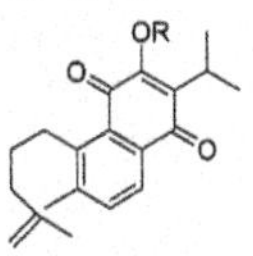

18: R= isopentenyl, 3-oxo
31: R=H, 1-oxo
32: R=H, Δ^2
41: R=H, 3-oxo
80: R=Ac
147: R=H

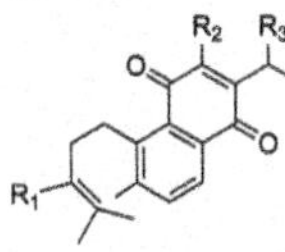

17: R_1=H, R_2=OH, R_3=Me
37: R_1=R_2=OH, R_3=Me, Δ^1
115: R_1=H, R_2=OH, R_3=CHO

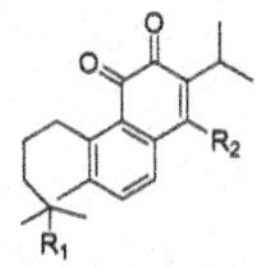

44: R_1=R_2=OH
154: R_1=R_2=H
155: R_1=OH, R_2=H, 3-oxo

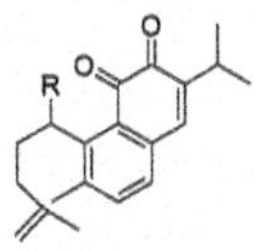

28: R=oxo
81: R=H

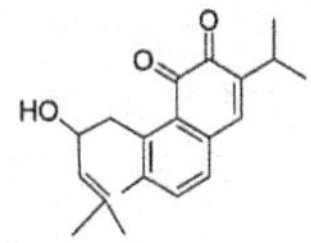

35

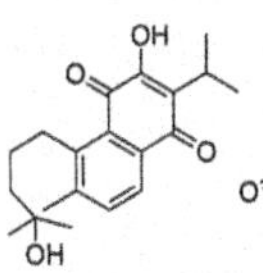

43

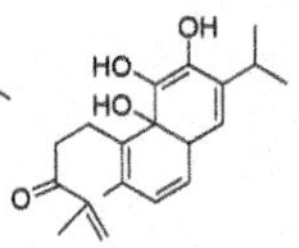

45

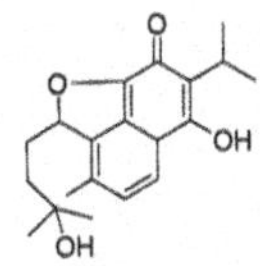

95

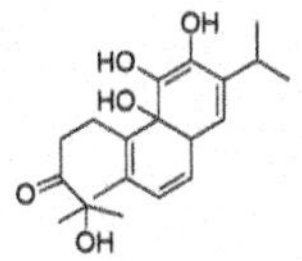

142

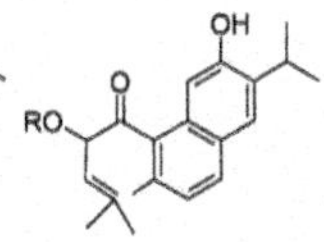

79: R=Ac
116: R=H

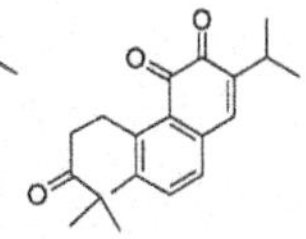

150

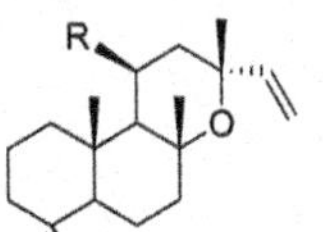

15: R=OH
118: R=H

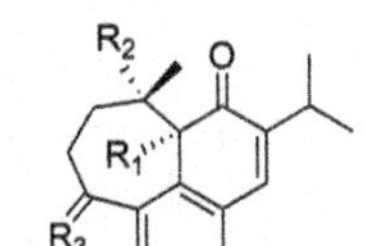

30: R_1=H, R_2=OH, R_3=O
119: R_1=OH, R_2=CH_3, R_3=O
140: R_1=H, R_2=OH, R_3=H_2

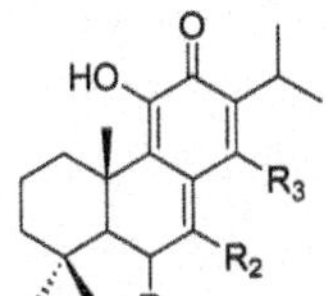

67: R_1=oxo, R_2=OH, R_3=H
159: R_1=oxo, R_2=R_3=H
164: R_1=α-OH, R_2=H, R_3=OMe

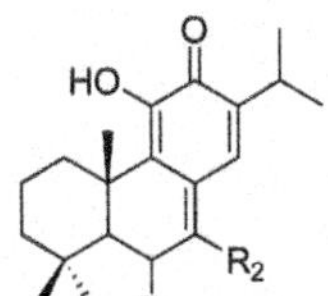

47: R_1=R_2=OH, Δ^5
60: R_1=αOH, R_2= H

Fig. 3 (continued)

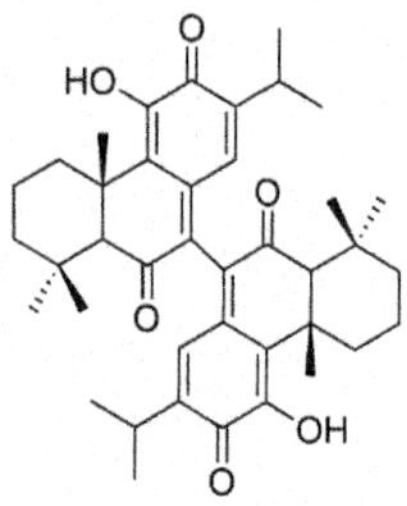

65

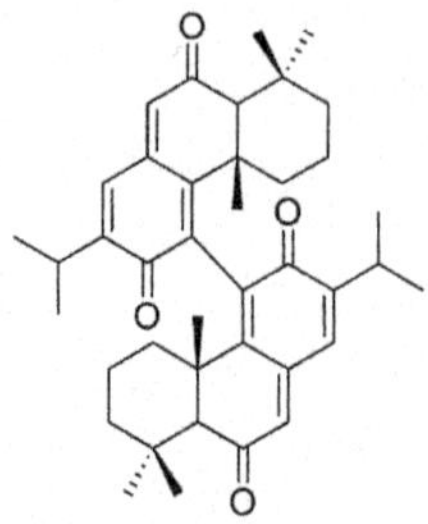

13

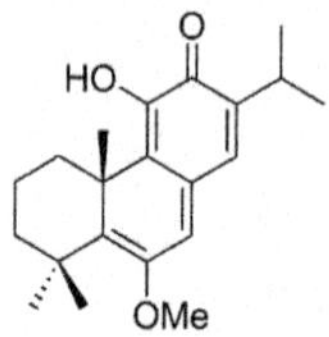

165

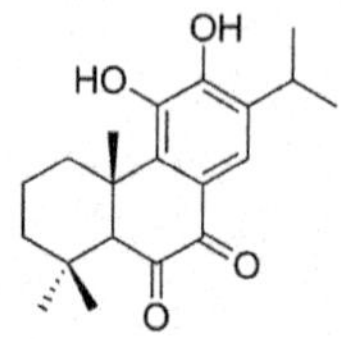

107

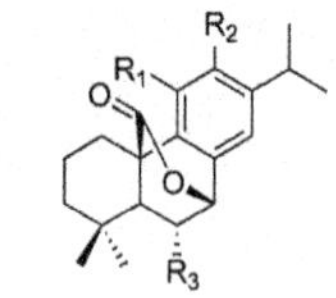

33: $R_1=R_2=OH$, $R_3=H$, 20-deoxo
89: $R_1=R_2=OH$, $R_3=H$
99: $R_1=R_3=H$, $R_2=OH$
113: $R_1=R_2=OH$, $R_3=OH$

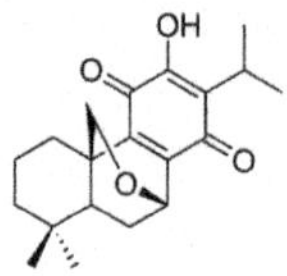

64

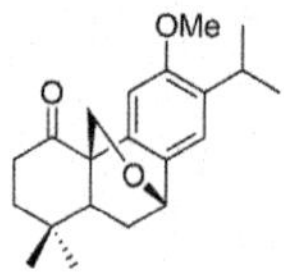

139

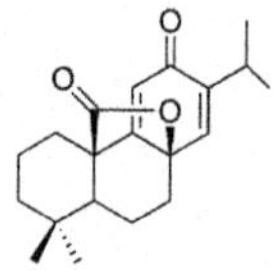

166

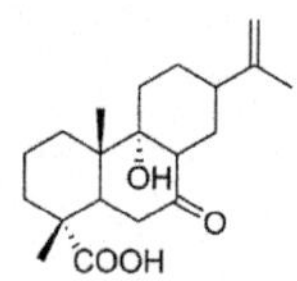

167

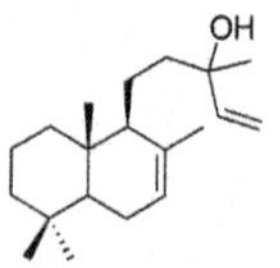

10

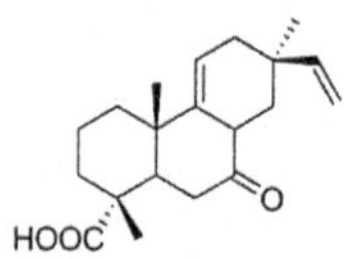

70

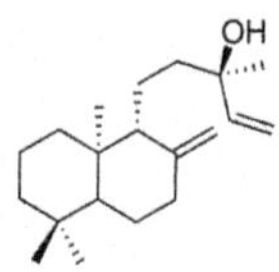

94

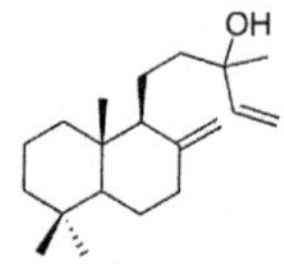

117

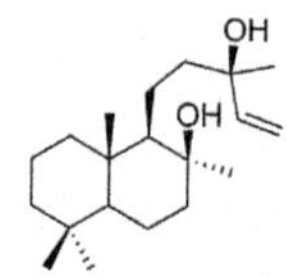

156

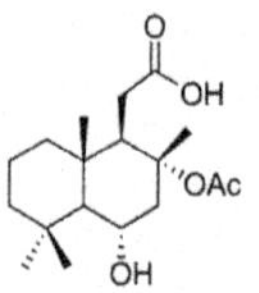

55

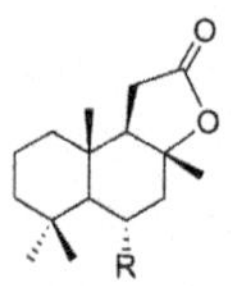

58: R=OH
125: R=H

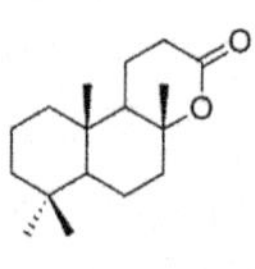

76

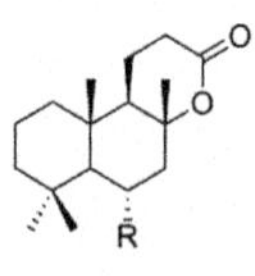

57: R=OH
82: R=H

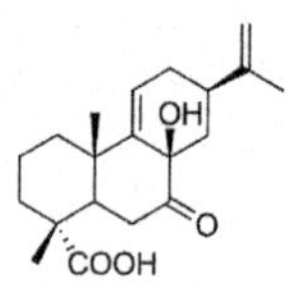

101

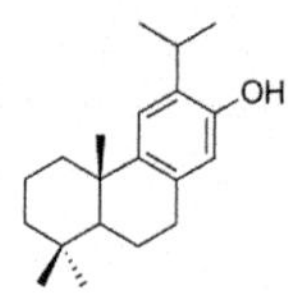

157

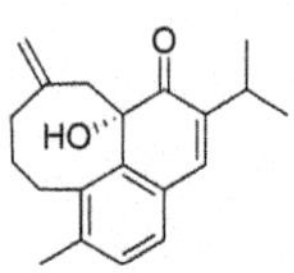

86

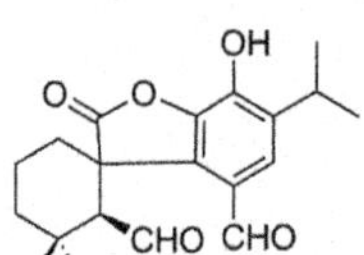

137

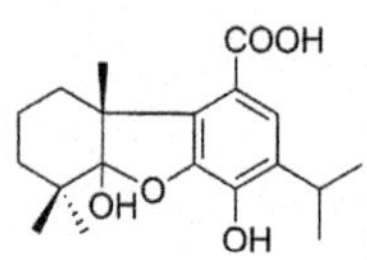

141

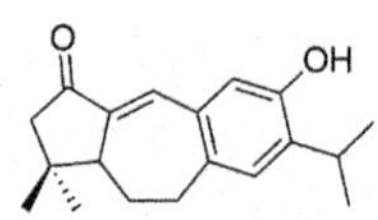

143

Fig. 3 (continued)

Table 2 Isolated diterpenes from Turkish *Salvia* species

No	Name	Plant Name	Ref
10	Δ^7-manool (isomanool)	*S. kronenburgii* Rech. f.	[49]
		S. sclarea L.	[50]
11	11,12-dioxoabieta-8,13-dien	*S. napifolia* Jacq.	[51]
		S. hypargeia Fisch. et Mey	[34]
		S. candidissima Vahl subsp. *candidissima*	[52]
12	10-acetylferruginol	*S. microstegia* Boiss. et Bal.	[53]
13	11,11′-didehydroxy-7,7′-dihydroxytaxodione	*S. montbretii* Benth.	[15]
14	11-hydroxy-12-methoxyabieta-8,11,13-trien	*S. candidissima* Vahl subsp. *candidissima*	[52]
		S. candidissima Vahl. subsp. *occidentalis* Hedge	[54]
15	11β -hydroxymanoyl oxide	*S. candidissima* Vahl subsp. *candidissima*	[52]
		S. candidissima Vahl. subsp. *occidentalis* Hedge	[41]
		S. cyanescens Boiss. & Balansa	[42]
16	12-demethylmulticauline	*S. multicaulis* Vahl.	[19]
17	12-hydroxysapriparaquinone	*S. kronenburgii* Rech. f.	[49]
		S. limbata C. A. Meyer	[55]
		S. eriophora Boiss. et Kotschy	[56]
18	12-isopentenyl-3-oxosalvipisone	*S. cyanescens* Boiss. & Balansa	[42]
19	12-methoxypisiferic acid	*S. wiedemannii* Boiss.	[57]
20	12-methyl-5-dehydroacetylhorminone	*S. multicaulis* Vahl.	[19]
21	12-methyl-5-dehydrohorminone	*S. multicaulis* Vahl.	[19]
		S. recognita Fisch. et Mey	[58]
22	12-methylferruginol	*S. bracteata* Banks and Sol	[59]
23	12-*O*-methylhorminone	*S. bracteata* Banks and Sol	[59]
24	14-hydroxyferruginol	*S. montbretii* Benth.	[60]
25	14-oxo-pimaric acid	*S. candidissima* Vahl. subsp. *occidentalis* Hedge	[54]
		S. wiedemannii Boiss.	[61]
26	18-hydroxy-8,11,13 abietatrien-7-one	*S. candidissima a Vahl.* subsp. *occidentalis Hedge*	[54]
27	1-oxo-5-hydroxy-abieta-8,11,13-trien-18-oic acid	*S. candidissima* Vahl subsp. *candidissima*	[52]
28	1-oxoaethiopinone	*S. candidissima* Vahl. subsp. *occidentalis* Hedge	[54]
		S. sclarea L.	[50]

(continued)

Table 2 (continued)

No	Name	Plant Name	Ref
29	1-oxoferruginol	*S. multicaulis* Vahl.	[19]
		S. napifolia Jacq.	[51]
		S. viridis L.	[62]
30	1-oxosalvibretol	*S. kronenburgii* Rech. f.	[49]
		S. montbretii Benth.	[15]
		S. candidissima Vahl subsp. *candidissima*	[52]
		S. staminea Montbret et Aucher ex Benth.	[45]
31	1-oxosalvipisone	*S. candidissima* Vahl. subsp. *occidentalis* Hedge	[54]
32	2,3-dehydrosalvipisone	*S. candidissima* Vahl subsp. *candidissima*	[52]
		S. sclarea L.	[44, 50]
		S. verticillata L.	[63]
33	20-deoxocarnosol	*S. candidissima* Vahl. subsp. *occidentalis* Hedge	[64]
34	2-demethylmultiorthoquinone	*S. multicaulis* Vahl.	[19]
		S. blepharochlaena Hedge & Hub. Mor	[18]
35	2-hydroxysaprorthoquinone	*S. limbata* C. A. Meyer	[55]
36	2β-hydroxyroyleanone	*S. cryptantha* Montbret and Aucher ex. Bentham	[14]
37	3,12-dihydroxy sapriparaquinone-1-ene	*S. limbata* C. A. Meyer	[55]
		S. eriophora Boiss. et Kotschy	[56]
38	3-oxo-12-methoxy-14-hydroxyabieta-8, 11,13-triene	*S. multicaulis* Vahl.	[65]
39	3-oxo-8,11,13-abietatrien-20-oic acid	*S. ceratophylla* L.	[66]
40	3-oxo-abieta-8,11,13-triene	*S. wiedemannii* Boiss.	[57]
41	3-oxosalvipisone	*S. candidissima* Vahl. subsp. *occidentalis* Hedge	[41]
42	3β-hydroxy-8,11,13,15-abietatetraen-18-oic acid	*S. tomentosa* Mill.	[67]
43	4,12-dihydroxy-sapriparaquinone	*S. eriophora* Boiss. et Kotschy	[56]
44	4,14-dihydroxy-saprorthoquinone	*S. eriophora* Boiss. et Kotschy	[56]
45	4-dehydrosalvilimbinol	*S. limbata* C. A. Meyer	[55]
		S. syriaca L.	[68]

(continued)

Table 2 (continued)

No	Name	Plant Name	Ref
46	5,11,12-trihydroxy-abieta-8,11,13-triene	*S. microstegia* Boiss. et Bal.	[53]
47	5,6-didehydro-7-hydroxytaxodone	*S. hypargeia* Fisch. et Mey	[20]
48	6,7-dehydroroyleanone	*S. tchihatcheffii* (Fisch. et Mey.) Boiss.	[47]
		S. bracteata Banks and Sol	[59]
		S. eriophora Boiss. et Kotschy	[56]
49	6,12,14-trihydroxyabieta-6,8,11,12-tetraen	*S. napifolia* Jacq.	[51]
50	6-hydroxysalvinolone	*S. hypargeia* Fisch. et Mey	[20]
		S. kronenburgii Rech. f.	[49]
		S. montbretii Benth.	[15]
51	6-oxo-12-methylroyleanone-18-oic acid	*S. divaricata* Montbret et Aucher ex Bentham	[39]
52	6-oxo-12-peroxyabieta-8,11,13-triene	*S. multicaulis* Vahl.	[65]
53	6-oxoferruginol	*S. multicaulis* Vahl.	[65]
		S. napifolia Jacq.	[51]
54	6-oxoroyleanone-18-oic acid	*S. divaricata* Montbret et Aucher ex Bentham	[39]
55	6α -hydroxy-8α-acetoxy-13,14,15, 16-tetranorlabdan-12-oic- acid	*S. yosgadensis* Freyn. et Bornm.	[69]
56	6α-hydroxy-11,12-dioxo-8,13-abieta-diene	*S. recognita* Fisch. et Mey	[58]
57	6α-hydroxyambreinolide	*S. yosgadensis* Freyn. et Bornm.	[69]
58	6α-hydroxynorambreinolide	*S. yosgadensis* Freyn. et Bornm.	[69]
59	6α-hydroxysalvinolone	*S. hypargeia* Fisch. et Mey	[34]
60	6α-hydroxytaxodone	*S. hypargeia* Fisch. et Mey	[34]
61	6β-hydroxy-8,15-pimaradien-1-one	*S. ceratophylla* L.	[66]
62	6β-hydroxy-isopimaric acid	*S. caespitosa* Montbret & Aucher ex Benth.	[70]
63	6β-hydroxyroyleanone	*S. fruticosa* Mill. (*S. triloba* L.)	[71]
64	7,20-epoxyroyleanone	*S. blepharochlaena* Hedge & Hub. Mor	[72]
		S. candidissima Vahl subsp. *candidissima*	[52]
		S. napifolia Jacq.	[51]
65	7,7′-bistaxodione	*S. montbretii* Benth.	[15]
66	7α-acetoxyroyleanone	*S. kronenburgii* Rech. f.	[49]
67	7-hydroxytaxodione	*S. montbretii* Benth.	[15]

(continued)

Table 2 (continued)

No	Name	Plant Name	Ref
68	7-*O*-acetylhorminone	*S. amplexicaulis* Lam.	[73]
		S. blepharochlaena Hedge & Hub. Mor	[18]
		S. cryptantha Montbret and Aucher ex. Bentham	[14]
		S. napifolia Jacq.	[51]
		S. bracteata Banks and Sol	[59]
		S. candidissima Vahl. subsp. *occidentalis* Hedge	[54]
		S. eriophora Boiss. et Kotschy	[56]
		S. nemorosa L.	[74]
		S. wiedemannii Boiss.	[57]
69	7-oxo-11,12,14-trihydroxy-abieta-5,8, 11,13-tetraene-20-oic acid methyl ester	*S. candidissima* Vahl subsp. *candidissima*	[52]
70	7-oxo-13-epi-pimara-9(11),15-dien-18-oic acid	*S. heldreichiana* Boiss. ex Bentham	[75]
71	7-oxo-abieta-9,12,14-triene	*S. amplexicaulis* Lam.	[73]
72	7-oxoferruginol-18-al	*S. sclarea* L.	[44]
73	7-oxoroyleanone	*S. sclarea* L.	[44]
74	7α-acetylroyleanone	*S. fruticosa* Mill. (*S. triloba* L.)	[71]
75	7β-hydroxysandarcopimaric acid	*S. wiedemannii* Boiss.	[57]
		S. heldreichiana Boiss. ex Bentham	[75]
76	8,13-diepimanoyl oxide	*S. candidissima* Vahl. subsp. *occidentalis* Hedge	[41]
77	abieta-8,11,13-triene	*S. limbata* C. A. Meyer	[43]
78	acetylhorminone	*S. multicaulis* Vahl.	[65]
79	acetyllimbinol	*S. limbata* C. A. Meyer	[43]
80	acetylsalvipisone	*S. sclarea* L.	[50]
81	aethiopinone	*S. candidissima* Vahl. subsp. *occidentalis* Hedge	[64]
		S. ceratophylla L.	[66]
		S. ceratophylla L.	[66]
		S. cyanescens Boiss. & Balansa	[42]
		S. eriophora Boiss. et Kotschy	[56]
		S. hypargeia Fisch. et Mey	[34]
		S. sclarea L.	[50]
		S. viridis L.	[62]

(continued)

Table 2 (continued)

No	Name	Plant Name	Ref
82	ambreinolide	*S. yosgadensis* Freyn. et Bornm.	[69]
83	antiochic acid	*S. tomentosa* Mill.	[67]
84	blephaein	*S. blepharochlaena* Hedge & Hub. Mor	[18]
85	bractealine	*S. bracteata* Banks and Sol	[59]
86	candidissiol	*S. candidissima* Vahl subsp. *candidissima*	[52]
		S. candidissima Vahl. subsp. *occidentalis* Hedge	[64]
		S. ceratophylla L.	[66]
		S. ceratophylla L.	[66]
		S. sclarea L.	[44, 50]
		S. syriaca L.	[68]
		S. verticillata L.	[63]
87	carnosic acid	*S. fruticosa* Mill. (*S. triloba* L.)	[30]
88	carnosic acid 12-*O*-methyl ether	*S. fruticosa* Mill. (*S. triloba* L.)	[30]
		S. microphylla Kunth.	[38]
89	carnosol	*S. fruticosa* Mill. (*S. triloba* L.)	[30, 76]
90	cryptanol	*S. blepharochlaena* Hedge & Hub. Mor	[72]
		S. candidissima Vahl. subsp. *occidentalis* Hedge	[54]
		S. cryptantha Montbret and Aucher ex. Bentham	[14]
		S. fruticosa Mill. (*S. triloba* L.)	[71]
		S. hypargeia Fisch. et Mey	[77]
		S. multicaulis Vahl.	[65]
		S. napifolia Jacq.	[51]
		S. wiedemannii Boiss.	[57]
91	cryptojaponol	*S. candidissima* Vahl subsp. *candidissima*	[52]
		S. napifolia Jacq.	[51]
		S. sclarea L.	[50]
		S. staminea Montbret et Aucher ex Benth.	[45]
92	dehydroabietic acid	*S. tchihatcheffii* (Fisch. et Mey.) Boiss.	[47]
		S. tomentosa	[67]
93	demethylcryptojaponol	*S. hypargeia* Fisch. et Mey	[20]
		S. montbretii Benth.	[60]
94	*ent*-sclareol	*S. candidissima* Vahl subsp. *candidissima*	[52]

(continued)

Table 2 (continued)

No	Name	Plant Name	Ref
95	eriophoroxide	*S. eriophora* Boiss. et Kotschy	[78]
96	ferruginol	*S. amplexicaulis* Lam.	[73]
		S. blepharochlaena Hedge & Hub. Mor	[18]
		S. bracteata Banks and Sol	[59]
		S. caespitosa Montbret & Aucher ex Benth.	[70]
		S. candidissima Vahl subsp. *candidissima*	[52]
		S. candidissima Vahl. subsp. *occidentalis* Hedge	[54]
		S. ceratophylla L.	[66]
		S. cyanescens Boiss. & Balansa	[42]
		S. eriophora Boiss. et Kotschy	[56]
		S. fruticosa Mill. (*S. triloba* L.)	[71, 30]
		S. hypargeia Fisch. et Mey	[34]
		S. kronenburgii Rech. f.	[49]
		S. limbata C. A. Meyer	[43]
		S. microstegia Boiss. et Bal.	[53]
		S. montbretii Benth.	[60]
		S. multicaulis Vahl.	[65]
		S. napifolia Jacq.	[51]
		S. recognita Fisch. et Mey	[58]
		S. sclarea L.	[44, 50]
		S. staminea Montbret et Aucher ex Benth.	[45]
		S. syriaca L.	[68]
		S. viridis L.	[62]
		S. wiedemannii Boiss.	[57]
97	ferruginol-18-al	*S. multicaulis* Vahl.	[19, 65]
98	ferruginol-12-methyl ether	*S. montbretii* Benth.	[60]
		S. pomifera L.	[79]
99	formosanolide	*S. blepharochlaena* Hedge & Hub. Mor	[72]
100	forskalinone	*S. forskahlei* L.	[80]
101	heldrichinic acid	*S. heldreichiana* Boiss. ex Bentham	[75]

(continued)

Table 2 (continued)

No	Name	Plant Name	Ref
102	hinokione	*S. multicaulis* Vahl.	[19]
103	horminone	*S. amplexicaulis* Lam.	[73]
		S. blepharochlaena Hedge & Hub. Mor	[18]
		S. bracteata Banks and Sol	[59]
		S. candidissima Vahl. subsp. *occidentalis* Hedge	[54]
		S. cryptantha Montbret and Aucher ex. Bentham	[14]
		S. eriophora Boiss. et Kotschy	[56]
		S. fruticosa Mill. (*S. triloba* L.)	[71]
		S. hypargeia Fisch. et Mey	[77]
		S. multicaulis Vahl.	[19]
		S. napifolia Jacq.	[51]
		S. nemorosa L.	[74]
		S. wiedemannii Boiss.	[57]
104	horminone-18-oic acid	*S. divaricata* Montbret et Aucher ex Bentham	[39]
105	hypargenin A	*S. hypargeia* Fisch. et Mey	[34, 77]
106	hypargenin B	*S. hypargeia* Fisch. et Mey	[77]
107	hypargenin C	*S. hypargeia* Fisch. et Mey	[77]
108	hypargenin D	*S. hypargeia* Fisch. et Mey	[34, 77]
109	hypargenin E	*S. hypargeia* Fisch. et Mey	[77]
110	hypargenin F	*S. hypargeia* Fisch. et Mey	[77]
		S. montbretii Benth.	[60]
111	isopimaradienal	*S. caespitosa* Montbret & Aucher ex Benth.	[70]
112	isopimaric acid	*S. caespitosa* Montbret & Aucher ex Benth.	[70]
		S. heldreichiana Boiss. ex Bentham	[75]
		S. wiedemannii Boiss.	[57]
113	isorosmanol	*S. fruticosa* Mill. (*S. triloba* L.)	[30]
114	kronenquinone	*S. kronenburgii* Rech. f.	[49]
115	limbinal	*S. limbata* C. A. Meyer	[43]
116	limbinol	*S. limbata* C. A. Meyer	[55]

(continued)

Table 2 (continued)

No	Name	Plant Name	Ref
117	manool	*S. candidissima* Vahl subsp. *candidissima*	[52]
		S. fruticosa Mill. (*S. triloba* L.)	[30]
		S. limbata C. A. Meyer	[55, 43]
		S. multicaulis Vahl.	[19]
		S. sclarea L.	[44]
		S. verticillata L.	[63]
118	manoyl oxide	*S. cyanescens* Boiss. & Balansa	[42]
		S. candidissima Vahl subsp. *candidissima*	[52]
		S. candidissima Vahl. subsp. *occidentalis* Hedge	[64]
		S. staminea Montbret et Aucher ex Benth.	[45]
119	microstegiol	*S. candidissima* Vahl subsp. *candidissima*	[52]
		S. candidissima Vahl. subsp. *occidentalis* Hedge	[54]
		S. hypargeia Fisch. et Mey	[34]
		S. microstegia Boiss. et Bal.	[81]
		S. napifolia Jacq.	[51]
		S. sclarea L.	[44, 50]
		S. staminea Montbret et Aucher ex Benth.	[45]
		S. viridis L.	[62]
120	montbretol	*S. montbretii* Benth.	[60]
121	montbretyl 12-*O*-methyl ether	*S. montbretii* Benth.	[60]
		S. candidissima Vahl. subsp. *occidentalis* Hedge	[54]
122	multicaulin	*S. blepharochlaena* Hedge & Hub. Mor	[18]
		S. multicaulis Vahl.	[19]
123	multiorthoquinone	*S. blepharochlaena* Hedge & Hub. Mor	[18]
		S. multicaulis Vahl.	[19]
124	nemorosin	*S. nemorosa* L.	[74]
125	norambreinolide	*S. yosgadensis* Freyn. et Bornm.	[69]

(continued)

Table 2 (continued)

No	Name	Plant Name	Ref
126	*O*-methylpisiferic acid methyl ester	*S. blepharochlaena* Hedge & Hub. Mor	[18]
127	pachystazone	*S. candidissima* Vahl subsp. *candidissima*	[52]
		S. candidissima Vahl. subsp. *occidentalis* Hedge	[64]
		S. napifolia Jacq.	[51]
		S. nemorosa L.	[74]
128	pisiferal	*S. microstegia* Boiss. et Bal.	[53]
		S. multicaulis Vahl.	[19]
		S. wiedemannii Boiss.	[61]
129	pisiferic acid	*S. blepharochlaena* Hedge & Hub. Mor	[18]
130	pomiferin A	*S. pomifera* L.	[79]
131	pomiferin B	*S. pomifera* L.	[79]
132	pomiferin C	*S. pomifera* L.	[79]
133	pomiferin D	*S. pomifera* L.	[79]
134	pomiferin E	*S. pomifera* L.	[79]
135	pomiferin F	*S. pomifera* L.	[82]
136	pomiferin G	*S. pomifera* L.	[82]
137	rosmadial	*S. fruticosa* Mill. (*S. triloba* L.)	[30]
138	royleanone	*S. blepharochlaena* Hedge & Hub. Mor	[72]
		S. kronenburgii Rech. f.	[49]
139	salvibracteone	*S. bracteata* Banks and Sol	[59]
140	salvibretol	*S. montbretii* Benth.	[15]
141	salvicanaric acid	*S. hypargeia* Fisch. et Mey	[20]
142	salvilimbinol	*S. eriophora* Boiss. et Kotschy	[56]
		S. limbata C. A. Meyer	[55]
143	salvimultine	*S. multicaulis* Vahl.	[83]
144	salvinolone	*S. bracteata* Banks and Sol	[59]
		S. candidissima Vahl subsp. *candidissima*	[52]
		S. sclarea L.	[50]

(continued)

Table 2 (continued)

No	Name	Plant Name	Ref
145	salvinolone-12-*O*-methyl ether	*S. candidissima* Vahl subsp. *candidissima*	[52]
		S. verticillata L.	[63]
146	salvipimarone	*S. multicaulis* Vahl.	[19]
147	salvipisone	*S. candidissima* Vahl subsp. *candidissima*	[52]
		S. candidissima Vahl. subsp. *occidentalis* Hedge	[54]
		S. ceratophylla L.	[66]
		S. cyanescens Boiss. & Balansa	[42]
		S. eriophora Boiss. et Kotschy	[56]
		S. kronenburgii Rech. f.	[49]
		S. nemorosa L.	[74]
		S. sclarea L.	[44, 50]
		S. verticillata L.	[63]
148	salvirecognine	*S. recognita* Fisch. et Mey	[58]
149	salvirecognone	*S. recognita* Fisch. et Mey	[58]
150	salvisyrianone	*S. syriaca* L.	[68]
151	salvitchihatine	*S. tchihatcheffii* (Fisch. et Mey.) Boiss.	[47]
152	salviviridinol	*S. viridis* L.	[62]
153	sandaracopimaric acid	*S. caespitosa* Montbret & Aucher ex Benth.	[70]
154	saprorthoquinone	*S. hypargeia* Fisch. et Mey	[34]
155	sclareapinone	*S. sclarea* L.	[50]
156	sclareol	*S. chrysophylla* Stapf.	[84]
		S. limbata C. A. Meyer	[43]
		S. palaestina Benth.	[85]
		S. poculata Nab.	[86]
		S. sclarea L.	[44]
157	sempervirol	*S. multicaulis* Vahl.	[19]
158	sugiol	*S. amplexicaulis* Lam.	[73]
		S. blepharochlaena Hedge & Hub. Mor	[72]
		S. bracteata Banks and Sol	[59]
		S. napifolia Jacq.	[51]
		S. recognita Fisch. et Mey	[58]
		S. viridis L.	[62]

(continued)

Table 2 (continued)

No	Name	Plant Name	Ref
159	taxodione	*S. kronenburgii* Rech. f.	[49]
		S. montbretii Benth.	[60]
		S. staminea Montbret et Aucher ex Benth.	[45]
		S. hypargeia Fisch. et Mey	[34]
160	tchihatine	*S. tchihatcheffii* (Fisch. et Mey.) Boiss.	[47]
161	trilobinol	*S. fruticosa* Mill. (*S. triloba* L.)	[71]
162	trilobinone	*S. fruticosa* Mill. (*S. triloba* L.)	[71]
163	virgatol	*S. fruticosa* Mill. (*S. triloba* L.)	[71]
164	viridinol	*S. viridis* L.	[62]
165	viridone	*S. syriaca* L.	[68]
		S. viridis L.	[62]
166	wiedelactone	*S. heldreichiana* Boiss. ex Bentham	[75]
		S. multicaulis Vahl.	[65]
		S. recognita Fisch. et Mey	[58]
		S. wiedemannii Boiss.	[87]
		S. blepharochlaena Hedge & Hub. Mor	[72]
167	wiedemannic acid	*S. heldreichiana* Boiss. ex Bentham	[75]
		S. wiedemannii Boiss.	[57]

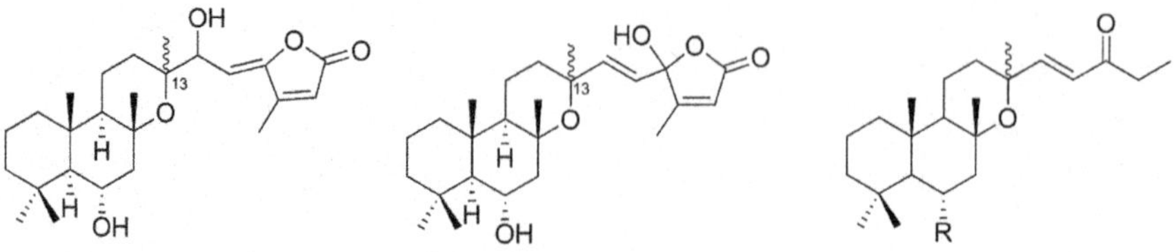

168

169

170: R=OH, 13β–Me
171: R=OH, 13α–Me
172: R=H

Fig. 4 Chemical structures of the isolated Anatolian *Salvia* sesterterpenes, in agreement with Table 3

Table 3 Isolated sesterterpenes from Turkish *Salvia* species

168	yosgadensolide A and its 13-epimer (*sesterterpene lactone*)	*S. yosgadensis* Freyn. et Bornm.	[40]
169	yosgadensolide B and its 13-epimer (*sesterterpene lactone*)	*S. yosgadensis* Freyn. et Bornm.	[40]
170	yosgadensonol	*S. yosgadensis* Freyn. et Bornm.	[69]
171	13-epi-yosgadensonol	*S. yosgadensis* Freyn. et Bornm.	[69]
172	6-dehidroxy-yosgadensonol	*S. limbata* C.A. Mey.	[43]

the first studies on *S. tomentosa* Mill., a new diterpene acid, 3*β*-hydroxy-8,11,13,15-abietatetraen-18-oic acid was obtained along with a known diterpene dehydroabietic acid [67]. Since the plant material was collected from the Mediterranean region of Turkey (Antakya-old Antioch) the new diterpene was named as antiochic acid.

Another study was carried out on *S. cryptantha* (syn = *S. absconditiflora*, endemic) which afforded two new abietane diterpenoids, 2*β*-hydroxyroyleanone and cryptanol along with the known abietanes horminone and 7-*O*-acetylhorminone [14].

Salvia fruticosa Mill. is the medicinal *Salvia* plant of Turkey, called Anatolian sage which is also named Greek sage [2] or Dalmatian sage, even Eastern Mediterranean sage, and its synonym name *S. triloba* has been used much more commonly in Anatolia due to its trilobed leaves, in general. Salvigenin is a new flavone was isolated from *S. triloba* and for the first time from nature [76], and later on it became a characteristic flavone for Anatolian *Salvia* species. In this study, an abietane diterpenoid carnosol was also obtained along with salvigenin (5-hydroxy-6,7,4′-trimethoxyflavone). In a continuing study with the roots of *S. triloba*, six known abietane diterpenoids; cryptanol, virgatol, 6*β*-hydroxyroyleanone, 7*α*-acetylroyleanone, horminone, ferruginol as well as two new ones were described. The structures of the new compounds trilobinol and trilobinone were established by based on spectral methods, namely NMR [71].

From the aerial parts of *Salvia wiedemannii* (endemic), two new abietane diterpenoids, wiedemannic acid and 3-oxo-abieta-8,11,13-triene were isolated along with the known four abietane diterpenoids; cryptanol, horminone, 7-*O*-acetylhorminone, ferruginol, and three known pimarane diterpenoids; 12-methoxypisiferic acid, isopimaric acid and 7β-hydroxysandracopimaric acid [57]. In addition, from the same extract, a new pimarane diterpene 14-oxo-pimaric acid and a known abietane diterpene pisiferal, and a known flavone salvigenin were isolated [61]. Continuing study on *S. wiedemannii* has revealed a new abietane diterpene 8-hydroxy-12-oxoabieta-9(11),13-dien-20-oic acid 8,20-lactone [87].

Four abietane diterpenoids were isolated from the aerial parts of *Salvia microstegia* Boiss. et Bal. Two of them (ferruginol and pisiferal) are known compounds. The other two, 10-acetylferruginol and 5,11,12-trihydroxy-abieta-8,11,13-triene are new compounds [53].

Microstegiol, a diterpene with a new carbon skeleton, was obtained from *Salvia microstegia* which is a first isolated rearranged abietane diterpene from Anatolian *Salvia* species. Determination of the carbon framework and substitution pattern of the isolate was made by series of selective INEPT, COSY and NOE experiments, which also permitted the unambiguous assignment of the C-13 NMR spectrum [81].

A second new rearranged abietane diterpene from Anatolian *Salvia* species was candidissiol with an eight-membered ring A which was isolated from the roots of *Salvia candidissima* together with four known diterpenoids manoyloxide, pachystazone, 20-deoxocarnosol and aethiopinone. In fact, there are two subspecies of *S. candidissima* growing in Turkey. Isolation of candidissiol was from *S. candidissima* Vahl. subsp. *occidentalis* Hedge which is noteworthy, because, appearance of an eight-membered ring structure in terpenoids is very rare in nature [64]. In a continuing study on the same subspecie, two new abietane diterpenoids 11-hydroxy-12-methoxyabieta-8,11,13-triene, 1-oxosalvipisone, and nine known diterpenoids; ferruginol, horminone, 7-*O*-acetylhorminone, cryptanol, montbretyl 12-methyl ether (its structure was corrected to salvinolonyl 12-methyl ether in a later study), microstegiol, 1-oxoaethiopinone and salvipisone were isolated from the roots of *S. candidissima*. The structures of the compounds were established by 1D- and 2D NMR and mass techniques. So, it can be concluded that *S. candidissima* is one of the *Salvia* species very rich in diterpenoids, and except 14-oxopimaric acid and manoyloxide, all the isolated 16 diterpenoids have either abietane or rearranged abietane skeleton [54].

From the aerial parts of *Salvia candidissima* Vahl. subsp. *occidentalis* Hedge, a sesquiterpene spathulenol and a new abietane diterpene 3-oxosalvipisone were obtained together with two labdane diterpenes. The known was identified as labdane 8,13-diepimanoyl oxide. The structure of the new labdane 11β-hydroxymanoyl oxide was elucidated based on extensive 1D and 2D NMR studies and confirmed by X-ray analysis [41].

The roots of *S. candidissima* Vahl. subsp. *candidissima* afforded 20 terpenoids, two of them being new, 7-oxo-11,12,14-trihydroxy-20-methoxyhydroxy-methylene-abieta-5,8,11,13-tetraene, and abietic acid derivative 1-oxo-5-hydroxy-abieta-8,11,13-trien-18-oic acid along with a new steroidal ester 1-oxostigmasterol 3β-palmitate, the triterpene α-amyrin acetate, and the flavone salvigenine. Eight diterpenes had been isolated before from *S. candidissima* subsp. *occidentalis* Hedge: 11-hydroxy-12-methoxy-abieta-8,11,13-triene, salvipisone, microstegiol, ferruginol, candidissiol, manoyl oxide, pachystazone and 11β-hydroxymanoyl oxide. The additional diterpenoids obtained from this subspecies were 7,20-epoxyroyleanone, 11,12-dioxo-abieta-8,13-dien, 2,3-dehydrosalvipisone, salvinolone, salvinolone 12-methyl ether, cryptojaponol, manool, 1-oxo-salvibretol and ent-sclareol [52].

From the root extracts of *Salvia hypargeia* Fisch. et Mey., collected from Sivas (central Turkey), six new abietane diterpenoids hypargenins (A–F) were isolated in addition to the known abietanes cryptanol and horminone. The structures of the new and the known compounds were established by spectral data. The new compounds, hypargenins A, B, C, D, and F showed antibacterial activity against standard bacteria while hypargenin E did not exhibit antibacterial activity. Hypargenin F was also found to be active against *Mycobacterium tuberculosis*. [77]. In a second study on the root extract of *S. hypargeia*, collected from Adana (south of Turkey), ten diterpenoids; 6α-hydroxysalvinolone, 6α-hydroxytaxodone, aethiopinone, microstegiol, ferruginol, saprorthoquinone, 11,12-dioxoabieta-8,13-diene, taxodione, hypargenin A and hypargenin D were isolated [34]. The last three abietanes were also isolated in the previous study.

The bioactivity-guided fractionation of this plant extract has yielded four abietane-type diterpenes (5,6-didehydro-7-hydroxytaxodone, 6-hydroxysalvinolone (=14-deoxycoleon U), dimethyl-cryptojaponol and salvicanaric acid), two triterpenes (lupeol and lupeol-3-acetate), and a fatty acid mixture consisting of mainly palmitic acid (51.6%) and palmitoleic acid (6.4%) and their cytotoxic activity was investigated against A2780 human ovarian cell lines [20].

Acid group containing abietane diterpenes are not very common in Anatolian *Salvia* species, even the simplets one; abietic acid was not isolated from any *Salvia* species. However, a few abietane diterpenes having isolated from some *Salvia* species, as an acid group containing abietane, dehydroabietic acid was isolated for the first time from *S. tomentosa*, and subsequently pisiferic acid and carnosic acid and derivatives from *S. triloba* and from a few Anatolian *Salvia* species. However, *S. divaricata* Montbret et Aucher ex. Bentham afforded three new diterpenic acids; 6-oxoroyleanone-18-oic acid, 6-oxo-12-methylroyleanone-18-oic acid, and horminone-18-oic acid, besides a new linear sesquiterpene salvinine, together with a group of known aromatic acids [39].

Five new and two known abietane diterpenoids were isolated from the aerial parts of *Salvia pomifera*. The new abietanes were established as pomiferin A-E and the two known ones as ferruginyl 12-methyl ether and 18-hydroxy-8,11,13-abieta-trien-7-one. Six known triterpenoids were also isolated [79]. Two new diterpenoids, pomiferin F and G and a new triterpenoid, 23-hydroxygermanicone and three known triterpenoids were obtained in addition to the previously isolated compounds from the same extract of *S. pomifera* [82].

A new diterpenoid nemorosin and known abietanes pachystazone, horminone, 7-*O*-acetylhorminone and salvipisone were isolated from the aerial parts of *Salvia nemorosa* L. The structures of the known and new compounds were established by spectral analysis [74].

From the acetone extract of the whole plant of *Salvia sclarea*, two new abietanes 2,3-dehydrosalvipisone and 7-oxoferruginol-18-al were obtained in addition to seven known diterpenes consist of two labdanes sclareol, manool, and three abietanes salvipisone, ferruginol, 7-oxoroyleanone, and two rearranged abietanes microstegiol and candidissiol were obtained [44].

From the acetone extract of the roots of *Salvia napifolia* Jacq., eight known abietanes; horminone, 7-acetyl-horminone, ferruginol, pachystazone, cryptanol, cryptojaponol, sugiol and microstegiol and five new abietanes; 6,12,14-trihydroxyabieta-6,8,11,12-tetraen, 7,20-epoxyroyleanone, 1-oxoferruginol, 6-oxoferruginol and 11,12-dioxo-abieta-8,13-dien were obtained. The structures of the new and the known compounds were established by extensive spectral methods [51].

From the roots of *S. heldrichiana*, a new abietane diterpene heldrichinic acid and five known diterpenes were isolated. From five known diterpenes, three of them have pimarane skeleton which were identified as isopimaric acid, 7β-hydroxysandracopimaric acid, 7-oxo-13-epi-pimara-9(11),15-dien-18-oic acid, and other two with abietane skeleton were identified for the firs time from native and named as wiedelactone and wiedemannic acid [75] isolated from *S. wiedemannii* [61, 87].

One of the *Salvia* species rich in abietane and rearranged abietane diterpenoids is *Salvia montbretii*, from its roots extract, five known compounds ferruginol, ferruginyl 12-methyl ether, taxodione, hypargenin F and demethylcryptojaponol, and one new abietane 14-hydroxyferruginol and two new rearranged abietane-type diterpenoids, montbretol (6,12-dihydroxyabieta-5,8,11,13-tetraen-7-one) and montbretyl 12-methyl ether (6-hydroxy-12-methoxyabieta-5,8,11,13-tetraen-7-one) were isolated [60]. However, in a subsequent study, Burnell et al. synthesized montbretyl 12-methyl ether through oxygenation of sugiol methyl ether and mentioned that the synthetic montbretyl 12-methyl ether [106] showed some spectral differences from the natural one. Therefore, re-investigation of the structure of montbretol and montbretyl 12-methyl ether by a co-group of Cordell and Ulubelen using the selective INEPT NMR technique verified their identification as salvinolone and salvinolonyl 12-methyl ether [107]. In a continuation study on a new collection of the roots of *Salvia montbretii*, additional six new diterpenoids to the previously reported ones [60] were isolated, and two of them were established as new rearranged abietane diterpenoids salvibretol, and 1-oxosalvibretol, others were abietanes; 6-hydroxysalvinolone, 7-hydroxytaxodione, 7,7′-bistaxodione, and 11,11′-didehydroxy-7,7′-dihydroxytaxodione, the latter two were dimeric abietanes. Structures of all the diterpenoids were established by IR, UV, and extensive NMR and mass spectroscopic techniques [15].

From the roots of *Salvia forskahlei*, a new diterpenoid forskalinone, two new aromatic compounds, the octanol esters of cis- and trans-4-*O*-methyl-caffeic acid dimers, were isolated together with the known steroids and a known triterpene α-amyrin [80].

Sesterterpenes are found fairly rare in nature. In fact, the first sesterterpene lactone salvisyriacolide from a *Salvia* species has been isolated from the aerial parts of *S. syriaca* by Rustaiyan et al. in Iran [108]. An endemic plant to Central Anatolia *S. yosgadensis* Freyn. et Bornm. afforded sesterterpenoids and sesterterpenoid lactones for the first time in Turkey beside other compounds as given below.

From the aerial parts of *Salvia yosgadensis* Freyn. et Bornm. two new sesterterpenes lactones, yosgadensolide A (6α,14-dihydroxymanoyloxide-15,17-dien-16,19-olide) and yosgadensolide B (6a, 16-dihydroxymanoyloxide-14,17-dien-16,19-olide) and their epimers with rare skeleton were obtained. In addition, a known sesquiterpene, three known sesquiterpene lactones, eight known flavonoids and two known aromatics were isolated [40].

In continuing study on *S. yosgadensis* yielded seven terpenes with unusual skeletons, two were known diterpenes, ambreinolide and norambreinolide. Two new norditerpenes were elucidated as 6α-hydroxynorambreinolide and 6α-hydroxy-8α-acetoxy-13,14,15,16-tetranorlabdan-12-oic acid, a new diterpene 6α-hydroxyambreinolide. In addition, two new 19,20-dinorsesterterpenes were obtained and named as yosgadensonol and 13-epi-yosgadensonol [69].

Six new rearranged abietane diterpenes were isolated from the roots of *Salvia limbata* C. A. Meyer (syn. *S. chrysadenia* Freyn). Their structures were established as new rearranged abietane diterpenoids as 12-hydroxysapriparaquinone, 3,12-dihydroxysapriparaquinone-1-ene, 2-hydroxysaprorthoquinone, limbinol, salvilimbinol and 4-dehydrosalvilimbinol by 1D- and 2D NMR techniques including spin decoupling, COSY, APT, HETCOR and COLOC experiments. In addition, a known labdane diterpene manool was also obtained [55].

From the aerial parts of *Salvia limbata*, two new diterpenes limbinal and acetyllimbinol and four known diterpenes ferruginol, abieta-8,11,13-triene, sclareol and manool were obtained. Two new unusual dinorsesterterpenes 6-dehydroxy-yosgadensonol and 6-dehydroxy-13-epiyosgadensonol were also isolated. From Anatolian *Salvia* species this is the second isolation of sesterterpenes which were first isolated from *S. yosgadensis* in addition to a sesquiterpene spathulenol [43].

The roots of *S. sclarea* yielded 12 diterpenoids, two of which are new, named sclareapinone and acetylsalvipisone. The known diterpenoids were identified as ferruginol, salvipisone, microstegiol, candidissiol, 2,3-dehydrosalvipisone, aethiopinone, 1-oxoaethiopinone, salvinolone, cryptojaponol and Δ^7-manool [50].

From the roots of *Salvia tchihatcheffii* (endemic), two new hydroxyabieta-tetraene diterpenes tchihatine and salvitchihatine, and a new triterpene, erithrodiol diacetate, were isolated in addition to the known compounds. The known abietanes were identified as dehydroabietic acid and 6,7-dehydroroyleanone, and a sesquiterpene (+)-ledol [47].

Abietane diterpenoids microstegiol, 2,3-dehydrosalvipisone, salvipisone, candidissiol, salvinolonyl 12-methyl ether, ferruginol and a labdane diterpene manool, and a linear alcohol nonacosan-1-ol were isolated from the roots of *S. verticillata* [63].

The aerial parts of *Salvia cyanescens* afforded a new diterpenoid 12-isopentenyl-3-oxosalvipisone, five known diterpenoids manoyl oxide, 11β-hydroxymanoyl oxide, ferruginol, aethiopinone and salvipisone in addition to the two sesquiterpenes mentioned sesquiterpene part before [42].

From the roots of *Salvia multicaulis*, four new aromatic norabietane diterpenoids, multicaulin, 12-demethylmulticauline, multiorthoquinone, and 12-demethylmultiorthoquinone, two new abietane diterpenoids, 12-methyl-5-dehydrohorminone and 12-methyl-5-dehydro-*O*-acetylhorminone, as well as a new pimarane diterpenoid, salvipimarone were isolated, the presence of norabietanes is fairly rare in *Salvia* species. As known diterpenoids; six abietanes 1-oxoferruginol, 18-oxoferruginol, pisiferal, hinokione, horminone and sempervirol and a labdane as diterpenes and two known triterpenes α-amyrin and lupeol were also obtained [19].

In a subsequent study on *S. multicaulis*, two new abietanes, 3-oxo-12-methoxy-14-hydroxyabieta-8,11,13-triene and 6-oxo-12-peroxyabieta-8,11,13-triene were obtained in addition to a new aromatic ester 4,4′-bisbenzoic acid heptyl ester from the roots of *Salvia multicaulis*. Six known abietanes; ferruginol, ferruginol-18-al, 6-oxoferruginol, *O*-acetylhorminone, cryptanol, wiedelactone were also isolated [65]. A new noricetexane diterpene, named salvimultine, was obtained as a unique icetexane isolated from nature for the first time. However, several icetexane were isolated from some *Salvia* species except for Anatolian *Salvia* species, has been isolated from the polar fractions of *Salvia multicaulis*. Its structure was established as 1(10)-*seco*-2(10)-cyclo-icetexane by using 1D- and 2D NMR spectral methods, including COSY, HETCOR, COLOC, and NOESY experiments as well as HRMS [83]. Totally, from *S. multicaulis* extract 23 diterpenoids were isolated consisting of 10 new diterpenoids; 4 abietanes, 4 nor-abietanes, a pimarane, a nor-icetexane were obtained besides 13 known diterpenoids consisting of 12 abietanes and a labdane diterpenes.

The roots of *Salvia recognita* Fisch. et Mey. (endemic) yielded seven diterpenoids, three of them being new, salvirecognone, salvirecognine, 6α-hydroxy-11,12-dioxo-8,13-abieta-diene along with four known triterpenoids. Four known diterpenoids ferruginol, wiedelactone, sugiol, 12-methyl-5-dehydrohorminone. The structures of the new compounds were deduced by 1D and 2D NMR techniques including APT, DEPT, COSY and HETCOR [58].

From the roots of *Salvia bracteata* Banks and Sol., two new diterpenoids salvibracteone and bractealine have been isolated together with eight known diterpenoids. The known compounds are 7-*O*-acetylhorminone, horminone, 12-*O*-methylhorminone, 6,7-dehydroroyleanone, ferruginol, 12-*O*-methylferruginol, salvinolone, and sugiol [59].

Salvia kronenburgii Rech. Fil., is one of the most rich and diverse plants in both di- and triterpenoid compounds. In the first study on the root extract of

S. kronenburgii, an endemic specie to the Eastern Anatolia, a new norditerpenoid named kronenquinone was obtained along with eight known abietane diterpenoids; taxodione, ferruginol, salvipisone, 12-hydroxysapriparaquinone, royleanone, 7-acetoxyroyleanone, 6-hydroxysalvinolone, 1-oxosalvibretol, and a labdane Δ^7-manool [49].

From the roots of *Salvia viridis* L. (syn: *S. horminum*), Ulubelen et al. isolated three new abietane diterpenes, salviviridinol, viridinol, viridone, together with five known diterpenoids. Three known compounds were sugiol, 1-oxoferruginol, ferruginol which have abietane skeleton, the other two diterpenoids have rearranged abietane skeleton, aethiopinone and microstegiol. Structures of the compounds were established by 1D and 2D NMR spectroscopic techniques [62].

A new rearranged abietane diterpene, salvisyrianone, isolated from the root extract of *Salvia syriaca L.* with four known abietanes ferruginol, 4-dehydrosalvilimbinol, viridone and candidissiol [68].

The roots of *Salvia blepharochlaena* Hedge and Hub. Mor. (endemic) have yielded two new diterpenoids, blephaein and *O*-methylpisiferic acid methyl ester together with eight known diterpenoids; pisiferic acid, 12-*O*-methylpisiferic acid, multicaulin, multiorthoquinone, 2-demethylmultiorthoquinone, ferruginol, horminone, and 7-*O*-acetylhorminone. The structures of the new compounds were established by spectroscopic analysis and by some chemical reactions [18].

In the second study on *S. blepharochlaena*, six abietane diterpenoids, 8-hydroxy-12-oxo-abieta-9(11),13-dien-20-oic acid-8,20-lactone, sugiol, 7,20-epoxyroyleanone, formosanolide, royleanone and cryptanol were obtained [72].

From the roots of *Salvia amplexicaulis* Lam., five diterpenoids, four triterpenoids, three steroids and a flavonoid salvigenin were isolated. Isolated diterpenoids were identified as 7-oxo-abieta-9,12,14-triene, ferruginol, horminone, 7-*O*-acetylhorminone and sugiol [73].

Another Anatolian *Salvia* specie *S. caespitosa* Montbret and Aucher ex. Bentham was found to be rich in pimarane diterpenoids. From its roots extract, a new pimarane diterpene 6β-hydroxyisopimaric acid has been isolated together with three known pimaranes; isopimaradienal, isopimaric acid, sandaracopimaric acid, and an abietane diterpene ferruginol [70].

Salvia ceratophylla L. is a lemon-scented biennial herbaceous plant. Its roots afforded two new compounds, 6β-hydroxy-8,15-pimaradien-1-one and 3-oxo-8,11,13-abietatrien-20-oic acid and four known abietanes, ferruginol, salvipisone, aethiopinone, candidissiol. From four known diterpenoids, aethiopinone and candidissiol are rearranged diterpenoids [66].

Salvia staminea Montbret et Aucher ex Benth. methanolic extract yielded six known diterpenoids as ferruginol, taxodione, microstegiol, manoyloxide, 1-oxo-salvibretol and cryptojaponol [45].

The roots of *S. eriophora* Boiss. et Kotschy. acetone extract afforded eleven abietane and rearranged abietanes; 4,14-dihydroxysaprorthoquinone, aethiopinone, salvipisone, salvilimbinol, 4,12-dihydroxy-sapriparaquinone, 12-hydroxysapriparaquinone, 3,12-dihydroxysapriparaquinone-1-ene, ferruginol, 6,7-dehydroroyleanone, horminone, and

7-*O*-acetylhorminone [56]. In a later study, a new rearranged abietane diterpenoid eriophoroxide was obtained from the root extract of *S. eriophora* Boiss. et Kotschy, and its structure was elucidated by spectroscopic methods and semi-empirical/ab-initio calculations [78].

Methanolic extract of the aerial parts of *S. chrysophylla* (endemic) yielded sclareol. The common triterpenoids ursolic and oleanolic acids and *β*-sitosterol were also obtained [84].

Recently, the Anatolian *Salvia S. fruticosa* L. (syn = *S. triloba* Mill.) hexane and dichloromethane extracts obtained by re-extraction of its MeOH extract, which was studied by Topcu et al., and isolated seven bioactive diterpenoids; carnosol, carnosic acid, carnosic acid 12-methyl ether, rosmadial, isorosmanol, ferruginol, and manool [30]. Among them, only carnosol was obtained in the first study on the same specie [76] while ferruginol only in second study besides the other seven abietanes.

Di-and triterpenoids of Anatolian *Salvia* species due to inhibiting both butyryl- and acetyl-cholinesterases enzymes may help for cognitive brain function and neuroprotection in some neurodegenerative disorders, as some other Lamiaceae family plants as well [109].

3.1.3 Triterpenoids and Steroids

One of the first chemical studies on Anatolian *Salvia* species was published on *S. virgata* in 1975 and 1976 by Ulubelen group. The former publication was about flavonoids of *S. virgata* [104], the latter one was about isolation of a new triterpene [110] from the same plant, its structure was elucidated as 3*β*-hydroxy-1-oxo-olean-12-en-28-oic acid and named virgatic acid (vergatic acid).

Seventy percent aqueous ethanolic extract of the *S. triloba* yielded ursolic acid oleaonolic acid and steroid *β*-sitosterol [76].

S. aethiopis yielded two steroids *β*-sitosterol and sitosteryl-3*β*-D-glucoside [94].

In a the first study on the chloroform ($CHCl_3$) extract of *S. horminum* yielded triterpenoids micromeric, ursolic and oleanolic acids [111]. A further investigation of the plant has led to the isolation of three triterpenic alcohols, including a new oleanane type triterpene olean-(13)18-ene-2*β*,3*β*-diol and two lupane triterpenes lupeol and lup-(20)29-ene-2*α*,3*β*-diol from the petroleum ether extract [112].

The first study on *S. tomentosa* Mill., was carried out by Ulubelen et. al. which afforded triterpenic acids; ursolic, oleanolic and crateagolic acids and a steroidal glycoside sitosteryl 3*β*-glucoside [67]. In a subsequent study, two known triterpenes uvaol and *β*-amyrin were also isolated from this species [90].

From the isolated triterpenic acids, oleanolic acid and ursolic acid are very common triterpenic acids in the genus *Salvia* and some other genera of Lamiaceae family and other family plants while virgatic and crataegolic acids were not found commonly in *Salvia* species. Virgatic acid was first isolated from *S. virgata* as mentioned above, then from *S. pinnata* L. [91], both by Ulubelen group. Crataegolic acid was isolated from some other *Salvia* species, including from *S. officinalis* by Brieskorn group in Europe

[113], subsequently from *S. tomentosa* in Turkey [67]. Another bioactive triterpene with lupane skeleton betulinic acid was found in *S. triloba* for the first time by Abdel-Moneim et al. [114], then in some Anatolian *Salvia* species including *S. verticillata* L. subsp. *verticillata.* In addition, the latter plant extract also afforded four known triterpenoids crataegolic acid, oleanolic acid, ursolic acid and vergatic acid and steroids β- sitosterol and sitosteryl 3β-D-glucoside [91].

A new and four known triterpenoids have been isolated from the leaves of *Salvia pinnata* L. Their structures have been established as a new trihydroxylated triterpene olean-13(18)-ene-2α,3β,11α-triol, and four known ones lupeol, lup-20(29)-ene-3β,11α-diol, oleanolic acid, and virgatic acid. The structure of the new compound was elucidated based on spectral data and by some chemical reactions [92].

From the leaves of *S. palaestina* Benth., triterpenoids vergatic acid, ursolic acid, crataegolic acid, lupane-3β,11α,20-triol, and sitosteryl 3β-glucoside and a diterpene sclareol were isolated [85].

From the roots of *S. potentillifolia* (Boiss. et Heldr.) ex Bentham the triterpenoids crataegolic, oleanolic, and vergatic acids, β-amyrin, lupan-3β,11α,20-triol were isolated as well as β-sitosterol [48].

Six known triterpenoids oleanolic acid, α-amyrin, lupeol, oleanolic acid methyl ester, taraxasterol and β-sitosterol were isolated from *S. pomifera* [79].

A new triterpenoid, 23-hydroxygermanicone and three known triterpenoids moronic acid, moradiol, erithrodiol were obtained in addition to the previously isolated compounds from the same extract of *Salvia pomifera* [82].

From the aerial parts of *S. montbretii*, a triterpene monogynol A and its two new esters were isolated besides known triterpenes α-amyrin, lupeol, ursolic and oleanolic acids and β-sitosterol. The structures of the new triterpene esters were established as 3β-*O*-*cis*-*p*-coumaroylmonogynol A and 3β-*O*-*trans*-*p*-coumaroylmonogynol A by NMR and mass spectroscopic techniques. The structure of monogynol A was further confirmed by X-ray diffraction analysis [89].

From *S. nemorosa* a new triterpenoid, salvinemorol and known triterpenes; α-amyrin, ursolic and oleanolic acids, steroids; stigmast-7-en-3-one, stigmast-4-en-3-one, β-sitosterol, stigmast-7-enol, a steroidal triterpene 24-methylene-cycloartanol, as well as flavonoids were isolated [74].

From the roots of *S. limbata* two triterpenic acids ursolic and oleanolic acid and two steroids stigmasterol and β-sitosterol were obtained [43, 55].

A known triterpene α-amyrin and steroids β-sitosterol, stigmast-3-one were obtained from the roots of *S. forskahlei* [80].

The roots of *S. candidissima* Vahl. subsp. *candidissima* afforded a new steroidal ester 1-oxostigmasterol 3β-palmitate, a known triterpene α-amyrin acetate, and isolated 19 diterpenoids mentioned in the diterpenoids part [52].

A new sterodial compound 1-oxo-7α-hydroxysitosterol was isolated from the whole plant of *Salvia glutinosa* L. in addition to 11 known triterpenoids and three steroids. The known triterpenoids were α-amyrin, α-amyrin acetate, 11-oxo-α-amyrin (3β-hydroxy-11-oxo-ursan-12-ene), 11-oxo-β-amyrin (3β-hydroxy-11-oxo-olean-12-ene), 3β-acetoxyolean-9,11-diene, ursolic and oleanolic acids, ursolic and oleanolic acid methyl esters, lupeol, erythrodiol 28-acetate and the known steroid compounds stigmasterol, sitosterol, 7α-hydroxysitosterol [88].

From an acetone extract of the whole plant *S. sclarea*, a triterpenoid α-amyrin and a steroid β-sitosterol were obtained [44]. 3-oxo-oleanolic acid was also obtained from a new collection of *S. sclarea* along with α-amyrin and β-sitosterol [50].

From the roots of *S. verticillata* L., five known triterpenoids α-amyrin, β-amyrin, α-amyrin-28-al, ursolic acid, acetyloleanolic acid, and two steroids β-sitosterol and 3β-*p*-pentyl-benzoic acid 7-oxositosteryl ester were isolated [63].

The aerial parts of *S. cyanescens* yielded known triterpenes α-amyrin, α-amyrin acetate and lupeol 3-acetate and a steroid β-sitosterol [42].

From the roots of *S. tchihatcheffii*, a new triterpene erithrodiol diacetate was isolated in addition to the known compounds. The known triterpenoids were identified as 3-acetylerithrodiol, 28-acetylerithrodiol, the acetate of 3β-hydroxyolean-12-en-28-al, and two steroids sitosterol and stigmasterol [47].

Three triterpenoids lupeone, 3β-hydroxyolean-12-en-28-al, 3-acetyloleanolic acid isolated from *S. recognita* root extract [58].

Four known triterpenoids; lupeone, 3-acetyl-beta-amyrin, 3α-acetylerythrodiol, hopanone, were obtained in addition to a new steroid brassicasterone, and two known steroids β-sitosterol and stigmast-4-en-3-one from the roots of *S. multicaulis* [65]. Lupeol and α-amyrin were also isolated in another work on the same plant [19]. As a result of the studies on *S. multicaulis* which is an endemic specie to Turkey 12 new and 21 known compounds were isolated.

Root extract of *S. syriaca* L. yielded two triterpenes ursolic and oleanolic acids, and two steroids 3β-hydroxy-stigmast-5-en-7-one and β-sitosterol [68].

From the roots of *S. caespitosa* Montbret and Aucher ex. Bentham, a new triterpenoid, 3-acetylvergatic acid, as well as five known triterpenoids; 3-oxo-oleanolic acid, α-amyrin, lupeol, oleanolic acid, and ursolic acid, and two steroids; sitosterol and sitosteryl 3β-glucoside were obtained along with five diterpenoids [70].

From *S. amplexicaulis* Lam., known compounds as four triterpenoids α-amyrin, ursolic acid, oleanolic acid, acetyloleanolic acid and three steroids stigmast-4-en-3-one, stigmasterol, β-sitosterol were isolated [73].

Callus, cell and root culture systems of *S. amplexicaulis* were established to study secondary metabolites of the plant. The roots of *Salvia amplexicaulis* cultured in Schenk-Hildebrand medium supplemented with 1 mg/L indol butyric acid which have shown hairy root characteristics. Root cultures produced two triterpenes oleanolic and ursolic acids and a steroid β-sitosterol [115].

From *S. ceratophylla*, common triterpenoids ursolic acid and oleanolic acid, and β-sitosterol were obtained [66].

From the methanol (MeOH) extract of *Salvia staminea*, a new ursane-type triterpenoid, salvistamineol and a steroid sitosterol 3β-D-glucoside and from acetone extract a triterpenoid oleanolic acid 3β-acetate and two steroids β-sitosterol and stigmast-4-en-3-one have been isolated [45].

Highly hydroxylated triterpenoids have been obtained from a few *Salvia* species, one of them is *S. kronenburgii* which is an endemic specie to the eastern part of Turkey, collected from Van. In the first study on the root extract of *S. kronenburgii*, six new triterpenoids were isolated. The new triterpenoids were elucidated as 1β,2α,3β,11α-tetrahydroxy-olean-12-ene, 1β,2α,3β,11α-tetrahydroxy-urs-12-ene, 2α-acetoxy-urs-5,12-diene-3β,11 α-diol, 3α-acetoxy-urs-12-ene-1β,11α-diol, 1β,3α,11α-trihydroxy-urs-12-ene, 2α,3β,11α-trihydroxy-urs-12-ene, and the known triterpenes 3β-acetoxy-urs-12-ene-1β,2α,11α-triol and 3β-acetoxyolean-12-ene-1β,2α,11α-triol, based on extensive NMR and MS analyses. Nine diterpenoids were also obtained as indicated in the diterpenoids part [49].

In a continuing search, three new and five known triterpenoids and a sterol were isolated from the acetone extract of *S. kronenburgii*. The structures of the new triterpenoids were established as 1β,2α-dihydroxy-3β-acetoxy-11-oxours-12-ene, 2α,20β-dihydroxy-3β-acetoxyurs-9(11),12-diene, and 1β,2α-dihydroxy-3β-acetoxyurs-9(11),12-diene on the basis of spectral analyses, including 1D and 2D NMR and mass spectroscopy. However, the latter two compounds probably are artifacts formed from dehydration of the corresponding allylic alcohols. The known triterpenoids were identified as 1β,2α-dihydroxy-3β-acetoxyolean-9(11),12-diene, 1β,2α,3β,11α-tetrahydroxyurs-12-ene, 3β-acetoxy-2α,11α-dihydroxyurs-12-ene, salvinemorol, salvistamineol and β-sitosterol. Among them, 1β,2α,3β,11α-tetrahydroxyurs-12-ene was the most abundant compound in the extract, was found to be highly cytotoxic to renal, non-small cell lung, and breast cancer cell lines [16].

In the second study on *S. blepharochlaena,* in addition to isolated six abietane diterpenoids, four triterpenoids, 24-methylenecycloartanol, erythrodiol-3-acetate, α-amyrin and oleanolic acid-3-acetate and four steroids, 3β-hydroxystigmast-5-en-7-one, stigmast- 4-en-3-one, β-sitosterol and stigmasterol were obtained [72].

Acetone extract of *S. microphylla* afforded three triterpenes erithrodiol 3-acetate, oleanolic acid, lupeol and a steroid β-sitosterol as known compounds in addition to a new and a known phenolic esters two sesquiterpenes. A diterpene carnosic acid 12-methyl ether (12-methoxycarnosic acid) was also isolated from the same plant extract [38].

Salvia macrochlamys Boiss. et Kotschy. was also one of the rich species in triterpenoids, and the MeOH extract yielded different type of terpenic compounds, being three new and seven known ones. The new triterpenoids were identified as monogynol A derivatives consisting of 3β-acetylmonogynol A, 3β-acetyl,22β-hydroxymonogynol A, 3β-acetyl,21β,22β-dihydroxymonogynol A. Their structures were determined by using spectroscopic methods including extensive 1D and 2D NMR techniques (DEPT, COSY, HMQC, HMBC and NOESY). Seven of the known terpenoids were triterpenes including germanicol, germanicol acetate, lupeol, lupeol acetate, monogynol A, ursolic acid and a sesquiterpene caryophyllene oxide [46].

Two triterpenoids oleanolic acid and betulinic acid were isolated besides two phenolic compounds from *S. cedronella* [103].

Lupeol and lupeol-3-acetate were isolated from *S. hypargeia* besides 4 diterpenoids [20].

Beside a diterpenoid sclareol; two triterpenoids, namely 2α,3α-dihydroxy-24-nor-4(23),12-olean-dien-28-oic acid, ursolic acid and a steroid β-sitosterol were isolated from the aerial parts of *S. poculata* Nab., an endemic *Salvia* species to Turkey [86].

The aerial parts of *S. chrysophylla* Stapf, which is an endemic species to south-western Anatolia. Structures of the isolated compounds were elucidated as triterpenoids oleanolic and ursolic acid, and a steroid β-sitosterol [84].

The hexane and dichloromethane (CH_2Cl_2) extracts, obtained by re-extraction of the MeOH extract of *S. fruticosa* Mill., afforded four triterpenoids (amyryl-tetracosanoate, oleanolic acid, ursolic acid, and erythrodiol), a steroid (3-acetyl-sitosterol) beside seven diterpenoids [30].

From the aerial parts of *S. trichoclada* Bentham two lupan type triterpene aglycones, lupeol, and 30-hydroxylup-20(29)-en-3-on, oleanolic acid and ursolic acid were isolated [93].

The MeOH extract of the aerial parts of *S. trichoclada* Bentham (Lamiaceae) was studied for bioactive non-volatile secondary metabolites, and nine compounds were isolated. Structures of the isolated compounds were elucidated as lupeol, lupenone, glochidone, monogynol A, oleanolic acid, ursolic acid, β-sitosterol, apigenin-7-*O*-β-D-rhamnoside, and a phenolic rosmarinic acid by using 1D and 2D-NMR spectroscopic techniques. A lupane triterpene glochidone was isolated from a *Salvia* species and Lamiaceae family for the first time in this study [21] (Figs. 5 and 6).

3.1.4 Flavonoids and Other Phenolics

Since polar extracts of Anatolian *Salvia* species were not investigated for their flavonoids and other phenolics intensively, the number of isolated flavonoids is limited as well as phenolics. In fact most of phenolics, particularly characteristic phenolic acids of *Salvia* species, such as salvianolic acids have been isolated from Chinese *Salvia*, *Salvia miltthorrhiza* (Danshen).

The majority of flavonoids belong to the flavones, particularly 6-hydroxylated (or 6-methoxylated) flavones which are accepted as useful taxonomic markers for the genus. Flavonols are less common when compared to flavones [26].

The characteristic flavone for Anatolian *Salvia* species is salvigenin which was isolated from most of the investigated *Salvia* species growing in Turkey. Salvigenin was isolated as a unique flavone from *S. caespitosa* [70], *S. candidissima* [52], *S. ceratophylla* [66], *S. chrysophylla* [84], *S. fruticosa* [30], *S. heldrichiana* [75], *S. syriaca* [68], *S. tchihatcheffii* [47], *S. wiedemannii* [61] species.

The first chemical study on the flavonoids of Anatolian *Salvia* species was published on *S. virgata* in 1975, and salvigenin, salvigenin-5-glucoside and luteolin-7,3′,4′-trimethyl ether were isolated [104].

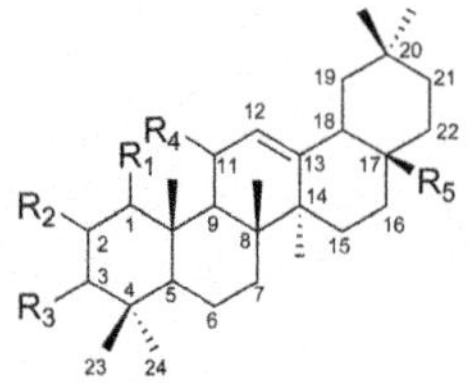

177: R_1=R_2=R_4=H, R_3=βOH, R_5=Me
179: R_1=R_2=H, R_3=βOH, R_4=oxo, R_5=Me
189: R_1= oxo, R_2=R_4=H, R_3=βOAc, R_5=COOH
190: R_1=R_2=R_4=H, R_3=oxo, R_5=COOH
193: R_1=βOH, R_2=R_4=αOH, R_3=βOAc, R_5=Me
194: R_1=R_2=R_4=H, R_3=βOAc, R_5=Me, $\Delta^{9(11)}$
199: R_1=R_2=R_4=H, R_3=βOAc, R_5=CHO
202: R_1=R_2=R_4=H, R_3=βOAc, R_5=COOH
204: R_1=R_4=H, R_2=αOH, R_3=βOH, R_5=COOH
205: R_1=R_2=R_4=H, R_3=βOH, R_5=CH_2OH
206: R_1=R_2=R_4=H, R_3=βOAc, R_5=CH_2OH
207: R_1=R_2=R_4=H, R_3=βOAc, R_5=CH_2OAc
208: R_1=R_2=R_4=H, R_3=βOH, R_5=CH_2OAc
223: R_1=R_2=R_4=H, R_3=βOH, R_5=COOH
224: R_1=R_2=R_4=H, R_3=βOAc, R_5=COOH
225: R_1=R_2=R_4=H, R_3=βOH, R_5=COOMe
226: R_1= H, R_2=H, R_3=βOH, R_4=21=αOH, R_5=Me
232: R_1= oxo, R_2=R_4=H, R_3=βOH, R_5=COOH

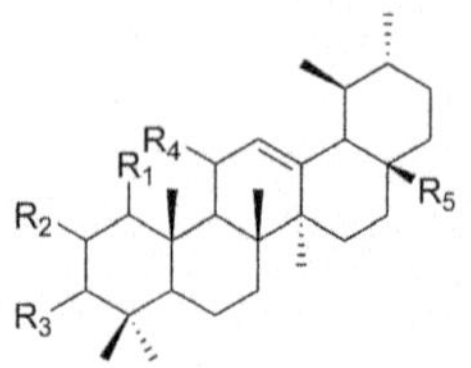

173: R_1=R_2=R_4=H, R_3=βOH, R_5=Me
174: R_1=R_2=R_4=H, R_3=βOAc, R_5=Me
175: R_1=R_2=R_4=H, R_3=βOH, R_5=CHO
176: R_1=R_2=R_4=H, R_3=β$C_{24}H_{49}COO$, R_5=Me
178: R_1=R_2=H, R_4=oxo, R_3=βOH, R_5=Me
180: R_1=R_3=βOH, R_2=R_4=αOH, R_5=Me
181: R_1=R_3=βOH, R_2=H, R_4=αOH, R_5=Me
182: R_1=βOH, R_2=H, R_3=R_4=αOH, R_5=Me
183: R_1=βOAc, R_2=H, R_3=R_4=αOAc, R_5=Me, Δ^5
184: R_1=βOH, R_2=αOH, R_3=βOAc, R_4=oxo, R_5=Me
185: R_1=βOH, R_2=αOAc, R_3=βOAc, R_4=oxo, R_5=Me
186: R_1=βOH, R_2=αOH, R_3=βOAc, R_4=H, R_5=Me, $\Delta^{9(11)}$
187: R_1=H, R_2=αOAc, R_3=βOH, R_4=H, R_5=Me, Δ^5
188: R_1=R_4=H, R_2=αOH, R_3=βOAc, R_5=Me, 20-βOH, $\Delta^{9(11)}$
191: R_1=βOH, R_2=H, R_3=αOAc, R_4=αOH, R_5=Me
192: R_1=H, R_2=R_4=αOH, R_3=βOAc, R_5=Me
195: R_1=βOH, R_2=R_4=αOH, R_3=βOAc, R_5=Me
227: R_1=R_2=H, R_3=βOAc, R_4=αOH, R_5=Me, 21-αOH
229: R_1=R_2=R_4=H, R_3=βOH, R_5=COOH
230: R_1=R_2=R_4=H, R_3=βOH, R_5=COOMe
231: R_1=R_2=R_4=H, R_3=βOH, R_5=CH_2OH

Fig. 5 Chemical structures of the triterpenes isolated from Anatolian *Salvia* species, in agreement with Table 4

From *Salvia aethiopis* two flavonoids 5-hydroxy-7,3′,4′-trimethoxyflavone and salvigenin were obtained besides a hydrocarbon n-eicosane [94].

One of the earlier studies was carried out on the leaves of *Salvia tomentosa* Mill. (syn, *Salvia grandiflora* Etling) which yielded seven flavonoids; 5-hydroxy-6,7,3′,4′-tetramethoxyflavone, cirsimaritin, jaceosidin, luteolin, luteolin 7-*O*-β-D-glucoside, 6-methoxyluteolin and a new one 6-methoxyluteolin 7-*O*-β-D-glucoside [95]. In addition, in a second study on *S. tomentosa* four known flavonoids; cirsilineol, eupatilin, diosmetin, 6-hydroxyluteolin 7-*O*-β-D-glucoside and a new glycoside 6-hydroxyluteolin 5-*O*-β-D-glucoside were obtained as *O*-glucosides by Ulubelen et al. [67].

Salvia verticillata L. is one of the most common *Salvia* species in Turkey. From its different collections in Anatolia, several flavonoids and phenolic compounds were obtained. In the first study, a simple coumarin esculetin and a known compound luteolin-7-*O*-β-D-glucoside were obtained from its subspecies; *S. verticillata* subsp. *verticillata*, and the latter was also obtained from *S. pinnata* along with apigenin, luteolin, 6-methoxyluteolin, apigenin 7-*O*-β-D-glucoside, and quercetin 3-*O*-β-D-glucoside in the same study [91]. A rare flavone glucoside 6-hydroxyluteolin

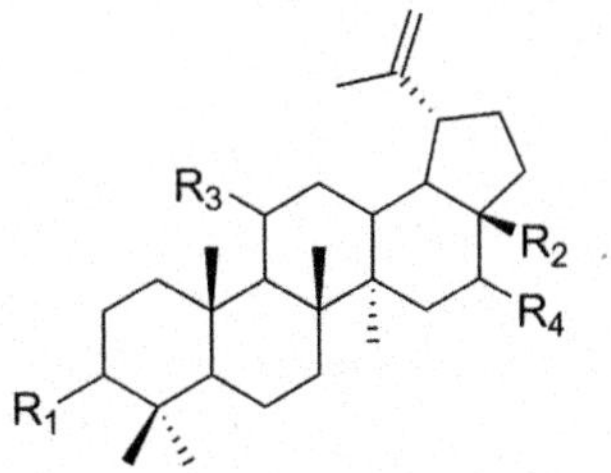

203: R_1=βOH, R_2=COOH, R_3=R_4=H
211: R_1=oxo, R_2=Me, R_3=R_4=H, Δ^1
213: R_1=βOH, R_2=R_4=H, R_3=αOH
215: R_1=oxo, R_2=Me, R_3=R_4=H
216: R_1=βOH, R_2=Me, R_3=R_4=H
217: R_1=βOAc, R_2=Me, R_3=R_4=H

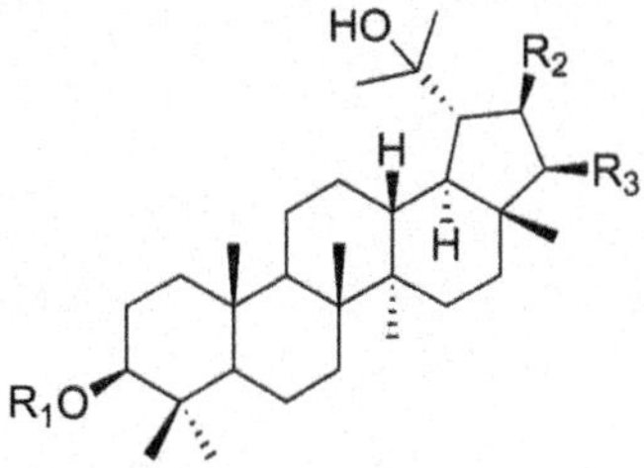

218: R_1=R_2=R_3=H
198: R_1=Ac, R_2=R_3=H
197: R_1=Ac, R_2=OH, R_3=H
196: R_1=Ac, R_2=OH, R_3=OH
214: R_1=R_2=R_3=H, 11-αOH
200: R_2=R_3=H, R_1=

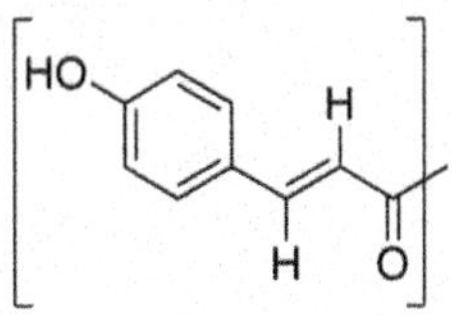

201: R_2=R_3=H, R_1=

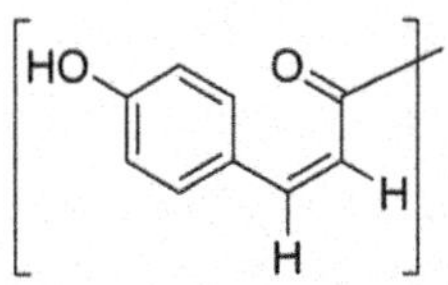

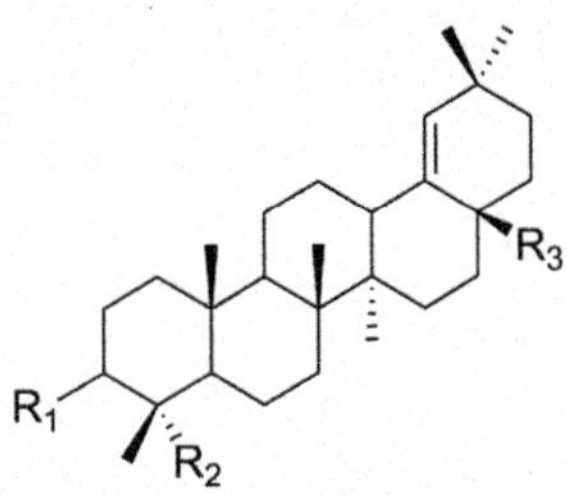

219: R_1=βOH, R_2=Me, R_3=CH_2OH
220: R_1=oxo, R_2=Me, R_3=COOH
209: R_1=βOH, R_2= R_3=Me
210: R_1=βOAc, R_2= R_3=Me

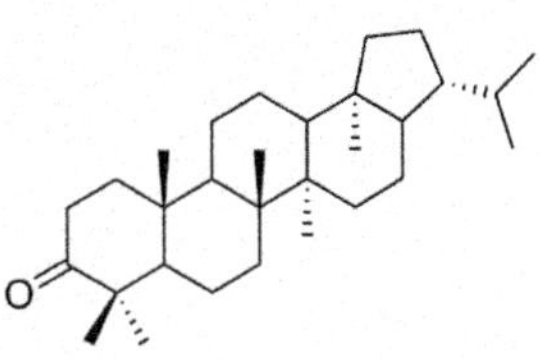

212

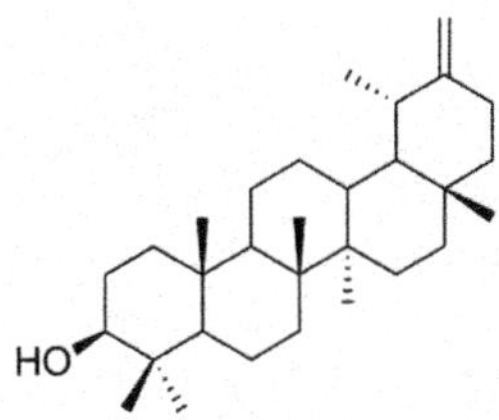

228

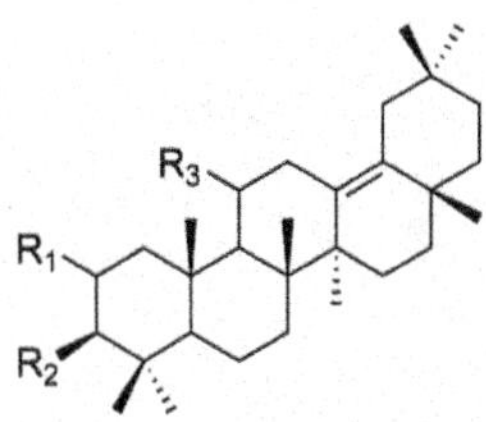

221: R_1=R_3=αOH,R_2=βOH
222: R_1=R_3=H,R_2=βOAc

Fig. 5 (continued)

5-*O*-β-D-glucoside and cirsimaritin were also obtained from *S. verticillata* subsp. *verticillata* [91].

From the leaves of *S. palaestina* Bentham, 16 known flavonoids; salvigenin, eupatilin, apigenin 7,4′-dimethyl ether, luteolin 7,4′-dimethyl ether, genkwanin, 6,7,3′,4′-tetramethoxyflavone, cirsimaritin, chrysoeriol, apigenin, luteolin, apigenin 7-*O*-β-D-glucoside, apigenin 7-*O*-β-D-glucuronide, luteolin 7-*O*-β-D-glucoside,

Table 4 Isolated triterpenes from Turkish *Salvia* species

No	Name	Plant Name	Ref
173	α-amyrin	*S. amplexicaulis* Lam.	[73]
		S. caespitosa Montbret & Aucher ex Benth.	[70]
		S. cyanescens Boiss. & Balansa	[42]
		S. forskahlei L.	[80]
		S. glutinosa L.	[88]
		S. monbretii Benth.	[89]
		S. multicaulis Vahl.	[19]
		S. nemorosa L.	[74]
		S. pomifera L.	[79]
		S. sclarea L.	[44, 50]
		S. verticillata L.	[63]
174	α-amyrin 3-acetate	*S. cyanescens* Boiss. et Bal.	[42]
		S. candidissima Vahl subsp. *candidissima*	[52]
		S. glutinosa L.	[88]
		S. multicaulis Vahl.	[65]
175	α-amyrin-28-al	*S. verticillata* L.	[63]
176	α-amyryl-tetracosanoate	*S. fruticosa* Mill. (*S. triloba* L.)	[30]
177	β-amyrin	*S. tomentosa* Mill.	[90]
		S. verticillata L.	[63]
		S. potentillifolia (Boiss. et Heldr.) ex Bentham	[48]
178	11-oxo-α-amyrin	*S. glutinosa* L.	[88]
179	11-oxo-β-amyrin	*S. glutinosa* L.	[88]
180	1β,2α,3β,11α-tetrahydroxy-urs-12-ene	*S. kronenburgii* Rech. f.	[16]
181	1β,3β,11α-trihydroxy-urs-12-ene	*S. kronenburgii* Rech. f.	[16]
182	1β,3α,11α-trihydroxy-urs-12-ene	*S. kronenburgii* Rech. f.	[16]
183	1β,3α,11α-triacetoxy-urs-5,12-diene	*S. kronenburgii* Rech. f.	[16]
184	1β,2α-dihydroxy-3β-acetoxy-urs-12-ene-11-on	*S. kronenburgii* Rech. f.	[16]
185	1β-hydroxy-2α,3β-diacetoxy-urs-12-ene-11-on	*S. kronenburgii* Rech. f.	[16]
186	1β,2α-dihydroxy-3β-acetoxy-urs-9(11),12-diene	*S. kronenburgii* Rech. f.	[16]
187	2α-acetoxy-3β-hydroxy-urs-5,12-diene	*S. kronenburgii* Rech. f.	[16]
188	2α,20β-dihydroxy-3β-acetoxy-urs-9(11),12-diene	*S. kronenburgii* Rech. f.	[16]
189	3-acetylvergatic acid	*S. caespitosa* Montbret & Aucher ex Benth.	[70]
190	3-oxo-oleanolic acid	*S. caespitosa* Montbret & Aucher ex Benth.	[70]
		S. sclarea L.	[50]

(continued)

Table 4 (continued)

No	Name	Plant Name	Ref
191	3α-acetoxy-urs-12-ene-1β,11α-diol	*S. kronenburgii* Rech. f.	[49]
		S. multicaulis Vahl.	[65]
192	3β-acetoxy-2α,11α-dihydroxyurs-12-ene	*S. kronenburgii* Rech. f.	[16]
193	3β-acetoxyolean-12-ene-1β,2α,11α-triol	*S. kronenburgii* Rech. f.	[49]
194	3β-acetoxyolean-9,12 diene	*S. glutinosa* L.	[88]
195	3β-acetoxy-urs-12-ene-1β,2α,11α-triol	*S. kronenburgii* Rech. f.	[49]
196	3β-acetyl,21β,22β-dihydroxymonogynol A	*S. macrochlamys* Boiss. et Kotschy.	[46]
197	3β-acetyl,22β-hydroxymonogynol A	*S. macrochlamys* Boiss. et Kotschy.	[46]
198	3β-acetylmonogynol A	*S. macrochlamys* Boiss. et Kotschy.	[46]
199	3β-hydroxyolean-12-en-28-al	*S. tchihatcheffii* (Fisch. et Mey.) Boiss.	[47]
200	3β-*O*-cis-p-coumaroyl monogynol A	*S. montbretii* Benth.	[89]
201	3β-*O*-trans-p-coumaroyl monogynol A	*S. montbretii* Benth.	[89]
202	acetyloleanolic acid	*S. amplexicaulis* Lam.	[73]
		S. verticillata L.	[63]
203	betulinic acid	*S. verticillata* L.	[91]
204	crataegolic acid	*S. palaestina* Benth.	[85]
		S. potentillifolia (Boiss. et Heldr.) ex Bentham	[48]
		S. tomentosa Mill.	[67]
		S. verticillata L.	[91]
205	erythrodiol	*S. pomifera* L.	[82]
		S. fruticosa Mill. (*S. triloba* L.)	[30]
206	erythrodiol 3-acetate	*S. microphylla* Kunth.	[38]
		S. blepharochlaena Hedge and Hub. Mor.	[72]
207	erythrodiol diacetate	*S. tchihatcheffii* (Fisch. et Mey.) Boiss.	[47]
208	erythrodiol 28- acetate	*S. glutinosa* L.	[88]
209	germanicol	*S. macrochlamys* Boiss. et Kotschy.	[46]
210	germanicol acetate	*S. macrochlamys* Boiss. et Kotschy.	[46]
211	glochidone	*S. trichoclada* Benth.	[21]
212	hopanone	*S. multicaulis* Vahl.	[65]
213	lup-20(29)-ene-3β,11α-diol	*S. pinnata* L.	[92]
214	lupan-3 β,11α,20-triol	*S. potentillifolia* (Boiss. et Heldr.) ex Bentham	[48]
		S. palaestina Benth.	[85]
215	lupenone	*S. trichoclada* Benth.	[21]
		S. multicaulis Vahl.	[65]
		S. recognita Fisch. et Mey.	[58]

(continued)

Table 4 (continued)

No	Name	Plant Name	Ref
216	lupeol	*S. caespitosa* Montbret & Aucher ex Benth.	[70]
		S. glutinosa L.	[20, 88]
		S. microphylla Kunth.	[38]
		S. montbretii Benth.	[89]
		S. multicaulis Vahl.	[19]
		S. pinnata L.	[92]
		S. pomifera L.	[79]
		S. trichoclada Benth.	[21, 93]
217	lupeol 3-acetate	*S. cyanescens* Boiss. et Bal.	[42]
		S. macrochlamys Boiss. et Kotschy.	[46]
		S. hypargeia Fisch. et Mey.	[20]
218	monogynol A	*S. macrochlamys* Boiss. et Kotschy.	[46]
		S. montbretii Benth.	[89]
		S. trichoclada Benth.	[21]
219	moradiol	*S. pomifera* L.	[82]
220	moronic acid	*S. pomifera* L.	[82]
221	olean-13(18)-ene-2α,3β,11α-triol	*S. pinnata* L.	[92]
222	olean-13(18)-en 3-acetate	*S. pinnata* L.	[92]
223	oleanolic acid	*S. pinnata* L.	[91]
		S. amplexicaulis Lam.	[73]
		S. caespitosa Montbret & Aucher ex Benth.	[70]
		S. ceratophylla L.	[66]
		S. chrysophylla Stapf.	[84]
		S. fruticosa Mill. (*S. triloba* L.)	[30, 76]
		S. glutinosa L.	[88]
		S. limbata C.A. Meyer	[43]
		S. microphylla Kunth.	[38]
		S. montbretii Benth.	[89]
		S. nemorosa L.	[74]
		S. pinnata L.	[92]
		S. potentillifolia (Boiss. et Heldr.) ex Bentham	[48]
		S. pomifera L.	[79]
		S. syriaca L.	[68]
		S. tomentosa Mill.	[67]
		S. trichoclada Benth.	[21, 93]
		S. verticillata L.	[91]

(continued)

Table 4 (continued)

No	Name	Plant Name	Ref
224	oleanolic acid 3β-acetate	*S. staminea* Montbret et Aucher ex Benth.	[45]
		S. blepharochlaena Lam.	[72]
225	oleanolic acid methyl ester	*S. glutinosa* L.	[88]
		S. pomifera L.	[82]
226	salvinemorol	*S. kronenburgii* Rech. f.	[16]
		S. nemorosa L.	[74]
227	salvistamineol	*S. kronenburgii* Rech. f.	[16]
		S. staminea Montbret et Aucher ex Benth.	[45]
228	taraxasterol	*S. pomifera* L.	[79]
229	ursolic acid	*S. amplexicaulis* Lam.	[73]
		S. caespitosa Montbret & Aucher ex Benth.	[70]
		S. ceratophylla L.	[66]
		S. chrysophylla Stapf.	[84]
		S. fruticosa Mill. (*S. triloba* L.)	[30, 76]
		S. glutinosa L.	[88]
		S. limbata C.A. Meyer	[43]
		S. macrochlamys Boiss. et Kotschy.	[46]
		S. montbretii Benth.	[89]
		S. nemorosa L.	[74]
		S. palaestina Benth.	[85]
		S. pinnata L.	[91]
		S. poculata Nab.	[86]
		S. syriaca L.	[68]
		S. tomentosa Mill.	[67]
		S. trichoclada Benth.	[21, 93]
		S. verticillata L.	[63, 91]
230	ursolic acid methyl ester	*S. glutinosa* L.	[88]
231	uvaol	*S. tomentosa* Mill.	[90]
232	vergatic acid	*S. palaestina* Benth.	[85]
		S. pinnata L.	[91, 92]
		S. potentillifolia (Boiss. et Heldr.) ex Bentham	[48]
		S. verticillata L.	[91]

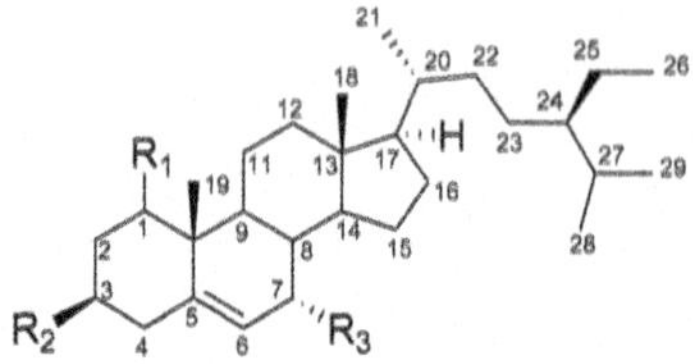

233: R_1=oxo, R_2=OH, R_3=αOH
234: R_1=oxo, R_2=[OOC$(CH_2)_{14}CH_3$], R_3=H, Δ^{22}
236: R_1=R_3=H, R_2=OAc
237: R_1=H, R_2=OH, R_3=oxo, Δ^{22}
238: R_1=H, R_3=oxo, R_2= $pC_5H_{11}C_6H_4COO$
239: R_1=H, R_2=OH, R_3= αOH
240: R_1=R_3=H, R_2=OH
241: R_1=H, R_2=oxo, R_3=H, 5,6-dihydro, $\Delta^{4,22}$
242: R_1=R_3=H, R_2=-O-β-D-glucoside
243: R_1=H, R_2=oxo, R_3=H, 5,6-dihydro, Δ^{22}
244: R_1=H, R_2=oxo, R_3=H, 5,6-dihydro, $\Delta^{4,22}$
245: R_1=H, R_2=oxo, R_3=H, 5,6-dihydro, $\Delta^{7,22}$
246: R_1=H, R_2=OH, R_3=H, 5,6-dihydro, $\Delta^{7,22}$
247: R_1=R_3=H, R_2=OH, Δ^{22}

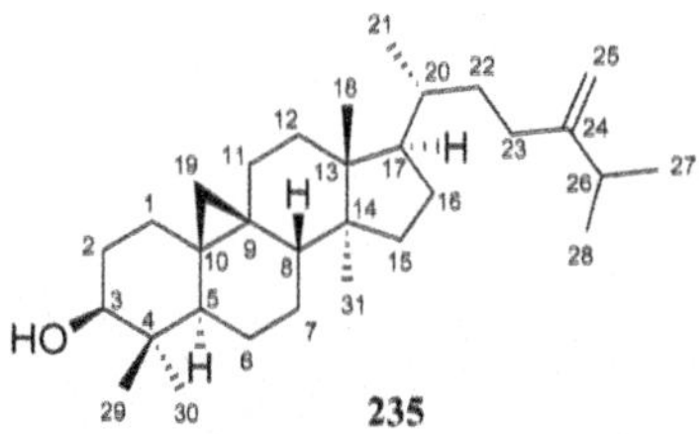

235

Fig. 6 Chemical structures of the steroids isolated from Anatolian *Salvia* species, in agreement with Table 5

luteolin 7-*O*-β-D-glucuronide, chrysoeriol 7-*O*-β-D-glucoside, chrysoeriol 7-*O*-β-D-glucuronide were isolated [96].

Although none of the flavonoids was obtained from *S. divaricata* Montbret et Aucher ex Bentham, a group of known aromatic acids; salicylic acid, 3-methoxysalicylic asid, *p*-hydroxybenzoic acid, *cis*- and *trans*-*p*-coumaric acid were isolated from its aerial parts [39] in addition to the three new diterpenoids, and a new linear sesquiterpene salvinine.

From the aerial parts of *S. montbretii*, known flavonoids apigenin, luteolin, cirsiliol were obtained [89].

A new aromatic compound di(4,4′-hexyloxycarbonylphenyl) ether were isolated from the roots of *S. heldrichiana* together with a large amount of salvigenin [75].

Two new aromatic compounds, the octanol esters of cis- and trans-4-*O*-methyl-caffeic acid dimers, were isolated from *S. forskahlei* [80].

Table 5 Isolated steroids from Turkish *Salvia* species

No	Name	Plant Name	Ref.
233	1-oxo-7α-hydroxy-β-sitosterol	*S. glutinosa* L.	[88]
234	1-oxostigmasterol 3β-palmitate	*S. candidissima* Vahl subsp. *candidissima*	[52]
235	24-methylene-cycloartanol	*S. nemorosa* L.	[74]
236	3-acetyl-sitosterol	*S. fruticosa* Mill. (*S. triloba* L.)	[30]
237	3β-hydroxy-stigmast-5-en-7-one	*S. blepharochlaena* Hedge & Hub. Mor	[18]
		S. syriaca L.	[68]
238	3β-*p*-pentyl-benzoic acid 7-oxositoseryl ester	*S. verticillata* L.	[63]
239	7α-hydroxy-β-sitosterol	*S. glutinosa* L.	[88]
240	β-sitosterol	*S. limbata* C. A. Meyer	[55]
		S. poculata Nab.	[86]
		S. multicaulis Vahl.	[65]
		S. potentillifolia (Boiss. et Heldr.) ex Bentham	[48]
		S. divaricata Montbret et Aucher ex Bentham	[39]
		S. verticillata L.	[63, 91]
		S. ceratophylla L.	[66]
		S. aethiopis L.	[94]
		S. amplexicaulis Lam.	[73]
		S. blepharochlaena Hedge & Hub. Mor	[18]
		S. caespitosa Montbret & Aucher ex Benth.	[70]
		S. cyanescens Boiss. & Balansa	[42]
		S. fruticosa Mill. (*S. triloba* L.)	[76]
		S. glutinosa L.	[88]
		S. kronenburgii Rech. f.	[16]
		S. limbata C. A. Meyer	[55]
		S. montbretii Benth.	[89]
		S. nemorosa L.	[74]
		S. pomifera L.	[79]
		S. sclarea L.	[44, 50]
		S. staminea Montbret et Aucher ex Benth.	[45]
		S. syriaca L.	[68]
		S. tchihatcheffii (Fisch. et Mey.) Boiss.	[47]
		S. trichoclada Benth.	[21]
		S. microphylla Kunth.	[38]
		S. forskahlei L.	[80]
		S. montbretii Benth.	[89]

(continued)

Table 5 (continued)

No	Name	Plant Name	Ref.
241	brassicasterone	*S. multicaulis* Vahl.	[65]
242	sitosterol 3-*O*-*β*-D-glucoside	*S. staminea* Montbret et Aucher ex Benth.	[45]
		S. caespitosa Montbret & Aucher ex Benth.	[70]
		S. verticillata L.	[91]
		S. tomentosa Mill.	[67]
		S. aethiopis L.	[94]
243	stigmast-3-one	*S. forskahlei* L.	[80]
244	stigmast-4-en-3-one	*S. blepharochlaena* Hedge & Hub. Mor	[18]
245	stigmast-7-en-3-one	*S. nemorosa* L.	[74]
246	stigmast-7-enol	*S. nemorosa* L.	[74]
247	stigmasterol	*S. blepharochlaena* Hedge & Hub. Mor	[18]
		S. glutinosa L.	[88]
		S. limbata C. A. Meyer	[55]
		S. tchihatcheffii (Fisch. et Mey.) Boiss.	[47]

Studies on another endemic *Salvia* species *S. cyanescens* were resulted in isolation of four dimethyl ether containing flavones; salvigenin (6-methoxyapigenin 4′,7-dimethyl ether) 6-hydroxyapigenin-7,4′-dimethyl ether, 6-hydroxykaempferol-3,6-dimethyl ether and kaempferol-3,7-dimethyl ether [42].

Flavones salvigenin, eupatilin, apigenin and luteolin were isolated from the aerial parts of *S. nemorosa* growing in North Eastern part of Turkey [74].

Flavonoids apigenin, luteolin, 4′-methylapigenin, 6-hydroxyluteolin-6,7,3′,4′-tetramethyl ether, 6-hydroxy apigenin-7,4′-dimethyl ether were obtained from *S. sclarea* [44].

From the aerial parts of *Salvia candidissima* subsp. *occidentalis* afforded three flavonoids salvigenin, crysoeriol, diosmetin and a phenolic acid *o*,*p*-dimethoxybenzoic acid with the mentioned terpenoids and steroids in the previos page [41]. However, *Salvia candidissima* subsp. *candidissima* yielded only salvigenin [41, 54].

From the roots of *Salvia limbata*, the flavonoids pectolinarigenin and salvigenin were obtained [55] while from the aerial parts of *Salvia limbata* four flavonoids salvigenin, luteolin, eupatilin and quercetin 3-methyl ether were isolated [43].

Three new compounds were isolated from *S. verticillata* L. collected from central Anatolia (Çorum). Two of them were elucidated as esters; 1-hydroxy-3,5,7-trien-nonanoic acid isobutyl ester and 4-acetyl-4′-palmityl-1,1′ oxybiphenyl (its name was corrected to 4-acetyl-4′-hexadecanyl-1,1′oxybiphenyl) beside a steroid ester and 14 other type of known compounds [63].

From *S. yosgadensis* eight flavonoids; luteolin, apigenin, apigenin 7-methyl ether, apigenin 4′-methyl ether, apigenin 7,4′-dimethyl ether, apigenin 6,4′-dimethyl ether, salvigenin, kaempferol 3-methyl ether and two simple phenolics *p*-hydroxyquinone and *p*-acetoxyphenol were isolated [40].

S. staminea MeOH extract yielded three flavones apigenin, luteolin and apigenin 7-*O*-*β*-D-glucoside, acetone extract yielded only apigenin 7,4′-methyl ether [45].

S. blepharochlaena afforded an aromatic ester, 4,4 -bisbenzoic acid heptyl ester as a sole phenolic compound [72].

The acetone extract of *S. microphylla* yielded two phenolic esters one being new compound 2-(*p*-hydroxyphenyl) ethyl eicosaheptanoic acid ester, and a known one eicosahexyltrans-4-hydroxy-3-methoxycinnamic acid ester (hexacosylferulate) [38].

A phytochemical analysis on *S. cedronella* Boiss. (endemic) yielded a new coumarin, 3-methoxy-4-hydroxymethyl coumarin, together with *p*-hydroxyphenylethyl docosanoate [103].

Four flavonoids; namely 5-hydroxy-7,4′-dimethoxyflavone, cirsimaritin, eupatilin, and salvigenin were also isolated from the aerial parts of *S. poculata* Nab [86].

Megastigmane glycosides were rarely seen in *Salvia* species, they were isolated for the first time from an Anatolian *Salvia* species *S. dichroantha* Stapf. [13]. Isolation of two megastigmane glycosides, premnaionoside, and salvionoside B, and an aliphatic alcohol glycoside, *(3R)*-1-octen-3-ol-3-*O*-*β*-D-xylopyranosyl-(1->6)-*O*-*β*-D-glucopyranoside, kaempferol 3,7,4′-trimethyl ether, rosmarinic acid and 3-*O*-methyl-rosmarinic acid as well as sucrose was carried out from the aerial parts of *S. dichroantha* (endemic) [98].

Analysis of methanolic extract of *S. verticillata* showed the presence of eight phenolic compounds, among them the highest quantity was observed for ferulic acid, hesperidin and quercetin [101]. In a study by Demirezer et al., two *Salvia* species were investigated. From the aerial parts of *S. verticillata* L., a new phenolic acid 3-(3′,4′-dihydroxyphenyl)-2-hydroxymethyl propionic acid, two flavones apigenin 4′-methyl ether 7-*O*-*β*-D-glucuronide, and luteolin 7-*O*-*β*-D-glucuronide were obtained. At the same study, from *S. trichoclada* two phenolic acids 3-(3′,4′-dihydroxyphenyl) lactic acid and rosmarinic acid were isolated besides four triterpenoids [93]. In a further study on methanolic extract of *S. trichoclada* afforded also rosmarinic acid beside apigenin-7-*O*-*β*-rhamnoside [21].

S. triloba (syn: *S. fruticosa*) is a medicinal plant of the genus *Salvia* growing in Turkey and Mediterranean countries. Salvigenin as a new flavone was isolated for the first time from *S. triloba* by Ulubelen group [76], and it was subsequently obtained from many *Salvia* species and became a chemotaxonomic marker of Turkish *Salvia* species. A later study, during the isolation of terpenoid compounds from *S. triloba*, collected from Gökova Bay, Muğla (South Aegean region of Turkey), salvigenin was only flavonoid obtained due to its nonpolar character (5-hydroxy-6,7,4′-trimethoxyflavone) [30]. But, MeOH or other more polar extracts of the plant were not particularly investigated for isolation of flavonoids and other phenolics.

In another study, detection of polar compounds of both cultivated and wild *S. fruticosa* by HPLC showed the presence of seventeen different phenolic compounds, consist of seven phenolic acids and ten flavonoids. Rosmarinic, *p*-coumaric and caffeic acids were determined to be the principal phenolic acids while the major flavonoids were found to be myricetin, morin and luteolin. However, in the cultivated plant extract the quantitity of phenolic compounds (vanillic acid, gallic acid, chlorogenic acid, *p*-coumaric acid, myricetin and morin) was increased while the quantity of flavonoids (catechin, epicatechin, luteolin, apigenin) and caffeic acid was decreased [99].

And analysis of rosmarinic acid among a series of *Salvia* extracts, the MeOH extract of *S. ekimiana* (153.50 mg/100) was revealed to be the richest in terms of rosmarinic acid quantity [116].

As a hydroxycinnamic acid derivative rosmarinic acid was usually determined as the main phenolic compound in many *Salvia* species. It was also determined by HPLC quantitatively as the main component of *S. tomentosa* (cultivated and wild), and followed by caffeic acid, morin, *p*-coumaric acid, and myricetin. Chlorogenic acid, *p*-coumaric acid, morin, kaempferol, hesperetin, and apigenin were increased through cultivation; gallic acid, caffeic acid, ferulic acid, rutin, catechin, and epicatechin were decreased [100].

In a study reported in 2014 on *S. fruticosa* extract, apigenin, rutin, hesperetin, luteolin as flavonoids and caffeic, *p*-coumaric, ferulic and rosmarinic acids as simple phenolic compounds were detected [102].

Phenolic and flavonoid contents of ethyl acetate (EtOAc), MeOH and water extracts of *Salvia cadmica* Boiss. were investigated by RP-HPLC quantitatively. Results indicate that MeOH extract contained considerable amounts of phenolic acids including caffeic acid, chlorogenic acid, ferulic acid, *p*-coumaric acid, protocatechuic acid, and rosmarinic acid, and flavonoids apigenin, hesperidin, kaempferol, luteolin [97].

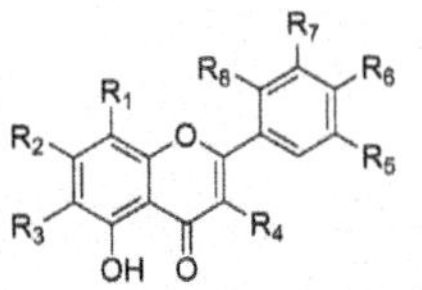

248: $R_1=R_3=R_4=R_5=R_7=R_8=H$, $R_2=R_6=OMe$
249: $R_1=R_4=R_5=R_8=H$, $R_2=R_3=R_6=R_7=OCH_3$
250: $R_1=R_3=R_4=R_5=R_8=H$, $R_2=R_6=OCH_3$
251: $R_1=R_3=R_4=R_5=R_7=R_8=H$, $R_2=R_6=OCH_3$
252: $R_1=R_4=R_5=R_7=R_8=H$, $R_3=OH$, $R_2=R_6=OCH_3$
253: $R_1=R_5=R_7=R_8=H$, $R_3=R_4=OCH_3$, $R_2=R_6=OH$
254: $R_1=R_4=R_5=R_8=H$, , $R_2=R_3=R_6=R_7=OCH_3$
255: $R_1=R_4=R_5=R_8=H$, $R_3=R_6=R_7=OH$, R_2=-*O*-*β*-D-glucoside
256: $R_3=OCH_3$, $R_1=R_4=R_5=R_8=H$, $R_2=R_6=R_7=OH$
257: $R_3=OCH_3$, $R_1=R_4=R_5=R_8=H$, $R_7=R_6=OH$, R_2=-*O*-*β*-D-glucoside
258: $R_1=R_3=R_4=R_5=R_7=R_8=H$, $R_2=R_6=OH$
259: $R_1=R_3=R_4=R_5=R_7=R_8=H$, $R_2=OH$, $R_6=OCH_3$
260: $R_1=R_3=R_4=R_5=R_7=R_8=H$, $R_6=OCH_3$, R_2=-*O*-*β*-D-glucoside
261: $R_1=R_4=R_5=R_7=R_8=H$, $R_2=OH$, $R_3=R_6=OCH_3$
262: $R_1=R_3=R_4=R_5=R_7=R_8=H$, $R_2=R_6=OCH_3$
263: $R_1=R_3=R_4=R_5=R_7=R_8=H$, $R_2=OCH_3$, $R_6=OH$
264: $R_1=R_3=R_4=R_5=R_7=R_8=H$, R_2=-*O*-*β*-D-glucoside, $R_6=OH$
265: $R_1=R_3=R_4=R_5=R_8=H$, $R_2=R_6=OH$, $R_7=OCH_3$
266: $R_1=R_4=R_5=R_8=H$, $R_2=R_3=OCH_3$, $R_6=R_7=OH$
267: $R_1=R_4=R_5=R_7=R_8=H$, $R_2=R_3=OCH_3$, $R_6=OH$
268: $R_1=R_3=R_4=R_5=R_8=H$, $R_2=OH$, $R_6=OCH_3$, $R_7=OH$
270: $R_1=R_4=R_5=H$, $R_2=OH$, $R_3=R_6=R_7=OCH_3$
271: $R_1=R_3=R_4=R_5=H$,$R_7=OH$, $R_6=OCH_3$, R_2=-*O*-rutinoside
272: $R_1=R_4=R_5=R_8=H$, $R_2=R_6=OH$, $R_3=R_7=OCH_3$
273: $R_1=R_3=R_5=R_7=R_8=H$, $R_2=R_4=R_6=OH$
274: $R_1=R_3=R_5=R_7=R_8=H$, $R_2=R_4=R_6=OCH_3$
275: $R_1=R_3=R_5=R_7=R_8=H$, $R_4=OCH_3$, $R_2=R_6=OH$
276: $R_1=R_3=R_5=R_7=R_8=H$, $R_2=R_4=OCH_3$, $R_6=OH$
277: $R_1=R_3=R_4=R_5=R_8=H$, $R_2=R_6=R_7=OH$
278: $R_1=R_3=R_4=R_5=R_8=H$, $R_6=R_7=OH$, R_2=-*O*-*β*-D-glucoside
279: $R_1=R_3=R_5=R_7=H$, $R_2=R_4=R_6=R_8=OH$
280: $R_1=R_3=R_8=H$, $R_2=R_4=R_5=R_6=R_7=OH$
281: $R_1=R_3=R_7=R_8=H$, $R_2=R_4=R_5=R_6=OH$
282: $R_1=R_3=R_5=R_8=H$, $R_4=OCH_3$, $R_2=R_5=R_6=OH$
283: $R_1=R_3=R_7=R_8=H$, R_4=-*O*-*β*-D-glucose, $R_2=R_5=R_6=OH$
284: $R_1=R_3=R_7=R_8=H$, $R_2=R_5=R_6=OH$, R_4= -*O*-rutinoside
285: $R_1=R_4=R_5=R_7=R_8=H$, $R_2=R_3=R_6=OCH_3$

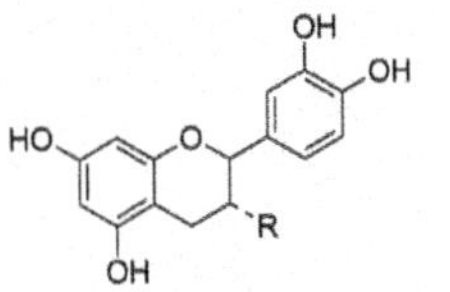

286 R=*β*OH
287 R=*α*OH

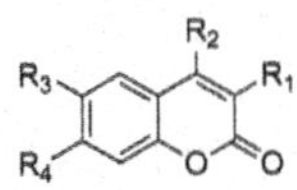

269 $R_1=R_2=H$, $R_3=R_4=OH$
288 $R_1=OCH_3$, $R_2=OH$, $R_3=H$, $R_4=CH_3$

Fig. 7 Chemical structures of flavonoids the isolated from Anatolian *Salvia* species, in agreement with Table 6

The identification and quantification of the phenolic contents of methanolic extracts of *S. brachyantha* (Bordz.) Pobed (endemic), *S. aethiopis* L., and *S. microstegia* Boiss. and Bal. were evaluated by using RP-HPLC/MS. The results showed that there is a total of 18 detectable phenolics, the most abundant of which was kaempferol in *S. microstegia* and rosmarinic acids in *S. brachyantha* and *S. aethiopis*. The other major phenolics were found to be apigenin, luteolin as flavones, and *p*-coumaric acid, and chlorogenic acid as phenolic acids [117].

As a result, salvigenin as a marker flavone along with other two flavones apigenin and luteolin derivatives which were found as the most common flavonoids in Anatolian *Salvia* extracts. All isolated flavonoids and other phenolics from Anatolia *Salvia* species are shown in Figs. 7 and 8.

Table 6 Isolated flavonoids from Turkish *Salvia* species

No	Name	Plant Name	Ref
248	4′-methylapigenin	*S. sclarea* L.	[44]
249	5-hydroxy-6,7,3′,4′-tetramethoxyflavone	*S. tomentosa* Mill.	[95]
250	5-hydroxy-7,3′,4′-trimethoxyflavone	*S. aethiopis* L	[94]
251	5-hydroxy-7,4′-dimethoxyflavone	*S. poculata* Nab.	[86]
252	6-hydroxy apigenin-7,4′-dimethyl ether	*S. sclarea* L.	[44]
		S. cyanescens Boiss. & Balansa	[42]
253	6-hydroxy-kaempferol-3,6-dimethyl ether	*S. cyanescens* Boiss. & Balansa	[42]
254	6-hydroxyluteolin -6,7,3′,4′-tetramethyl ether	*S. sclarea* L.	[44]
255	6-hydroxyluteolin 7-*O*-*β*-D-glucoside	*S. tomentosa* Mill.	[90]
256	6-methoxyluteolin	*S. pinnata* L.	[91]
		S. tomentosa Mill.	[95]
257	6-methoxyluteolin-7- *O*-*β*-D-glucoside	*S. tomentosa* Mill.	[95]
258	apigenin	*S. montbretii* Benth.	[89]
		S. nemorosa L.	[74]
		S. palaestina Benth.	[96]
		S. pinnata L.	[91]
		S. sclarea L.	[44]
		S. staminea Montbret et Aucher ex Benth.	[45]
		S. yosgadensis Freyn. et Bornm.	[40]
259	apigenin 4′-O-methyl ether	*S. yosgadensis* Freyn. et Bornm.	[40]
260	apigenin 4′-O-methyl ether 7-*O*-*β*-D-glucuronide	*S. verticillata* L.	[93]
261	apigenin 6,4′-dimethyl ether	*S. yosgadensis* Freyn. et Bornm.	[40]
262	apigenin 7,4′-dimethyl ether	*S. staminea* Montbret et Aucher ex Benth.	[45]
		S. palaestina Benth.	[96]
		S. yosgadensis Freyn. et Bornm.	[40]
263	apigenin 7-*O*-methyl ether	*S. yosgadensis* Freyn. et Bornm.	[40]
264	apigenin-7-*O*-*β*-D-glucoside	*S. pinnata* L.	[91]
265	chrysoeriol	*S. palaestina* Benth.	[96]
266	cirsiliol	*S. montbretii* Benth.	[89]
267	cirsimaritin	*S. palaestina* Benth.	[96]
		S. poculata Nab.	[86]
		S. tomentosa Mill.	[95]

(continued)

Table 6 (continued)

No	Name	Plant Name	Ref
268	diosmetin	*S. candidissima* Vahl. subsp. *occidentalis* Hedge	[41]
		S. tomentosa Mill.	[90]
269	esculetin	*S. pinnata* L.	[91]
270	eupatilin	*S. limbata* C. A. Meyer	[43]
		S. nemorosa L.	[74]
		S. palaestina Benth.	[96]
		S. poculata Nab.	[86]
		S. tomentosa Mill.	[90]
271	hesperidin	*S. cadmica* Boiss.	[97]
272	jaceosidin	*S. tomentosa* Mill.	[95]
273	kaempferol	*S. cadmica* Boiss.	[97]
274	kaempferol 3,7,4′-trimethyl ether	*S. dichroantha* Stapf.	[98]
275	kaempferol 3-methyl ether	*S. yosgadensis* Freyn. et Bornm.	[40]
276	kaempferol-3,7- dimethyl ether	*S. cyanescens* Boiss. & Balansa	[42]
277	luteolin	*S. montbretii* Benth.	[89]
		S. nemorosa L.	[74]
		S. palaestina Benth.	[96]
		S. pinnata L.	[91]
		S. sclarea L.	[44]
		S. staminea Montbret et Aucher ex Benth.	[45]
		S. tomentosa Mill.	[95]
		S. yosgadensis Freyn. et Bornm.	[40]
		S. limbata C. A. Meyer	[43]
278	luteolin 7-*O*-*β*-D-glucoside	*S. pinnata* L.	[91]
		S. tomentosa Mill.	[95]
		S. palaestina Benth.	[96]
		S. verticillata L.	[91]
279	morin	*S. fruticosa* Mill.	[99]
280	myricetin	*S. tomentosa* Mill.	[100]
		S. fruticosa Mill.	[99]
281	quercetin	*S. verticillata* L.	[101]
282	quercetin 3-*O*-methyl ether	*S. limbata* C. A. Meyer	[43]
283	quercetin 3-*O*-*β*-D-glucoside	*S. pinnata* L.	[91]
284	rutin	*S. fruticosa* Mill.	[102]
		S. tomentosa Mill.	[100]

(continued)

Table 6 (continued)

No	Name	Plant Name	Ref
285	salvigenin	*S. aethiopis* L.	[94]
		S. amplexicaulis Lam.	[73]
		S. caespitosa Montbret & Aucher ex Benth.	[70]
		S. candidissima Vahl subsp. *candidissima*	[52]
		S. candidissima Vahl. subsp. *occidentalis* Hedge	[41]
		S. ceratophylla L.	[66]
		S. cyanescens Boiss. & Balansa	[42]
		S. fruticosa Mill. (*S. triloba* L.)	[76]
		S. fruticosa Mill. (*S. triloba* L.)	[30]
		S. heldreichiana Boiss. ex Bentham	[75]
		S. limbata C. A. Meyer	[55]
		S. limbata C. A. Meyer	[43]
		S. nemorosa L.	[74]
		S. palaestina Benth.	[96]
		S. poculata Nab.	[86]
		S. syriaca L.	[68]
		S. wiedemannii Boiss.	[61]
		S. yosgadensis Freyn. et Bornm.	[40]
		S. tchihatcheffii (Fisch. et Mey.) Boiss.	[47]
286	catechin	*S. fruticosa* L.	[99]
		S. tomentosa Mill.	[100]
287	epicatechin	*S. fruticosa* Mill.	[99]
		S. tomentosa Mill.	[100]
288	3-methoxy-4-hydroxymethyl coumarin	*S. cedronella* Boiss.	[103]

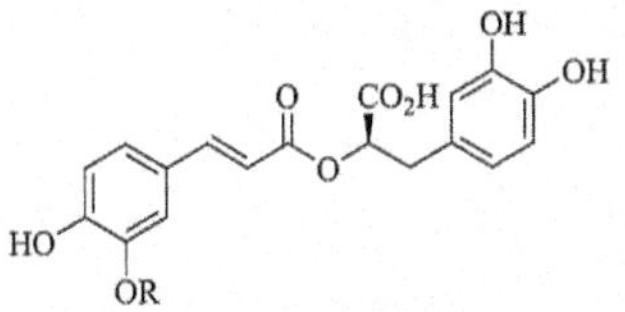

291: R=CH_3
307: R=H

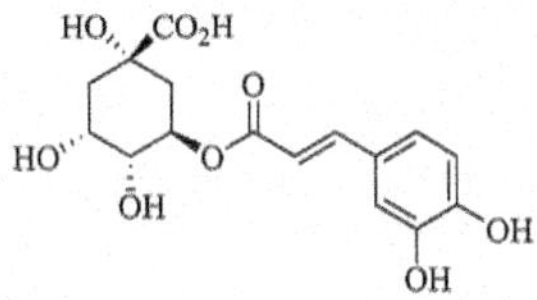

294

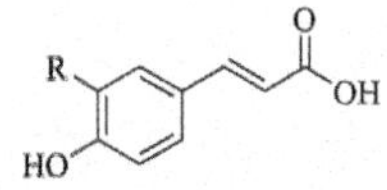

293: R=OH
296: R=H
298: R=OCH3

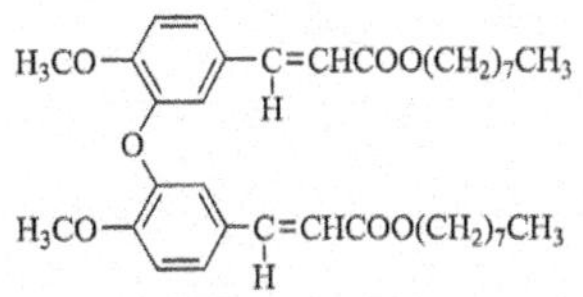

295: *cis-*
311: *trans-*

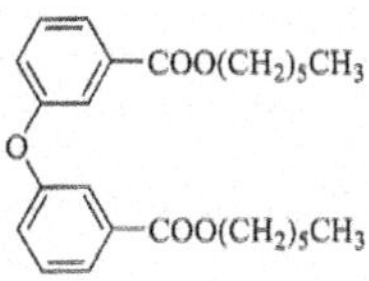

297

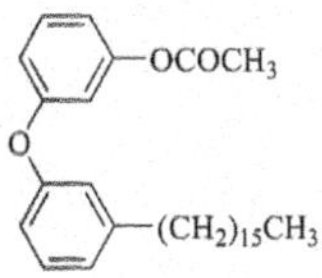

317

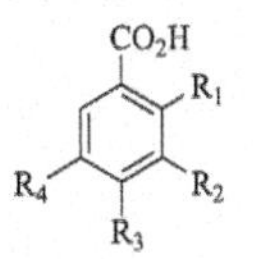

304: R_1=R_2=R_4=H, R_3=OH
308: R_1=OH, R_2= R_3=R_4=H
312: R_1=OH, R_3=OCH_3, R_2=R_4=H
313: R_1 =H, R_2= R_3=R_4=OH
314: R_1=R_3=OCH_3, R_2=R_4=H
315: R_1=R_2=H, R_3=R_4=OH
316: R_1=R_2=H, R_3=OH, R_4=OCH_3

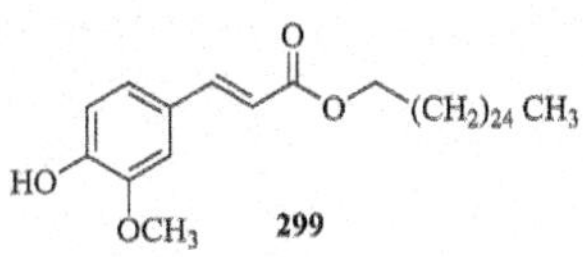

299

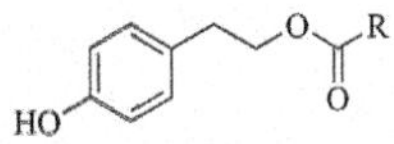

290: R= $C_{26}H_{53}$
303: R= $C_{21}H_{43}$

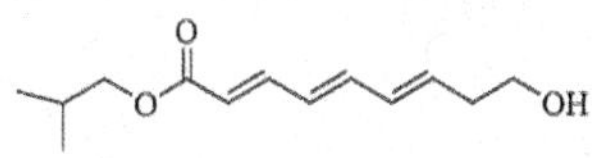

289

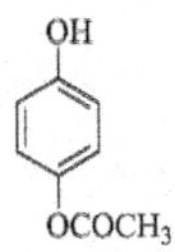

302

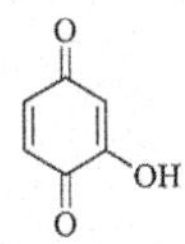

305

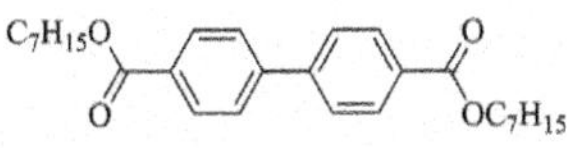

292

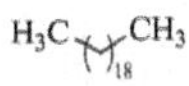

300

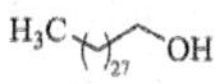

301

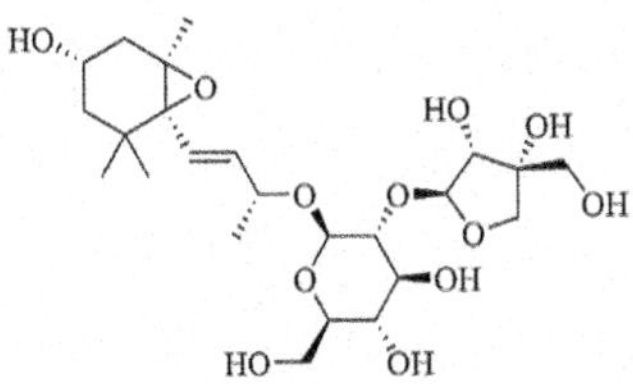

306

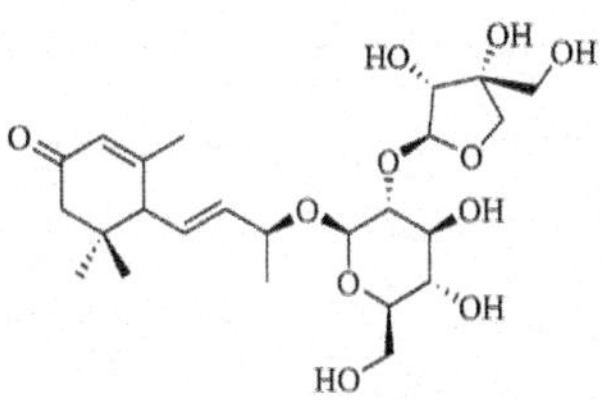

309

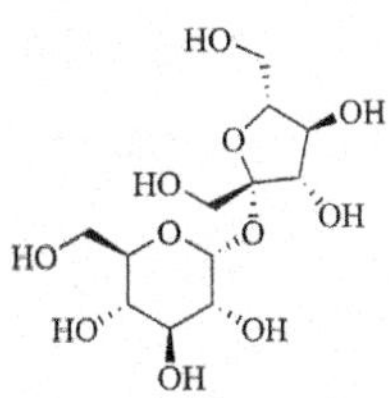

310

Fig. 8 Chemical structures of the isolated phenolic and other compounds from Turkish *Salvia* species, in agreement with Table 7

Table 7 Isolated phenolics and other compounds from Turkish *Salvia* species

No	Name	Plant Name	Ref
289	1-hydroxy-3,5,7-trien-nonanoic acid-isobutylester	*S. verticillata* L.	[63]
290	2-(*p*-hydroxy-phenyl)ethyl eicosaheptanoic acid ester	*S. microphylla* Kunth.	[69]
291	3-*O*-methyl-rosmarinic acid	*S. dichroantha* Stapf.	[98]
292	4,4′-bisbenzoic acid heptyl ester	*S. blepharochlaena* Hedge & Hub. Mor	[72]
		S. multicaulis Vahl.	[65]
293	caffeic acid	*S. fruticosa* Mill.	[99]
		S. tomentosa Mill.	[100]
		S. cadmica Boiss.	[97]
294	chlorogenic acid	*S. fruticosa* Mill.	[99]
		S. tomentosa Mill.	[100]
		S. cadmica Boiss.	[97]
295	*cis*-4-*O*-methyl-caffeic acid octanol ester	*S. forskahlei* L.	[80]
296	coumaric acid	*S. divaricata* Montbret & Aucher ex Benth.	[39]
297	di(4,4′-hexyloxy-carbonylphenyl) ether	*S. heldreichiana* Boiss. ex Bentham	[75]
298	ferulic acid	*S. verticillata* L.	[101]
299	hexacosylferulate	*S. microphylla* Kunth.	[69]
300	n-eicosane	*S. aethiopis* L.	[94]
301	nonacosan-1-ol	*S. verticillata* L.	[97]
302	*p*-acetoxyphenol	*S. yosgadensis* Freyn. et Bornm.	[71]
303	*p*-hydroxy phenylethyl docosanoate	*S. cedronella* Boiss.	[103]
304	p-hydroxy-benzoic acid,	*S. divaricata* Montbret et Aucher ex Benth.	[39]
305	*p*-hydroxyquinone	*S. yosgadensis* Freyn. et Bornm.	[71]
306	premnaionoside	*S. dichroantha* Stapf.	[98]
307	rosmarinic acid	*S. trichoclada* Benth.	[55] [93]
308	salicylic acid	*S. divaricata* Montbret et Aucher ex Benth.	[39]
309	salvionoside B	*S. dichroantha* Stapf.	[98]
310	sucrose	*S. dichroantha* Stapf.	[98]
311	*trans*-4-*O*-methyl-caffeic acid octanol ester	*S. forskahlei* L.	[80]

(continued)

Table 7 (continued)

No	Name	Plant Name	Ref
312	3-methoxysalicylic acid	*S. divaricata* Montbret et Aucher ex Benth.	[39]
313	gallic acid	*S. tomentosa* Mill.	[100]
314	*o,p*-dimethoxybenzoic acid	*S. candidissima* subsp. *occidentalis*	[41]
315	protocatechuic acid	*S. cadmica* Boiss	[97]
316	vanillic acid	*S. fruticosa* Mill.	[99]
317	4-acetyl- 4′-palmityl- 1,1′oxybiphenyl	*S. verticillata* L.	[63]

4 Biological Activities

The crude plant extracts of *Salvia* species and its essential oils as well as pure compounds have many biological activities including antimicrobial, antioxidant, cytotoxic/antitumor, vasodepressant, anti-inflammatory, antispasmodic, analgesic, immunomodulatory, hypoglycemic and some enzyme inhibitory properties including cholinesterase, α-amylase, α-glucosidase and tyrosinase aldehyde oxidase, xanthine oxidase (Table 9).

The activities on their triterpenoid constituents reported by Topcu in a review article [27] as well as several chapters, publications by other authors, particularly the major triterpenoids oleanolic acid and ursolic acid. Since these two compounds have been isolated from many Lamiaceae and Boraginaceae family plants and their activities studied extensively, they were not displayed herein in detail. In fact, they have anti-inflammatory, anti-cancer, anti-viral, anti-aging anti-cholinestarese especially ursolic acid has stronger activity rather than oleanolic acid which has promising therapeutic potential [16, 27, 45]. Lupane triterpenoids have been also investigated for some biological activities, particularly for antioxidant and anticholinesterase activities [21, 46, 84] (Table 9).

4.1 Antimicrobial Activity

Among antimicrobial activity studies on *Salvia* species, antibacterial activity takes place the main part, along with antifungal, yeast and antiviral test assays.

4.1.1 Antibacterial, Antifungal and Yeast Activities

Ulubelen and Topcu group have investigated several *Salvia* species for their antibacterial activity against same standard bacterial strains and a yeast; namely

Table 8 All the chemical constituents isolated from each Anatolian *Salvia* plant species

	Chemical Constituents						
Plant name	Sesquiterpenes	Diterpenoids	Triterpenoids	Steroids	Flavonoids	Phenolics and Others	Ref.
Salvia aethiopis L				• β-sitosterol • sitosteryl-3β-D-glucoside	• 5-hydroxy-7,3',4'-trimethoxyflavone • salvigenin	• n-eicosane	[94]
Salvia amplexicaulis Lam.		• 7-acetylhorminone • 7-oxo-abieta-9,12,14-triene • ferruginol • horminone • sugiol	• acetyloleanolic acid • oleanolic acid • ursolic acid • α-amyrin	• β-sitosterol • stigmast-4-en-3-one, stigmasterol	• salvigenin		[73]
Salvia blepharochlaena Hedge & Hub. Mor (endemic)		• 12-*O*-methylpisiferic acid • 2-demethyl-multiorthoquinone • 7-acetylhorminone • blephacin • ferruginol • horminone • multicaulin • multiorthoquinone • *O*-methylpisiferic acid methyl ester • pisiferic acid					[18]
		• 7,20-epoxyroyleanone • 8-hydroxy-12-oxo-abieta-9(11), 13-dien-20-oic acid 8,20-lactone • cryptanol • formosanolide • royleanone • sugiol	• 24-methylene-cycloartanol • erythrodiol-3-acetate • oleanolic acid-3-acetate • α-amyrin	• 3β-hydroxy-stigmast-5-en-7-one • stigmast-4-en-3-one • β-sitosterol • stigmasterol		• 4,4'-bisbenzoic acid heptyl ester	[72]
Salvia bracteata Banks and Sol		• 12-methylferruginol • 12-*O*-methylhorminone • 6,7-dehydroroyleanone • 7-*O*-acetylhorminone • bractealine • ferruginol • horminone • salvibracteone • salvinolone • sugiol					[59]

(continued)

Table 8 (continued)

	Chemical Constituents						
Salvia caespitosa Montbret & Aucher ex Benth. (endemic)		• 6β-hydroxy-isopimaric acid • isopimaradienal • ferruginol • isopimaric acid • sandaracopimaric acid	• 3-acetylvergatic acid • α-amyrin • lupeol • 3-oxo-oleanolic acid • oleanolic acid • ursolic acid	• sitosterol • sitosterol 3β-glucoside	• salvigenin		[70]
Salvia candidissima Vahl subsp. *candidissima*		• 11,12-dioxo-abieta-8,13-dien • 11-hydroxy-12- methoxyabieta-8,11,13-trien • 11β -hydroxymanoyl oxide • 1-oxo-5-hydroxy-abieta-8,11,13-trien-18-oic acid • 1-oxo-salvibretol • 2,3-dehydrosalvipisone • 7,20-epoxyroyleanone • 7-oxo-11,12,14-trihydroxy-20-methoxyhydroxy-methylene-abieta-5,8,11,13-tetraene • candidissiol • cryptojaponol • *ent*-sclareol • ferruginol • manool • manoyloxide • microstegiol • pachystazone • salvinolone • salvinolone-12-methyl ether • salvipisone	• α-amyrin acetate	• 1-oxostigmasterol 3β-palmitate	• salvigenin		[52]

(continued)

Table 8 (continued)

	Chemical Constituents						
Salvia candidissima Vahl. subsp. *occidentalis* Hedge		• 11-hydroxy-12- methoxyabieta-8,11,13-trien • 14-oxo-pimaric acid • 1-oxoaethiopinone • 1-oxosalvipisone • 7-*O*-acetylhorminone • cryptanol • ferruginol • horminone • microstegiol • montbretyl 12-methyl ether • salvipisone					[54]
		• 20-deoxocarnosol • aethiopinone • candidissiol • manoyloxide • pachystazone					[64]
	• spathulenol	• 11β -hydroxymanoyl oxide • 3-oxosalvipisone • 8,13-diepimanoyl oxide			• crysoeriol • diosmetin • salvigenin		[41]
Salvia cedronella Boiss. (endemic)			• betulinic acid • oleanolic acid		• 3-methoxy-4-hydroxymethyl coumarin	• *p*-hydroxy phenylethyl docosanoate	[103]
Salvia ceratophylla L.		• 6β-hydroxy-8,15-pimaradien-1-one • 3-oxo-8,11,13-abietatrien-20-oic acid • ferruginol • salvipisone • aethiopinone • candidissiol • aethiopinone • candidissiol	• oleanolic acid • ursolic acid	• β-sitosterol	• salvigenin		[66]
Salvia chrysophylla Stapf. (endemic)		• sclareol	• oleanolic acid • ursolic acid				[84]

(continued)

Table 8 (continued)

	Chemical Constituents						
Salvia cryptantha Montbret and Aucher ex. Bentham (endemic)		• 2β-hydroxyroyleanone • 7-acetylhorminone • cryptanol • horminone					[14]
Salvia cyanescens Boiss. & Balansa (endemic)	• caryophyllene oxide • spathulenol	• 11β-hydroxymanoyl oxide • 12-isopentenyl-3-oxosalvipisone • aethiopinone • ferruginol • manoyl oxide • salvipisone	• α-amyrin 3-acetate • α-amyrin • lupeol 3-acetate	• β-sitosterol	• 6-hydroxy-apigenin-7,4′-dimethyl ether • 6-hydroxy-kaempferol-3,6-dimethyl ether • kaempferol-3,7- dimethyl ether • salvigenin		[42]
Salvia dichroantha Stapf. (endemic)					• 3-*O*-methyl-rosmarinic acid • kaempferol 3,7,4′-trimethyl ether rosmarinic acid	• premnaionoside • salvionoside B • (3R)-1-octen-3-ol-3-*O*-β-D-xylopyranosyl-(1->6)-*O*-β-D-glucopyrano-side • sucrose	[98]
Salvia divaricata Montbret et Aucher ex Bentham (endemic)	• salvinine	• 6-oxo-12-methylroyleanone-18-oic acid horminone-18-oic acid • 6-oxoroyleanone-18-oic acid,		• β- sitosterol		• 3-methoxy- • p-hydroxy-benzoic acid, • *p*-coumaric acid • salicylic acid	[39]
Salvia eriophora Boiss. et Kotschy (endemic)		• eriophoroxide					[78]
		• 12-hydroxy-sapriparaquinone • 3,12-dihydroxy-sapriparaquinone-1-ene • 4,12-dihydroxy-sapriparaquinone • 4,14-dihydroxy-saprorthoquinone • 6,7-dehydroroyleanone • 7-*O*-acetylhorminone • vaethiopinone • ferruginol • horminone • salvilimbinol					[56]

(continued)

Table 8 (continued)

	Chemical Constituents						
		• salvipisone					
Salvia forskahlei L.		• forskalinone	• α-amyrin	• stigmast-3-one • sitosterol		• trans-4-*O*-methyl-caffeic acid octanol ester • cis-4-*O*-methyl-caffeic acid octanol ester	[80]
Salvia fruticosa Mill. (*S. triloba* L.)		• carnosol	• oleanolic acid • ursolic acid	• β-sitosterol	• salvigenin		[76]
		• 6β-hydroxyroyleanone • 7α-acetylroyleanone • cryptanol • ferruginol • horminone • trilobinol • trilobinone • virgatol					[71]
		• carnosic acid • carnosic acid 12-methyl ether • carnosol • ferruginol • isorosmanol • manool • rosmadial	• α-amyryl-tetracosanoate • erythrodiol • oleanolic acid • ursolic acid	• 3-acetylsitosterol	• salvigenin		[30]
Salvia glutinosa L.			• α-amyrin • α-amyrin acetate • 11-oxo-α-amyrin • 11-oxo-β-amyrin • 3β-acetoxyolean-9,11 diene • erythrodiol 28- acetate • lupeol • oleanolic acid • oleanolic acid methyl esters • ursolic acid • ursolic acid methyl esters	• 1-oxo-7α-hydroxy-sitosterol, • stigmasterol • sitosterol			[88]

(continued)

Table 8 (continued)

	Chemical Constituents						
Salvia heldreichiana Boiss. ex Bentham (endemic)		• 7-oxo-13-epi-pimara-9(11), 15-dien-18-oic acid • 7β-hydroxysandracopimaric acid • heldrichinic acid • isopimaric acid • wiedelactone • wiedemannic acid			• salvigenin	• di(4,4′-hexyloxy-carbonyl-phenyl) ether	[75]
Salvia hypargeia Fisch. et Mey (endemic)		• cryptanol • horminone • hypargenin A • hypargenin B • hypargenin C • hypargenin D • hypargenin E • hypargenin F					[77]
		• 11,12-dioxoabieta-8,13-diene taxodione • 6α-hydroxysalvinolone • 6α-hydroxytaxodone • aethiopinone • ferruginol • hypargenin A • hypargenin D • microstegiol • saprorthoquinone					[34]
		• 5,6-didehydro-7-hydroxytaxodone • 6-hydroxysalvinolone • demethylcryptojaponol • salvicanaric acid	• lupeol • lupeol-3-acetate				[20]
Salvia kronenburgii		• Δ^7-manool • 12-hydroxysapriparaquinone	• 1β,2α,3β,11α-tetra hydroxy-olean-12-ene,				[49]

(continued)

Table 8 (continued)

	Chemical Constituents						
Rech. f. (endemic)		• royleanone • 1-oxosalvibretol • 6-hydroxysalvinolone • 7-acetoxyroyleanone • ferruginol • kronenquinone • salvipisone • taxodione	• 1β,2α,3β,11α-tetra hydroxy-urs-12-ene • 1β,3α,11α-trihydroxy-urs-12-ene • 2α,3β,11α-trihydroxy-urs-12-ene • 2α-acetoxy-urs-5,12-diene-3β,11α-diol • 3α-acetoxy-urs-12-ene-1β,11α-diol • 3β-acetoxyolean-12-ene-1β,2α,11α-triol • 3β-acetoxy-urs-12-ene-1β,2α,11α-triol				
			• 1β,2α,3β,11α-tetra hydroxyurs-12-ene, • 1β,2α-dihydroxy-3β-acetoxy-11-oxours-12-ene • 1β,2α-dihydroxy-3β-acetoxyurs-9(11),12-diene • 1β,2α-dihydroxy-3β-acetoxyolean-9(11),12-diene • 2α,20β-dihydroxy-3β-acetoxyurs-9(11),12-diene • 3β-acetoxy-2α,11α-dihydroxyurs-12-ene • salvinemorol • salvistamineol	• β-sitosterol			[16]

(continued)

Table 8 (continued)

	Chemical Constituents						
Salvia limbata C. A. Meyer		• 12-hydroxysapriparaquinone • 2-hydroxysaprorthoquinone • 3,12-dihydroxy sapriparaquinone-1-ene • 4-dehydrosalvilimbinol • limbinol • manool • salvilimbinol		• β-sitosterol • stigmasterol	• pectolinarigenin • salvigenin		[55]
	• 6-dehydroxy-13-epi-yosgadensenol • spathulenol • yosgadensenol	• abieta-8,11,13-triene • acetyllimbinol • ferruginol • limbinal • manool • sclareol	• oleanolic acid • ursolic acid	• β -sitosterol	• eupatilin • luteoline • quercetin 3-methyl ether • salvigenin		[43]
Salvia macrochlamys Boiss. et Kotschy.	• caryophyllene oxide		• 3β-acetyl,21β,22β-dihydroxymonogynol A lupeol • 3β-acetyl,22β-hydroxymonogynol A • 3β-acetylmonogynol A • germanicol • germanicol acetate • lupeol acetate • monogynol A • ursolic acid				[46]
Salvia microphylla Kunth.	• 8α-hydroxy-β-eudesmol • β-eudesmol	• carnosic acid 12-methyl ether	• erithrodiol 3-acetate • lupeol • oleanolic acid	• β-sitosterol		• 2-(*p*-hydroxy-phenyl)ethyl eicosahep-tanoic acid ester • hexaco sylferulate	[38]

(continued)

Table 8 (continued)

	Chemical Constituents						
Salvia microstegia Boiss. et Bal.		• 10-acetylferruginol • 5,11,12-trihydroxy-abieta-8,11,13-triene • ferruginol • pisiferal					[53]
		• microstegiol					[81]
Salvia montbretii Benth.		• 14-hydroxyferruginol • demethylcryptojaponol • ferruginol • ferruginyl 12-methyl ether • hypargenin F • montbretol • montbretyl 12- methyl ether • taxodione					[60]
			• 3*β*-*O*-cis-p-coumaroyl monogynol A • 3*β*-*O*-trans-p-coumaroyl monogynol A • lupeol • monogynol A • oleanolic acid • ursolic acid • α-amyrin	• sitosterol	• apigenin • cirsiliol • luteolin		[89]
		• 11,11′-didehydroxy-7,7′-dihydroxytaxodione • 1-oxosalvibretol • 6-hydroxysalvinolone • 7,7′-bistaxodione • 7-hydroxytaxodione • salvibretol					[15]

(continued)

Table 8 (continued)

	Chemical Constituents						
Salvia multicaulis Vahl.		• 12-demethylmulticauline • 2-demethylmultiorthoquinone • 12-methyl-5-dehydroacetylhorminone • 12-methyl-5-dehydrohorminone • 18-oxoferruginol • 1-oxoferruginol • hinokione • horminone • manool • multicaulin • multiorthoquinone • pisiferal • salvipimarone • sempervirol	• α-amyrin • lupeol				[19]
		• 3-oxo-12-methoxy-14-hydroxyabieta-8,11,13-triene • 6-oxo-12-peroxyabieta-8,11,13-triene • 6-oxoferruginol • acetylhorminone • cryptanol • ferruginol • ferruginol-18-al • wiedelactone	• lupeone • 3-acetyl-β-amyrin *(incorrectly written as 3α-acetylamyrin in original article* • 3-acetyl-erythrodiol • hopanone	• brassicasterone • β –sitosterol • stigmast-4-en-3-one		• 4,4'-bisbenzoic acid heptyl ester	[65]
		• salvimultine					[83]
Salvia napifolia Jacq.		• 1,12-dioxoabieta-8,13-dien • 1-oxoferruginol • 6,12,14-trihydroxyabieta-6,8,11,12-tetraen • 6-oxoferruginol • 7,20-epoxyroyleanone • 7-acetyl-horminone • cryptanol • cryptojaponol • ferruginol • horminone • microstegiol • pachystazone • sugiol					[51]

(continued)

Table 8 (continued)

	Chemical Constituents						
Salvia nemorosa L.		• 7-*O*-acetylhorminone • horminone • nemorosin • pachystazone • salvipisone	• α-amyrin • oleanolic acid • salvinemorol • ursolic acid	• 24-methylene-cycloartanol • stigmast-4-en-3-one • stigmast-7en-3-one • stigmast-7-enol • β-sitosterol	• apigenin • eupatilin • luteolin • salvigenin		[74]
Salvia palaestina Benth.		• sclareol	• crataegolic acid • lupane-3β,11α, 20-triol • ursolic acid • vergatic acid	• sitosteryl 3β-glucoside			[85]
					• 6,7,3′,4′-tetra methoxyflavone • apigenin • apigenin-7-*O*-β-D-glucuronide • apigenin-7,4′-dimethyl ether • apigenin-7- *O*-β-D-glucoside • chrysoeriol • chrysoeriol-7-*O*-β-D-glucoside • chrysoeriol-7-*O*-β-D-glucuronide • cirsimaritin • eupatilin • genkwanin • luteolin • luteolin-7,4′-dimethyl ether • luteolin-7-*O*-β-D-glucoside • luteolin-7-*O*-β-D-glucuronide • salvigenin		[96]

(continued)

Table 8 (continued)

	Chemical Constituents						
Salvia pinnata L.			• oleanoic acid • ursolic acid • vergatic acid		• 6-methoxyluteolin • apigenin • apigenin-7-*O*-β-D-glucoside • esculetin • luteolin • luteolin 7-*O*-β-D-glucoside • quercetin 3-*O*-β-D-glucoside		[91]
			• lup-20(29)-ene-3β,11α-diol • lupeol • olean-13(18)-ene-2α,3β,11α-triol • oleanolic acid • vergatic acid				[92]
Salvia poculata Nab.		• sclareol	• 2α,3α-dihydroxy-24-nor-4(23),12-oleanadien-28-oic acid • ursolic acid	• β -sitosterol	• 5-hydroxy-7,4′-dimethoxyflavone • cirsimaritin • eupatilin • salvigenin		[86]
Salvia potentillifolia (Boiss. et Heldr.) ex Bentham (endemic)	• 1-acetoxy-8-hydrox yeudesm-4(15),7(11)-dien-8,12-olide		• β-amyrin • crataegolic acid • lupan-3 β,11α,20-triol • oleanolic acid • vergatic acid	• β -sitosterol			[48]
Salvia pomifera L.		• 18-hydroxy-8,11,13 abietatrien-7-one • Ferruginyl-12-methyl ether • pomiferin A • pomiferin B • pomiferin C • pomiferin D • pomiferin E	• α-amyrin • lupeol • oleanolic acid • oleanolic acid methyl ester • taraxasterol	• β-sitosterol			[79]
		• pomiferin F • pomiferin G	• 23-hydroxy-germanicone • erithrodiol • moradiol • moronic acid				[82]

(continued)

Table 8 (continued)

	Chemical Constituents						
Salvia recognita Fisch. et Mey. (endemic)		• 12-methyl-5-dehydrohorminone • 6α-hydroxy-11,12-dioxo-8,13-abieta-diene • ferruginol • salvirecognine • salvirecognone • sugiol • wiedelactone	• 3β-hydroxyolean-12-en-28-al *(incorrectly written as 3α-acetylamyrin-28-al)* • 3-acetyloleanolic acid • lupeone				[58]
Salvia sclarea L.	• caryophyllene oxide • spathulenol	• 2,3-dehydrosalvipisone • 7-oxoferruginol-18-al • 7-oxoroyleanone • candidissiol • ferruginol • manool • microstegiol • salvipisone • sclareol	• α-amyrin	• β-sitosterol	• 4′-methylapigenin • apigenin • luteolin • 6-hydroxyluteolin - 6,7,3′,4′-tetramethyl ether • 6-hydroxy apigenin-7,4′-dimethyl ether		[44]
		• Δ^7-manool • 1-oxoaethiopinone • 2,3-dehydrosalvipisone • acetylsalvipisone • aethiopinone • candidissiol • cryptojaponol • ferruginol • microstegiol • salvinolone • salvipisone • sclareapinone	• 3-oxo-oleanolic acid • α-amyrin	• β-sitosterol			[50]
Salvia staminea Montbret et Aucher ex Benth.	• caryophyllene oxide • spathulenol	• 1-oxo-salvibretol • cryptojaponol • ferruginol • manoyloxide • microstegiol • taxodione	• oleanolic acid 3β-acetate • salvistamineol	• β-sitosterol • sitosterol 3β-D-glucoside • stigmast-4-en-3-one	• apigenin • apigenin 7,4′-methyl ether • apigenin 7-*O*-β-D-glucoside • luteolin		[45]

(continued)

Table 8 (continued)

	Chemical Constituents						
Salvia syriaca L.		• 4-dehydrosalvilimbinol • candidissiol • ferruginol • salvisyrianone • viridone	• oleanolic acid • ursolic acid	• β-sitosterol • 3β-hydroxy-stigmast-5-en-7-one	• salvigenin		[68]
Salvia tchihatcheffii (Fisch. et Mey.) Boiss. (endemic)	• (+)-ledol	• 6 7-dehydroroyleanone • dehydroabietic acid • salvitchihatine • tchihatine	• 28-acetylerithrodiol • 3-acetylerithrodiol • 3β-hydroxy olean-12- en-28-al • erithrodiol diacetate	• β-sitosterol • stigmasterol	• salvigenine		[47]
Salvia tomentosa Mill.					• 5-hydroxy-6,7,3′,4′-tetramethoxyflavone • 6-methoxyluteolin • 6-methoxyluteolin-7- *O*-β-D-glucoside • cirsimaritin • jaceosidin • luteolin • luteolin 7-*O*-β-D-glucoside		[95]
		• 3β-hydroxy-8,11,13,15-abietatetraen-18-oic acid • dehydroabietic acid	• crateagolic acid • oleanolic acid • ursolic acid	• sitosteryl 3β-glucoside			[67]
			• β-amyrin • uvaol		• 6-hydroxyluteolin 5-*O*-β-D-glucoside • 6-hydroxyluteolin 7-*O*-β-D-glucoside • cirsilineol • diosmetin • eupatilin		[90]

(continued)

Table 8 (continued)

	Chemical Constituents						
Salvia trichoclada Benth.			• 30-hydroxylup-20 (29)-en-3-on • lupeol • oleanolic acid • ursolic acid			• 3-(3′,4′-dihydroxy-phenyl) lactic acid • rosmarinic acid	[93]
			• glochidone • lupenone • lupeol • monogynol A • oleanolic acid • ursolic acid	• *β*-sitosterol	• apigenin-7-*O*-*β*-D-rhamnoside	• rosmarinic acid	[21]
Salvia verticillata L			• betulinic acid • crateagolic acid • oleanolic acid • ursolic acid • vergatic acid	• *β*- sitosterol • sitosteryl 3*β*-D-glucoside	• 6-hyroxyluteolin 5-*O*-*β*-D-glucoside • luteolin-7-*O*-*β*-D-glucoside		[91]
		• 2,3-dehydrosalvipisone • candidissiol • ferruginal • manool • salvinolonyl 12-methyl ether • salvipisone	• *α*-amyrin • *α*-amyrin-28-al • *β*-amyrin • acetyl-oleanolic acid • ursolic acid	• *β*- sitosterol • 3*β*-p-pentyl-benzoic acid 7-oxositoseryl ester	• 4-acetyl-4′-palmityl-1,1′oxybiphenly • microstegiol	• 1-hydroxy-3,5,7-trien-nonanoic acid-isobutylester • nonacosan-1-ol	[63]
					• apigenin 4′-methyl ether 7*O*-*β*-D-glucuronide • luteolin 7-*O*-*β*-D-glucuronide	• 3-(3′,4′-dihydroxy-phenyl)-2-hydroxymethyl propionic acid	[93]
Salvia viridis		• 1-oxoferruginol • aethiopinone • ferruginol • microstegiol • salviviridinol • sugiol • viridinol • viridone					[62]

(continued)

Table 8 (continued)

	Chemical Constituents						
Salvia wiedemannii Boiss. (endemic)		• 12-methoxypisiferic acid • 3-oxo-abieta-8,11,13-triene • 7-*O*-acetylhorminone • 7β-hydroxysandaracopimaric acid • cryptanol • ferruginol • horminone • isopimaric acid • wiedemannic acid					[57]
		• 14-oxo-pimaric acid • pisiferal			• salvigenin		[61]
		• 8-hydroxy-12-oxoabieta-9(11),13-dien-20-oic acid 8,20-lactone					[87]
Salvia yosgadensis Freyn. et Bornm. (endemic)	• 1-acetoxy-8-hydroxy-eudesm-4(15),7(11)-dien-8,12-olide • 1-acetoxyeudes-m-4(15),7-dien-8,12-olide • istanbulin D • spathulenol	• yosgadensolide A and its 13-epimer (*sesterterpene lactone*) • yosgadensolide B and its 13-epimer (*sesterterpene lactone*)			• apigenin • apigenin 4′-methyl ether • apigenin 6,4′-dimethyl ether • apigenin 7,4′-dimethyl ether • apigenin 7-methyl ether • kaempferol 3-methyl ether • luteolin • salvigenin	• *p*-acetoxy-phenol • *p*-hydroxy-quinone	[40]
		• 6α -hydroxy-8 alpha-acetoxy-13,14,15,16-tetranorlabdan-12-oic- acid • 6α-hydroxyambreinolide • 6α-hydroxynorambreinolide • ambreinolide • norambreinolide yosgadensonol (*sesterterpene*) • 13-epi-yosgadensonol (*sesterterpene*)					[69]

Table 9 Bioactivity studies of the Anatolian *Salvia* extracts

Plant name	Activities	Ref.
S. adenocaulon	Antioxidant, Anticholinesterase	[22]
S. adenophylla	Antioxidant, Anticholinesterase	[22]
S. aethiopis	Antioxidant Antimicrobial Anticholinesterase Cytotoxic	[22, 119, 120, 117, 121] [117] [22] [121]
S. albimaculata	Antioxidant, Anticholinesterase	[122]
S. amplexicaulis	Antioxidant, Anticholinesterase, Enzyme Inhibitory Vasodepressor	[116] [73]
S. anatolica	Antioxidant, Anticholinesterase	[22]
S. aramiensis	Antioxidant, Anticholinesterase	[22]
S. argentea	Antioxidant, Anticholinesterase	[32]
S. atropatana	Antioxidant, Anticholinesterase	[22]
S. aucheri	Antioxidant Anticholinesterase	[22, 122, 123] [22, 122]
S. aytachii	Antioxidant, Anticholinesterase, Enzyme Inhibitory	[116]
S. blepharochlaena	Antibacterial Antioxidant, Anticholinesterase	[18] [22]
S. brachyantha	Antioxidant, Anticholinesterase Antimicrobial	[22, 117] [117]
S. bracteata	Antimicrobial Antioxidan, Anticholinesterase	[59] [32]
S. cadmica	Antioxidant Anticholinesterase Enzyme inhibitory	[97, 22, 120, 116] [97, 22, 116] [97, 116]
S. caespitosa	Antibacterial Antioxidant Amoebicidal, Cytotoxic Anticholinesterase	[70] [119, 32] [36] [32]
S. candidissima subsp. *candidissima*	Antioxidant Anticholinesterase	[22, 119] [22]
S. candidissima Vahl. ssp. *occidentalis*	Antioxidant activity, Anticholinesterase	[122]
S. cassia	Antioxidant, Anticholinesterase, Enzyme Inhibitory	[116]
S. cedronella	Antioxidant Anticholinesterase Antistaphylococcal	[103, 22] [22] [124]
S. cerino-pruinosa	Antioxidant, Anticholinesterase, Enzyme Inhibitory	[116]
S. ceratophylla	Antibacterial Antioxidant	[66] [121, 122]

(continued)

Table 9 (continued)

Plant name	Activities	Ref.
	Anticholinesterase Cytotoxic	[122] [121]
S. chionantha	Antioxidant activity Anticholinesterase	[22, 125] [22]
S. chrysophylla	Anticholinesterase Antioxidant	[84, 22, 8] [84, 22]
S. cilicica	Antioxidant, Anticholinesterase	[22]
S. cryptantha	Antimicrobial Anticholinesterase Antioxidant Cytotoxic	[33] [122, 32] [122, 32, 33] [126]
S. cyanescens	Antioxidant, Anticholinesterase	[122]
S. dicroantha	Antioxidant, Anticholinesterase	[22]
S. divaricata	Antioxidant, Anticholinesterase	[22]
S. ekimiana	Antioxidant, Anticholinesterase, Enzyme Inhibitory	[116]
S. eriophora	DNA damaging, Cytotoxic and Antioxidant	[78]
	Vasodepressor	[56, 35]
	Antioxidant, Anticholinesterase, Enzyme Inhibitory	[116]
S. euphratica	Antioxidant Anticholinesterase	[22, 119, 127] [22]
S. forskahlei	Antimicrobial Antioxidant, Anticholinesterase	[80] [122]
S. freyniana	Antioxidant, Anticholinesterase, Enzyme Inhibitory	[116]
S. frigida	Antioxidant, Anticholinesterase	[122]
S. fruticosa (*S. triloba* L.)	Antistaphylococcal Anti-inflammatory Antimycobacterial Antioxidant Anticholinesterase Antiamnesic Enzyme inhibitory Cytotoxic	[124] [128] [129, 130] [30, 99, 22, 131–137] [30, 22, 131, 134, 137] [137] [138] [138–140]
S. glutinosa	Cytotoxic Anticholinesterase Antioxidant	[88] [32] [88, 32]
S. halophila	Antibacterial Antioxidant Anticholinesterase	[141] [122, 141, 142] [122]
S. hedgeana	Antioxidant, Anticholinesterase	[22]
S. heldreichiana	Antimicrobial Antioxidant, Anticholinesterase	[143] [22]

(continued)

Table 9 (continued)

Plant name	Activities	Ref.
S. huberi	Antioxidant, Anticholinesterase	[22]
S. hydrangea	Antioxidant, Anticholinesterase	[22]
S. hypargeia	Antioxidant Anticholinesterase Cytotoxic	[22, 119] [22] [34, 20]
S. indica	Anticholinesterase, Antioxidant	[32]
S. kronenburgii	Anticholinesterase, Antioxidant Cytotoxic	[22] [16]
S. limbata	Antioxidant Anticholinesterase	[22, 120] [22]
S. longipedicellata	Anticholinesterase, Antioxidant	[22]
S. macrochlamys	Antioxidant Anticholinesterase	[46, 22] [22]
S. marashica	Antioxidant, Anticholinesterase, Enzyme Inhibitory	[116]
S. microphylla	Antimicrobial	[38]
S. microstegia	Anticholinesterase Antioxidant Antimicrobial	[122, 32] [120, 117, 122, 32, 144] [117, 144]
S. modesta	Anticholinesterase, Antioxidant	[22]
S. montbrettii	Anticholinesterase, Antioxidant	[22]
S. multicaulis	Antituberculous Antimicrobial Anticholinesterase Antioxidant	[19] [33] [122, 32] [122, 32, 33, 145]
S. napifolia	Anticholinesterase, Antioxidant	[22]
S. nemorosa	Antioxidant Anticholinesterase	[22, 120, 146] [22]
S. nutans	Antioxidant, Anticholinesterase, Enzyme Inhibitory	[116]
S. nydeggeri	Anticholinesterase, Antioxidant	[22]
S. officinalis	Antimycobacterial	[129]
S. pachystachys	Antioxidant Anticholinesterase	[22, 120] [22]
S. palaestina	Antimicrobial Anticholinesterase, Antioxidant	[96] [22]
S. pilifera	Antioxidant, Anticholinesterase, Enzyme Inhibitory	[116]
S. pinnata	Antioxidant, Anticholinesterase	[32]
S. pisidica	Antioxidant, Anticholinesterase	[22]
S. poculata	Antioxidant Anticholinesterase	[86, 22] [86, 22, 8]
S. pomifera	Antioxidant	[22, 133]

(continued)

Table 9 (continued)

Plant name	Activities	Ref.
S. potentillifolia	Antioxidant Anticholinesterase Enzyme Inhibitory	[22, 116, 37] [22, 116] [116]
S. quezelii	Antioxidant, Anticholinesterase	[32]
S. recognita	Antioxidant, Anticholinesterase	[22]
S. rosifolia	Antioxidant, Anticholinesterase	[22]
S. russellii	Antioxidant, Anticholinesterase	[22]
S. sericeo-tomentosa	Antioxidant Anticholinesterase, Enzyme Inhibitory	[116, 133] [116]
S. sclarea	Antimicrobial Antifungal Antioxidant Anticholinesterase	[44] [147] [119, 122, 147] [122]
S. smyrnaea	Anticholinesterase, Antioxidant	[22]
S spinosa	Anticholinesterase, Antioxidant	[22]
S. staminea	Cytotoxic Antioxidant Amoebicidal, Cytotoxic Anticholinesterase	[45] [22, 148] [36] [22, 8]
S. syriaca	Antihypertensive, Vasodepressor Antioxidant, Anticholinesterase	[68, 35] [122, 32, 149]
S. suffruticosa	Anticholinesterase, Antioxidant, Enzyme Inhibitory	[22, 116]
S. tchiatcheffii	Anticholinesterase, Antioxidant, Enzyme Inhibitory	[116]
S. tigrina	Antifungal	[150]
S. tobeyi	Anticholinesterase, Antioxidant	[32]
S. tomentosa	Antibacterial Antimycobacterial Anticholinesterase, Antioxidant	[151] [129, 130] [22]
S. trichoclada	Anticholinesterase Antioxidant Anti-inflammatory Antimicrobial	[21, 22, 134] [21, 93, 22, 134] [128] [152]
S. verbenaca	Antioxidant	[148]
S. verticillata	Antioxidant Anti-inflammatory Antibacterial Anticholinesterase	[93, 101, 22, 120, 122, 32, 134, 153] [128] [101, 151] [22, 122, 32, 134]
S. virgata	Antioxidant Anticholinesterase	[22, 120, 148] [22]
S. viridis (syn: *S. horminum*)	Antibacterial Antioxidant Anticholinesterase	[62] [22, 145] [22]

(continued)

Table 9 (continued)

Plant name	Activities	Ref.
S. viscosa	Antioxidant, Anticholinesterase	[32]
S. wiedemannii	Analgesic Antiviral Antioxidant, Anticholinesterase	[154] [155] [22]
S. xanthocheila	Antioxidant, Anticholinesterase	[22]
S. yosgadensis	Antioxidant, Anticholinesterase	[22]

Bacillus subtilis ATCC 6633, *Staphylococcus aureus* 6538P, *S. epidermidis* ATCC 12228, *Proteus mirabilis* ATCC 14153, *Escherichia coli* ATCC 8739, *Klebsiella pneumonia* ATC 4352, *Pseudomonus aeruginosa* ATCC 9027, *Enterococcus faecalis* ATCC 29212, *β*-haemolytic *Streptococcus* and *Candida albicans* ATCC 10231.

Isolated flavonoids from *S. palaestina;* 6,7,3′,4′-tetra-methoxyflavone, cirsimaritin, luteolin, luteolin 7-glucoside were tested against *S. aureus*, *S. epidermis*, *E. coli*, *P. aeruginosa*, *P. vulgaris*, *and K. pneumoniae*; only cirsimaritin gave significant activity [96].

Salvia sclarea (clary sage) has been used as a flavoring agent in Europe due to strong, warm and aromatic taste and odour. It has been used in the treatment of coughs, colds, sore throat, and sore eyes, and wounds as well as diuretic and blood cleaning. Its herb has been used for stomachic and digestive disorders and in kidney diseases. From the acetone extract of the whole plant *S. sclarea*, isolated diterpenes (sclareol, manool, salvipisone, ferruginol, microstegiol, candidissiol, 7-oxoroyleanone, 2,3-dehydrosalvipisone and 7-oxoferruginol-18-al) and two sesquiterpenes (caryophyllene oxide and spathulenol) were tested for antimicrobial activity against a series bacteria as mentioned above. Among them, the new diterpene 2,3-dehydrosalvipisone, sclareol, manool, 7-oxoroyleanone, spathulenol and caryophyllene oxide were found to be active against *Staphylococcus aureus*, while the first and third compound against *Candida albicans* and the last compound against *Proteus mirabilis* [44].

Isolated diterpenoid forskalinone and aromatic compounds (octanol esters of cis- and trans-4-*O*-methyl-caffeic acid dimers) from the roots of *Salvia forskahlei* were also tested against the standard bacterial strains and a yeast. Forskalinone showed moderate resistance against *S. epidermidis* (670 µg/mL) and slight activity against *E. faecalis* (168 µg/mL). Trans-4-*O*-methyl-caffeic acid dimer octanol ester was found to be inactive while the cis isomer showed a slight activity against *C. albicans* (156 µg/mL) [80].

Ulubelen and Topcu investigated a series abietane diterpenoids for standard bacterial strains including Gram (+) and Gram (−) bacteria as seen in Table 10.

Among them salvipisone was found to be highly active against *B. subtillus*, *S. aureus* and *S. epidermidis.* Pisiferic acid 12-methyl ether was shown activity

Table 10 MIC values of the selected diterpenoids from Anatolian *Salvia* Species against standard bacterial strains (μg/ml)

Compounds	*B. subtilis*	*S. aureus*	*S. epidermidis*	*P. mirabilis*	*E. coli*	*Kl. pneumonia*	*P. aeruginosa*	*E. faecalis*	*C. albicans*
microstegiol	-	-	-	-	125	125	-	125	-
salvibretol	125	-	-	125	125	-	125	-	-
1-oxosalvibretol	-	-	-	-	-	-	-	12.5	-
salvipisone	13.5	26.95	52.18	-	-	-	-	-	125
2,3-dehydro-salvipisone	-	10.5	-	-	-	-	-	NT	5.3
aethiopinone	125	>125	>125	-	125	50	-	-	125
salvinolone	125	-	125	-	-	125	-	125	125
saphrorthoquinone	100	100	100	-	-	-	>125	50	100
taxodione	100	25	-	-	-	-	-	50	-
horminone	-	-	-	-	-	16.91	-	-	-
pomiferin C	18.75	-	-	-	-	-	-	-	-
pisiferic acid 12-methyl ether	6.45	-	-	-	-	-	-	-	-
11-OH,12-OCH_3-abieta-8,11,13-triene	-	>125	>125	-	125	>125	-	NT	-
7-oxoroyleanone	-	54.0	-	-	-	-	-	NT	6.75

against *B. subtillus* while 2,3-dehydrosalvipisone and 7-oxoroyleanone were found strongly active against the yeast *C. albicans* [29].

S. bracteata and the crude extract were also tested against the standard bacterial strains. The crude extract, and a new abietane bractealine, and the known compound horminone showed activity against *B. subtilis*, *S. aureus*, and *S. epidermidis*. Ferruginol showed slight activity against these strains among tested the ten diterpenoids which are namely salvibracteone, 7-*O*-acetylhorminone, 12-methylhorminone, 6,7-dehydroroyleanone, ferruginol, 12-methylferruginol, salvinolone, sugiol, bractealine, and horminone including ferruginol [59].

Three new diterpenes and five known diterpenes mentioned diterpenoids part isolated from the roots of *S. viridis* were tested against the same standard series. 1-Oxoferruginol showed high activity against *B. subtilis*, *S. aureus*, *S. epidermidis* and a modest activity against *P. mirabilis*, while microstegiol showed low activity against *B. subtilis* [62].

The isolated two new abietanes (blephaein and *O*-methylpisiferic acid methyl ester) and eight known diterpenoids from *S. blepharochlaena* were tested for their antibacterial activity. Potent antibacterial activity was exhibited by the known compounds horminone and 7-*O*-acetylhorminone against *Staphylococcus aureus* ATCC 6538 P, *S. epidermidis* ATCC 12226, and *B. subtilis* ATCC 6633. Horminone was also found to be active against *E. faecalis* ATCC 29212 [18].

Antibacterial activity of the isolated compounds; a new pimarane diterpene 6*β*-hydroxyisopimaric acid, four known diterpenes, one new triterpenoid, 3-acetylvergatic acid, five known triterpenoids, two steroids and a flavone from the roots of *S. caespitosa* Montbret and Aucher ex. was evaluated testing against the standard bacterial strains. Only the new diterpene, 6*β*-hydroxyisopimaric acid demonstrated strong activity against *S. aureus, S. epidermidis* and *B. subtilis*, with minimum inhibitory concentration (MIC) values were found 9, 18 and 9 μg/mL, respectively [70].

Four known and two new diterpenoids together with two triterpenic acids, a steroid and a flavone isolated from *S. ceratophylla* L. were tested for their antibacterial activity. It is noteworthy that only a rearranged abietane diterpene which has an eight-membered ring structure, named candidissiol exhibited strong antibacterial activity against *S. epidermidis* and *P. mirabilis* [66].

The extract of *S. microphylla* and its constituents 2-(*p*-hydroxyphenyl)ethyl eicosaheptanoic acid ester and hexacosylferulate, a sesquiterpene 8*α*-hydroxy-*β*-eudesmol, a diterpene carnosic acid 12-methyl ether were tested for antimicrobial activity against the same standard bacterial strains, and only carnosic acid 12-methyl ether showed moderate antimicrobial activity *against S. aureus* at 78 μg/mL [38].

Another group studied antimicrobial activity of the of *S. halophila* Hedge extract against 13 bacteria and 2 yeasts (*E. coli* ATCC 25922, *P. mirabilis* BC 3624, *Aeromonas hydrophila* ATCC 7965, *B. subtilis* var. *niger* ATCC 10, *B. cereus* RSKK 863, *B. brevis* FMC 3, *S. aureus* ATCC 25923, *Morganella morganii*, *Salmonella typhimurium* NRRLE 4463, *Listeria monocytogenes* 1/2B, *K. pneumoniae* FMC 5, *P. aeruginosa* ATCC 27853, *Yersinia enterocolitica* ATCC 1501,

Saccharomyces cerevisiae BC 5461, and *C. albicans* ATCC 1223.) by agar diffusion method. The samples prepared from the dried ethanol (EtOH) extract at four concentrations (1, 2.5, 5 and 10%) showed no effect against the tested yeasts. However, they showed antibacterial activity against all of the tested Gram (+) bacteria at all concentrations except *L. monocytogenes* while the extract had no effect on the whole tested Gram (−) bacteria except for *A. hydrophila* and *P. aeruginosa* [141].

The EtOH extracts obtained from the leaves, rootstock and the combined formulation of endemic *Salvia tigrina* Hedge & Hub.-Mor. have been investigated for their antifungal activities. The antifungal activity of the extract was tested against *Candida* species (*C. albicans* ATCC 10231, *C. tropicalis* ATCC 13803 and *C. guilliermondii* ATCC 6260 and *Cryptococcus* species (*C. neoformans* ATCC 90112 and *C. laurentii* 34142), *Aspergillus flavus*, *Penicillium frequentans*, *Botrytis cinerea*, *Geotrichum candidum*, *Fusarium oxysporum* and *Alternaria alternara* by the visual broth macrodilution method. Ketoconazole was used as a positive reference standard to determine the sensitivity of the strains. All the extracts exhibited a strong antifungal effect, the MIC values ranging from 3.12 to 25 mg/mL, and higher antifungal effect was observed against *C. albicans*, *C. neoformans* and *B. cinarea* [150].

In vitro antimicrobial properties of various extracts (hexane, CH_2Cl_2, MeOH/$CHCl_3$, MeOH/water and direct MeOH) from *Salvia microstegia* Boiss. et Balansa were evaluated. Direct MeOH extract of *S. microstegia* exhibited more effective antibacterial activity on the Gram (−) and Gram (+) bacteria. None of the extracts exhibited antifungal activity on the *Candida albicans* ATCC 10231 [144].

Antimicrobial activity of $CHCl_3$ and EtOH extract of *Salvia trichoclada* L. was investigated by disk diffusion method. Active components extracted showed an average antimicrobial activity against microorganisms (*E. faecalis* ATCC 29212, *M. luteus* NRLL B-4375, *M. luteus* ATCC 9341, *E. aerogenes* ATCC 13048, *B. subtilis* ATCC 6633, *E. coli* ATCC 39628, *P. aeruginosa* ATCC 27853, *P. vulgaris* ATCC 6897, *L. monocytogenes* ATCC 7644, *Pseudomonas* sp. (clinic isolate), *E. cloacae* ATCC 13047D, *S. aureus* ATCC 25923, *S. epidermidis* ATCC 12228, *S. typhimurium* CCM 5445, *K. pneumonia* (clinic isolate), *C. albicans* (clinic isolate), *S. cerevisia* and *A. flavus*.) The results showed that the Gram (−) bacteria were more sensitive than Gram (+) bacteria. EtOH extract was more effective than the $CHCl_3$ extract. Soxhlet extraction was found to be more suitable than maceration for stronger inhibition on the test microorganisms [152].

Antibacterial activity of *Salvia verticillata* L. subsp. *amasiaca* (Frey & Bornm.) Bornm. and *Salvia tomentosa* Mill. was compared with other 14 different plants (*Clinopodium vulgare* L. subsp. *vulgare* L., *Mentha pulegium* L., *Melilotus officinalis* (L.) Desr., *Melilotus alba* Desr., *Medicago lupulina* L., *Galega officinalis* L., *Xeranthemum annuum* L., *Cichorium intybus* L., *Plantago lanceolate* L., *Plantago major* L. subsp. *major*, *Fumaria officinalis* L., *Galium palustre* L., *Echium vulgare* L., and *Sambucus nigra* L.). For each plant, 3 different extracts (aqueous, EtOH, and MeOH) were prepared, and a total of 48 extracts were evaluated. Antibacterial activity was carried out against 10 bacteria, including *S. pyogenes*, *S. aureus*, *S.*

epidermidis, *E. coli*, *P. aeruginosa*, *S. typhimurium*, *S. marcescens*, *P. vulgaris*, *Enterobacter cloacae*, and *K. pneumoniae* by disk diffusion method. All plants, except *M. alba*, *M. lupulina*, *X. annuum*, *G. palustre*, and *S. nigra* showed inhibitory activity against both Gram (+) and Gram (−) bacteria. The best inhibitory activity was observed with aqueous extract of *Melilotus officinalis* (22.5 mm); it performed better than all positive controls (erythromycin, ampicillin, carbenicillin, tetracycline, and chloramphenicol; 7–20 mm) against *P. aeruginosa*. The tested two *Salvia* plant extracts exhibited moderate-high activity against three Gram (−) bacteria including *P. vulgaris*, *K. pneumonia* and *E. cloacae* [151].

Antimicrobial activities of the aqueous herbal extracts and essential oils of *Salvia heldreichiana* Boiss. Ex Bentham (endemic) and *Ballota saxatilis* subsp. *brachyodonta* Boiss. P. H. Davis & Doroszenko were screened against nine human pathogenic bacteria and yeast [*E. coli* (ATCC 25293), *E. feacalis* (ATCC 29212), *B. subtilis* (ATCC 6633), *S. thyphimurium*, *S. aureus* (ATCC 25925), *S. epidermidis* (ATCC 12228), *K. pneumoniae* (10031), *C. albicans* (clinic strain), *C. parapsilosis* (ATCC 22019)]. Antimicrobial activity were determined by using macrodilution method. The aqueous extracts of the plants showed rather low antimicrobial activity against the studied microorganisms than their essential oils [143].

The ability of methanolic extract of *Salvia verticillata* L. to increase the permeability of multidrug resistant bacterial cells was conducted by flow cytometric assay on *Listeria innocua* and *E. coli*. Intense concentration of *S. verticillata* L. has destroyed 97% of living cells for *Listeria innocua* and 94.86% for *E. coli* [101].

In vitro antifungal activity of the *Salvia sclarea* L. from Munzur Valley in Tunceli, Turkey was evaluated. The activity of EtOH, hexane and aqueous extracts of *S. sclarea* against pathogen fungi *Epicoccum nigrum* and *Colletotrichum coccodes* was investigated. The best antifungal effect was observed for ethanolic *S. sclarea* extract [147].

Antimicrobial activities of *Salvia* species (*S. brachyantha* (Bordz.) Pobed, *S. aethiopis* L., and *S. microstegia* Boiss. and Bal.) were tested using the disc diffusion method against three Gram (+) and four Gram (−) microbial species, as well as three fungi species. The ethanolic extracts of *Salvia* species revealed a wide range of antimicrobial activity. *S. brachyantha* and *S. microstegia* showed the highest antimicrobial activities against *B. subtilis*, whereas *S. aethiopis* was more effective on *Y. lipolytica*. None of the extracts showed anti-fungal activity against *S. cerevisiae* [117].

There are several antimicrobial activity studies which make activity comparison among many different plant species having at least antiseptic folkloric usage, but their results were not presented herein due to a huge number of plant extracts either belonging to Lamiaceae family or not were tested besides only one or two *Salvia* species [156, 157].

4.1.2 Antituberculous Activity

Since antiquity, some *Salvia* species have been used to treat cough, cold and night sweating and tuberculosis [2, 5]. In China, a series fully aromatic abietane diterpenoids isolated from *S. miltiorrhiza* have screened for antituberculous activities and especially tanshinones were found to be active.

Therefore, antituberculous activity of the isolated abietane diterpenoids which have fully aromatic rings (multicaulin, 12-demethylmulticauline, multiorthoquinone, 12-demethylmultiorthoquinone), and other isolated diterpenes (12-methyl-5-dehydrohorminone, 12-methyl-5-dehydro-*O*-acetylhorminone, salvipimarone) from *S. multicaulis* was carried out against *Mycobacterium tuberculosis* strain H37Rv, and all compounds were found to be significantly active, particularly 12-demethylmulticauline, 12-demethylmultiorthoquinone, 12-methyl-5-dehydrohorminone and 12-methyl-5-dehydro-*O*-acetylhorminone being the most potent substances. Six of these novel compounds were further evaluated against a number of standard bacterial strains [19].

Antibacterial and antimycobacterial activity of the methanol extracts of five Labiatae plants were investigated including *Salvia fruticosa* Mill., *Salvia tomentosa* Mill., *Sideritis albiflora* Hub.-Mor. (endemic), *Sideritis leptoclada* O. Schwarz & P. H. Davis, (endemic), and *Origanum onites* L. Antibacterial activity was analyzed against *S.aureus*, *S. epidermidis*, *E. faecalis*, *B. cereus*, *E. coli*, *S. typhimurium*, *E. aerogenes,* and *K. pneumoniae.* Antimycobacterial activity was assayed against Mycobacterium tuberculosis. The best antibacterial activity (MIC 640 μg/mL) was shown against *S. typhimurium* and *E. aerogenes* by *S. fruticosa*; *E. coli*, and *S. typhimurium*, *E. aerogenes* by *S. tomentosa; S. typhimurium*, and *E. aerogenes* by *S. leptoclada* and *S. typhimurium*, *E. aerogenes* and *S. epidermidis* by *O. onites*, respectively. The best antimycobacterial activity was shown by *S. tomentosa* (MIC 196 μg/mL) while *S. fruticosa* (MIC 392 μg/mL) and *O. onites* (MIC 784 μg/mL) showed moderate activity against *M. tuberculosis*. [130].

The antimycobacterial activities of the EtOH extracts from the root and aerial parts of *Salvia cilicica*; the aerial parts of *Salvia fruticosa*, *Salvia officinalis* and *Salvia tomentosa* were investigated against three *Mycobacterium tuberculosis* (sensitive-, resistant-standard and multidrug resistance (MDR) clinical isolate) strains. The *S. officinalis* showed the highest antimycobacterial activity with MIC 25 μg/mL value against clinical isolate of MDR- *M.tuberculosis* and H37Rv *M. tuberculosis* (ATCC 27294 sensitive) and with MIC value of 100 μg/mL against H37Rv *M. tuberculosis* (ATCC 35838 R resistant) followed MIC 100 μg/mL by *S. fruticosa* and *S. tomentosa* against all the three tested *M. tuberculosis* strains. The antimycobacterial activity of the each aerial and root parts extracts of *S. cilicica* showed a MIC value of 400 μg/mL [129].

4.1.3 Antiviral Activity

Antiviral efficacy of *S. wiedemannii* Boiss. extracts on Herpes simplex type 1 (HSV-1) and Parainfluenza-3 (PI-3) viruses was evaluated by using Vero cell lines. Antiviral efficacy of these extracts, obtained from aerial parts of *S. wiedemannii*, was compared to that of acyclovir and oseltamivir. The H_2O, $CHCl_3$, and EtOH extracts of *S. wiedemannii* (16–0.0625 µg/mL) showed a significant antiviral activity on HSV-1 with a maximum nontoxic concentration 16 µg/mL. Only the butanol extract of *S. wiedemannii* demonstrated important antiviral activity on PI-3 with a range of 64–16 µg/mL of inhibitory concentration for cytopathic effect,[1] which was close to the anti PI-3 activity of oseltamivir [155].

4.2 Antioxidant Activity

Antioxidant activity tests are carried out for many Anatolian *Salvia* species in the last fifteen years. In most cases, lipid peroxidation inhibitory test (*β*-carotene-linoleic acid assay) and radical scavenging assays including DPPH (1,1-diphenyl-2-picrylhydrazyl) free radical, ABTS cation radical, superoxide anion radical tests were the most applied ones. Ferric (III) and cupric (II) reducing based assays were also applied in some studies. In the tests as control compounds butylated hydroxyanisole (BHA), butylated hydroxytoluene (BHT), *α*-tocopherol and quercetin have been used. Many spices and herbs are found to be potent sources of natural antioxidants. Among the spices sage, rosemary, thyme and oregano are well known with antioxidant activity. *Salvia* species have received particular attention as a source of natural antioxidants. The main antioxidant activity of *S. officinalis* and *S. fruticosa* (syn = *S. triloba*) was attributed to its mainly phenolic compounds, such as carnosic acid, carnosol, and rosmarinic acid [158].

Most of the antioxidant activity studies, particularly antioxidant activity-directed isolation studies on the plants, the initial step must cover the analysis of total phenolic and flavonoid contents quantitatively of the crude extracts, calculated as pyrocatechol and quercetin equivalents, respectively. Most of the chemical studies on Anatolian *Salvia* species were not carried out based on bioactivity-directed isolation procedure. However, recent isolation and structure elucidation studies are based on bioactivity directed way.

The methanolic extracts of *S. cryptantha* and *S. multicaulis* were examined for radical scavenging activity by using DPPH, hydroxyl radical-scavenging and lipid peroxidation test assays, and no or slight activity was observed in the polar and non-polar subfractions of the both plant extracts [33].

---

[1]Cytopathic effect or cytopathogenic effect (abbreviated CPE) refers to structural changes in host cells that are caused by viral invasion.

Antioxidant activities of the MeOH extracts of six *Salvia* species; *S. caespitosa* Montbret & Aucher ex Bentham (endemic), *S. hypargeia* Fisch. & Mey. (endemic), *S. euphratica* subsp. *euphratica* Montbret & Aucher ex Bentham (endemic), *S. sclarea* L., *S. candidissima* subsp. *candidissima* Montbret & Aucher ex Bentham and *S. aethiopis* L. from Turkey were investigated in vitro. The extracts were screened in DPPH and β-carotene/linoleic acid test systems. Non-polar subfractions of the MeOH extracts of *Salvia* species studied did not show any antioxidant activity in both test systems. The most active plant was found to be *S. euphratica* subsp. *euphratica* having an IC_{50} value of 20.7 $\pm$ 1.22 μg/mL, followed by *S. sclarea* (IC_{50} = 23.4 $\pm$ 0.97 μg/mL) among the polar subfractions. In the β-carotene-linoleic acid test system, polar extract of *S. hypargeia* was superior to the polar extracts of other *Salvia* species studied (69.2 $\pm$ 1.90%), it was followed by *S. sclarea* with 63.51 $\pm$ 4.24% inhibition [119].

The antioxidant activities of four Lamiaceae plants, *Salvia viridis* L., *S. multicaulis* Vahl., *Stachys byzantina* C. Koch and *Eremostachys laciniata* (L.) Bunge have been determined by using DPPH as well as by flow injection analysis-luminol chemiluminescence. All extracts were shown to possess a significant scavenger activity against DPPH radical and an inhibitory effect on $H_2O_2^-$ or $HOCl^-$ luminol chemiluminescence. Both the highest DPPH radical scavenging activity and the most potent $H_2O_2^-$ induced peak chemiluminescence was observed for *S. viridis* extract [145].

The crude MeOH extract of *Salvia macrochlamys* Boiss. et Kotschy. was tested in five different systems involving free radical scavenging activity by DPPH, β-caroten–linoleic acid, superoxide anion radical scavenging activity in PMS-NADH system, CUPRAC test for Cu (II) reducing capacity, and metal chelating activity with ferrozine for antioxidant activity. Particularly, in inhibition of superoxide anion radical generation, the extract showed better activity than standard BHT and a similar activity to that of ascorbic acid at 12.5–25 μg/mL concentrations, while in inhibition of lipid peroxidation it exhibited less activity. The extract and a lupane triterpenoid monogynol A and its three derivatives were found to be highly active particularly in the metal chelating test system on ferrous ions [46].

The 56 extracts prepared with petroleum ether (PE), $CHCl_3$, EtOAc and MeOH solvents, obtained from 14 *Salvia* species (*S. albimaculata* Hedge and Hub (endemic), *S. aucheri* Bentham var. *canescens* Boiss. and Heldr (endemic), *S. candidissima* Vahl. ssp. *occidentalis*, *S. ceratophylla* L., *S. cryptantha* Montbret and Bentham (endemic), *S. cyanescens* Boiss. and Bal. (endemic), *S. frigida* Boiss, *S. forskahlei* L., *S. halophila* Hedge (endemic), *S. microstegia* Boiss. and Bal., *S. multicaulis* Vahl., *Salvia sclarea* L., *S. syriaca* L., S. *verticillata* L. ssp. *amasiaca*) were screened for their antioxidant activities. The antioxidant activities were carried out by DPPH radical-scavenging and enzymatic xanthine/xanthine oxidase (XO) methods. Particularly, the EtOAc and MeOH extracts were observed to be highly active against both DPPH and XO [122].

A study was designed to examine rosmarinic acid levels of the MeOH extracts of *Salvia verticillata* subsp. *verticillata* and *S. verticillata* subsp. *amasiaca* and screening their antioxidant activity by two complementary test systems, DPPH free radical-scavenging and *β*-carotene/linoleic acid systems. In DPPH system, *S. verticillata* subsp. *verticillata* was superior to the subsp. *amasiaca* with an IC_{50} value of 14.5 μg/mg while in the *β*-carotene/linoleic acid test system. Inhibition capacity of *S. verticillata* subsp. *verticillata* was found to be 74.4% correlating with its slightly higher rosmarinic acid level (28.7 μg/mg) than the other subspecie [153].

Same study was carried out for *S. virgata*, *S. staminea* and *S. verbenaca*. In the DPPH system, the most active plant was found to be *S. verbenaca* (14.30 μg/mg), followed by *S. virgata* (65.70 μg/mg). The extract of *S. staminae* exhibited the weakest antioxidant activity in this test system of which IC_{50} value is 75.40 μg/mg. In *β*-carotene/linoleic acid test system, *S. verbenaca* extract was superior to the other extracts studied (inhibition value is 77.03%). According to the results obtained by spectrophotometric antioxidant/radical scavenging analysis and further supported by HPLC analysis, *S. verbenaca* has the highest rosmarinic acid level with a value of 29.30 μg/mg. Results showed that the rosmarinic acid and its derivatives are more likely to be responsible for most of the observed antioxidant activities of *Salvia* species [148].

The antioxidant activity of EtOH extract of *S. halophila* Hedge was evaluated by two antioxidant assays, including phosphomolybdenum reduction and DPPH test systems. The EtOH extract showed free radical scavenging activity with IC_{50} of 67.73 μg/mL in DPPH the assay. The antioxidant activity of the extract was found to be 84.87 ± 0.7 mg ascorbic acid equivalent (AAE)/g extract in phosphomolybdenum assay [141].

The acetone extract prepared from the aerial parts of *S. cedronella* Boiss., was evaluated for the antioxidant potential by using three different assays; *β*-carotene-linoleic acid test system for lipid peroxidation inhibition, DPPH for free radical scavenging activity, Fe^{2+}-ferrozine test system for metal chelation. Due to the presence of high phenolics content, high radical scavenging ability was observed and a moderate iron chelating effect. However, inhibition of lipid peroxidation of the extract was not significant [103].

The antioxidant activity of the methanolic extract of *S. microstegia* was examined with two complementary test systems; DPPH and *β*-carotene-linoleic acid assays. The IC_{50} value of free radical scavenging activity of the methanolic extract of *S. microstegia* was determined as 7.63 mg/mL while lipid peroxidation inhibition percentage was found to be 79.92% [144].

The crude extract and the isolated compounds from the aerial parts of *Salvia poculata* Nab., a Turkish endemic *Salvia* species, were investigated for antioxidant activity by using *β*-carotene bleaching, superoxide anion radical, and ABTS cation radical scavenging activity test methods. The crude extract of *S. poculata* exhibited a significant antioxidant effect. From eight isolated compounds, sclareol and *β*-sitosterol showed high total antioxidant activity in the *β*-carotene bleaching method [86].

Methanolic extracts of eight *Salvia* species including *S. aethiopis*, *S. candidissima. S. limbata*, *S. microstegia*, *S. nemorosa*, *S. pachystachys*, *S. verticillata*, and *S. virgata*, collected from Eastern Anatolia in Turkey, were screened for their possible antioxidant activities by two complementary test systems, including DPPH free radical scavenging and *β*-carotene/linoleic acid. In both systems, the most active plant was found to be *S.* verticillata with values of IC_{50} = 18.3 μg/mL and 75.8%, respectively [120].

The antioxidant activities of ethanol extract of *S. potentillifolia* were determined by using complementary tests, namely, DPPH radical-scavenging, *β*-carotene-linoleic acid and reducing power assays. The ethanol extract showed better activity (IC_{50} = 69.4 ± 0.99 μg/mL) than that of standard BHT in the DPPH system, and showed great lipid peroxidation inhibition in the *β*-carotene-linoleic acid system (IC_{50} = 30.4 ± 0.50 μg/mL) [37].

The antioxidant activity of the hexane extract and essential oil of the *S. chionantha* Boiss. (endemic) was determined by using four complementary test systems; namely, *β*-carotene-linoleic acid, DPPH scavenging, ABTS scavenging, and CUPRAC assays. In *β*-carotene-linoleic acid assay, the extract showed 81.2 ± 0.1% lipid peroxidation inhibition at 0.8 mg/mL concentration, while in ABTS assay the essential oil exhibited 77.4 ± 0.5% inhibition at same concentration [125].

The CH_2Cl_2, EtOAc, and MeOH extracts prepared from 55 *Salvia taxa* were screened for their scavenging effect against DPPH free radical and iron-chelating capacity at three concentrations (25, 50, 100 μg/mL). Only the CH_2Cl_2 and EtOAc extracts of *S. cilicica* Boiss. et Kotschy had a moderate iron-chelating capacity (54.71%) at 100 μg/mL. However, about fifteen *Salvia* extracts showed notable scavenging effect against DPPH free radical, particularly at 100 μg/mL [22].

Antioxidant activities of CH_2Cl_2, EtOAc, and EtOH extracts of three wildly grown samples (SFW) and a cultivated sample of *S. fruticosa* were assessed by determining DPPH radical-scavenging activity, iron-chelating capacity, and ferric-reducing antioxidant power (FRAP). All the extracts of the cultivated plant showed better scavenging activity against DPPH than did the wildly grown samples extracts. The EtOH extract of the cultivated plant also showed greater activity on the FRAP test than did the EtOH extract of wild plant. All the extracts exerted low or no chelation effect on ferric ion. The CH_2Cl_2 and EtOH extracts of wild plant had lower DPPH radical—scavenging activity and FRAP activity at all tested concentrations [131].

In the villages of Anatolia, some Sideritis species are also sold as sage. The species of 87 plant samples were sold as "sage", and 61 of them were identified as *S. fruticosa*, 20 samples as *Sideritis congesta*, the all rest samples corresponded to Sideritis species including *Sideritis pisidica* var. *termessi*, *Sideritis arguta*, *Sideritis perfoliata* and *Sideritis libanotica* subsp. *linearis* except a *Salvia* specie *Salvia tomentosa*. Also there was one unidentified species. Infusions from the 87 samples were prepared and the preliminary antioxidant activity of all infusions were determined using the DPPH radical scavenging test at a concentration of 1.0 mg/mL. Their radical scavenging activity against DPPH was found to change between 12.69 and 71.39%. Further antioxidant activity tests were performed for the infusions of the 7 extracts having best results, by the DPPH radical scavenging, ferrous ion-chelating and FRAP tests at concentrations of 0.25, 0.50, and 1.0 mg/mL. The investigated two *S. fruticosa* infusions showed highest scavenging effect (65.04%)

towards DPPH, as well as gave the best FRAP result. However, none of the infusions displayed a significant effect in the ferrous ion-chelating tests [132].

Aerial parts of *S. halophila* Hedge (endemic) were examined for its antioxidant activity and phenolic compositions. The *S. halophila* was extracted with different solvents including hexane, EtOAc, MeOH, and 50% MeOH in a Soxhlet apparatus. Water extract was also prepared from *S. halophila* by reflux. All the extracts were investigated for Fe(III) reductive free radical scavenging and the effect upon linoleic acid peroxidation activities. The peroxidation level was also determined by the TBA method. Polar fractions were found more active in free radical activity test system whereas non-polar fractions protected the peroxidation of linoleic acid. Rosmarinic acid was the most abundant component in the extracts [142].

Free radical scavenging activities of n-butanol, methanol water, acetone and chloroform extracts of *S. verticillata* Bentham and *S. trichoclada* L. were examined by DPPH test at concentrations of 1, 5, 10, 25, 50 and 100 μg/mL. The results were compared with ascorbic acid. The highest scavenging effect was observed at 50 μg/mL conc. for n-butanol extract and at 100 μg/mL conc. for both n-butanol and methanol extracts [93].

The EtOAc and MeOH extracts from 16 *Salvia* L. species (*S. amplexicaulis* Lam., *S. aytachii* Vural & Adiguzel (endemic), *S. cadmica* Boiss. var. *cadmica* (endemic), *S. cassia* Samuelss. ex. Rech. Fil, *S. cerino-pruinosa* Rech. Fil var. *cerino-pruinosa* (endemic), *S. ekimiana* F. Celep & Dogan (endemic), *S. eriophora* Boiss. & Kotschy. (endemic), *S. freyniana* Bornm. (endemic), *S. marashica* A. Ilcim (endemic), F. Celep & Dogan, *S. nutans* L., *S. pilifera* Montbret & Aucher ex Benth. (endemic), *S. potentillifolia* Boiss. & Heldr. ex Bentham (endemic), *S. sericeo-tomentosa* Rech. Fil var. *sericeo-tomentosa* (endemic), *S. sericeo-tomentosa* var. *hatayica* Celep & Dogan (endemic), *S. suffruticosa* Montbr. & Auch. ex Benth. *bracteata* Banks et Sol. (endemic), and *S. tchihatcheffii* (Fisch. & C.A. Mey.) Boiss) were screened for their antioxidant activity by using DPPH radical scavenging, metal chelation, and ferric-reducing antioxidant power (FRAP) assays. Among the extracts screened, the best scavenging activity was obtained with MeOH extract of *S. sericeo-tomentosa* var. *sericeo-tomentosa* (90.29%). In general, the MeOH extracts showed higher activity than the EtOAc extracts, ranging from 5.42 to 17.92%. The extracts exhibited weak metal-chelation capacity varying between 6.92 and 24.19% at 1 mg/mL. In the FRAP assay, the EtOAc extracts exhibited quite low activity, whereas the MeOH extracts were found to be highly active. The best FRAP value was observed for the MeOH extract of *S. cerino-pruinosa* var. *cerino-pruinosa*. Among the screened 16 *Salvia* extracts, the MeOH extract of *S. ekimiana* (153.50 mg/100 g) was revealed to be the richest in terms of rosmarinic acid [116].

The crude methanol extract of *S. fruticosa* and its sub-extracts hexane and dichloromethane as well as pure isolated triterpenoids were investigated for antioxidant activity by six complementary test methods including DPPH free radical scavenging, β-carotene bleaching, superoxide anion radical scavenging, ferric reducing power, and cupric reducing antioxidant capacity (CUPRAC) tests. The methanol extract exhibited the highest antioxidant activity surpassing the other tested extracts and pure isolated triterpenoids α-amyryltetracosanoate, oleanolic acid, and ursolic acid [30].

In Turkey, another group also studied on a cultivated *S. fruticosa* extract and investigated antioxidant activity by comparison of both wild and cultivated plant samples which were found to be higher for the second harvesting year whereas there were no significant differences between wild and cultivated *S. fruticosa* during storage period [99].

In vitro antioxidant activity was evaluated by DPPH and beta-carotene/linoleic acid assays for n-hexane, CH_2Cl_2, MeOH extracts obtained from two varieties of an endemic species *S. euphratica* (*S. euphratica* var. *euphratica* and *S. euphratica* var. *leiocalycina*) as well as their phenolic acid composition. MeOH extracts of both varieties exhibited strong antioxidant activity, probably their high rosmarinic acid content (39.4 and 55.8 μg/mg, respectively) [127].

As an additional study [116], antioxidant activities of the CH_2Cl_2, EtOAc, and EtOH extracts from the aerial parts and/or roots of a group of *Salvia* (sage) species (*S. argentea, S. bracteata, S. caespitosa* (endemic), *S. cryptantha* (endemic), *S. glutinosa, S. indica, S. microstegia, S. multicaulis, S. pinnata, S. quezelii* (endemic), *S. syriaca, S. tobeyi* (endemic), *S. verticillata* subsp. *amasiaca*, and *S. viscosa*) were investigated by DPPH free radical scavenging activity, metal-chelation capacity, and FRAP assays. As a result the EtOH extracts exhibited better FRAP and DPPH scavenging activity [32].

The pure compounds isolated from CH_2Cl_2 extract of the aerial parts of *S. chrysophylla* Stapf were studied for lipid peroxidation inhibitory and the DPPH free radical scavenging activity to establish their antioxidant potential. Among the tested five compounds (ursolic acid, oleanolic acid, β-sitosterol, sclareol, salvigenin), oleanolic acid showed better lipid peroxidation inhibitory activity than the others, and followed by ursolic acid. Except ursolic acid, none of the isolated compounds showed DPPH radical scavenging activity [84].

Methanolic extracts of 39 different populations of three species of *Salvia* (*S. fruticosa* Mill. 20 sample, *S. pomifera* Mill. 4 sample and *S. tomentosa* Mill. 15 sample) were analyzed for their antioxidant properties. Samples were collected from different natural ecological areas in Marmara Region in Turkey. Antioxidant activity was analyzed by DPPH radical scavenging method and expressed as Trolox equivalent (TE). The antioxidant values of the species ranged from 287.57 to 450.51 μmol TE/100 g dry weight (DW) [133].

In vitro antioxidant activity of the ethanol, hexane and aqueous extracts of *Salvia sclarea* L. was evaluated. Total antioxidant status was determined by using Rel assay diagnostics TAS assay kit (Lot. RL024) by Multiscan FC. The best antioxidant effect was seen in ethanolic *S. sclarea* extract [147].

Antioxidant capacity of the three *Salvia* species and their major molecule rosmarinic acid was evaluated, and the antioxidant capacity of different extracts of *S. trichoclada, S. verticillata* and *S. fruticosa* was determined with DPPH test assay. All the extracts and rosmarinic acid showed high radical scavenging capacity [134].

The antioxidant potential of the *S. trichoclada* methanolic extract and the isolated compounds (lupeol, lupenone, glochidone, monogynol A, oleanolic acid, ursolic acid, β-sitosterol, apigenin-7-*O*-β-D-rhamnoside, and rosmarinic acid) were carried out by using lipid peroxidation inhibitory (β-carotene bleaching method) and DPPH free radical scavenging activity test assays. Among all the tested compounds, rosmarinic acid exhibited highest antioxidant activity and strong anticholinesterase activity [21].

In a recent study, the antioxidant capacity of methanolic extracts of three *Salvia* L. species; *S. brachyantha* (Bordz.) Pobed, *S. aethiopis* L., and *S. microstegia* Boiss. and Bal., was evaluated through reducing assays of cupric ions (Cu^{2+}) and ferric ions (Fe^{3+}) (FRAP) beside DPPH assay, and they showed weak-moderate activity, in general [117].

Antioxidant activity of another endemic *Salvia* specie, *Salvia cadmica* Boiss. preparing EtOAc, MeOH and water extracts was evaluated through DPPH free radical scavenging, cupric ion reducing activity (CUPRAC), metal ions (Fe^{2+}) chelating effect and phosphomolybdenum assays. The MeOH extract exhibited strong activity in phosphomolybdenum, DPPH radical scavenging and CUPRAC assays, (311.96, 54.71, and 59.89 μmol TEs/g dry plant, respectively) and the water extract showed remarkable effect in Fe^{2+} chelating assay, ABTS radical scavenging and FRAP assays (8.82 μmol EDTAEs/g dry plant, 102.23, and 97.96 μmol TEs/g dry plant, respectively) [97].

In a very recent study, MeOH and EtOAc extracts of *Salvia aethiopis* L. and *S. ceratophylla* L. extracts were investigated for in vitro antioxidant activity beside cytotoxicity activity. HPLC results allowed to make a correlation between antioxidant capacity and quantity of phenolic acids in the both extracts which contain rosmarinic acid as the main phenolic acid having strong antioxidant activity [121].

In some studies, antioxidant activity of *Salvia* extracts were compared with those of other plant extracts. These type of studies were not presented herein [137], except a few notable ones. In one of them, antioxidant activity of the water extract of *S. triloba* L. was compared with water extracts of tilia (*Tilia argentea* Desf ex DC) and two Turkish black teas commercially called Rite tea and young shoot tea (*Camellia sinensis*) was compared. For this purpose, 100 μg portions of the extracts were added into test samples. All samples were able to show statistically significant antioxidant effect. Both of the tea extracts showed highest antioxidant activities, nevertheless, differences between tilia and sage and tilia and tea were not statistically significant. Unlike antioxidant activity, the highest reducing power activity was shown by the sage extract [135].

4.3 Enzyme Inhibitory Activity

In the treatment of Alzheimer's disease (AD) two types of drugs are used throughout the world; cholinesterase inhibitors and *N*-methyl D-aspartate (NMDA) receptor antagonists. There are four cholinesterase inhibitors approved by FDA including tacrine, donepezil, rivastigmine, and a natural drug galantamine which might cure patients with weak to moderate AD. Memantine, is the only NMDA receptor antagonist approved by the FDA in the treatment of patients with moderate to severe AD. Since none of them provides a satisfactory treatment for Alzheimer's disease studies are still going on to find new potential drugs from both synthetic chemicals and natural sources. Potential anti-Alzheimer activity of the extracts and constituents against two enzymes acetylcholinesterase (AChE) and butyrylcholinesterase (BChE) was measured by a spectrophotometric Ellman method in vitro.

In vitro anticholinesterase activity of 56 Anatolian sage extracts prepared with petroleum ether (PE), $CHCl_3$, EtOAc and MeOH obtained from 14 *Salvia* species (*S. albimaculata* Hedge and Hub, *S. aucheri* Bentham var. *canescens* Boiss. and Heldr, *S. candidissima* Vahl. ssp. *occidentalis*, *S. ceratophylla* L., *S. cryptantha* Montbret and Bentham, *S. cyanescens* Boiss. and Bal., *S. frigida* Boiss, *S. forskahlei* L., *S. halophila* Hedge, *S. microstegia* Boiss. and Bal., *S. multicaulis* Vahl., *S. sclarea* L., *S. syriaca* L., *S. verticillata* L. ssp. *amasiaca*) growing in Turkey. Anticholinesterase effect of the extracts was tested against both AChE and BChE enzymes at concentrations of 0.2 and 1 mg/mL using a microplate-reader assay based on the Ellman method. Against AChE, most of the extracts did not show any activity, while against BChE, the $CHCl_3$ extracts had moderate inhibition between 47.7 and 74.7% at 0.2 mg/mL concentration. The high inhibition on AChE enzyme was observed for the PE extract of *Salvia albimaculata* (endemic) (89.4%), and $CHCl_3$ extract of *S. cyanescens* (80.2%) at 1 mg/mL. The EtOAc extracts of *S. frigida* and *S. microstegia*, $CHCl_3$ extracts of *S. candidissima* ssp. *occidentalis* and *S. ceratophylla*, as well as PE extract of *S. cyanescens* strongly inhibited BChE enzyme (92.2, 89.6, 91.1, 91.3, and 91.8%, respectively) [122].

The anticholinesterase activity of the crude extract of *S. poculata* and the isolated compounds (sclareol, ursolic acid, β-sitosterol, and flavonoids 5-hydroxy-7,4′-dimethoxyflavone, cirsimaritin, eupatilin, and salvigenin) against the enzymes AChE and BChE was determined. The crude extract of *S. poculata* exhibited a significant BChE inhibitory activity and isolates ursolic acid and cirsimaritin possessed high BChE inhibitory activity [86].

The hydroalcoholic extracts of three traditionally used Lamiaceae species for memory-enhancement; *S. triloba* L., *Melissa officinalis* L., and *Teucrium polium* L., were assessed for their in vivo antiamnesic activity and anticholinesterase activity along with antioxidant activities. Scopolamine-induced antiamnesic activity was determined in mice by passive avoidance test, while anticholinesterase effect was measured by spectrophotometric Ellman method at 0.25, 0.50, 1.0, and 2.0 mg/mL. It should be emphasized that *Salvia triloba* was found to have the highest antiamnesic effect at 100, 200, and 400 mg/kg doses having 22.7, 57.1, and 71.4% of relative effects, respectively. However, in the anticholinesterase assay, the three extracts showed similar inhibition against AChE and *Teucrium polium* had the highest inhibition [137].

One of the most active Anatolian *Salvia* plants was found as *S. potentillifolia* showing high antioxidant, antimicrobial and anticholinesterase activities. The essential oil and the ethanol extract of *S. potentillifolia* exhibited moderate BChE activity at 200 μg/ml concentration, while they showed almost no activity against AChE enzyme. In contrast, the major compound of the essential oil, α-pinene, demonstrated AChE inhibitory activity with an IC_{50} value of 81.7 ± 0.96 μM, and showed almost no activity against BChE. Interestingly, β-pinene displayed no inhibitory activity on either enzyme at a concentration of 200 μM [37], which was supported the results of Miyazawa and Yamafuji indicating that α-pinene is more active than β-pinene [159].

Inhibitory potential of the CH_2Cl_2, EtOAc, and MeOH extracts prepared from 55 Turkish *Salvia taxa*, 28 of which are endemic, were tested for their AChE inhibitory activity at 25, 50, and 100 μg/mL using an ELISA microplate reader. Among the total 165 *Salvia* extracts at 100 μg/mL concentration, only the CH_2Cl_2 extract of *S. fruticosa* (*S. triloba*) could show inhibition (51.07%) towards AChE [22]. In fact, *S. fruticosa* is a medicinal plant of Anatolia, and numerous activity studies were carried out with it. In the studies making a comparision with other *Salvia* species, it showed highest effect in the analyzing of different activities (antioxidant, cytotoxic, anti-inflammatory and anticholinesterase).

The inhibitory activity against AChE and BChE enzymes of CH_2Cl_2, EtOAc, and EtOH extracts of 3 wild-grown and a cultivated samples of *S. fruticosa* was tested comparatively by the Ellman method. The CH_2Cl_2 extract of all fractions and subfractions obtained by activity-guided fractionation were tested, and inactive subfractions were discarded. Among the extracts, the most active ones against AChE and BChE were found to be the CH_2Cl_2 and EtOAc extracts of the cultivated plant with inhibition 35.97 and 35.78%, respectively. All the extracts of the wild-grown species had a lower inhibitory effect than those of the cultivated species [131].

The MeOH extract of *S. fruticosa* Mill., hexane and CH_2Cl_2 subextracts, obtained by re-extraction from the MeOH extract along with the isolated triterpenoids, were evaluated for the anticholinesterase potential against AChE and BChE enzymes. The MeOH extract exhibited the highest anticholinesterase activity, surpassing the other two tested extracts and all pure compounds α-amyryltetracosanoate, oleanolic acid, ursolic acid, and sitosterol acetate [30].

The EA and MeOH extracts from 16 *Salvia* L. species were screened for their inhibitory activity against AChE, BChE, as well as lipoxygenase, and tyrosinase enzymes linked to neurodegeneration. While the extracts exerted weak anticholinesterase activity (5.32% for AChE and 44.12% for BChE), the highest inhibition on BChE enzyme was observed for the EtOAc extract of *S. cassia* (44.12%). The extracts displayed either no or very low inhibition towards tyrosinase enzyme at 100 μg/mL. However, they showed high inhibition against lipoxygenase enzyme in general at 100 μg/mL (13.07–74.21%) and the best inhibition was seen for the EtOAc extract of *S. sericeo-tomentosa* var. *hatayica* [116].

Dichloromethane extract and its isolated compounds (sclareol, β-sitosterol, salvigenin, oleanolic acid and ursolic acid) of *S. chrysophylla* were evaluated for their anticholinesterase activity, carried out by the Ellman assay against both enzymes AChE and BChE. Diterpene sclareol exhibited fairly good activity against both the enzymes which is noticeable. Furthermore, the two triterpenoids oleanolic and ursolic acids exhibited selective activity against AChE [84].

The extracts (CH_2Cl_2, EtOAc, and EtOH) prepared from the aerial parts and/or roots of 14 *Salvia* species (*S. argentea*, *S. bracteata*, *S. caespitosa*, *S. cryptantha*, *S. glutinosa*, *S. indica*, *S. microstegia*, *S. multicaulis*, *S. pinnata*, *S. quezelii*, *S. syriaca*, *S. tobeyi*, *S. verticillata* subsp. *amasiaca*, and *S. viscosa*) were investigated for anticholinesterase potential using the Ellman method at 25, 50, and 100 μg/mL concentrations. Cholinesterase enzymes inhibition results of the tested extracts were not promising, only CH_2Cl_2 extract of the aerial parts of *S. cryptantha* was shown moderate inhibition on AChE (56.22%) [32].

In continuation of our studies on the genus *Salvia* fourteen alcoholic extracts of *Salvia* species were evaluated for their anticholinesterase activity. Among them, the most active extract at 200 μg/mL was found to be *S. staminea* with the inhibition of 55.17 and 79.75% against enzymes AChE and BChE, respectively. *S. chrysophylla* and *S. poculata* extracts followed it. Pure isolated compounds from *S. staminea* were investigated for AChE activity at 200 μM, and the abietane diterpenoids ferruginol and taxodione were found to be the most promising isolates, which were then further investigated at five different concentrations, and both diterpenoids exhibited high activity, particularly against BChE enzyme with IC_{50} values of 7.73 and 10.52 μM, respectively [8].

Isolated triterpenoids (lupeol, lupenone, glochidone, monogynol A, oleanolic acid, ursolic acid), and *β*-sitosterol, apigenin-7-*O*-*β*-D-rhamnoside, and rosmarinic acid from the MeOH extract of the aerial parts of *S. trichoclada* were studied for their anticholinesterase activity. The activity was investigated by the Ellman method against both AChE and BChE enzymes. Lupeol, glochidone, oleanolic acid, ursolic acid, apigenin-7-*O*-*β*-D-rhamnoside, rosmarinic acid showed fairly high anticholinesterase activity almost at all concentrations (25, 50, 100, 200 μM), particularly oleanolic acid and ursolic acid as well as glochidone. The highest and selective AChE inhibitory activity was observed for a flavone glycoside apigenin-7-*O*-*β*-D-rhamnoside. Glochidone was found to be the most active one among the isolated lupanes, even though it exhibited a close activity to that of standard compound galanthamine [21].

In another study, AChE inhibitory activity of three *Salvia* species including *S. trichoclada*, *S. verticillata*, and *S. fruticosa* was determined by in vitro Ellman and in vivo isolated guinea pig ileum methods. The AChE inhibitory activity of rosmarinic acid was also determined by in silico docking and isolated guinea pig ileum methods. The MeOH extract of *S. trichoclada* showed the highest inhibition on AChE and significant contraction responses on the isolated guinea pig ileum, rosmarinic acid as well. Rosmarinic acid showed promising results in both in vitro and in vivo methods tested docking results of rosmarinic acid verified these results showing high affinity to the selected target AChE [134].

Inhibitory properties on different enzymes of some Anatolian *Salvia* species were also investigated by several groups in Turkey and some recent ones were mentioned below shortly.

Salvia cadmica Boiss. EtOAc, MeOH and water extracts were analyzed for their inhibitory activities on the enzymes AChE, BChE, α-amylase, α-glucosidase and tyrosinase. According to the results of this assays, the extracts showed no activity on AChE, BChE and tyrosinase. In contrast, MeOH extract showed considerable activity on α-amylase and α-glucosidase enzymes (102.28 and 869.21 μmol ACEs/g dry plant, respectively) [97].

4.4 Cytotoxic Activity

Since 1990 Ulubelen and Topcu investigated some pure isolates for cytotoxic activity in vitro.

Cytotoxic activity of the isolated new compound 1-oxo-7α-hydroxysitosterol and known one 7α-hydroxysitosterol were tested against P-388 murine lymphocytic leukemia and KB systems. 7β-hydroxysteroids are reported to be highly active while their 7α-hydroxy counterparts have little or no cytotoxic activity. In order to confirm this finding the cytotoxic activity tests in P-388 and KB test systems were performed for two of compounds having a 7α-hydroxyl group and both showed marginal activity [88].

Cytotoxic activity of the abietanes (6α-hydroxysalvinolone, 6α-hydroxytaxodone, ferruginol, saprorthoquinone, 11,12-dioxoabieta-8,13-diene, taxodione, hypargenin A, hypargenin D) isolated from the roots of *Salvia hypargeia* and the crude extract were tested against a panel of human cancer cell lines [human breast cancer (BC 1), human lung cancer (LU 2), human colon cancer (COL 2), human epidermoidal carcinoma in mouth (KB), vinblastine-resistant KB-VI hormone-dependent human prostate cancer (LNCaP)] as well as P388 and ASK cells in culture. The crude extract was active in all the test systems investigated, except for KB. 6α-Hydroxysalvinolone and taxodione were also found to mediate a generalized cytotoxic response against the panel of human cancer lines with low ED_{50} values ranging 0.7–10.1 μg/mL [34].

S. staminea methanolic extract and *S. staminea* acetone extract and its five diterpenoids; ferruginol, taxodione, microstegiol, cryptojaponol, salvibretol, 1-oxosalvibretol, and four triterpenoids; salvistamineol, salvistamineol 3-acetate, salvinemorol, oleanolic acid 3-acetate were tested against a panel of BC1, LU1, COL2, KB, KB-VI, LNCaP, P388 and A2780 human cell lines. All the tested compounds and two extracts showed more or less cytotoxicity against A2780 cell lines except microstegiol and oleanolic acid 3-acetate. Taxodione was the only compound showed cytotoxic activity on all the tested cell lines with best IC_{50} value ranging between 0.3 and 9.0 μg/mL [45].

Phytochemical studies carried out by Topcu et al on an endemic Anatolian *Salvia* specie *S. kronenburgii*, afforded a new and nine known diterpenes along with nine new and seven known triterpenoids [16, 49]. Of the total of 15 triterpenes isolated from this plant, 11 were tested against the A2780 human ovarian cancer cell line, but all were found to be weakly cytotoxic to this cell line. The most active one was salvinemorol, with an IC_{50} of 8.6 μg/mL, consistent with its selective cytotoxicity. The most abundant triterpenoid was 1*β*,2*α*,3*β*,11*α*-tetrahydroxyurs-12-ene, and this was submitted for testing to the NCI developmental therapeutics program. It was found to be selectively active against only 3 of tested 60 cancer cell lines, showing growth inhibitions (log GI_{50}) of −7.66 in nonsmall cell lung cancer (EKVX), −6.44 in renal cancer (RXF 393), and −4.73 in breast cancer (T-47D) [16].

In a later study, a new rearranged abietane diterpenoid eriophoroxide was obtained from the root extract of *Salvia eriophora* Boiss. et Kotscshy, and its DNA damaging, cytotoxic and antioxidant activities of the plant extract were investigated [78].

Over 200 Turkish plant extracts from different families were investigated for their DNA damaging activity against three yeasts YCp50 Galactose, RAD52 Galactose and RAD52 Glucose. Among them, the most active four extracts were selected for further studies, one of them was the *S. hypargeia* root extract [167].

Among the screened 16 *Salvia* extracts, *S. hypargeia* Fisch. et Mey. root acetone extract showed the highest activity (IC_{50} = 15.5 μg/mL) against the A2780 human ovarian cancer cell line as well as *S. fruticosa* dichloromethane extract (IC_{50} = 17.2 μg/mL). Then, bioactivity-guided fractionation of *S. hypargeia* extract has yielded four abietane-type diterpenes which are 5,6-didehydro-7-hydroxytaxodone, 6-hydroxysalvinolone (14-deoxycoleon U), demethylcryptojaponol, and salvicanaric acid. Among dose dependently tested abietanes demethylcryptojaponol, 6-hydroxysalvinolone and taxodione were shown significant activity against A2780 human ovarian cancer cell lines with IC_{50} values of 1.2, 3.9 and 9.0 μg/mL, respectively. And surprisingly, fatty acid mixture of the obtained from the first fractions of the same extract was found to be most active part in the test system (IC_{50} = 0.6 μg/mL) [20].

The cytotoxic potential of five Lamiaceae family plants *Salvia tomentosa*, *S. sclarea*, *S. dichroantha* and *Nepeta nuda* subsp. *nuda* and *Marrubium astracanicum* subsp. *macrodon* extracts on L6 cells was evaluated. The extracts displayed low or no cytotoxicity towards mammalian L6 cells [160].

In another study, *Salvia cryptantha* extracts were studied for cytotoxicity against the breast cancer cell lines. In-vitro apoptosis studies of breast cancer cells were performed by Annexin V staining in flow cytometry analyses, in-vivo anticancer activity test was carried out by inhibiting the growth of mammary tumor in rats. The extract showed cytotoxicity to three cancer cell lines. Annexin-positive cells level in *S. cryptantha* extracts treated cell lines were higher than the untreated control cells. In vivo studies showed that mean tumor volume inhibition ratio in *S. cryptantha* extract treated group was 38% compared with the untreated rats [126].

The genotoxic and antigenotoxic effects of Anatolian (Mediterranian) *S fruticosa* leaf extract with the absence and presence of S9 mix using sister chromatid exchange, chromosome aberration and micronucleus formation test systems in human peripheral blood lymphocytes were treated with 1.5-, 3.0- and 6.0-μL/mL concentrations for 24- and 48-h treatment periods. The cytotoxicity of the leaf extract was also investigated by calculating the mitotic index, proliferation index and nuclear division index. Extract had no cytotoxic effect; however, it induced the cytotoxicity of cyclophosphamide [140].

In a study on cytotoxicity of HT-29 colon cancer cell lines, *S. fruticosa* extract, rosmarinic acid and caffeic acid treatments were investigated by XTT and TBE assays by dose and time dependent manner. Cell proliferation of HT-29 cells was inhibited moderately by *S. fruticosa* extract, rosmarinic acid and caffeic acid with an IC_{50} values of 185, 148 and 149 μg/mL for 48 h incubation, respectively and 229, 184 and 181 μg/mL for 72 h incubation, respectively in XTT assay, indicating that HT-29 cells increase their carcinogenic activity depending on the subjected time [138].

In another study on the same plant cytotoxic, apoptotic and anti-angiogenic effects of methanolic extract of *S. fruticosa* was investigated on prostate cancer cells. Cell viability was evaluated by XTT; apoptosis was investigated by DNA fragmentation and caspase 3/7 activity assays. Changes in the angiogenic cytokine levels were investigated by human angiogenesis antibody array. Scratch assay was used to determine the cell motility. The *S. fruticosa* extract induced cytotoxicity and apoptosis in a concentration-dependent manner in both cancer cells; however, it was not cytotoxic to normal cells. Cell motility was reduced in PC-3, DU-145 and HUVEC cells by STE treatment [139].

Cytotoxicity activities of MeOH and EtOAc extracts of *S. aethiopis* L. and *S. ceratophylla* L. were evaluated. The proliferation of the cells was assessed by the MTT assay. Viability percentage of the extracts was determined relative to controls and measured on 15.6–1000 μg/mL extract concentrations. The most toxic extract was found to be *S. aethiopis* EtOAc extract with high phenolic content [121].

4.5 Cardiovascular Activity

S. milthiorrhiza, is originally a Chinese medicinal herb, called Danshen or Tanshen and commonly used in China and Far East Asian Countries, especially for its cardiovascular effects since for hundred years.

About one hundred components are found in *S. miltiorrhiza* roots, which are classified into two groups: hydrophilic phenolic acids and lipophilic tanshinones. More than 200 individual compounds have been isolated and characterized from *S. miltiorrhiza*, which exhibited various pharmacological activities targeting different pathways for the treatment of cardiovascular diseases in various animal and cell models [161]. Among them, particulary tanshinones which have fully aromatic abietane structure can cause coronary vasodilatation, suppress thromboxane formation, inhibit platelet adhesion and aggregation, and scavenge free radicals [161].

Since Turkish *Salvia* species rich in abietane diterpenoids, Ulubelen et al., have studied cardiovascular effects of some *Salvia* species including mainly three species (*S. syriaca*, *S. amplexicaulis*, *S. eriophora*), their crude extracts and isolated compounds were found to have more or less active especially abietane diterpenoids and steroids.

Investigation of vasodepressor effect of the extracts and their compounds from these three *Salvia* species, an experimental model was applied. The prepared three extracts and isolated 26 compounds from them were tested for cardiovascular activity. Each sample was dissolved in the least amount of ethanol and the solution was diluted to 25% with saline adjusting to 4 mL. The experimental study was carried out on adult Wistar Albino rats (250–300 g) as two groups. The control group received ethanol diluted to 25% with saline, while the test group received single compounds intravenously in doses calculated from a dose response curve. A significant reduction in the direct pressure was observed, together with a slight increase in the heart rate, which did not reach a significant level.

The roots crude extract, the known terpenoids and a new rearranged diterpene salvisyrianone isolated from the roots of *S. syriaca* were tested for antihypertensive activity, and the extract as well as the two isolates ferruginol and 3*β*-hydroxystigmast-5-en-7-one showed antihypertensive activity. However, they had no significant effect on heart rate [35, 68].

S. amplexicaulis Lam. crude extract and its isolates, 7-oxo-abieta-9,12,14-triene, ferruginol, and stigmast-4-en-3-one showed a vasodepressor effect on Wistar Albino rats [73].

S. eriophora Boiss. et Kotschy. crude extract and isolated compounds; 4,14-dihydroxy-saprorthoquinone, aethiopinone, salvipisone, 4,12-dihydroxysapriparaquinone, 12-hydroxysapriparaquinone, 3,12-dihydroxysapriparaquinone-1-ene, 6,7-dehydroroyleanone, salvilimbinol, ferruginol, horminone, and 7-*O*-acetylhorminone were tested on Wistar Albino rats for their cardiovascular activities. Ferruginol, aethiopinone, 4,12-dihydrosapriparaquinone, 6,7-dehydroroyleanone, and 4,14-dihydroxysaprorthoquinone showed vasodepressor effect [56].

As a result of cardiovascular studies, ferruginol was found as one of the vasodepressor constituents, and antihypertensive activity was also observed for ferruginol and 3*β*-hydroxystigmast-5-en-7-one. The compounds 7-oxo-abieta-9,12,14-triene, stigmast-4-en-3-one, 4,14-dihydroxysaprorthoquinone, aethiopinone, 4,12 dihydroxysapriparaquinone and 6,7-dehydroroyleanone and the plant extracts showed antihypertensive activity, while the other samples were not found to be active [35].

In addition to terpenic compounds of Anatolian *Salvia* species, the vascular effect of salvigenin (6-hydroxyapigenin 6,7,4′-trimethyl ether), a natural flavone, was investigated in comparison with another flavone, 6-hydroxyluteolin 6,7,3′,4′-tetramethyl ether in rat aotic rings. The maximal relaxations induced by flavones were similar, however, based on their pEC_{50} values salvigenin displayed a higher potency than 6-hydroxyluteolin 6,7,3′,4′-tetramethyl ether. Endothelium removal markedly reduced the relaxations to salvigenin while the responses to 6-hydroxyluteolin 6,7,3′,4′-tetramethyl ether were partially affected. In addition, a significant decrease was observed in maximal responsiveness and sensitivity to flavones in the presence of L-NOARG, a NO synthase inhibitor. The cyclooxygenase inhibitor indomethacin significantly inhibited the relaxations to salvigenin, but not altered the responses to 6-hydroxyluteolin 6,7,3′,4′-tetramethyl ether [162].

It is concluded that a marker flavone of Anatolian *Salvia* species salvigenin with three methoxy group was found to be effective flavone in causing vasorelaxation which appears to be mediated by endothelium derived NO and prostacyclin. Whereas, the four methoxylated flavone 6-hydroxyluteolin 6,7,3′,4′-tetramethyl ether induced relaxant responses are partially endothelium, presumably NO mediated.

4.6 Analgesic Activity

In Turkish folk medicine, the aerial part of *S. wiedemannii* Boiss. has been used for treatment of peptic ulcers and pain relieving agent. To evaluate the analgesic effect of *S. wiedemannii*, tail flick and acetic acid-induced writhing tests were applied in mice. The $CHCl_3$ extract (500 mg/kg, i.p.) obtained from *S. wiedemannii* showed significant analgesic activity on tail flick assay, while water, EtOH and butanol extracts of the plant had no activity on the same test. $CHCl_3$ extract (500 mg/kg, i. p.) also inhibited number of writhings induced by acetic acid. As a result, $CHCl_3$ extract provided analgesic effects similar to morphine and its effect was quick and durable. This in vivo study demonstrates that *S. wiedemannii* has strong analgesic effect which verified its folkloric use [154].

Kappa Opioid Peptide Receptor (KOPr) activation produces analgesic, psychotomimetic, diuretic and antipruritic effects. KOPr ligands are investigated for their potential roles in the treatment of addiction, depression, feeding behavior, psychosis and schizophrenia. The methanolic extracts of a number of *Salvia* species native to Turkey (*S. tomentosa*, *S. tchihatcheffii*, *S. rosifolia*, *S. dichroantha* and *S. sclarea*) were tested for their potential binding to opioid receptors in rat brain membranes and Chinese Hamster Ovary Cells expressing human KOPr (CHO-KOPh). [^{3}H] Diprenorphine, an unselective opioid antagonist, was utilized in the radioligand receptor binding assays. All extracts (0.11 mg/mL) inhibited the [^{3}H] diprenorphine binding with ranging KOPr binding affinities. *S. dichroantha* and *S. sclarea* methanolic extracts showed more than 50% inhibition of diprenorphine binding both in rat brain membranes and CHO-KOPh membranes. Among all the extracts *S. sclarea* deserves further investigation for its active component(s) and its pharmacological characterization [163].

4.7 Other Activities

In vitro amoebic effectiveness of methanolic extract of *S. staminea* and *S. caespitosa* was investigated against *Acanthamoeba castellanii*, as well as their cytotoxicity on corneal cells in vitro. The effect of *Salvia* species, with concentrations ranging 1.0–32.0 mg/mL, on the proliferation of *A. castellanii* trophozoites and cysts were examined in vitro. According to the inverted light microscope analysis results, *S. staminea* showed remarkable amoebicidal effect on *A. castellanii*. For determining the cytotoxicity of *Salvia* species on corneal cells, agar diffusion tests were performed. In the case of the cytotoxic activity, methanolic extract of *S. staminea* showed no cytotoxicity on corneal cells with a concentration of 16 mg/ mL [36].

The in vitro antiprotozoal activities of crude methanolic extracts from the aerial parts of five Lamiaceae plants (*Salvia tomentosa, S. sclarea, S. dichroantha, Nepeta nuda* subsp. *nuda* and *Marrubium astracanicum* subsp. *macrodon*) were evaluated against four parasitic protozoa, i.e. *Trypanosoma brucei rhodesiense, T. cruzi, Leishmania donovani* and *Plasmodium falciparum*. Melarsoprol, benznidazole, miltefosine, chloroquine and podophyllotoxin were used as reference drugs. All crude methanol extracts showed antiprotozoal potential against at least three parasites, so they were dispersed in water and partitioned against n-hexane and to yield three subextracts that were screened in the same test systems. Chloroform extracts of *S. tomentosa* and *S. dichroantha* showed significant activity against *L. donovani*. All extracts displayed in vitro antimalarial and moderate trypanocidal activities against *T. cruzi* with the n-hexane extract of *S. sclarea* being the most active against the latter [160].

To evaluate wound healing effect of the ethanol extracts of *Salvia cryptantha and* Salvia *cyanescens* linear incision and circular excision wound models were employed on rats and mice. The wound healing effect was comparatively evaluated with the standard skin ointment Madecassol®. Inhibition of tyrosinase, a key enzyme in skin aging, was achieved using ELISA microplate reader. The ethanol extract of *S. cryptantha* treated groups of animals showed 56.5% contraction, whereas the reference drug Madecassol® showed 100% contraction. On the other hand, the same extract on linear incision wound model demonstrated a significant increase (33.2%) in wound tensile strength as compared to other groups. Antioxidant activity was further evaluated by DPPH and superoxide radical scavenger effect, ferrous ion-chelating ability, and FRAP tests [164].

Water, methanol *n*-butanol, acetone, and chloroform extracts from three *Salvia* species (*S. fruticosa, S. verticillata*, and *S. trichoclada*) were screened for their anti-inflammatory activity using in vivo experimental models in rats to confirm their ethnobotanical usage. For this purpose, a carrageenan-induced inflammatory paw edema model was used. All extracts demonstrated anti-inflammatory activities, n-butanol extract of *S. fruticosa* was found to be the most active one [128].

The aerial parts extracts of *S. halophila* and *S. virgata* were studied in *p*-benzoquinone induced writhing reflex for the assessment of antinociceptive activity and carrageenan-induced hind paw edema and 12-*O*-tetradecanoyl-13-acetate-induced ear edema models in mice for the anti-inflammatory activity. The aerial parts of *S. halophila* and *S. virgata* were subjected to Soxhlet extraction with different solvents such as n-hexane, ethyl acetate, methanol and aqueous methanol (50%) as well as water extraction under reflux. Results have shown that methanol extract of *S. virgata* significantly inhibited carrageenan-induced paw edema and *p*-benzoquinone-induced writhing reflex at 100 mg/kg dose, while this extract showed no effect in the TPA-induced ear edema. In contrast, the other extracts did not show any inhibitory antinociceptive and anti-inflammatory activities [165].

Another bioactivity study was carried out to investigate cytotoxic, antioxidant, iNOS, AChE, BChE inhibitory and antimicrobial activities of the ethanol extract of *Salvia veneris*, an endemic specie to the Northern Cyprus. The cytotoxicity of the extract was evaluated against PC3, HeLa, CaCo-2, MCF-7, U87MG, HEK293, mPanc-96 cell lines. Highest activity was observed against PC3, mPanc-96 and HeLa cell lines 6.158, 8.704 and 9.907 µg/mL (IC_{50}) respectively. The extract produced 11.00 µg/mL iNOS inhibitory activity which is higher concentration than the positive control parthenolide 0.6 µg/mL (IC_{50}) [166].

Until now, three patents were obtained related to the isolation of the new compounds and their activity from *Salvia* species by Topcu group, one was on "Rearranged norditerpenoids and diterpenoids as anti-tuberculosis agents" and the other two were on "Inhibition of NF-kB activity by triterpene compounds".

5 Conclusions and Future Prospects

60 Anatolian *Salvia* species have been investigated which afforded mainly abietane diterpenoids and oleanane and ursane triterpenoids as well as some flavonoids and other phenolics.

Sesquiterpenes have been isolated from only 9 *Salvia* species, and spathulenol was present in most of them.

The genus *Salvia* plants are very rich in abietane diterpenoids with high biodiversity, among them horminone and 7-*O*-acetylhorminone were dominant abietanes in Anatolian *Salvia* species. Abietane diterpenes followed by pimaran and labdane diterpenoids found in Anatolian *Salvia* species. Also, rearranged abietane diterpenoids were isolated. As representative examples to the rearranged abietanes with opened ring A (1-oxo and 3-oxosalvipisone), and seven membered ring A (microstegiol, salvibretol) and eight membered ring A (candidissiol) can be given. Also norabietanes which consist of fully aromatic three rings, such as multicaulin and multiorthoquinone and their derivatives, were isolated with remarkable anti-tuberculous activity from *S. multicaulis* and a few *Salvia* species growing in Turkey.

Sesterterpenoids are very rare in nature and were isolated only from a few Anatolian *Salvia* species, particularly *S. yosgadensis* afforded several norditerpenes and norsesterterpenes which is an endemic specie to the Central Anatolia. The new ones from this specie were two sesterterpene lactones, named yosgadensolide A and yosgadensolide B and two dinorsesterterpenes, named yosgadensonol and 13-epi-yosgadensonol.

Taxodione and 6-hydroxysalvinolone are the most active abietanes which were isolated from numerous *Salvia* species and they showed high antioxidant, cytotoxic and anticholinesterase activities. Horminone and derivatives also showed high antioxidant (especially radical scavenging) and cytotoxic activities. A series abietane diterpenoids were investigated for anticholinesterase activities. Taxodione and 6-hydroxysalvinolone were found to be fairly active, and cholinesterase inhibitory activity studies on 40 abietane diterpenoids are completed to submit for publication by Topcu group.

Phenolic abietane diterpenoids carnosol, carnosic acid and a phenolic compound rosmarinic acid revealed the highest antioxidant effects, particularly the two mentioned abietanes are promising lead drugs as neuroprotective agents due to strong blood brain circulatory properties.

Since sesterterpenoids were isolated from only a few Anatolian *Salvia* species, their activity have not been investigated yet.

As triterpenic secondary metabolites, ursolic acid and oleanolic acid were the most common triterpenoids in Anatolian *Salvia* species, and followed them with lupeol and its derivatives.

There are many activity studies on the first two types of triterpenoids, however, not enough activity studies were reported on lupanes, except betulinic acid. Glochidone, isolated from *S. trichoclada* was found to be the most active lupane triterpenoid, in terms of anticholinesterase activity, revealing a close activity to that of the commercial drug galanthamine.

Among a series oleanane and ursane triterpenoids was screened for cytotoxic activity against A2780 human ovarian cell lines, salvinemorol showed highest activity (8.6 μg/mL), salvistamineol followed it. A series highly hydroxylated oleanolic and ursolic acid derivatives were isolated from an endemic *Salvia* species *S. kronenburgii* and one of them 1β,2α,3β,11α-tetrahydroxyurs-12-ene was found to be the most active compound against renal, non-small cell lung, and breast cancer cell lines. Later on, the triterpenoids isolated from *S. kronenburgii* were investigated for NF-κappa B activity and among them, 3β-acetoxy-1β,2α,11α-trihydroxyurs-12-ene, and 3β-acetoxy-1β,2α,11α-trihydroyolean-12-ene were found to be most active compounds and they were patented in 2012.

As steroids, β-sitosterol was isolated from almost all *Salvia* species, and salvigenin with 3 methoxy group has a nonpolar structure was a marker flavone for Anatolian *Salvia* species and isolated from most of them.

Activity directed studies are focused by several groups, in the last years. But, unfortunately only a limited number of researchers is continuing to make isolation studies from Anatolian *Salvia* plants. However, activity studies are intensively going on, particularly on the extracts of *Salvia* species for various activities rather than pure isolates, especially antioxidant and anticholinesterase activities.

Until now, three patents were obtained on di- and triterpenoids of Anatolian *Salvia* species, recent one (2012) was a PCT patent entitled "Inhibition of NF-Kb activity by triterpene Compounds" (WO 2012130481 A1).

Flavonoids and phenolics of Anatolian *Salvia* species showed high antioxidant and antiradical activities. However, there are limited number of structure-activity relationship studies including both the extracts and their secondary metabolites. Some methoxylated flavones and flavonols, especially 6-methoxylated ones were found more cytotoxic than the others.

As a result, the future studies should be focused on diterpenoids and phenolic compounds as well as some triterpenoids such as ursolic acid and betulinic acid which are lead drug compounds of *Salvia* species and phase studies are going on them to complete.

References

1. Jash SK, Gorai D, Roy R (2016) Salvia genus and triterpenoids. Int J Pharm Sci Res 7:4710
2. Kintzios SE (2003) Sage: the genus Salvia. CRC Press, USA
3. Lin LZ, Wang X-M, Xiu-Lan H et al (1989) Sapriolactone, a cytotoxic norditerpene from *Salvia prionitis*. Phytochemistry 28:3542–3543
4. Zhang Y, Mao X, Su J et al (2017) A network pharmacology-based strategy deciphers the underlying molecular mechanisms of Qixuehe Capsule in the treatment of menstrual disorders. Chin Med 12:23
5. Chen W (1981) Acta Pharm Sinica 19:876
6. Perry NSL, Bollen C, Perry EK et al (2003) Salvia for dementia therapy: review of pharmacological activity and pilot tolerability clinical trial. Pharmacol Biochem Behav 75:651–659. https://doi.org/10.1016/s0091-3057(03)00108-4
7. Howes MJR, Perry NS, Houghton PJ (2003) Plants with traditional uses and activities, relevant to the management of Alzheimer's disease and other cognitive disorders. Phytotherapy Res 17:1–18
8. Topcu G et al (2013) Investigation of anticholinesterase activity of a series of Salvia extracts and the constituents of *Salvia staminea*. Nat Prod J 3(1):3–9
9. Azizkhani M, Tooryan F, Azizkhani M (2016) Inhibitory potential of *Salvia sclarea* and *Ocimum basilicum* against chemical and microbial spoilage in cheese. J Food Saf 36:109–119
10. Shibly O, Ciancio S, Kazmierczak M et al (1996) Clinical evaluation of the effect of a hydrogen peroxide mouth rinse, sodium bicarbonate dentifrice, and mouth moisturizer on oral health. J Clin Dent 8:145–149
11. Pitten F-A, Kramer A (1999) Antimicrobial efficacy of antiseptic mouthrinse solutions. Eur J Clin Pharmacol 55:95–100
12. Sezik E, Yesilada E (1999) Ucucu Yag Tasıyan Turk Halk Ilacları (Turkish folk medicine containing volatile oils). In: Kirimer N, Mat A (eds) Essential Oils—In Honour of Prof. Dr. K. Hüsnü Can Baser On his 50th Birthday, Anadolu University Press, Eskisehir, pp 98–131
13. Hedge IC (1982) Salvia L. In: Davis PH (ed) Flora of Turkey and the East Aegean Islands, vol 7, (pp 400–461). Edinburgh: Edinburgh University Press
14. Ulubelen A, Topcu G, Terem B (1987) Abietane diterpenoids from the roots of *Salvia cryptantha*. Phytochemistry 26(5):1534–1535
15. Topcu G, Ulubelen A (1996) Abietane and rearranged abietane diterpenes from *Salvia montbretii*. J Nat Prod 59(8):734–737
16. Topcu G et al (2004) Highly hydroxylated triterpenes from *Salvia kronenburgii*. J Nat Prod 67(1):118–121
17. Topcu G, Ulubelen A (2007) Structure elucidation of organic compounds from natural sources using 1D and 2D NMR techniques. J Mol Struct 834:57–73
18. Ulubelen A et al (2001) Antibacterial diterpenes from the roots of *Salvia blepharochlaena*. J Nat Prod 64(4):549–551
19. Ulubelen A, Topcu G, Johansson CB (1997) Norditerpenoids and diterpenoids from *Salvia multicaulis* with antituberculous activity. J Nat Prod 60(12):1275–1280
20. Topcu G et al (2008) Cytotoxic activity of some Anatolian Salvia extracts and isolated abietane diterpenoids. Pharm Biol 46(3):180–184

21. Çulhaoğlu B et al (2015) Antioxidant and anticholinesterase activities of lupane triterpenoids and other constituents of *Salvia trichoclada*. Med Chem Res 24(11):3831–3837
22. Senol FS et al (2010) Survey of 55 Turkish Salvia taxa for their acetylcholinesterase inhibitory and antioxidant activities. Food Chem 120(1):34–43
23. Hatipoglu SD, Zorlu N, Dirmenci T et al (2016) Determination of volatile organic compounds in fourty five Salvia species by thermal desorption-GC-MS technique. Rec Nat Prod 10:659
24. Baser HC, Buchbauer G (2015) Handbook of essential oils: science, technology, and applications, 2nd ed. CRC Press, USA
25. Kunduhoglu B, Kurkcuoglu M, Duru ME et al (2011) Antimicrobial and anticholinesterase activities of the essential oils isolated from Salvia dicroantha Stapf., *Salvia verticillata* L. subsp *amasiaca* (Freyn. and Bornm.) Bornm. and *Salvia wiedemannii* Boiss. J Med Plants Res 5:6484–6490. https://doi.org/10.5897/jmpr11.220
26. Lu Y, Foo LY (2002) Polyphenolics of salvia—a review. Phytochemistry 59:117–140
27. Topçu G (2006) Bioactive Triterpenoids from Salvia Species. J Nat Prod 69:482–487
28. Wu Y-B, Ni Z-Y, Shi Q-W et al (2012) Constituents from Salvia species and their biological activities. Chem Rev 112:5967–6026
29. Ulubelen A, Topçu G (1997) Chemical and biological investigations of Salvia species growing in Turkey. Stud Nat Prod Chem 20:659–718
30. Topçu G et al (2013) Terpenoids, essential oil composition, fatty acid profile, and biological activities of Anatolian *Salvia fruticosa* Mill. Turk J Chem 37(4):619–632
31. Kintzios SE (2003) III. Chemical constituents, 4. Ulubelen A (ed) Terpenoids in genus Salvia. Ser. vol 55, pp (58–71). CRC Press
32. Orhan IE et al (2013) Assessment of anticholinesterase and antioxidant properties of selected sage (Salvia) species with their total phenol and flavonoid contents. Ind Crop Prod 41:21–30
33. Tepe B et al (2004) Antimicrobial and antioxidative activities of the essential oils and methanol extracts of *Salvia cryptantha* (Montbret et Aucher ex Benth.) and Salvia multicaulis (Vahl). Food Chem 84(4):519–525
34. Ulubelen A et al (1999) Cytotoxic activity of diterpenoids isolated from *Salvia hypargeia*. Pharm Biol 37(2):148–151
35. Ulubelen A (2003) Cardioactive and antibacterial terpenoids from some Salvia species. Phytochemistry 64(2):395–399
36. Goze I et al (2009) In Vitro Amoebicidal Activity of *Salvia staminea* and *Salvia caespitosa* on Acanthamoeba castellanii and Their Cytotoxic Potentials on Corneal Cells. J Ocul Pharmacol Th 25(4):293–298
37. Kivrak İ et al (2009) Antioxidant, anticholinesterase and antimicrobial constituents from the essential oil and ethanol extract of *Salvia potentillifolia*. Food Chem 116(2):470–479
38. Aydoğmuş Z, Yeşilyurt V, Topcu G (2006) Constituents of *Salvia microphylla*. Nat Prod Res 20(8):775–781
39. Ulubelen A, Topcu G, Tuzlaci E (1992) New diterpenoids from *Salvia divaricata*. J Nat Prod 55(10):1518–1521
40. Topcu G et al (1996) Sesterterpenes and other constituents of *Salvia yosgadensis*. Phytochemistry 42(4):1089–1092
41. Topcu G et al (1995) Terpenoids and flavonoids from the aerial parts of *Salvia candidissima*. Phytochemistry 40(2):501–504
42. Gökdil G et al (1997) Terpenoids and flavonoids from *Salvia cyanescens*. Phytochemistry 46 (4):799–800
43. Ulubelen A et al (1996) Norsesterterpenes and diterpenes from the aerial parts of *Salvia limbata*. Phytochemistry 43(2):431–434
44. Ulubelen A et al (1994) Terpenoids from *Salvia sclarea*. Phytochemistry 36(4):971–974
45. Topcu G et al (2003) Studies on di-and triterpenoids from *Salvia staminea* with cytotoxic activity. Planta Med 69(05):464–467
46. Topcu G et al (2007) Antioxidant activity tests on novel triterpenoids from *Salvia macrochlamys*. Arkivoc 7:195–208

47. Topçu G, Kartal M, Ulubelen A (1997) Terpenoids from *Salvia tchihatcheffii*. Phytochemistry 44(7):1393–1395
48. Ulubelen A, Tuzlacı E (1987) Terpenoids from *Salvia potentillifolia*. Planta Med 53 (06):578–578
49. Topçu G, Ulubelen A (1999) Terpenoids from *Salvia kronenburgii*. J Nat Prod 62(12):1605–1608
50. Ulubelen A, Sonmez U, Topcu G (1997) Diterpenoids from the roots of *Salvia sclarea*. Phytochemistry 44(7):1297–1299
51. Ulubelen A et al (1995) Abietane diterpenes from *Salvia napifolia*. Phytochemistry 40 (3):861–864
52. Ulubelen A, Tan N, Topcu G (1997) Terpenoids from *Salvia candidissima* subsp. candidissima. Phytochemistry 45(6):1221–1223
53. Ulubelen A, Topcu G (1991) Abietane diterpenoids from *Salvia microstegia*. Phytochemistry 30(6):2085–2086
54. Ulubelen A, Topcu G, Tan N (1992) Rearranged abietane diterpenes from *Salvia candidissima*. Phytochemistry 31(10):3637–3638
55. Topcu G, Eriş C, Ulubelen A (1996) Rearranged abietane diterpenes from *Salvia limbata*. Phytochemistry 41(4):1143–1147
56. Ulubelen A et al (2002) Cardioactive diterpenes from the roots of *Salvia eriophora*. Planta Med 68(09):818–821
57. Topcu G, Ulubelen A (1990) Diterpenoids from *Salvia wiedemannii*. Phytochemistry 29 (7):2346–2348
58. Tan N, Topçu G, Ulubelen A (1998) Norabietane diterpenoids and other terpenoids from *Salvia recognita*. Phytochemistry 49(1):175–178
59. Ulubelen A et al (1999) Diterpenoids from the roots of *Salvia bracteata*. Phytochemistry 52 (8):1455–1459
60. Ulubelen A, Topcu G (1992) New abietane diterpenoids from *Salvia montbretii*. J Nat Prod 55(4):441–444
61. Topcu G, Ulubelen A (1991) Diterpenoids from *Salvia wiedemannii*. Phytochemistry 30 (7):2412–2413
62. Ulubelen A et al (2000) Antibacterial diterpenes from the roots of *Salvia viridis*. Planta Med 66(05):458–462
63. Sonmez U, Topcu G, Ulubelen A (1997) Constituents of *Salvia verticillata*. Turk J Chem 21 (4):376–382
64. Ulubelen A, Topcu G, Tan N (1992) Diterpenoids from *Salvia candidissima*. Tetrahedron Lett 33(47):7241–7244
65. Ulubelen A et al (1998) Diterpenoids and triterpenoids from *Salvia multicaulis*. Phytochemistry 47(5):899–901
66. Goren AC et al (2002) Diterpenoids from *Salvia ceratophylla*. Nat Prod Lett 16(1):47–52
67. Ulubelen A, Miski M, Mabry T (1981) A new diterpene acid from *Salvia tomentosa*. J Nat Prod 44(1):119–124
68. Ulubelen A et al (2000) Cardioactive terpenoids and a new rearranged diterpene from *Salvia syriaca*. Planta Med 66(07):627–629
69. Topcu G et al (1996) Norditerpenes and Norsesterterpenes from *Salvia yosgadensis*. J Nat Prod 59(2):113–116
70. Ulubelen A et al (2001) A new antibacterial diterpene from the roots of *Salvia caespitosa*. Nat Prod Lett 15(5):307–314
71. Ulubelen A (1990) New diterpenoids from the roots of *Salvia triloba*. Planta Med 56 (01):82–83
72. Kolak U et al (2005) Terpenoids and steroids from the roots of *Salvia blepharochlaena*. Turk J Chem 29(2):177–186
73. Kolak US et al (2001) Cardioactive diterpenoids from the roots of *Salvia amplexicaulis*. Planta Med 67(08):761–763

74. Ulubelen A et al (1994) Terpenoids from *Salvia nemorosa*. Phytochemistry 35(4):1065–1067
75. Ulubelen A, Topcu G, Tan N (1995) Diterpenoids from *Salvia heldrichiana*. Phytochemistry 40(5):1473–1475
76. Ulubelen A, Öztürk S, Işıldatıcı S (1968) A new flavone from *Salvia triloba* L.f (Labiatae). J Pharm Sci 57(6):1037–1038
77. Ulubelen A et al (1988) Diterpenoids from the roots of *Salvia hypargeia*. J Nat Prod 51 (6):1178–1183
78. Topcu G et al (2007) Structure elucidation of a new rearranged abietane diterpene from a biologically active plant. *Salvia eriophora*. 2:981–986
79. Ulubelen A, Topcu G (1992) Abietane diterpenoids from *Salvia pomifera*. Phytochemistry 31(11):3949–3951
80. Ulubelen A et al (1996) An abietane diterpene and two phenolics from *Salvia forskahlei*. Phytochemistry 42(1):145–147
81. Ulubelen A et al (1992) Microstegiol, a rearranged diterpene from *Salvia microstegia*. Phytochemistry 31(7):2419–2421
82. Topcu G, Ulubelen A, Eris C (1994) Diterpenoids and tritepenoids from *Salvia pomifera*. Phytochemistry 36(3):743–745
83. Ulubelen A, Topcu G (2000) Salvimultine, a new noricetexane diterpene from the roots of *Salvia multicaulis*. J Nat Prod 63(6):879–880
84. Çulhaoğlu B et al (2013) Bioactive constituents of *Salvia chrysophylla* Stapf. Nat Prod Res 27(4–5):438–447
85. Ulubelen A et al (1985) Terpenoids from *Salvia palaestina*. Phytochemistry 24(6):1386–1387
86. Kolak U et al (2009) Antioxidant and anticholinesterase constituents of *Salvia poculata*. Turk J Chem 33(6):813–823
87. Ulubelen A et al (1991) A new abietane diterpene from *Salvia wiedemannii* Boiss. J Org Chem 56(26):7354–7356
88. Topcu G et al (1997) Terpenoids from *Salvia glutinosa*. Phytochemistry 45(6):1293–1294
89. Ulubelen A, Topcu G, Lotter H et al (1994) Triterpenoids from the aerial parts of *Salvia montbretii*. Phytochemistry 36:413–415
90. Ulubelen A, Miski M, Mabry T (1981) Further flavones and triterpenes and the new 6-hydroxyluteolin 5-β-D-glucoside from *Salvia tomentosa*. J Nat Prod 44(5):586–587
91. Ulubelen A, Topçu G (1984) Flavonoids and terpenoids from *Salvia verticillata* and *Salvia pinnata*. J Nat Prod 47(6):1068
92. Ulubelen A, Topçu G (1984) Triterpenoids from *Salvia pinnata*. Phytochemistry 23(1):133–134
93. Demirezer OL et al (2012) Chemical constituents of two sages with free radical scavenging activity. Nat Prod Comm 7(2):187–190
94. Ulubelen A, Uygur I (1976) Flavonoidal and other compounds of *Salvia aethiopis*. Planta Med 29(04):318–320
95. Ulubelen A et al (1979) Flavonoids of *Salvia tomentosa* (Labiatae). J Nat Prod 42(3):261–263
96. Miski M et al (1983) Antibacterial activity studies of flavonoids from *Salvia palaestina*. J Nat Prod 46(6):874–875
97. Kocak MS et al (2016) *Salvia cadmica*: Phenolic composition and biological activity. Ind Crop Prod 85:204–212
98. Kirmizibekmez H et al (2012) Chemical constituents of *Salvia dichroantha*. Biochem Syst Ecol 42:18–20
99. Dincer C et al (2012) A comparative study on phenolic composition, antioxidant activity and essential oil content of wild and cultivated sage (*Salvia fruticosa* Miller) as influenced by storage. Ind Crop Prod 39:170–176
100. Dincer C et al (2013) Phenolic composition and antioxidant activity of *Salvia tomentosa* Miller: effects of cultivation, harvesting year, and storage. Turk J Agricul Forest 37(5):561–567

101. Tekeli Y et al (2014) Phenolic composition, Antioxidant capacity of *Salvia verticilata* and effect on multidrug resistant bacteria by flow-cytometry. Afr J Tradit Complem Altern Med 11(4):147–152
102. Torun M et al (2014) Concentration of sage (*Salvia fruticosa* Miller) extract by using integrated membrane process. Sep Purif Technol 132:244–251
103. Yeşilyurt V et al (2008) Antioxidant potential and phenolic constituents of *Salvia cedronella*. Food Chem 108(1):31–39
104. Ulubelen A, Ayanoglu E (1975) Flavonoids of *Salvia virgata*. Lloydia 38:446–447
105. Ulubelen A (1989) Virgatol, a new diterpene from the roots of *Salvia virgata*. Planta Med 55:397. https://doi.org/10.1055/s-2006-962043
106. Burnell RH, Desfossés S, Jean M (1993) Concerning montbretyl 12-methyl ether. J Nat Prod 56:627–628
107. Gil RR, Cordell GA, Topçu G et al (1994) Montbretol and salvinolone are identical. J Nat Prod 57:181–185
108. Rustaiyan A, Sadjadi A (1987) Salvisyriacolide, a sesterterpene from *Salvia syriaca*, vol 26, pp 3078–3079
109. Topcu G, Kusman T (2014) Lamiaceae family plants as a potential anticholinesterase source in the treatment of Alzheimer's disease Bezmialem. Science 2:1–25
110. Ulubelen A, Ayanoğlu E (1976) Vergatic acid, a new pentacyclic triterpene from *Salvia virgata*. Phytochemistry 15:309–311
111. Ulubelen A, Brieskorn C (1975) Pentacyclic triterpenic acids: micromeric acid from *Salvia horminum*. Phytochemistry 14:820–821
112. Ulubelen A, Brieskorn CH, Özdemir N (1977) Triterpenoids of *Salvia horminum*, constitution of a new diol. Phytochemistry 16:790–791. https://doi.org/10.1016/S0031-9422(00)89266-3
113. Brieskorn CH, Kapadia Z (1980) Bestandteile von *Salvia officinalis*. Planta Med 38:86–90
114. Abdel-Moneim F, Elgamal M, Fayez M et al (1967) Constituents of local plants—XI.: the triterpenoid acids of *Salvia lanigera* Poir. and *S. triloba* L. Phytochemistry 6:1035–1036
115. Arı Ş, Kolak U, Hasancebi S et al (2001) Characterization of two triterpenes and a steroid from the cultured roots of *Salvia amplexicaulis*. Biotechnol Biotechnol Equip 15:23–26
116. Orhan IE, Senol FS, Ozturk N et al (2012) Profiling of in vitro neurobiological effects and phenolic acids of selected endemic Salvia species. Food Chem 132(3):1360–1367. https://doi.org/10.1016/j.foodchem.2011.11.119
117. Tohma H et al (2016) RP-HPLC/MS/MS analysis of the phenolic compounds, antioxidant and antimicrobial activities of Salvia L. species. Antioxidants 5(4):38
118. Wang LL, Ma RF, Liu CY et al (2017) *Salvia miltiorrhiza*: a potential red light to the development of cardiovascular diseases. Curr Pharm Des 23:1077–1097. https://doi.org/10.2174/1381612822666161010105242
119. Tepe B et al (2006) Screening of the antioxidant potentials of six Salvia species from Turkey. Food Chem 95(2):200–204
120. Tosun M et al (2009) Antioxidant properties and total phenolic content of eight Salvia species from Turkey. Biol Res 42(2):175–181
121. Poyraz İE, Çiftçi GA, Öztürk N (2017) Phenolic contents, in vitro antioxidant and cytotoxicity activities of *Salvia aethiopis* L. and *S. ceratophylla* L. (Lamiaceae). Rec Nat. Prod 11(4):2
122. Orhan I et al (2007) Antioxidant and anticholinesterase evaluation of selected Turkish Salvia species. Food Chem 103(4):1247–1254
123. Ozkan G, Ozcan MM, Al Juhaimi FY (2012) Sage (*Salvia aucheri* Bentham var. canescens Boiss. and Heldr.): essential oil composition, phenolics and antioxidant activity. Asian J Chem 24(5):2225–2227
124. Arslan I, Celik A, Mercan N (2009) Chemical constituents and in vitro antistaphylococcal activities of endemic *Salvia cedronella* and *S. fruticosa* naturally distributed in Denizli (Turkey). Planta Med 75(9):962

125. Tel G et al (2010) Chemical composition of the essential oil and hexane extract of *Salvia chionantha* and their antioxidant and anticholinesterase activities. Food Chem Toxicol 48 (11):3189–3193
126. Ozer H et al (2013) Antitumoral effects of *Salvia absconditiflora* Greuter & Burdet syn. Salvia cryptantha Montbret & Aucher ex Benth. on Breast cancer. Indian J Tradit Knowl 12 (3):390–397
127. Yumrutas O et al (2012) Phenolic acid contents, essential oil compositions and antioxidant activities of two varieties of *Salvia euphratica* from Turkey. Nat Prod Res 26(19):1848–1851
128. Cadirci E et al (2012) Anti-inflammatory effects of different extracts from three Salvia species. Turk J Biol 36(1):59–64
129. Tan N et al (2016) Antimycobacterial and antifungal activities of selected sour Salvia species. Rec Nat Prod 10(5):593
130. Askun T et al (2009) Characterization of the phenolic composition and antimicrobial activities of Turkish medicinal plants. Pharm Biol 47(7):563–571
131. Senol FS et al (2011) Evaluation of cholinesterase inhibitory and antioxidant activities of wild and cultivated samples of sage (*Salvia fruticosa*) by activity-guided fractionation. J Med Food 14(11):1476–1483
132. Erdogan-Orhan I et al (2010) Sage-called plant species sold in Turkey and their antioxidant activities. J Serb Chem Soc 75(11):1491–1501
133. Erdoğan SS, Karık Ü, Başer KHC (2014) The determination of antioxidant activity of some sage populations of in the Marmara Region. Türk Tarım ve Doğa Bilimleri 7(7):1877–1882
134. Demirezer LÖ et al (2015) Molecular docking and ex vivo and in vitro anticholinesterase activity studies of Salvia sp. and highlighted rosmarinic acid. Turk. J Med Sci 45(5):1141–1148
135. Yıldırım A et al (2000) Comparison of antioxidant and antimicrobial activities of Tilia (Tilia argentea Desf ex DC), sage (*Salvia triloba* L.), and Black tea (*Camellia sinensis*) extracts. J Agricul Food Chem 48(10):5030–5034
136. Özcan MM, Özkan G (2015) Determination of antioxidant activity and total phenol contents of two Salvia extracts. Indian J Trad Knowl 14:226–230
137. Orhan I, Aslan M (2009) Appraisal of scopolamine-induced antiamnesic effect in mice and in vitro antiacetylcholinesterase and antioxidant activities of some traditionally used Lamiaceae plants. J Ethnopharmacol 122(2):327–332
138. Altay A (2015) Antioxidant and cytotoxic properties of *Salvia fruticosa* M. and its effects on gene expression of some CYP450 and antioxidant enzymes in HT-29 cell line. Middle East Technical University
139. Atmaca H, Bozkurt E (2016) Apoptotic and anti-angiogenic effects of *Salvia triloba* extract in prostate cancer cell lines. Tumor Biol 37(3):3639–3646
140. Sevindik N, Rencuzogullari E (2014) The genotoxic and antigenotoxic effects of *Salvia fruticosa* leaf extract in human blood lymphocytes. Drug Chem Toxicol 37(3):295–302
141. Albayrak S, Aksoy A, Hamzaoğlu E (2008) Determination of antimicrobial and antioxidant activities of Turkish endemic *Salvia halophila* Hedge. Turk J Biol 32(4):265–270
142. Kosar M, Goger F, Baser KHC (2011) In vitro antioxidant properties and phenolic composition of *Salvia halophila* Hedge from Turkey. Food Chem 129(2):374–379
143. Erdogan EA, Everest A, Kaplan E (2013) Antimicrobial activities of aqueous extracts and essential oils of two endemic species from Turkey. Indian J Tradit Knowl 12(2):221–224
144. Ozcan B et al (2009) In vitro antimicrobial and antioxidant activities of various extracts of *Salvia microstegia* (Boiss.) Et. Bal. from Antakya, Turkey. Fresenius Environmental Bulletin 18(5a):658–662
145. Erdemoglu N et al (2006) Antioxidant activities of some Lamiaceae plant extracts. Phytother Res 20(1):9–13
146. Bahadori MB et al (2017) *Salvia nemorosa* L.: A novel source of bioactive agents with functional connections. LWT-Food Sci Technol 75:42–50

147. Yuce E et al (2014) Essential oil composition, antioxidant and antifungal activities of *Salvia sclarea* L. from Munzur Valley in Tunceli, Turkey. Cell Mol Biol 60(2):1–5
148. Tepe B (2008) Antioxidant potentials and rosmarinic acid levels of the methanolic extracts of *Salvia virgata* (Jacq), *Salvia staminea* (Montbret & Aucher ex Bentham) and *Salvia verbenaca* (L.) from Turkey. Biores Technol 99(6):1584–1588
149. Bahadori MB et al (2017) Functional components, antidiabetic, anti-Alzheimer's disease, and antioxidant activities of *Salvia syriaca* L. Int J Food Prop 20(8):1761–1772
150. Dulger B, Hacioglu N (2008) Antifungal activity of endemic *Salvia tigrina* in Turkey. Trop J Pharm Res 7(3):1051–1054
151. Karakaş FP, Yildirim A, Türker A (2012) Biological screening of various medicinal plant extracts for antibacterial and antitumor activities. Turk J Biol 36(6):641–652
152. Karcioglu L et al (2011) Antimicrobial activity of *Salvia trichoclada* in southern Turkey. Int J Agric Biol 13:134–136
153. Tepe B et al (2007) Antioxidant potentials and rosmarinic acid levels of the methanolic extracts of *Salvia verticillata* (L.) subsp. *verticillata* and *S. verticillata* (L.) subsp. *amasiaca* (Freyn & Bornm.) Bornm. Food Chem 100(3):985–989
154. Ustun O, Sezik E (2011) Analgesic activity of *Salvia wiedemannii* Boiss. used in Turkish folk medicine. Rec Nat. Prod 5(4):328
155. Ustun O, Ozcelik B (2011) In vitro antiviral activity and cytotoxicity of the extracts of *Salvia wiedemannii* Boiss. Planta Med 77(12):PL23
156. Yiğit D, Yiğit N, Özgen U (2009) An investigation on the anticandidal activity of some traditional medicinal plants in Turkey. Mycoses 52:135–140
157. Orhan DD, Ozcelik B, Hosbas S et al (2012) Assessment of antioxidant, antibacterial, antimycobacterial, and antifungal activities of some plants used as folk remedies in Turkey against dermatophytes and yeast-like fungi. Turk J Biol 36:672–686. https://doi.org/10.3906/biy-1203-33
158. Cuvelier M-E, Richard H, Berset C (1996) Antioxidative activity and phenolic composition of pilot-plant and commercial extracts of sage and rosemary. J Am Oil Chem Soc 73:645–652
159. Miyazawa M, Yamafuji C (2005) Inhibition of acetylcholinesterase activity by bicyclic monoterpenoids. J Agric Food Chem 53:1765–1768
160. Kirmizibekmez H, Atay I, Kaiser M et al (2011) In vitro antiprotozoal activity of extracts of five Turkish Lamiaceae species. Nat Prod Commun 6:1697–1700
161. Chen F, Li L, Tian DD (2017) *Salvia miltiorrhiza* roots against cardiovascular disease: consideration of herb-drug interactions. Biomed Res Int. https://doi.org/10.1155/2017/9868694
162. Uydeş-Doğan BS, Takır S, Özdemir O et al (2005) The comparison of the relaxant effects of two methoxylated flavones in rat aortic rings. Vascul Pharmacol 43:220–226. https://doi.org/10.1016/j.vph.2005.07.002
163. Çınar ÖG, Kırmızıbekmez H, Akaydın G et al (2011) Investigation of in vitro opioid receptor binding activities of some turkish Salvia species. Rec Nat Prod 5(4)
164. Suntar I, Akkol EK, Senol FS et al (2011) Investigating wound healing, tyrosinase inhibitory and antioxidant activities of the ethanol extracts of *Salvia cryptantha* and *Salvia cyanescens* using in vivo and in vitro experimental models. J Ethnopharmacol 135:71–77. https://doi.org/10.1016/j.jep.2011.02.022
165. Küpeli E, Göger F, Kosar M et al (2007) Anti-inflammatory and antinociceptive activities of *Salvia halophila* and *Salvia virgata* from Turkey. Planta Med 73:P_042
166. Polatoglu K, Yucel YY, Yalcin HTO et al (2015) Cytotoxic, antioxidant, iNOS, AChE. BChE inhibitory and antimicrobial activities of ethanol extract of Cyprus endemic plant *Salvia veneris*. Planta Med 81:1458
167. Topçu G, Tümen G, Kiliç T et al (2009) Bioactive Turkish plant extracts and their constituents in Innovations. Chem Biol Ed: Şener, B. 61–81

Composition of Aroma Compounds in *Salvia* spp. from Bulgaria and Their Potential Practical Applications

Velizar Gochev, Yana Hristova, Tanya Girova and Albena Stoyanova

Abstract The aim of present chapter is to gather the information about chemical composition, biological activities and possible applications of essential oils from seven different species of Genus *Salvia* growing in Bulgaria (*S. officinalis* L., *S. sclarea* L., *S. tomentosa* Mill, *S. scabiosifolia* Lam., *S. ringens* Sibth. et Sm., *S. amplexicaulis* Lam. and *S. aethiopis* L.). The chapter is focused on the variations in chemical composition of the essential oil caused by different climate and agro-industrial factors. Different forms of bioactivities of essential oils from *Salvia* spp. growing in Bulgaria and their possible applications are also discussed.

Keywords Salvia · Essential oils · Chemical composition · Application

1 Introduction

Genus *Salvia* L. is the largest in Lamiaceae family comprises approximately 900 species and is widely distributed in various geographic regions [1–3].

Based on its wide variety of bioactivities (antibacterial, antimycotic, antiinflamatory, antioxidant and other) infusions and essential oils from plants belonging to this genus are used in ethnopharmacology and folk medicine for treatment of infections and diseases of cardiovascular, nervous and respiratory systems. Different forms of bioactivities depends mainly on the chemical composition of essential oils and

V. Gochev (✉) · T. Girova
Department of Biochemistry and Microbiology, "Paisii Hilendarski" University of Plovdiv, Plovdiv, Bulgaria
e-mail: vgochev@uni-plovdiv.bg

Y. Hristova
Food Research and Development Institute at Agricultutal Academy, Plovdiv, Bulgaria

A. Stoyanova
Department of Tobacco, Sugar, Vegetable and Essential Oils, University of Food Technologies, Plovdiv, Bulgaria

V. Georgiev and A. Pavlov (eds.), *Salvia Biotechnology*,
https://doi.org/10.1007/978-3-319-73900-7_3

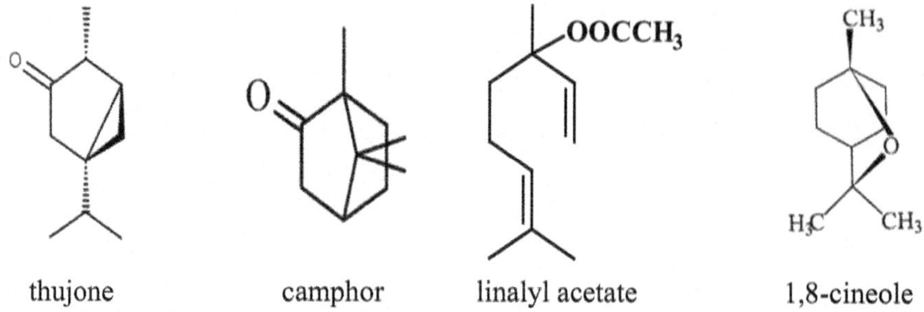

Fig. 1 Chemical structure of the main aromatic compounds in sage oils

infusions. The main aromatic compounds in sage oils, are thujone, camphor, linalyl acetate and 1,8 cineole [4–6] (Fig. 1).

About 18–19 native species of the genus *Salvia* L. is to be found in Bulgaria [7].

Among these species *S. officianlis* L. (garden sage) and *S. sclarea* L. (clary sage) are the most important from industrial point of view for cultivation and production of commercial essential oils and their application in a food industry and cosmetics as flavouring agents [8, 9].

The rest species are wild plants distributed in several isolated habitats. Four species are currently under protection: *S. scabiosifolia* L. and *S. pinnata* L. are classified as "critically endangered", *S. forskaohlei* L. is listed as "near threatened" and *S. verbenaceae* L. is listed as "near threatened" [10, 11]. Two species *S. amplexicaulis* Lam. and *S. aethiopis* L. are classified as rare and threatened on Bulgarian territory [12].

The aim of this brief review is to gather and systematized the results of the carried out researches focused on the chemical composition and bioactivities of the Bulgarian essential oils from seven species of genus *Salvia* L. and to expand their potential for practical applications.

2 *Salvia officinalis* L.

The species *S. officinalis* L. originated from the Mediterranean region and as wild growing plant it is widely distributed on the west coast of the Balkan peninsula, Italy, Greece, Spain and some island in Adriatic sea. For production of essential oil the plant is cultivated in many countries in Central and Eastern Europe, Turkey, Algeria and some of the United States [9, 13–16]. In Bulgaria several different cultivars were cultivated such as "*C-786*", "*Iskara*", "*Trakiya*" and "*Boyana*" but for an industrial production of essential oil only the cultivar "*Dessislava*" yielding 2% essential oil is used. Essential oil form *S. officinalis* L. demonstrates antiseptic, anti-inflammatory, antitumor, spasmolytic and antioxidant activities [17–19]. Its pharmacological properties depend on the chemical composition of the essential oil.

About 50 different compounds are identified in the sage oil from various countries but *cis*-thujone, camphor, 1,8-cineole, *trans*-thujone, β-caryophyllene, α- and β-pinene are the dominant. Qualitative characteristics of the essential oils from different countries are similar but great quantitative variations of the content of major compounds are determined, which means that these oils probably demonstrate different biological activities [17, 20–26].

First data about chemical composition of Bulgarian sage oil are published by Staikov and Zolotovitch during the distant 1961 [27]. The authors calculated the total ketone content of the oil as thujone and determine that the thujone content varied between 43.1 and 51%, depending on the moisture content of the raw plant material and the parts of the plant (whole plants or leaves) used for distillation.

More than 20 years later Balinova-Tzvetkova et al. [28] carried out a detailed research on the changes in the chemical composition of the oil during the different vegetation stages of *S. officianlis* L. growing in the experimental field of Institute of Roses, Essential and Medicinal Cultures—Kazanlik (Central Bulgaria, 380 m elev.). The authors reported that content of the main compound α-thujone dramatically decreases from 30 to 35% at the beginning of vegetation period to the 12.8% on the end of the vegetation period. The content of β-thujone and hydrocarbons also decreases but the content of β-pinene and 1,8-cineole increases. In accordance with the first published research [13] Balinova-Tzvetkova et al. [28] also determined that both the essential oil yield and its chemical composition depend on the parts of the plant used for distillation (whole plant or leaves). The major constituents of the essential oil are α-thujone (20–31.6%), camphor (7–18.3%), β-pinene (11–18%), 1,8-cineol (8.2–14%), and β-thujone (4–9.4%).

Tsankova et al. [29] studied the chemical composition of essential oil from sage, growing in the experimental field of the Institute of Botany, Sofia (West Bulgaria, 500–699 m elev.). The air-dried leaves were hydrodistilled in a Clevenger-type apparatus for 2 h reaching essential oil yield of 1.4% (w/w). By GS and GC/MS 39 constituents were identified corresponding to 96% of the total oil content. The oil has the typical high content of oxygenated compounds (79%), among which α-thujone (29.4%), β-thujone (17.4%) and camphor (11.7%) are the most abundant. The percentage of mono- and sesquiterpene hydrocarbons is only 13% and 4%, respectively. The compound selin-11-en-4-ol has been for the first time reported in the oil.

In their efforts to develop technologies for decontamination and utilization of heavy metals polluted areas some Bulgarian authors studied the effect of Zn, Cu, Cd and Pb on the chemical composition and yield of essential oil distilled from *S. officinalis* L. grown on polluted soils. Zheljazkov et al. [30] studied the essential oil yield of sage growing in metal contaminated soil (Zn–Cu smelter) near Plovdiv (Central Bulgaria) at three sites situated at 0.8, 3 and 9 km (control) away from the source of pollution. The results obtained demonstrated that high concentration of heavy metals in soil did not result in metal transfer into the essential oil but caused decreasing in the essential oil yield.

One year later, Stancheva et al. [31] evaluated the effects of heavy metals polluted soil on the growth, essential oil composition and accumulation of Cd, Pb,

Cu and Zn in the plant of *S. officinalis* L. The soil was collected from the vicinities (1 km) of a Non-Ferrous Metals Combine near the town of Plovdiv (Central Bulgaria, 164 m elev.) with pH 7.35 and the following content of heavy metals (μg/gDW): Cd 9.02 ± 0.98, Cu 82.10 ± 3.69, Pb 301.75 ± 21.03, Zn 641.60 ± 35.89. For the control non-polluted leached cinnamonic forest soil was used with the following agrochemical characteristics: pH (H2) 6.2, 8 mg kg^{-1} soil total mobile nitrogen ($N–NO_3^-$ + $N–NH_4^+$), 30 mg kg^{-1} soil P_2O_5, 120 mg k^{-1} soil K_2O. The obtained results show that the main compounds in the essential oil decreased as a result of heavy metals pollution as follows: α-thujon (−27%), β-thujon (−13%), β-caryophyllene (−54%) and viridoflorol (−21%), while camphor (+31%), borneol (+34%), 1.8-cineole (+38%) and bornyl acetate (+25%) significantly increased. Observed decrease of the levels of α- and β-thujones and elevated camphor level in the leaves of sage grown on heavy metals polluted soil indicated a deterioration of the essential oil composition.

Zheljazkov et al. [32] also studied the effect of distillation period on the sage oil yield and its chemical composition. For the experiments the authors used sage plants growing at the North Mississippi Research and Extraction Center at Verona but using certified seeds of commercially grown Bulgarian cultivar "*Dessislava*". The essential oil yield reached at different distillation times was from 0.06 to 0.28%. Monoterpenes represented the major percentage (58.2–84.1%) of oil composition followed by sesquiterpenes (4.03–16.12%) and diterpenes (0.31–7.61%). *Cis*-thujone (23.5–29.1%) and camphor (18.3–27.3%) were the major constituents.

Hristova [33] investigated the chemical composition of commercial sample of sage oil purchased from Vigalex Ltd, town of Gurkovo (Central Bulgaria, 380 m elev.). By GS and GC/MS 74 different compounds were identified corresponding to 97.43% of the total oil content. The number of detected and identified compounds is much higher in comparison with those published in previous researches by other Bulgarian authors. The evaluated sage oil sample characterized with more complex and rich chemical composition in comparison with sage oils from Central and Eastern Europe, Near East and some of the United States [4, 13, 18, 21–23, 34, 35]. The major components of sage oil sample are: *α*-thujone (22.34%), 1,8-cineole (10.55%), *β*-thujone (9.56%), camphor (8.52%), *trans*-*β*-caryophyllene (7.70%), limonene (6.59%), *β*-pinene (4.71%), *α*-humulene (4.68%), *α*-pinene (4.05%) and camphene (3.78%). According to its chemical composition and content of major compounds the studied Bulgarian sage oil sample can not be categorized into neither of the main five sage oil chemotypes, published in the literature [9].

ISO 9909 [36] standard regulates the content of the following compounds in the sage oil: *α*-thujone (18–43%), *β*-thujone (3–8.5%), %), camphor (4.5–24.5%), 1,8-cineole (5.5–13%), humulene (0–12%), *α*-pinene (1–6.5%), camphene (1.5–7%), limonene (0.5–3%), linalool ($\leq$ 1%) and bornyl acetate ($\leq$ 2.5%). All of the studied Bulgarian sage oil samples are in accordance with ISO 9909 [36].

New perspective method for production of aromatic product is extraction with liquid gases. The installations for extraction with liquefied gases characterized with high working pressure (0.6 MPa when butane is used, 1.5 MPa when propane is used and 4.7 MPa when CO_2 is used) and increased capital investments are needed.

Table 1 The content of main compounds in the essential oil and freon extract from sage

Compounds, %	Essential oil	1,1,1,2-Tetrafluoroethane extract
Camphor	21.8	4.8
cis-Thujone	20.2	6.7
trans-Thujone	11.3	5.4
1.8-Cineole	9.9	25.2
Ledene	4.9	–
α-Pinene	4.3	3.6
β-Pinene	–	5.4
Borneol	4.3	3.7
Bornyl acetate	–	3.6
Camphene	4.2	–
α-Humulene	3.0	6.1
E-β-Caryophyllene	–	7.5
allo-Aromadendrene	–	4.6

There are one laboratory installations with liquefied $C_2H_2F_4$, situated in University of Food Technologies in Plovdiv, and two industrial working with liquefied CO_2 and one with liquefied $C_2H_2F_4$ These installations are used for processing of different essential oil bearing plants in Bulgaria, for example rose, lavender, coriander, fennel and etc. For the first time in Bulgaria Atansova et al. [37] applied liquefied gas $C_2H_2F_4$ (1,1,1,2-tetrafluoroethane) to produce extract from *S. officinalis* L. in laboratory-extractor under following conditions: pressure 0.70–0.780 MPa, temperature 25–30 °C, time 60 min. The authors also produced essential oil from the same batch of the raw plant material by classical water distillation for 4 h in laboratory glass apparatus according to British Pharmacopoeia, modified as described elsewhere [38]. Chemical composition of both aroma products is shown in Table 1. As seen chemical composition of the obtained aroma products differs and strongly depends on the technological regiments for processing of the plant material. Because the 1,1,1,2-tetrafluoroethane extract is rich of 1,8-cineole and poor of thujone the authors proposed that it could be applied as a natural supplement in food products.

As we already mentioned biological activity of essential oil depends on its chemical composition. Many authors studied antimicrobial activity of sage oil from various geographic regions mainly against food spoilage bacteria and phytopathogenic fungi and less are focused on the clinical isolates of dimorphic yeasts. Most of the authors determined the sage oil as strong according to its antimicrobial activity, nevertheless that inhibitory zone diameters and minimal inhibitory concentration values published by the authors varied in a wide range [39–43]. Also there is no clearance which one of the major constituents (*cis*-thujone, camphor, 1.8-cineole, pinene) of the oil or combination among them is responsible for its activity.

The rich chemical composition of Bulgarian essential oil from *S. officinalis* L. is prerequisite for high levels of bioactivity. For this reason bioactivity, in particular antimicrobial activity of Bulgarian sage oil is intensively evaluated.

Nestorova et al. [44] determined antimicrobial activity of sage oil against 20 different strains of food spoilage bacteria, yeasts and molds. The oil demonstrated high levels of activity against all of the used test microorganisms and according to the authors it can be used as a natural food preservative.

Gurgulova et al. [45] tested antibacterial and antimycotic activity of sage oil focusing on the plant and insect pathogens. The studied oil sample demonstrated moderate antibacterial activity with MIC 0.1% and antimycotic activity (MIC 0.2%) against the best-known ascomycetous pathogen for honey bee larvae. Based on the obtained results the authors concluded that the oil is prospective for treatment of plant and insect pathogens.

Hristova [33] studied antimicrobial activity of Bulgarian sage oil against 60 strains of clinical isolates belonging to six species of genus *Candida—C. albicans, C. glabrata, C. tropicalis, C. parapsilosis, C. krusei and C. kefyr.* The oil was tested both by Disc Diffusion Test M44-A2 (2009) and Serial Broth Dilution Test M27-A3 (2008) in accordance with CLSI [47]. The author determined that the sage oil demonstrated antimicrobial activity against both fluconazole-sensitive and fluconazole-resistant strains, which means that the essential oil of *S. officinalis* L. could be used as an alternative antimycotic agent. According to its antimycotic activity Bulgarian sage oil exceeded sage oils from various geographic regions published in the literature [4, 24, 48].

In addition Amalieva [49] determined the beneficial effect of massage with combination of sage and lavender oil for treatment of depression, nervousness and anxiety.

The above mentioned studies demonstrated that Bulgarian essential oil from *S. officinalis* L. is appropriate for application in food industry, medicine and cosmetics.

3 *Salvia sclarea* L.

The species *S. sclarea* L., originated from the Mediterranean region and as wild growing plan it is widely distributed on the Balkan peninsula, and Middle Asia. For production of essential oil the plant is cultivated in many European countries (Italy, France, Spain, Serbia, Bulgaria, Moldova and Russia), in Uzbekistan, Tajikistan and some of the United States [9, 50]. About 80% from the world production of clary sage oil belongs to Russia. *S. sclarea* L. is cultivated In Bulgaria for industrial production of different aroma products—essential oil, concrete and absolute [9] and it is applied mainly in perfumery and cosmetics. About 60 different compounds are identified in a clary sage oils from Italy, Serbia, Tajikistan, Iran, Tunisia, Russia, Greece and other [51–56]. Linalyl acetate, linalool, 1,8-cineole, β-caryophyllene, geraniol, germacrene D and sclareole are determined. The content even of the both major compounds linalyl acetate and linalool, which correspond to 80% of the total oil content, varies in a wide range from 19.7 to 72.9% and from 3.0 to 30.4%, respectively. Many factors as climate conditions, blossoming stage, but mainly the way and duration of plant storage before distillation influenced the chemical composition of the essential oil. Diverse composition of clary sage oil determines varieties of bioactivities as antimicrobial, antimalarial, antioxidant activities and other [51–56].

Bulgarian clary sage oil has been intensively studied. The first data about essential oil yield, its linalyl acetate content and comparison with essential oils from Russia, Germany and France were published by Ilieva et al. [57] and Dimitrova et al. [58]. They also studied the influence of flowering stage and type of the raw plant material (whole plant, leaves or flowers) used for distillation on the essential oil yield.

Later Ilieva et al. [59] selected new forms of *S. sclarea* L., which characterized with erected bush, vertical bloom carrying stem disposition with compact floccule and big flowers and are cold resistant. The new forms are distinguished by higher essential oil yield and high content of linalyl acetate. The obtained clary sage oil characterized with fine, agreeable and long-lasting aroma [60].

Seven years later, Ilieva et al. [51] selected a new cultivar "Mechta", which was cultivated near the city of Varna (East Bulgari, 20–40 elev.). The yield of essential oil was 0.4% and the content of linalyl acetate was 65–78%.

Currently in Bulgaria six different cultivars from the species *S. sclarea* L. are selected. One of them cultivar "Trakiyka" is the dominant with 0.45% essential oil yield. From the flowers two aroma products are obtained. The first is concrete produced by solvent extraction with main compound sclareol and the second is essential oil produced by steam distillation with main compounds linalyl acetate and linalool.

Kovatcheva and Zhelhazkov [62] compared the oil composition from plants with different flower colours. The major components in the oils were as follows:

- from white flowers—α-pinene (2.5%, camphene (0.5%), β-pinene (17.9%), α-thujone (17.4%), β-thujone (1.8%), camphor (2.4%) and β-caryophyllene (17.1%);
- from pink flowers—α-pinene (2.9–4.2%, camphene (1.9–4.4%), β-pinene (13.8–17.6%), α-thujone (13.9–14.24%), β-thujone (2.7–9.4%), camphor (1.7–2.3%) and β-caryophyllene (5.2–7.41%);
- from violet flowers—α-pinene (2.5%, camphene (0.7–1.0%), β-pinene (16.7–17.8%), α-thujone (10.0–33.3%), β-thujone (2.5–4.0%), camphor (0.3–0.4%) and β-caryophyllene 7.9–9.8%);
- from dark violet flowers—α-pinene (0.1–0.2%), camphene (1.8–3.8%), β-pinene (1.0–3.9%), α-thujone (19.7–24.7%), β-thujone (0.7–1.7%), camphor (0.1–11.2%) and β-caryophyllene (7.6–11.7%).

According to Lawrence [63] there is no correlation between flower colour and chemical composition of the clary sage oils.

Hristova et al. [64] evaluated the commercial oil sample of clary sage produced by Vigalex Ltd., town of Gurkovo (Central Bulgaria, 380 m elev.) and by GC and GC/MS detected and identified 52 different compounds representing 98.25% of total oil content. Linalyl acetate (56.88%) and linalool (20.75%) were determined as major constituents, followed by germacrene D (5.08%) and β-caryophyllene (3.41%). The essential oil belongs to the linalyl acetatae/linalool chemotype [9, 50]. Qualitative composition of Bulgarian oil from *S. sclarea* L. is in accordance with clary sage oils from Serbia [51], Iran [65], Italy [52], Russia [53] and other countries. Some essential differences according to the quantitative content of major compounds linalyl acetate and linalool in Bulgarian oil and other oils are detected. These differences could be attributed mainly to the specific agrometeorological conditions in Bulgaria.

Investigations on bioactivity of clary sage oil are focused mainly on its antimicrobial properties.

Gurgulova et al. [66] tested antibacterial activity of essential oil from clary sage against 3 strains of *Bacillus subtilis* and 3 strains of *Bacillus cereus* isolated from honey bee samples. The antibacterial activity of the clary sage oil was categorized by the authors as "*relatively high*" (MIC from 0.05 to 0.1%) but lower in comparison with activity of thyme, savory and marjoram.

Rusenova and Parvanov [67] tested the antimicrobial activity of commercial oil sample obtained from Lavena Ltd, Shumen, Bulgaria. The oil demonstrated good antimicrobial activities against tested Gram-positive bacteria (*Staphylococcus aureus* ATCC 25923, *Bacillus lichenformis* ATCC 14575, *Listeria monocytogenes* (clinical isolate), *Enterococcus faecalis* ATCC 29212 and Gram-negative bacteria (*Proteus vulgaris* (clinical isolate), *Proteus mirabilis* (clinical isolate), *Escherichia coli* ATCC 25922), and against yeasts *Candida albicans* (clinical isolate) and *Malassezia pachydermatis* (clinical isolate).

Two years later Roussenova [68] reported that the same essential oil has highest activity (MIC 0.06% v/v) against the etiological agent of American foulbrood disease (*Paenibacillus larvae*). According to the author these results are reason for possible inclusion of the clary sage oil as an alternative in the measures for prevention and control of American foulbrood without the use of antibiotics. Unfortunately in both papers published by Roussenova [66, 67] chemical composition of the tested clary sage oil is unknown and more detailed discussions on the antimicrobial activity of the oil are impossible.

Antifungal activities of clary sage oil from Bulgaria and its major compounds linalyl acetate and linalool against 30 clinical isolates, belonging to species *Candida albicans*, *C. tropicalis*, *C. krusei*, *C. glabrata* and *C. parapsilopsis* were evaluated by Hristova et al. [64]. The studied essential oil characterized with stronger antimicrobial activity against *Candida* spp. in comparison with pure compounds. The activity of essential oil can not be attributed only to the single major compounds. Probably different compounds in the essential oil act synergic and additional activity of some minor compounds cannot be excluded. The obtained results revealed possibilities for application of Bulgarian clary sage oil as a natural antimycotic and alternative of azole preparations.

4 *Salvia tomentosa* Mill.

The species popular as "balsamic sage" is traditionally used for preparation of high quality sage tea. In folk medicine, the plant is popular for treatment of stomachache, asthma, cold and inflamed wounds [69].

Tsankova et al. [29] analyzed the chemical composition of essential oil from the plants, which were collected near the town of Krumovgrad (Rodopa mountain, 670 m elev.). The air-dried leaves, after hydrodistillation in a Clevenger-type apparatus for 2 h gave a yield of 1.8% (w/w). 39 constituents corresponding to 96% of the total oil content were identified. The content of monoterpene hydrocarbons in the oil was much higher (49%) and the amount of β-pinene was found to reach 29.1%. Among the oxygenated compounds present in this oil, borneol (19.4%) and camphor (4.9%) were the most abundant. Furthermore, the sesquiterpenoids α-copaene, γ- and δ-cadinene, spathulenol, ledol and valeranone were identified for the first time in the oil.

Marchev et al. [70] investigated the essential oil from fresh leaves from experimental field-growing collection of Institute for Biodiversity and Ecosystem Research, Bulgarian Academy of Sciences, Sofia (West Bulgaria, 500–600 m elev.). The essential oil was obtained by water distillation for 2.3 h in laboratory glass apparatus according to British Pharmacopoeia, modified as described elsewhere [38]. The oil yield was 0.96% v/w (67% moisture). By GS and GS/MS 60 compounds were detected and identified, representing 98% of the total oil content. The major compounds (above 3%) were: borneol (10.3%), β-pinene (9%), camphor (7.9%), α-pinene (6%), camphene (4%), 1.8-cineole (3.8%), α-limonene (3.5%) and β-caryophyllene (3%). The major monoterpenes include: hydrocarbons (47.5%), alcohols (23.0%), aldehydes and ketones (17.1%), ethers (8.3%), and esters (4.1%). Sesquiterpenes are presented by hydrocarbons (67.1%), alcohols (23.6%) ethers (6.4%), aldehydes and ketones (2.9%). The essential oil showed considerable acetylcholinesterase inhibitory activity ($IC_{50} = 0.28 \pm 0.06$ μg/mL), comparable with that of galanthamine. Study of antioxidant activity strongly suggested that the hydrogen atom transfer reaction was preferable over the electron transfer (ORAC = 175.0 ± 0.40 μM Trolox equivalents/g oil and FRAP = 1.45 ± 0.21 mM Trolox equivalents/g oil). The essential oil shoved moderate antifungal and antibacterial activities against *Candida albicans* and Gram-positive bacteria, whereas it was almost inactive against the investigated Gram-negative stains. The results suggested that the essential oil of *S. tomentosa* could be considered as a prospective active ingredient for prevention of oxidative stress-related and neurodegenerative disorders in aromatherapy. Because of the high antioxidant capacity, the oil according to the authors could be considered as natural supplement or antioxidant in cosmetics and food products. The effect of essential oil on in vitro acetylcholinesterase inhibitory activity of galanthamin was investigated and existence of cumulative interaction was demonstrated and discussed for the first time.

5 *Salvia scabiosifolia* Lam.

The species grows on dry and rocky soils. It could be distinguished from the other plants by its 20–50 cm in length, woody at base; stems—much branched, ascending, patent—villous, glandular; leaves pinnate or pinnatisect, petiolate, with many pairs of linear or oblong lateral segments and a large terminal segment; verticillasters with 6–10 flowers; calux 10–15 mm, accrescent; cololla 25–30 mm, violet-blue or violet [7].

Tsankova et al. [29] investigated the chemical composition of essential oil from the plants, which were collected near the town of Bjala (North Bulgaria, 200 m elev.). The air-dried leaves, after hydrodistillation in a Clevenger-type apparatus for 2 h gave a yield of 0.3% (w/w). 39 constituents according for 90% of the oil were identified. The major constituent was camphor (48.7%). Among them camphene (6.1%), *p*-cymene (3.1%), α-humulene (2.7%), β-boubonene (2.3%) and δ-cadinene (2.3%) were dominant. In the oil was identified 12 monoterpene (19.3%), 8 sesquiterpene hydrocarbons (12.4%) and 10 oxygenated terpenes (9.6%).

Marchev et al. [71] determined the chemical composition of the essential oil from the fresh leaves of the plants collected from the natural habitat in the municipality Byala, region the town of Ruse (North-East Bulgaria, 45.5 m. elev.). The essential oil was obtained by water distillation for 2.3 h in laboratory glass apparatus according to British Pharmacopoeia, modified as described elsewhere [38]. The yield of the oil was 0.14% (moisture 67.3%). 60 components representing 97.5% of the total oil were identified. The major compounds (above 3%) were: camphor (19. 2%), borneol (7.8%), camphene (7.0%), α-caryophyllene (4.2%), β-pinene (4.0%), γ-terpinene (3.5%), *p*-cymene (3.4%), eucalyptol (3.3%), limonene (3.2%) and α-pinene (3.0%). The oil consisted mainly of monoterpenes (68.9%), followed by sesquiterpenes (27.4%), aromatic (3.5%) and other compounds (0.2%). Distribution of the components according to the functional groups revealed that in monoterpenes the major classes of substances were hydrocarbons (43.4%) followed by aldehydes and ketones (31.7%), alcohols (15.8%), ethers (7.9%) and esters (1.2%). In sesquiterpenes again dominant compounds were hydrocarbons (62.5%) followed by alcohols (32.2%), ethers (3.4%), aldehydes and ketones (1.9%). According to the authors high camphor plants have found broader application for non-perfumery uses.

6 *Salvia ringens* Sibth. et Sm.

The plant is Balkan endemic species growing in several isolated habitats in Bulgaria. The aerial parts of plant, including leaves and flowers, were widely used in ethnomedicine for drugs and infusions preparation by the locals.

It was found that the flowers of Bulgarian species contain higher amoints of flavonoids compared to the leaves [72].

Table 2 Comparative percentage composition of essential oils from leaves and flowers

Compounds, %	Leaves	Flowers
Camphor	17.2	18.8
Borneol	7.2	8.7
β-Pinene	6.0	4.0
Camphene	–	5.0
β-*trans*-Ocimene	4.0	–
Germacrene D	3.5	–
o-Cymene	3.1	3.3
Eucalyptol	3.1	3.4
α-Caryophyllene	–	3.4
α-Pinene	–	3.2
α-Limonene	–	3.1

For first time Georgiev et al. [73] obtained the essential oil from different aerial parts (leaves and flowers) of the plants which were collected near to the village Nevsha, district of town of Varna (East Bulgari, 20–40 elev.). The yield of leaves and flowers oils were similar—0.03% v/w (moisture 67.3%) and 0.04% v/w (moisture 65.5%) respectively. 60 compounds were identified in both oils, which correspond to 97.7% and 97.3% pf the total oils constituents from leaves and flower oils, respectively. Both oils showed similar chemical compositions and the major compounds (above 3%) and the results can be seen summarized in Table 2.

Compared to the essential oil from leaves, a slight increase in relative percentages of monoterpenes was observed in the oil from flowers compared to that from leaves (68.4% vs. 64.3%). The relative percentage of sesquiterpenes in flowers oil was reduced with 4.3% compared to the sesquiterpenes levels in leaves oil (28.0% vs. 32.3%). There were no differences in the percentages of the other cycloaliphatic compounds (0.2% in both oils), and aromatic compounds (3.1% vs. 3.3% in oils from leaves and flowers, respectively). The data according the authors demonstrates that the oil could be used as potential source for obtaining of camphor reach oil with stable chemical composition.

7 *Salvia amplexicaulis* Lam.

In Bulgaria there is no data on the use of the plant or its oil and extracts in folk medicine.

For the first time Bozov et al. [74] investigated the chemical composition of the oil from the plants collected from the natural habitat in the town of Krichim (Central Bulgaria, 100–199 m. elev.) in the stage of flowering. The essential oil was obtained by water distillation for 2.3 h in laboratory glass apparatus according to British Pharmacopoeia, modified as described elsewhere [38]. The yield of the oil was 0.08% (v/w) (moisture 11.80%). 26 components representing 85.5% of the

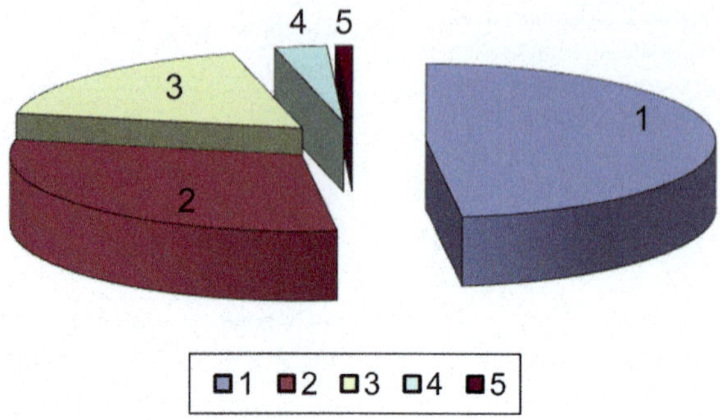

Fig. 2 Group of components in essential oil from *S. amplexicaulis*. 1 sesquiterpene hydrocarbons, 2 hydrocarbons, 3 oxygen sesquiterpenes, 4 phenyl propanoids, 5 oxygen hydrocarbons

total content were identified. The major compounds (above 3%) were: γ-muurolene (29.20%), nonadecane (5.58%), heneicosane (5.48%), spathulenol (5.14%), aromadendrene oxide-(2) (4.09%), ledene oxide (3.69%) and octadecane (3.57%). Distribution of major groups of aroma substances in the essential oil is shown in Fig. 2. Sesquiterpene hydrocarbons (48.01%) are the dominant group in the oil from *S. amplexicaulis*, followed by hydrocarbons (30.11%), oxygen sesquiterpenes (18.01%), phenyl propanoids (2.92%) and oxygen hydrocarbons (0.95%). According the authors the difference in chemical composition of this oil and the reported data [75, 76] may be due to environmental conditions under which the plant has grown as well as the variation in conditions of analysis.

The low essential oil yield characterized *S. amplexicaulis* as inappropriate for industrial application as essential oil bearing plant.

8 *Salvia aethiopis* L.

In Bulgaria there is no data on the use of the plant or its oil and extracts in folk medicine.

For the first time Damianova et al. [77] obtained essential oils by hydrodistillation from flowers and leaves from wild plants, grown in the region of the town of Plovdiv (South Bulgaria, 164 m elev.). The essential oil was obtained by water distillation for 2 h in laboratory glass apparatus according to British Pharmacopoeia, modified as described elsewhere [66]. The yield of essential oil, % in abs. dry mass was 0.19 and 0.03%, respectively. The main compounds of essential oils of flowers and leaves were as follows, respectively: germacrene D (29.37–24.19%), β-caryophyllene (23.55–21.91%), α-copaene (13.35–17.24%), β-cubebene (7.02–9.71%), δ-cadinene (5.56–6.69%) and α-caryophyllene (5.46–6.79%). The total sesquiterpene hydrocarbons constituted the highest percentage of the components of the essential oil constituting 94.0 and 95.82% in the oil of flowers and leaves, respectively. The results indicated that Bulgarian essential oil obtained from flowers and leaves of *S. aethiopis* is also a germacrene D and β-caryophyllene chemotype. The classification of the identified compounds, based on functional groups, is summarized in Fig. 3.

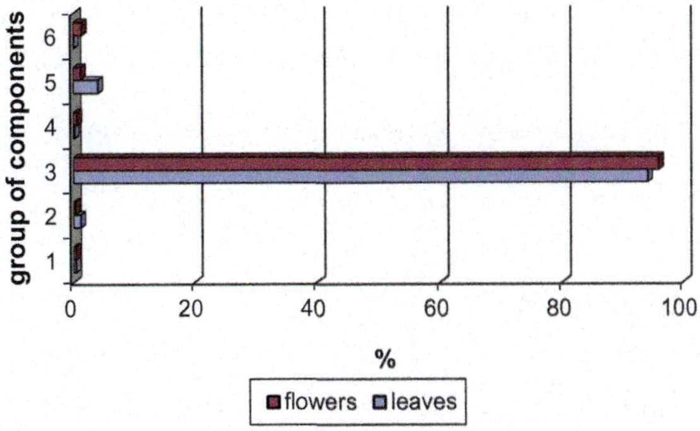

Fig. 3 Group of components in essential oils from *S. aethiopis*. 1 monoterpene hydrocarbons, 2 oxygenated monoterpenes, 3 sesquiterpene hydrocarbons, 4 oxygenated sesquiterpenes, 5 hydrocarbons, 6 others

The total sesquiterpene hydrocarbons constituted the highest percentage of the components of the essential oil constituting 94.0 and 95.82% in the oil of flowers and leaves, respectively. The oil of flowers consisted above 1% concentration hydrocarbons (4.08%). The oil of leaves consisted above 1% concentration hydrocarbons (1.29%) and others (1.46%). The percentage of monoterpene hydrocarbons, oxygenated monoterpenes and oxygenated sesquiterpenes are under 1%. The sesquiterpenes were found as a major constituents of oils produced on Yugoslavia [78], Serbia [79], Iran [80, 81], Spain [82] and Turkey [83]. The difference in chemical composition of the Bulgarian oil and the reported data according the authors may be due to environmental conditions under which the plant has grown as well as the variation in conditions of analysis.

The low essential oil yield characterized *S. amplexicaulis* as inappropriate for industrial application as essential oil bearing plant.

9 Conclusions Future Prospects

Essential oils from Bulgarian *Salvia* spp. characterized with rich chemical composition which could be attributed mainly to the unique geographic and climate conditions. From industrial point of view only species *S. officinalis* L. and *S. sclarea* L. are important for production of essential oils and aroma compounds. Bulgarian essential oils from *Salvia* spp. are rich of bioactive substances which determine broad spectrum of activities such as antibacterial, antifungal, antioxidant and acetylcholinesterase inhibitory action. These forms of bioactivities expand the possibilities for application of Bulgarian essential oils from *Salvia* spp. not only as flavouring agents in a food industry, perfumery and cosmetics but even as preparations with pharmacological importance. Probably the most prospective application of sage oils is as natural food biopreservatives. To reach real innovative application of sage oils as a component of so called “active packaging systems” future studies on these oils must be redirected, because currently most of the published studies are focused only on the chemical composition and biological activities of the oils. The possibilities for including of sage oils in polymer films or gel beads for coating of foods and food packages would be of a great interest.

Acknowledgements Experimental work and studies of the authors was financially supported by Project NI15/FK012 by "P. Hilendarski" University of Plovdiv.

References

1. Bown D (1995) Encyclopedia of herbs and their uses. Dorling Kindersley, London
2. Chevallier A (2009) The Encyclopedia of medicinal plants. Dorling Kindersley, London
3. Kintzios S (2000) Sage—the genus Salvia. Harwood Amsterdam, Netherland
4. Abu-Darwish M, Cabral C, Ferreira I et al (2013) Essential oil of common sage (*Salvia officinalis* L.) from Jordan: assessment of safety in mammalian cells and its antifungal and Anti-inflammatory potential. Bio Med Res Int, Article ID 538940:1–9
5. Dobetsberger C, Buchbauer G (2011) Actions of essential oils on the central nervous system: an updated review. Flavour Fragr J 26(5):300–316
6. Kamatou G, Vilijoen A, Gono-Bwalya AB (2005) The in vitro pharmacological activities and a chemical investigation of three South African species. J Ethnopharmacol 102(3):382–390
7. Markova M (1989) Genus *Salvia* in Flora Bulgaricae. Sofia Acad Sci Bulg 9:442
8. Dikova B (2011) Tomato spotted wilt virus on some medicinal and essential oil-bearing plants in Bulgaria. Bulg J Agric Sci 17(3):306–313
9. Georgiev E, Stoyanova A (2006) A guide for the specialist in aromatic industry. Plovdiv, Bulgaria
10. Petrova A, Vladimirov V (2009) Red list of Bulgarian vascular plants. Phytologia Balcanica 15:63–94
11. Tashev A, Tsavkov E (2008) medicinal plants of the Bulgarian dendroflora. Phytologia Balcanica 14:269–278
12. Fägäras M, Anastasiu P, Gavril N (2010) Rare and threatened plants in the Black Sea coastal area between Cape Midia (Romania) and Cape Kaliakra (Bulgaria). Botanica Serb 34:37–43
13. Badiee P, Nasirzadeh A, Motaffal M (2012) Comparison of *Salvia officinalis* L. essential oil and antifungal agents against *candida* species. J Pharm Technol Drug Res 1(7):1–6
14. Başer H, Kirimer N (2006) Essential oils of Lamiaceae plants of Turkey (Conference Paper). Acta Hort 723:163–171
15. Mirijalili M, Salehi P, Sonboli A, Vala M (2006) Essential oil variation of *Salvia officinalis* aerial parts during its phenological cycle. Chem Nat Compd 42(1):19–23
16. Raal A, Orav A, Arak E (2007) Composition of the essential oil of *Salvia officinalis* L., from varios European countres. Nat Prod Res 21(5):406–411
17. Abu-Darwish M, Cabral C, Ferreira I, Gonçalves M, Cavaleiro C, Cruz M, Al-Bdour T, Salgueiro I (2013) Essential oil of Common Sage (*Salvia officinalis* L.) from Jordan: assessment of safety in mammalian cells and its antifungal and Anti-inflammatory potential. Bio Med Res Int, Article ID 538940:1–9
18. Fu Z, Wang H, Hu X, Sun Z, Han C (2013) The Pharmacological properties of *Salvia* essential Oil. J Appl Pharm Sci 3(07):122–127
19. Loizzo R, Tundis R, Menichini F, Saab M, Statti A, Menichini F (2007) Cytotoxic activity of essential oils from Labiatae and Lauraceae families against in vitro human tumor models. Anticancer Res 27:3293–3300
20. Mahboubi E, Ahmadi A (2013) Evalution of *Salvia officinalis* antifungal properties on the growth and morphogenesis of *Alternaria alternata* under in-vitro conditions. Tech J Eng Appl Sci 3(17):2062–2069

21. Djeddi S, Bouchenak N, Settar I, Halli L (2012) Screening of chemical composition and antimicrobial potential of Algerian sage essential oil. Glob J Med Plant Res 1(1):46–49
22. Lakhal H, Ghorab H, Chibani S, Kabauche A, Semra Z, Smati Z, Smati F, Abuhamdah S, Kabouche Z (2013) Chemical composition and biological activities of the essential oil of *Salvia officinalis* from Batna (Algeria). Scholars Res Libr 5(3):310–314
23. Orhan I, Özçelik B, Kartal M, Kan Y (2012) Antimicrobial and antiviral effects of essential oils from selected *Umbeliferae* and *Labiatae* plants and individual essential oil components. Turk J Biol 36:239–246
24. Pinto E, Salguero L, Cavaleiro C, Palmeira A, Gonzalves M (2007) In vitro susceptibility of some species of yeasts and filamentous fungi to essential oils of *Salvia officinalis*. Ind Crops Prod 26(2):135–141
25. Soković M, Clamoćlija J, Ćirić A, Kataranovski D, Marin PD, Vukojević J, Brkić D (2012) Antifungal activity of the essential oils and components in vitro and in vivo on experimentally induced dermatomycoses at rats. Dig J Nanomater Biostruct 7(3):959–966
26. Miladinović D, Mihajilov-Krstev S, Nikolić M, Miladinović D, Cvetković C (2012) Investigation of the chemical composition-antibacterial activity relationship of essential oils by chemometric methods. Anal Bioanal Chem 403(4):1007–1018
27. Gildemeister E, Hoffman F (1961) Die Ätherischen Öle. VII. Akademie Verlag, Berlin
28. Balinova-Tzvetkova A, Tsoneva D, Tsankova P et al (1987) Essential oil from garden sage. In: Scientific conference "80 years scientific work with essential oil bearing plants in Bulgaria", 21–22 Oct 1987
29. Tsankova E, Konaktchiev A, Genova E (1994) Constituents of essential oils from three *Salvia* species. J Essent Oil Res 6:375–378
30. Zheljazkov V, Craker L, Xing B et al (2008) Aromatic plant production on metal contaminated soils. Sci Total Environ 395:51–62
31. Stancheva I, Geneva M, Hriztozkova M et al (2009) Essential oil variation of *Salvia officinalis* (L.), grown on heavy metals polluted soil. Biotechnol Equip 23:373–376
32. Zheljazkov V, Astatide T, Shiwakoti S et al (2014) Essential oil yield and composition of garden sage as a function of different steam distillation times. HoartScience 49(6):785–790
33. Hristova Y (2015) Inhibitory activity of essential oils from Bulgarian essential oil bearing plants against medically important fungi from genus *Candida*. Ph.D. thesis, "Paisii Hilendarski" University of Plovdiv, Bulgaria
34. Pozzatti P, Loreto É, Nunes M et al (2010) Activities of essential oils in the inhibition of *Candida albicans* and *Candida dubliniensis* germ tube formation. J Mycol Méd 20:185–189
35. Tullio V, Nostro A, Mandras N et al (2006) Antifungal activity of essential oils against filamentous fungi determined by broth microdilution and vapour contact methods. J Appl Microbiol 102:1544–1550
36. ISO 9909:1997 (1997) Oil of Dalmatian sage (*Salvia officinalis* L.). International Standard Organization
37. Atanasova T, Nenov N, Dimitrova-Djulgerova I et al (2012) Chemical composition of aromatic products, 1. Sage (*Salvia officinalis* L.). Sci Works Univ Food Technol 61:275–278
38. Balinova A, Diakov G (1974) On improved apparatus for microdistillation of rose flowers. Plant Sci:79–85
39. Tabassum N, Vidiasagar G (2013) Antifungal investigations on plant essential oils: a review. Int J Pharm Pharm Sci 5(2):19–28
40. Alizadeh A, Shaabani M (2012) Essential oil composition, phenolic content, antioxidant and antimicrobial activity in *Salvia officinalis* L. cultivated in Iran. Adv Environ Biol 6(1): 221–226
41. Hosseinzadeh H, Sadeghnia HR, Imenshahidi M, Fazly B (2009) Review of the pharmacological and toxicological effects of *Salvia leriifolia*. Iran J Basic Med Sci 12:1–8
42. Naeni A, Khosravi A, Chitsaz M, Shokri H, Kamlnejad M (2009) Anti-*Candida albicans* activity of some Iranian plants used in traditional medicine. J Mycol Méd 19:168–172

43. Zuzarte M, Gonçalves M, Canhoto J, Salgueiro L (2011) Antidermatophytic activity of essential oils, In: Méndez-Vilas A (ed) Science against microbial pathogens: communicating current research and technological advances. FORMATEX, Badajoz, pp 1167–1178
44. Nestorova V, Georgiev E, Balinova A (1989) About the antimicrobial effect of some food spices. 4. Essential oils obtained from spearmint, rosemary, sage, milfoil, monarda and marjoram. In: VIIth congress on Microbiology. Varna, Bulgaria, 27–29 October
45. Gurgulova K, Nenchev P, Zhelyazkova I et al (2006) Study on the antibacterial and antimycotic activity of some ethereal oils to strains of microorganisms causing diseases in bees and brood. Ecol Fut - J Agric Sci For Sci 5:32–36
46. Clinical Laboratory Standard Institute (2009) Reference method for antifungal disk diffusion susceptibility testing of yeasts; Approved guideline-second edition. M44-A2, vol 29, no 7. Wayene, USA
47. Clinical Laboratory Standard Institute (2008) Reference method for Broth dilution antifungal susceptibility testing of yeasts. Approved standard-third edition. M27-A 28 vol 28, no 14. Wayene, USA
48. el Hayouni A, Chraief I, Abedrabba M et al (2008) Tunisian *Salvia officinalis* L. and *Schinus molle* L. essential oils: their chemical compositions and their preservative effects against *Salmonella* inoculated in minced beef meat. Int J Food Microbiol 125:242–251
49. Amalieva K (2011) Investigation on the influence of essential oils and their mixture in aroma preparations. Ph.D. Thesis, University of Food Technologies Plovdiv, Bulgaria
50. Sharopov F, Setzer N (2012) The essential oil of Salvia sclarea L. from Tajikistan. Rec Nat Prod 6(1):75–79
51. Džamić A, Soković M, Ristić M, Grujić-Jovanović S, Vukojević J, Marina P (2008) Chemical composition and antifungal activity of *Salvia sclarea* (Lamiaceae) essential oil. Arch. Biol. Sci. Belgrade 60(2):233–237
52. Fratenale D, Glampert L, Bucchini A, Ricci D, Epifano F, Genovese S, Curini M (2005) Composition and antifungal activity of essential oil of *Salvia sclarea* from Italia. Chem Nat Compd 41(5):604–606
53. Jirovets L, Wlcek K, Buchbauer G, Gochev V, Girova T, Stoyanova A, Schmidt E (2007) Antifungal activities of essential oils of *Salvia lavandulifolia*, *Salvia officinalis* and *Salvia sclarea* against various Pathogenic *Candida* species. J Essent Oil Bear Plants 10(5):430–439
54. Pitarokili D, Couladis M, Petsikos-Panayotarou N, Tzakou O (2002) Composition and antifungal activity on soil-borne pathogens of the essential oil of *Salvia sclarea* from Greece. J Agric Food Chem 6(50):6688–6691
55. Taârit MB, Msaada K, Hosni K, Marzouk B (2014) GC analyses of salvia seeds as valuable essential oil source. Adv Chem, Article ID 838162, 6
56. Lorenzo D, Paz D, Davies P (2004) Characterization and enantiomeric distribution of some terpenes in the essential oil of a Uruguayan biotype of *Salvia sclarea* L. Flavour Fragr J 19 (4):303–307
57. Ilieva S, Dimitrova E, Zolotovitch G (1955) The production of essential oil of *Salvia sclarea* L. in Bulgaria (first announcement). C R l'Acad Bulg Sci 8(1): 49–52
58. Dimitrova E, Ilieva S, Zolotovitch G (1955) La production d'hule de *Salvia sclarea* en Bulgarie. C R l'Acad Bulg Sci 8(3):13–16
59. Ilieva S, Nikolov N, Ivanov I et al (1978) New modifications of *Salvia sclarea* L. with a high essential oil content. In: Third National Conference "Problems of cosmetics and perfumery". Varna, Bulgaria, 15–17 October
60. Ilieva S (1979) New *Salvia sclarea* cultivars developed by hybridization. Herba Hungarica 18 (3):197–203
61. Ilieva S, Grigorova P, Tsutsulova A (1985) Cultivar *Salvia sclarea* with higher yield and quality of essential oil. In: Five National Conference of aromatic "Problems of cosmetics and perfumery". Varna, Bulgaria, 17–19 October
62. Kovatcheva N, Zheljazkov V (1996) Essential oil content and components of *Salvia officinalis* L. from Bulgaria. In: 27th International Symposium on Essential Oils. Vienna
63. Lawrence B (2001) Progress in essential oil. Sage oil Perfum Flavour 26(3):66–79

64. Hristova Y, Gochev V, Wanner J et al (2013) Chemical composition and antifungal activity of essential oil of *Salvia sclarea* L. from Bulgaria against clinical isolates of *Candida* species. J BioSci Biotechnol 2(1):39–44
65. Yousefzadia M, Sonbolib A, Karimic F et al (2007) Antimicrobial activity of some *Salvia* species essential oils from Iran. Z Naturforsch 62:514–518
66. Gurgulova K, Nenchev P, Zhelyazkova I et al (2006) Study on biological activity of some essential oils to microorganisms *Bacillus cereus* and *Bacillus subtilis*. J Anim Sci 43:42–45
67. Rusenova N, Parvanov P (2009) Antimicrobial activities of twelve essential oils against microorganisms of veterinary importance. Trakia J Sci 7(1):37–43
68. Roussenova N (2011) Antibacterial activity of essential oils against the etiological agent of American foulbrood disease (*Paenibacillus larvae*). Bul J Veter Med 14(1):17–24
69. Georgiev V, Marchev A, Haas C et al (2011) Production of oleanolic and ursolic acids by callus cultures of *Salvia tomentosa* Mill. Biotechnol Biotechnol Equip 25(4):34–38
70. Marchev A, Ivanov I, Denev P et al (2015) Acetylcholinesterase inhibitory, antioxidant, and antimicrobial activities of *Salvia tomentosa* Mill. essential oil. J BioSci Biotechnol 4(2): 219–229
71. Marchev A, Georgiev V, Nikolova M et al (2012) Chemical composition of essential oil from *Salvia scabiosifolia* from Bulgaria. J Essent Oil Bear Pl 15(6):908–914
72. Nikolova M, Janicsak Genova E et al (2006) Comparative analysis of extremal flavonoids of Bulgarian and Hungarian samples of *Salva* species. Acta Bot Hung 48:361–367
73. Georgiev V, Marchev A, Nikolova M et al (2013) Chemical composition of essential oil from leaves and flowers of *Salvia ringens* Sibth. et Sm. growing wild in Bulgaria. J Essent Oil Bear Pl 16(5):624–629
74. Bozov P, Atanasova T, Merdzhanov P et al (2016) Chemical composition of the essential oils of *Salvia amplexicaulis* Lam. and *Ajuga laxmanii* Benth. from Bulgaria. J Food Packag Sci Technol Technol 5:14–17
75. Petrovic S, Pavlovic M, Tzakou O et al (2009) Composition and antimicrobial activity of *Salvia amplexicaulis* Lam. essential oil. J Essent Oil Res 21:563–566
76. Velickovic D, Ristic M, Milosavlievic N et al (2012) Chemical composition of the essential oils of *Salvia austriaca* Jacq. and *Salvia amplexicaulis* Lam. from Serbia. Agro Food Ind Tech 5:56–58
77. Damyanova S, Stoyanova A, Bozov P et al (2016) Chemical composition of essential oil from *Salvia aethiopis* L. from Bulgaria. Ukr Food J 5:497–504
78. Chalchat J-C, Gorunovic M, Petrovic S et al (1999) Chemical composition of two wild species of the genus *Salvia* L. from Yugoslavia: *Salvia aethiopis* and *Salvia verticillata*. J Essent Oil Res 13(6):416–418
79. Velickovic D, Ristic M, Velickovic A (2003) Chemical composition of the essential oils obtained from the flower, leaf and stem of *Salvia aethiopis* L. and *Salvia glutinosa* L. originating from the southeast region of Serbia. J Essent Oil Res 15 (5):346–349
80. Morteza-Semnani K, Goodarzi A, Azadcakht M (2005) The essential oil of *Salvia aethiopis* L. J Essent Oil Res 17(3):274–275
81. Tajbakhsh M, Rineh A, Khalilzadeh M et al (2007) Chemical constituents of the essential oils from leaves, flowers, stem and aerial parts of *Salvia aethiopis* L. from Iran. J Essent Oil Res 19(6):569–571
82. Torres E, Velasco-Negueruela A, Perez-Alonso M et al (1997) Volatile constituents of two *Salvia* species grown wild in Spain. J Essent Oil Res 9(1):27–33
83. Güllüce M, Ozer H, Bariş O et al (2006) Chemical composition of the essential oil of *Salvia aethiopis* L. Turk J Biol 30:231–233

Salvia spp.: An Updated on Antioxidant Activity and Pharmacological Uses

María Porres-Martínez, Elena González-Burgos, M. Emilia Carretero and M. Pilar Gómez-Serranillos

Abstract *Salvia* spp. (sage) is the largest genus of the family *Lamiaceae* (formerly *Labiatae*). This important genus includes about 900 species widespread throughout the world. Their species have been cultivated worldwide due to their excellent medicinal values, pharmaceutical applications and for culinary purposes. Studies on the chemical constituents of *Salvia* have revealed the presence of different active compounds, including terpenoids, flavonoids and phenolic compounds; the interesting pharmacological properties of sage have been attributed mainly to them. As example, some of the phenolic compounds of plants belonging to this genus have shown activity of active-oxygen scavengers such as superoxide anion, hydroxyl radical and singlet oxygen and activity to inhibit lipid peroxidation. Sage has been credited with a long list of medicinal uses. *Salvia officinalis*, *Salvia lavandulifolia*, *Salvia miltiorrhiza* and *Salvia leriifolia*, among others, have demonstrated potential antioxidant properties, an effect currently relevant to the prevention and treatment of Alzheimer's disease and other neurodegenerative diseases. In recent years, there is an increasing evidence to suggest that many degenerative diseases, such as brain dysfunction, cancer, heart diseases, ageing processes and immune system decline, could be the result of cellular damage caused by free radicals. Antioxidants may play an important role in preventing diseases caused by oxidative stress. The aim of this chapter is to update of antioxidant activity and pharmacological uses of species of *Salvia* to provide a better understanding of their therapeutic application.

Keywords *Salvia* species · Antioxidant · Pharmacological uses

María Porres-Martínez and Elena González-Burgos authors are contributed equally to this work.

M. Porres-Martínez (✉) · E. González-Burgos · M. Emilia Carretero
M. Pilar Gómez-Serranillos
Department of Pharmacology, Faculty of Pharmacy, Complutense University of Madrid, Plaza de Ramón Y Cajal S/N, 28040 Madrid, Spain
e-mail: m.porres@farm.ucm.es

V. Georgiev and A. Pavlov (eds.), *Salvia Biotechnology*,
https://doi.org/10.1007/978-3-319-73900-7_4

List of Abbreviations

ROS	Reactive oxygen species
MDA	Malondialdehyde
IsoPs	Isoprostanes
HNE	4-Hydroxynonenal
CAT	Catalase
SOD	Superoxide dismutase
GR	Glutathione reductase
GPx	Glutathione peroxidase
Se-GPx	Selenium-dependent glutathione peroxidase
NS-GPx	Selenium-independent glutathione peroxidase
GSSG	Oxidized glutathione
LDL	Low-density lipoprotein
DPPH	2,2-Diphenyl-1- picrylhydrazyl hydrate
ORAC	Oxygen radical absorbance capacity
ABTS	2,2′-Azino-bis(3-ethylbenzothiazoline-6-sulphonic acid)

1 Introduction

Under physiological conditions, there is a redox cell human body balanced between Reactive Oxygen Species (ROS) production and endogenous antioxidant defense systems activity [1]. An excessive production of ROS (i.e. hydroxyl radical, superoxide anion, hydrogen peroxide) may cause oxidative stress situation which produces oxidative cellular dysfunction, leading even to cell death [2]. Strong evidences support there is a relationship between ROS overproduction and the development of many diseases associated with age such as neurodegenerative diseases including Alzheimer's disease, Parkinson's disease, Amyotrophic lateral sclerosis, and Huntington's disease [3, 4].

However, the use of natural antioxidants may prevent and protect against the progression of these diseases associated with oxidative stress. In this context, the genus *Salvia* (Lamiaceae family) is an important source of diverse phytochemicals with antioxidant properties [5] and they may have beneficial effects for the prevention and treatment of these common age-related diseases associated with oxidative stress.

The aim of this chapter is to update of antioxidant activity and pharmacological uses of species of *Salvia* to provide a better understanding of their therapeutic application.

2 Oxidative Stress

Oxidative stress is a pathological event that occurs when there is an overproduction and accumulation of free radicals and reactive oxygen species (ROS) and these chemically reactive molecules cannot be neutralized or destroyed by antioxidant defenses of the body (enzymatic and non-enzymatic antioxidants). As consequence of this redox imbalance, there is an oxidative damage to biomolecules (lipids, proteins and DNA) and can cause excitotoxicity and cell death, mainly by apoptosis. Oxidative stress has been involved as a cause or consequence of more than 100 different diseases including cancer, heart diseases, neurodegenerative conditions (Alzheimer's disease, Parkinson's disease and amyotrophic lateral sclerosis), inflammatory diseases and metabolic disorders, among others [6–8] (Fig. 1).

2.1 *Reactive Oxygen Species (ROS)*

The term reactive oxygen species (ROS) includes both free radical species [i.e. superoxide radical ($^{\bullet}O^{-}{}_{2}$); hydroxyl radical ($^{\bullet}OH$)] and non-radical species derived from molecular oxygen which are oxidizing and/or easily converted to free radicals [i.e. hydrogen peroxide (H_2O_2)]. These ROS are found at low concentrations under physiological conditions and they have a key role in regulation as they participate in signal transduction as second messengers, they mediate cell proliferation, migration, and differentiations as well as they regulate gene expression. However, because endogenous (i.e. increased mitochondrial leakage, inflammation) and

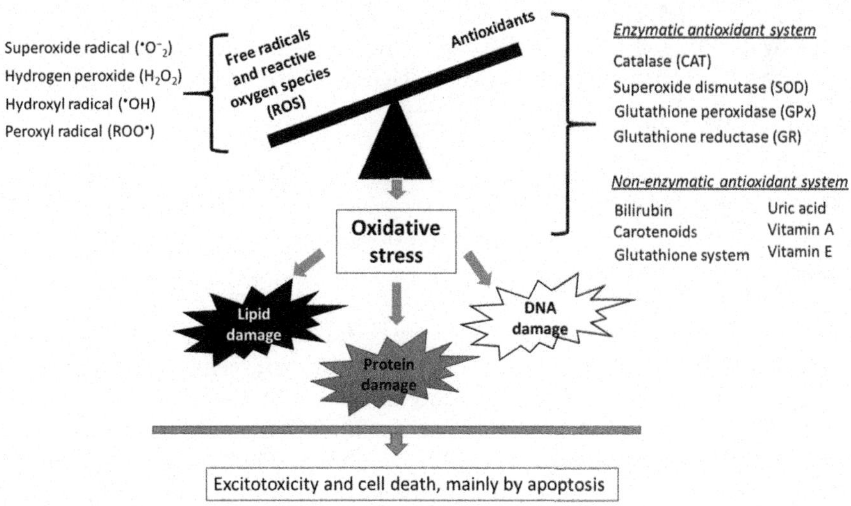

Fig. 1 Graphical summary of oxidative stress process

exogenous (i.e. smoking, stress, nutrition, environmental pollution) reasons, there could be an excessive amount of intracellular ROS which can react with macromolecules by chain reactions leading to a redox imbalance and cell damage (oxidative stress) [8, 9].

Examples of reactive oxygen species (ROS):

- Superoxide radical ($\cdot O^{-}_2$): This reactive species is the most abundant in the body at physiological pH. It is mainly originated in the respiratory chain through mitochondrial I and III complexes. Although this ROS is not so reactive as others because its inability to pass across cellular membranes, superoxide radical is involved in many oxidation reactions to produce more reactive and damaging free radicals such as hydroxyl radical ($\cdot OH$) [10].
- Hydrogen peroxide (H_2O_2): This non-radical oxygen species can be produced by the bivalent oxygen reduction and by the enzymatic action of superoxide dismutase and oxidases (i.e. monoamine oxidase, L-amino acid oxidase, glycolate oxidase). Hydrogen peroxide has less selective reactivity than free radicals because it has not unpaired electrons in the outer shell; however, unlike superoxide anion, the hydrogen peroxide is able to pass across cell membranes and inside the cell it is involved in numerous reactions of free radical formation [11].
- Hydroxyl radical ($\cdot OH$): This is the most known reactive and toxic oxygen-derived species. This free radical attacks DNA modifying their nucleotide bases by oxidation and it also causes the rupture in the double-stranded DNA molecule. Moreover, hydroxyl radical attacks lipid membranes and proteins (especially cysteine and methionine residues). This free radical can be formed through different reactions: (i) by H_2O_2 and H_2O lysis, (ii) by Haber-Weiss reaction (this reaction consists on producing hydroxyl radicals from hydrogen peroxide and anion superoxide) and (iii) by H_2O_2 reduction catalyzed metal ions reaction [12].
- Peroxyl radical ($ROO\cdot$): This is one of the most abundant free radicals. It is formed from hydroperoxide organic (ROOH) for loss of hydrogen atom [13].

Oxidative damage to biomolecules

Reactive oxygen species (ROS) can react with biomolecules (proteins, lipids and DNA), causing accumulative damage that leads to structural and functional cell alterations. This loss of redox homeostasis can cause cell death, mainly by apoptotic mechanisms [14, 15].

At DNA level, the major ROS that attacks DNA are $HO^{\bullet-}$ and $H^{\bullet-}$; these ROS cause breakage of DNA strands, and oxidize both purine bases and pyrimidine and sugar deoxyribose inducing mutations that lead to permanent functional genetic material associated with cancer and aging diseases. Mitochondrial DNA is more susceptible to oxidative damage than nuclear DNA because its lower antioxidant repair capacity and its proximity to the electron transport chain [16, 17].

Regarding lipids, the reactive oxygen species ($^{\cdot}OH$, $\cdot HOO$, $^1O_2$) can pull off a hydrogen atom from phospholipids giving to a saturated lipid radical ($R\cdot$) which reacts with oxygen and forms in last term peroxyl radicals (ROO·). This is a chain reaction called as lipid peroxidation and provokes structural and functional alterations of cell membranes (loss of fluidity and secretory functions, increase of porosity and permeability). There have been identified several markers of lipid peroxidation including malondialdehyde (MDA), isoprostanes (IsoPs) and 4-hydroxynonenal (HNE) [18, 19].

Proteins, main components of biological systems, are also susceptible to ROS attack. Particularly, it has been reported that hydroxyl radical attacks sulfurized or aromatic amino acids in protein structures lead to protein denaturation and modifications in metabolic and genetic regulation, signal transmission and enzyme activity [15, 20].

Carbohydrates are the biomolecules that react with ROS in lesser extent. Polysaccharides can suffer depolymerization and loss its structure and function by hydroxyl radical attack. The oxidative carbohydrate damage is related to diabetes mellitus pathogenesis [21].

2.2 *Antioxidants*

Under physiological conditions, there is a balance between ROS generation and effective cellular antioxidant defense mechanisms. Antioxidants (enzymatic and non-enzymatic system) are defined as molecules which at low concentrations (compared with the oxidizable substrate) protect cells from ROS damage by scavenging action, up-regulation of antioxidant enzymes and chelating metal-induced oxidative stress, maintaining redox state and homeostasis. However, when there is a ROS overproduction and intracellular accumulation, antioxidants cannot face up ROS damage. Therefore, the use of exogenous antioxidants has become one of the most promising strategies to minimize oxidative stress [22].

Enzymatic antioxidant system

Antioxidant enzymes including catalase (CAT), superoxide dismutase (SOD), glutathione reductase (GR) and glutathione peroxidase (GPx), neutralize intracellular ROS action transforming them into less harmful molecules and repairing their oxidative damage.

- Catalase (CAT): This antioxidant enzyme, located mainly in cytoplasm, mitochondria and peroxisomes of all aerobic organisms, decompose hydroperoxides such as hydrogen peroxide to water and oxygen molecules. Structurally, catalase is a tetrameric enzyme (60 kDa/each protein subunit) and it has a heme group (ferro [Fe (III)]-protoporphyrin) [23, 24].
- Superoxide dismutase (SOD): This enzyme, composed of 2 subunits of similar size, catalyzes the conversion of superoxide radical into hydrogen peroxide and oxygen. There have been identified three different SOD isoforms in humans [25, 26]:

- SOD 1: cytosolic Cu Zn SOD (32 kDa dimer).
- SOD 2: Mn SOD mitochondrial (96 kDa tetramer).
- SOD 3: Cu Zn SOD extracellular (132 kDa tetramer).

- Glutathione peroxidase (GPx): This enzyme is the main antioxidant defense system against low levels of oxidative stress. There have been identified two different isoforms of GPx [27]:
  - Selenium-dependent glutathione peroxidase (Se-GPx): Structurally, Se-GPx consists of four identical subunits (80 kDa), each one with a selenocysteine residue (necessary for its enzymatic activity) and it is presented in cytosol. This isoform catalyzes the conversion of hydrogen peroxide and organic hydroperoxides (ROOH).
  - Selenium-independent glutathione peroxidase (NS-GPx): It is a dimer (19–25 kDa) located in cytoplasm. This enzyme converts organic peroxides (ROOH) to hydroxides (ROH) using GSH as cofactor.
- Glutathione reductase (GR): This is a homodimeric enzyme (120 kDa) with a flavin group in its active center and it is mainly located in cytosol. This antioxidant enzyme reduces oxidized glutathione (GSSG) using nicotinamide adenine dinucleotide phosphate (NADPH) as cofactor [28]:

Non-enzymatic antioxidant system

The non-enzymatic antioxidant defense system is the first line of defense against ROS and complement antioxidant enzymes activity. It consists on molecules with low molecular weight which slow oxidation process or scavenging ROS, preserving redox balance and restoring oxidative damage [29].

- Bilirubin: This lipid-soluble antioxidant, formed during the heme degradation process, inhibits lipid peroxidation by scavenging peroxyl radicals [30].
- Carotenoids: These natural pigments quench singlet oxygen through its long conjugated double bonds chains, blocking lipid peroxidation process. The combination of carotenoids and vitamin E increase synergically their antioxidant action [31].
- Glutathione system: Glutathione (L-γ-glutamyl-L-cysteinylglycine) is the major endogenous antioxidant molecule in the body. Glutathione can be found as reduced glutathione (GSH) and oxidized glutathione (GSSG). The reduced glutathione (GSH) has antioxidant activity through its sulfhydryl active group of cysteine residue, exchanging electrons, ensuring the reduction of organic peroxides and hydrogen peroxide. GSH is then converted into GSSG [32, 33].
- Uric acid: This end product of purine metabolism acts as an effective antioxidant by chelating metal ion (iron and copper) and scavenging superoxide radical, hydroxyl radical and other ROS [34].

- Vitamin A (retinol): It is another important lipid-soluble antioxidant which protects membranes against lipid peroxidation [35].
- Vitamin E (α-tocopherol): Like vitamin A does, vitamin E protects lipids from oxidation by blocking reactive oxygen species reactions and consequently, lipid peroxidation chain reactions [35].

3 Oxidative Stress Related Diseases

Oxidative stress is implicated in the pathogenesis of many chronic diseases such as neurodegenerative diseases, cancer, heart diseases, ageing processes and immune system decline.

Regarding neurodegenerative diseases, brain is especially vulnerable to oxidative stress because its low glutathione levels and antioxidant enzymatic content, its richness in polyunsaturated fatty acids (highly susceptible to lipid peroxidation) and its large oxygen metabolism requirements. In vitro, in vivo and post mortem studies of human brains from patients with different neurodegenerative diseases (i.e. Alzheimer's, Parkinson's disease, lateral sclerosis amyotrophic and Huntington's disease) have detected high levels of reactive oxygen species, high amounts of lipid peroxidation biomarkers such as 4-hydroxy-2-nonenal and malondialdehyde, and an imbalance in endogenous enzymatic and non-enzymatic antioxidants [36, 37].

Oxidative stress has also been linked to wide variety of different cancers including brain tumor, breast cancer, gastric cancer and lung cancer, among others. Cancer cells are more sensitive to oxidative stress than normal cells. The redox imbalance leads to oxidative cellular damage (DNA, protein and lipids) and defects in antioxidant and DNA repair mechanisms, which in last term causes mutation and genomic instability and alterations in the normal cell growth [38, 39].

The involvement of oxidative stress has been also reported for several cardiovascular diseases such as hypertension, ischemic heart disease and atherosclerosis. Experimental and animal models have demonstrated that injured cardiac and vascular myocytes present low levels of reduced glutathione and erythrocyte glutathione-peroxidase-1, impaired mitochondrial reduction of molecular oxygen, high content in oxidized low-density lipoprotein (LDL) and alterations in calcium homeostasis [40, 41].

Oxidative stress is also associated with normal aging process. During ageing, antioxidant defenses start to function more poorly and there is an accumulation of ROS as a result of the years, which lead to a functionality decrease of the cells and make them more vulnerable to oxidative stress-related pathological conditions [42].

4 Salvia

Salvia L. spp. belongs to the Lamiaceae family and it is the genre that has the greatest species diversity and the largest number of species with nearly 900 species. This genus is native to the Mediterranean area and nowadays, it is widely distributed in Southeast Asia, Central and South America and Euro-Mediterranean region [43, 44].

Salvia spp. are aromatic herbaceous and perennial plants. They have simple and compound leaves, and verticillaster inflorescence with calyx and corolla campanulate bilabiate. They grow preferably in mountainous terrain (400 and 2000 m above sea level) with calcareous and sandy-calcareous soils (pH approximately 6.5). Moreover, *Salvia* spp. can grow spontaneously in all types of soil in temperate and warm areas and occasionally, these plants type are cultivated [45, 46].

From a phytochemical point of view, *Salvia* spp. has a complex chemical composition being terpenes and polyphenols their main compounds. Moreover, sages are rich in essential oils; there have found great qualitative and quantitative variations in essential oil composition intra and interspecies, which is attributed to soil type, daylight hours, geographic distribution and water regime, among other factors [47, 48].

Traditionally, *Salvia* spp. have been widely used by Egyptians, Greeks, Romans, American Indians and Chinese medicine for its properties as enhancer memory, anti-ulcerogenic, anti-epileptic, vulnerary agent, astringent, digestive, antipyretic and expectorant agent. *Salvia* has been used in treatment of pain, for protecting the human body against oxidative stress, free radical damages, inflammation or infections. Additionally species of this genus treat common illnesses such as depression, dementia, obesity, diabetes, lupus, heart disease, autism and cancer. Essential oils are very important sources for the treatment of different illness such as cancer, antimicrobial, antioxidant, and free radical scavenging agents [49]. *Salvia officinalis*, *Salvia miltiorrhiza* and *Salvia lavandulifolia* are some of the most extensive sage plants used in medicine [50–52]. In addition to its pharmacological properties, sages have been widely employed in food industry and cosmetic formulations to enhance the aroma and flavor and as conservator for its antioxidant properties [44, 53].

4.1 Salvia *Antioxidant Species*

Salvia genus is a source of phytochemical compounds with various beneficial effects on health. However, there are many differences in pharmacological effects among species. *Salvia* genus has been subject of intensive study in the last years for its antioxidant properties related to their active components in a large number of *Salvia* species [54]. Some examples of *Salvia* antioxidant species are:

- *Salvia bicolor* Desf. This specie grows in Mexico and North Africa. The extracts from its aerial parts exhibited antioxidant activity. *S. bicolor* contains unsaturated fatty acids, sterols, triterpenes, phenolic acids and flavonoids [55].
- *Salvia chloroleuca* Rech. f. and Allen. is one of Iran's endemic species. The antioxidant activity of this plant could be in part related to flavonoids and sterols [56].
- *Salvia desoleana* Atzei & Picci is native to the island of Sardinia in the Mediterranean. It is used in folk medicine to treat diseases of the central nervous system and it has showed antioxidant activity [57].
- *Salvia divinorum* is endemic to Mexico and it is characteristic due to its mental effects. The most prevalent diterpene, salvinorin A, and other active components produce psychoactive properties, for example sedative effects [58, 59].
- *Salvia fruticosa* Mill. is native to the eastern Mediterranean and it is widely used in folk medicine for the treatment of different disorders. It is also named as *S. libanotica* and formerly named as *S. triloba*. This endemic Lebanese plant has also been used for improving memory and as hypoglycemic agent with antioxidant potential [54, 60, 61].
- *Salvia haematodes* Wall. This name is a synonym of *S. sclarea* L. It is native to the northern Mediterranean, along with some areas in North Africa and Central Asia. It has significant antioxidant activity and it also possesses central nervous system depressant (anticonvulsant) properties [62].
- *Salvia hispanica* L. commonly named as Chia. Chia is a rich source of polyunsaturated fatty acids. Chia seeds and oil should be considered as functional ingredients with high potential antioxidant activity in food products due to its composition (tocopherols, phytosterols, carotenoids and phenolic compounds, including chlorogenic acid, caffeic acid, myricetin, quercetin and kaempferol) [63, 64].
- *Salvia lanigera* Poir. Its essential oil has mainly thymol, hexadecanoic acid, carvacrol, α-thujone. Its essential oil possesses a good antioxidant activity and also shows an antimicrobial activity [65].
- *Salvia lavandulifolia* Vahl. commonly known as "Spanish sage", is an aromatic plant native to the Iberian Peninsula. *S. lavandulifolia* has been commonly used in Mediterranean folk medicine as antioxidant. Furthermore, experimental studies have shown the potential value of *S. lavandulifolia* in dementia therapy attributed to its sedative, antioxidant, anti-inflammatory, estrogenic and anti-cholinesterase activities [53, 66].
- *Salvia leriifolia* Benth. is native to Iran and Afghanistan. This plant showed antiproliferative activity in cancer cell lines [67] and its essential oil has antioxidant effect [68].
- *Salvia macilenta* Boiss. which is endemic to Iran, is a member of the *Salvia* genus whose antioxidant activity and neuroprotective effect has been demonstrated [69].

- *Salvia miltiorrhiza* Bunge is a traditional Chinese Herbal. It has been used for more than a millennium in Asian countries, especially in China, Japan and Korea. It presents sedative, anti-inflammatory, anticancer, cardiovascular and antioxidant properties [70–72].
- *Salvia mirzayanii* Rech. f. & Esfand is an endemic herbaceous plant of Iran and is broadly used in folk medicine. *S. mirzayanii* has antioxidant, anti-inflammatory, immunomodulatory, antimicrobial, and neuroprotective activities. Its main components are essential oil, flavonoids (chrysoeriol, cirsimaritin, luteolin, salvigenin), catechin and rosmarinic acid [73, 74].
- *Salvia officinalis*, L. is native to the Mediterranean region and it is now extensively cultivated all over the world. It is one of the most widespread herbal species used in the area of human health because its antioxidant properties [75]. The essential oil of *S. officinalis* is characterized by a high amount of α- and β-thujone. The non-volatile fraction of sage is mainly composed of various diterpenes, phenolic acids and flavonoids [76]. Rosmarinic acid and carnosol were the main compounds of all the antioxidant phenolic extracts isolated from this plant [77]. It is showed pharmacological effects of *S. officinalis* on the central nervous system including neuroprotective and memory enhancing effects [78].
- *Salvia palaestina* Benth. is a plant from Turkey, Iran, Egypt, Syria, Lebanon, Jordan and Palestine. It has antioxidant activity due to active phenolic compounds [79].
- *Salvia plebeia* R. Br. is widely distributed in many provinces of China. As a traditional Chinese medicine, it has been used to treat various diseases. Current pharmacological studies have shown that it exhibits antioxidant activities. Phytochemical studies of *S. plebeia* revealed the presence of flavonoids, terpenoids and lignans [80].
- *Salvia reuterana* Boiss. is a medicinal herb with various therapeutic usages. It possesses neurological and antioxidant properties [81]. The antioxidant property of methanolic extract of *S. teuterana* was in vitro assessed [82].
- *Salvia verbenaca* L. is a potential source of antioxidants. Rosmarinic acid is the most abundant phenolic compound [83]. The plant extracts exhibit in vitro antioxidant activity by inhibition of lipid peroxidation. The antioxidant activity of the plant extract may be due to the polyphenolic compounds [84, 85].
- *Salvia verticillata* L. specie from Iran, whose extracts presents high antioxidant activity [86] and exhibited a potent free radical scavenging activity. Chrysoeriol is one isolated active compound presented in this plant specie [87].

4.2 Salvia *Antioxidant Chemical Composition*

Salvia is a rich source of phytochemicals including phenolic acids, polyphenols, flavonoids, anthocyanins, terpenoids and essential oils [79, 88, 89]. The antioxidant properties of *Salvia* are mainly due to its active constituents that are known to be a rich source of antioxidants [77, 79] in enzyme-dependent and enzyme-independent systems [90, 91]. The plants of this genus synthesize several types of secondary metabolites with radical scavenging, useful in pharmaceuticals industry due to its potential therapeutic use [92].

Many *Salvia* species present antioxidant capacity related to the presence of rosmarinic acid and carnosic acid [79]. However, other components play also a very important role in this antioxidant activity: carnosol, methyl carnosate, and monoterpenoid α-pinene or carvacrol [75, 93].

The main secondary metabolites in *Salvia* plants are classified in terpenoids and phenolic compounds [94, 95].

- *Terpenoids*

Terpenoids are lipids synthesized from acetyl-CoA via the mevalonic acid pathway.

The most common antioxidant terpenoids are:

- Carnosic acid is an active diterpene found in *Salvia* species such as *S. officinalis* L. with antioxidant, cytotoxic and antimicrobial properties [96]. Carnosic acid protects against glutamate toxicity in rat cortical cultures and against cerebral ischemia in mice by activating the nuclear factor E2-related factor 2 (Nrf2)/ Kelch-like ECH-associated protein 1 (Keap1) pathway [97].
- Carnosol is a phenolic diterpene presented in *Salvia* species such as *S. officinalis*. Carnosol has antioxidant, anti-microbial, anti-cancer, anti-inflammatory, and lipid lowering properties. Recently a number of investigations have demonstrated that these compounds also exhibit biological activity which may be useful against neurodegenerative diseases [88, 97, 98].
- Tanshinone is a group of active diterpenes present in the root of *Salvia miltiorrhiza* [99]. Traditionally it is used due to its anti-inflammatory activity [100] and antioxidant effects by activating signaling pathways associated with the redox state, especially in Chinese medicine. In addition, recently tanshinone activities have been studied due to its clinical applications. Tanshinone I protects mitochondria against prooxidants agents [101]. As examples: regulation of immune system [102], cardioprotective [103] and antitumor activity [72, 104]. Tanshinones have been identified as a novel class of Nrf2-inducers for antioxidant protection. The NF-E2 p45-related factor 2 (Nrf2) signaling pathway represents an important cellular defense system that suppresses damage caused by exposure to toxicants [70].
- The volatile oils are considered a good source of natural bioactive compounds [105]. In sage are chemically complex mixtures where terpenoid molecules predominate. Volatile oils can also act as antioxidants due to their redox properties, which play an important role in adsorbing and neutralizing free radicals, or by decomposing peroxides, quenching singlet and triplet oxygen, or by chelating the metal ions [106–108]. In example, there are studies with the major selected monoterpenes α-pinene and 1,8-cineole found in *Salvia lavandulifolia* essential oils that show their activity as regulators of cellular redox balance [109].

- *Phenolics*

Phenolic compounds are aromatic substances mainly formed via the shikimic acid or the malonic acid pathway in various ways. Phenolic compounds perform a potent antioxidant activity due to its effectiveness in scavenging radicals [110]. The higher amount of flavonoids and phenolic compounds correspond to their greater antioxidant activity [111, 112]. Some of the main antioxidant phenolics compounds are:

- Carvacrol, is a monoterpenic phenol presents in many essential oils. Carvacrol has a pronounced antioxidant effect against free radicals generated in vitro since it is able to scavenge peroxyl radicals [113].
- Luteolin is one of the most common flavonoids present in plants. Some studies suggest a potential therapeutic use as anticancer and antioxidant [66, 114].
- Rosmarinic acid is a phenylpropanoid with antioxidant properties. Rosmarinic acid and its by-products are present in several *Salvia* species, such as *S. officinalis* and *S. fruticosa*. Rosmarinic acid shows a great intensity for free radical scavenging and antioxidant activity [115]. This compound may be a potential therapeutic option in the management of learning and memory [78, 116].
- Salvianolic acids such as salvianolic acids A, B and K present protective action against peroxidative damage to biomembranes. They have anti-oxygen radicals and antioxidant activity. They are found in roots of *S. miltiorrhiza* and in *S. deserta*. Salvianolic acid B (Sal B) is one of effective compounds in the root and rhizome of *S. miltiorrhiza* Bunge and it is officially listed in the Chinese Pharmacopoeia [117]. Sal B shows great free radical scavenging and antioxidant activity [118]. It has demonstrated that Sal B inhibits lipid peroxidation and scavenges superoxide anions and hydroxyl radicals [119].

The main antioxidant compounds of *Salvia* spp. are represented in the Table 1.

4.3 *Salvia Antioxidant Activity in* In Vitro *Studies*

The genus *Salvia* has attracted great interest and it has been the topic of many pharmacological studies. Some species belonging to this genus have been investigated recently for their potent antioxidant and protective effects using different in vitro methods, including screening of the antioxidant potential of some extracts and essential oils [76]. The antioxidant compounds of this genus play a defensive role by preventing the generation of free radicals and consequently are extremely beneficial to improve many diseases caused by oxidative stress [120].

The present chapter has focused to evaluate in vitro antioxidant potential activity of *Salvia* and free radical scavenging effects. For these purpose, most of the studies assess the potential antioxidant activity of sage by spectrophotometric methods as 2,2-diphenyl-1- picrylhydrazyl hydrate (DPPH) free radical scavenging assay [89]. The total antioxidant capacity of *Salvia* species can be determinate also by phosphomolybdenum quantitative method—number of equivalents of gallic acid—[55],

Table 1 Main components from *Salvia* spp. with antioxidant properties

Compounds	Chemical structure	Examples	Sources	References
Carnosic acid	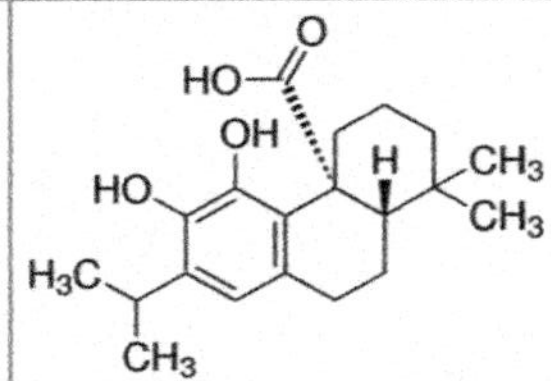	*S. fruticosa* *S. officinalis*	Leaves	[54, 122]
Carnosol	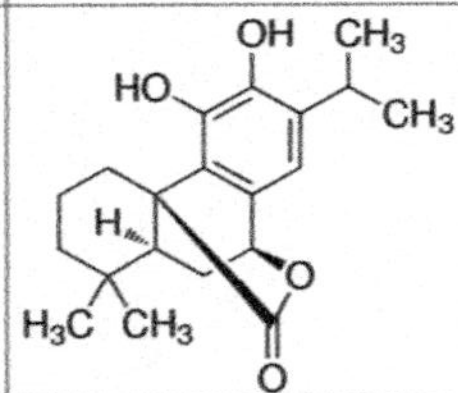	*S. fruticosa* *S. officinalis*	Plant	[54, 122]
Carvacrol	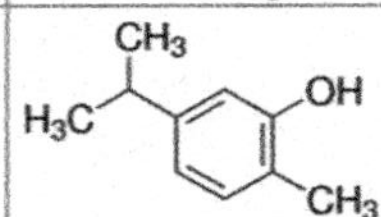	*S. lanigera* *S. lavandulifolia*	Essential oil	[65, 123, 124]
Luteolin	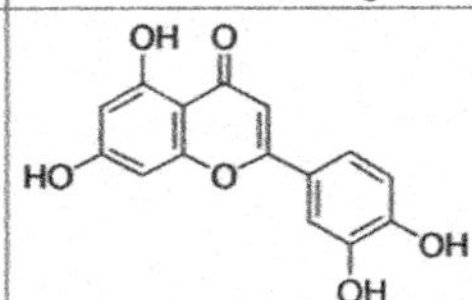	*S.palaestina* *S. plebeia* *S. sclarea*	Plant	[79, 80]
Salvianolic acid B	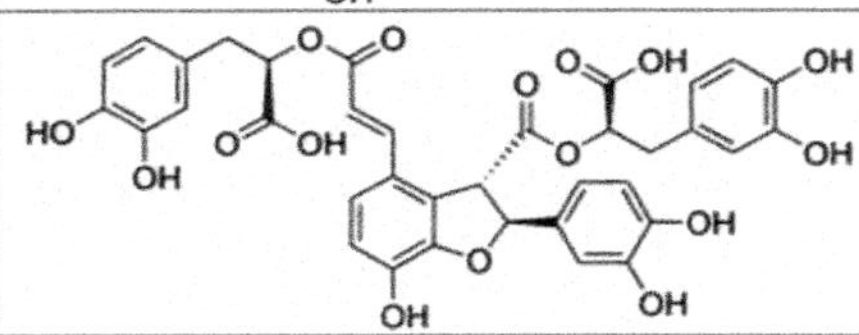	*S. miltiorrhiza*	Root and rhizome	[120]

(continued)

Table 1 (continued)

Compounds	Chemical structure	Examples	Sources	References
Rosmarinic acid	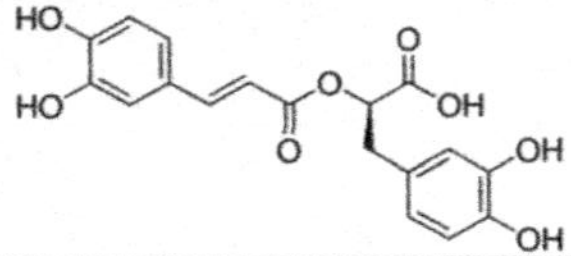	*S. fruticosa* *S. lavandulifolia* *S. miltiorrhiza* *S.officinalis* *S. verbenaca*	Plant	[54, 66, 85, 120]
Tanshinone	I 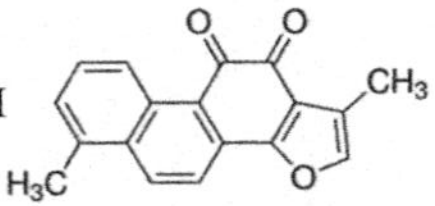 IIA 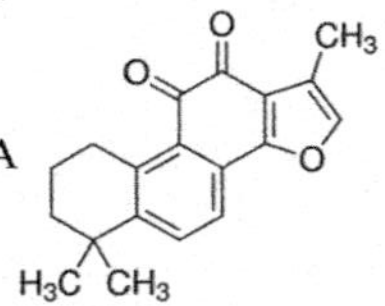 Dihydrotanshinone I 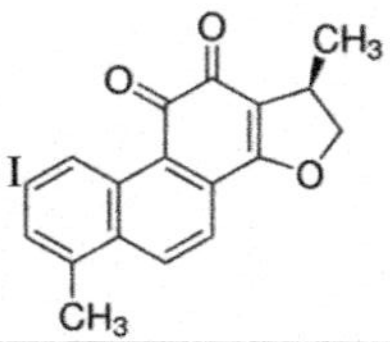	*S. miltiorrhiza*	Root	[101]
1,8-cineole	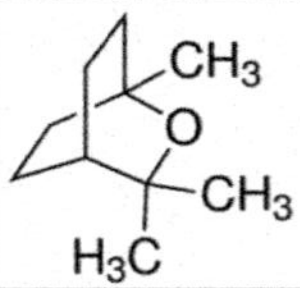	*S. lavandulifolia*	Essential oil	[66]
α-Pinene	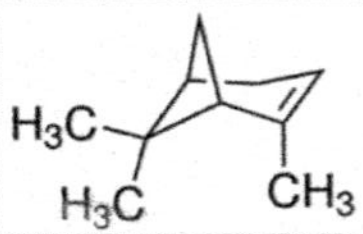	*S. lavandulifolia*	Essential oil	[66]

ABTS free radical scavenging assay [107] or oxygen radical absorbance capacity (ORAC) assay, total oxidant scavenging capacity assay [111].

An update on recent *Salvia* antioxidant studies is shown in Table 2.

4.4 Salvia *Pharmacological Uses*

The name *Salvia*, from the Greek word "*Salvere*", refers to the curative properties of the plants of this genus [121]. Many *Salvia* species have been used for the treatment of several disorders in traditional medicine all around the world [75, 77]. *Salvia* species have been reported to possess a wide variety of biological activities, and a wide diversity of phytochemicals. *Salvia* species have been traditionally used in medicine due to its anxiolytic, sedative and memory-enhancing properties. Furthermore, *Salvia* is a natural source of flavonoids and polyphenolics compound with antioxidant properties, protecting organism against oxidative stress and free radicals damages. Other important activities of sage are anti-inflammatory, antibacterial, analgesic, oestrogenic and antidepressive activities. *Salvia* also treats depression, obesity, diabetes, cardiovascular diseases and cancer [49, 93].

These pharmacological activities, especially antioxidant activity, highlight the potential use in the development of new drugs. *Salvia* species are a rich source of diverse phytochemicals antioxidants that could play a beneficial role in the treatment of mental and nervous conditions. Antioxidants of *Salvia* play a very important role to protect the body against the oxidative stress and free radical damages which are the cause of diseases such as cardiovascular diseases, cancer or central nervous system dysfunction [79]. *Salvia* species are known for the beneficial effects on memory disorders, depression and cerebral ischemia [93]. Many species are found to be rich in phenolic acids as active constituents that significantly contribute to their neuroprotective properties due to its antioxidant and free radical scavenging effect. The antioxidant properties of sage are found to be related to the presence of rosmarinic acid and carnosic acid. Several studies showed that sage play an active role to protect against a number of diseases related to oxidative stress and free radical-induced damage. These studies show the potential protective effects of several *Salvia* species against oxidative damage related to neurodegenerative diseases. For example, in treatment of Alzheimer's, Parkison disease, and other related diseases [76, 116].

There are some explanations to this antioxidant activity. Fischedick et al. (2013) described the structure activity relationship of phenolic diterpenes from *Salvia officinalis* as activators of the nuclear factor E2 (erythroid-derived 2)-related factor 2 (Nrf2) signaling pathway [97]. Nrf2 regulates the expression of antioxidant enzymes response such as Mn-superoxide dismutase (Mn-SOD), glutathione peroxidase (GPx), Hemo-oxygenase 1 (HO-1) and the metabolism of glutathione and is a transcription factor known to activate cytoprotective genes which may be useful in the treatment of neurodegenerative diseases [119].

Table 2 Recent studies of *Salvia* species on their antioxidant properties

Salvia specie	Relevant compounds/ plant area	Type of study	Effects	References
S. aegyptiaca *S. aethiopis* *S. atropatana* *S. eremophila* *S. hypoleuca* *S. limbata* *S. nemorosa* *S. santolinifolia* *S. sclarea* *S. syriaca* *S. xanthocheila*	Rosmarinic acid, carnosol, carnosic acid	Antioxidant potential of plant extracts. DPPH assay	*S. nemorosa*, *S. atropatana*, *S. santolinifolia*, and *S. eremophila* are the most active radical scavengers with higher total phenol contents	[92]
S. bicolor	A petroleum ether extract (PEE) and a methanolic extract (ME) of aerial parts	Antioxidant activities of extracts. DPPH assay Total antioxidant capacity	The higher antioxidant activity is for the ME. It may be attributed to its content of flavonoids and phenolic acids, as protocatechuic, ferulic and caffeic acids that have antioxidant activity	[55]
S. cryptantha	Carnosic and rosmarinic acids	DPPH radical-scavenging method and xanthine/xanthine oxidase (XO) system generated superoxide anion radical inhibition assay	High antioxidative property against DPPH. EtOAc extracts had high antioxidant activity against Xanthine oxidase	[125]
S. fruticosa	Root and aerial plant extracts	Antioxidant activity. DPPH assay Total phenolic contents. Folin-Ciocalteu reagent Phenolic constituents. High-performance liquid chromatography (HPLC)	Strong radical scavenging activity, the highest value in the roots and the lowest in the aerial parts. This antioxidant activity is correlated to the total phenolic content of different extracts, where rutin and luteolin are the most abundant constituents	[54]

(continued)

Table 2 (continued)

Salvia specie	Relevant compounds/ plant area	Type of study	Effects	References
S. hispanica	Extracts	DPPH free radical scavenging activity FRAP Assay (ferric reducing antioxidant power) Lipophilic and hydrophilic ORAC assay (oxygen radical absorbance capacity)	High antioxidant capacity and low levels of lipid autoxidation in chia seed extract The consumption of chia seed and oil could be interesting for health improvement	[64]
S. lanigera	Essential oil	Ferric reducing/antioxidant power (FRAP method). The ability of extracts to scavenge the DPPH radical was measured using the method of Brand-Williams	*S. lanigera* essential oil possess a good antioxidant activity	[65]
S. lavandulifolia	Essential oil	In vitro neuroprotective activity against H_2O_2-induced oxidative stress in PC12 cells. The effect on cell viability and morphology, lipid peroxidation, GSH/ GSSG ratio, intracellular ROS levels, antioxidant enzymes (CAT, SOD, GR, GPx, HO-1) and apoptotic enzymes is investigated	*S. lavandulifolia* essential oil has therapeutic value for the prevention and treatment of neurodegenerative diseases associated with oxidative stress-induced neuronal injury	[126]
S. lavandulifolia	Essential oil	ORAC, Intracellular ROS generation, enzyme activity assays, expression of the antioxidant enzymes CAT, SOD, GPx, GR and HO-1 and the Nrf2 transcription factor is determined by Western blot analysis in a H_2O_2-induced oxidative stress model in astrocytes (U373 MG cell line)	Essential oil samples exhibited a potent antioxidant capacity by ROS scavenging, enhancing the endogenous antioxidant system and inhibiting lipid peroxidation	[123, 124]
S. macilenta	Methanolic extract	Differentiated PC12 cells are exposed to oxidative stress. Western blot analysis of caspase-3, Bax, Bcl-2 and PARP	It exhibits antiglycating activity and protects PC12 cells against oxidative stress-induced apoptotic cell death. *S. macilenta* has hydroxyl radical scavenging capacity	[69]

(continued)

Table 2 (continued)

Salvia specie	Relevant compounds/ plant area	Type of study	Effects	References
S. microphylla	Shoots and leaves	Antioxidant activity is determined using butylated hydroxytoluene and vitamin C as reference standards by DPPH assay	Effective antioxidant activity	[127]
S. miltiorrhiza	Tanshinone I	Human neuroblastoma SH-SY5Y cells treated with paraquat mitochondrial toxic agent	Induction of mitochondrial protection through an Nrf2-dependent mechanism	[101]
S. miltiorrhiza	Salvianolic acid B	Human hepatoma cell line HepG2. Determination whether the protective effect of SalB against liver injury is associated with Nrf2 and phase II enzymes protein expression in vitro. Western blot analysis and mRNA expression	Sal B pretreatment increased the expression of nuclear factor-erythroid-2-related factor 2 (Nrf2), HO-1 and glutamate-L-cysteine ligase catalytic subunit (GCLC)	[119]
S. miltiorrhiza	Salvianolic acid B	ORAC assay and DPPH assay are conducted to explore the antioxidant activities	Strong antioxidant activity	[111]
S. miltiorrhiza	Salvianolate	H_2O_2-treated cardiomyocytes in vitro via down regulation of Smad2/3 and TGFβ1 expression. Total antioxidant capacity (TAOC) of culture medium	Salvianolate inhibits ROS and iNOS production and increases TAOC and nitric oxide (NO) levels	[128]
S. miltiorrhiza	Tanshinone I	ROS detection. Antibodies for Nrf2, Keap1, NQO1, c-GCS and b-actin, immunoblot analysis, ubiquitination assay and protein half-life	Tanshinone I activates the Nrf2-dependent antioxidant response. T-I activates Nrf2 and its downstream genes in HBE cells and protects HBE cells from As(III)-induced oxidative stress and cytotoxicity	[70]
S. miltiorrhiza	Ethanol and water extracts of plant material	Free radical DPPH scavenging assay. Antioxidant activity screening in assay using *Saccharomyces cerevisiae*	Significantly high free radical scavenging ability to inhibit the H_2O_2 induced yeast oxidative stress. Significant antioxidant activity	[120]

(continued)

Table 2 (continued)

Salvia specie	Relevant compounds/ plant area	Type of study	Effects	References
S. officinalis	Tartaric, malic, citric, rosmarinic acids, Eucalyptol	Evaluate in vitro ability to scavenge the free radical DPPH· and 2,2′-azino-bis (3-ethylbenzothiazoline-6-sulfonic acid) (ABTS· +), catalase (CAT-*like*) and superoxide dismutase (SOD-*like*) activity	Verification of the antioxidant properties of hydroalcoholic and waterbased extracts	[129]
S. officinalis	Essential oil (EO)	DPPH is used for determination of free radical-scavenging activity of the EO, ABTS free radical scavenging assay, β-carotene/linoleic acid bleaching assay	EOs from *S. officinalis* showed in general moderate to weak antioxidant activity. The antioxidant activity observed could be explained partially by the presence of *trans*-thujone, camphor, 1,8-cineole, and α-pinene	[130]
S. officinalis	Carnosic acid, carnosol, epirosmanol, rosmanol, 12-methoxy-carnosic acid, sageone, and carnosaldehyde	Screening in vitro for their ability to active the Nrf2 and general cellular toxicity using mouse primary cortical cultures	All compounds except 12-methoxy-carnosic acid are able to activate the antioxidant response element Furthermore both carnosol and carnoasldehyde are able to induce Nrf2-dependent gene expression as well as protect mouse primary cortical neuronal cultures from H_2O_2 induced cell death	[97]
S. officinalis	Rosmarinic acid, diterpenoids (carnosic acid and carnosol) and phenolic compounds	Scavenging of the free radicals using DPPH, transition metal reduction in phosphomolybdenum assay and inhibition of lipid oxidation	Plant extracts: effects on reducing Mo and DPPH radical scavenging	[122]
S. palaestina	Phenolics salpalaestinin methyl 3-O-methylrosmarinate	DPPH scavenging activities of compounds	Detection the highest antioxidant activities of these compounds	[79]

(continued)

Table 2 (continued)

Salvia specie	Relevant compounds/ plant area	Type of study	Effects	References
S. plebeia	Diterpenoids	Antioxidant assay: DPPH radical scavenging activity, Intracellular ROS assay with RAW 264.7, a murine macrophage cell line	Significant inhibitory effects on reactive oxygen species (ROS) production in lipopolysaccharide (LPS)-induced macrophages	[80]
S. sharifii	Ent-13-epi-manoyloxide, 6-hydroxy-6,7,4′-trimethoxyflavone, Ladanein	The antioxidant activates of the isolated compounds are evaluated using DPPH free radical scavenging assays	Phenolic compounds, ladanein and 6-hydroxy-5,7,4′-methoxy-flavone show higher antioxidant activities in comparison with the non-phenolic diterpene, *ent*-13-*epi*-manoyloxide	[121]
S. runcinata, S. repens, S. stenophylla	Rosmarinic acid	Quantitative DPPH assay	*S. runcinata* exhibited the best activity compared with *S. stenophylla* and *S. repens*	[131]
S. sclareoides	Linalool, linalyl acetate, *trans*-caryophyllene, germacrene, bicyclogermacrene	Inhibition of 2,2-diphenyl-1-picrylhydrazyl hidrate radical method	Antioxidant potential activity of essential oil	[89]
S. verbenaca	Extracts	DPPH activity according to the method described by Brand-Williams. Ferric Reducing Antioxidant Power (FRAP)	Different polyphenolic composition due to different phenological stage influence in antioxidant properties	[85]
S. verticillata	Chrysoeriol-	To identify the major antioxidant compound(s) from the methanol extract using activity-guided fractionation	The crude extract showed strong antioxidant activities in DPPH and β-carotene/linoleic acid tests. The ethyl acetate fraction also exhibited a potent free radical scavenging activity	[87]

5 Conclusions and Future Prospects

Salvia spp. are an important source of antioxidant bioactive compounds; these different *Salvia* spp. and their main components have demonstrated to be of interest for maintaining human redox balance and consequently, for the prevention and treatment of those pathologies associated with oxidative stress such as common neurodegenerative diseases like Parkinson's disease and Alzheimer's disease. Future studies should be directed to delve into the mechanism of action of these compounds in in vitro and in vivo models as well as to investigate the ability of these bioactive components to pass blood brain barrier.

This will offer considerable scope in the discovery and development of novel agents for preventing the progression of neurological disorders.

References

1. Halliwell B (2003) Oxidative stress in cell culture: an underappreciated problem? Fed Soc Biochem Mol Biol Lett 540:3–6
2. Halliwell B (1992) Reactive oxygen species and the central nervous system. J Neurochemi 59:1609–1623
3. Valko M, Leibfritz D, Moncol J et al (2007) Free radicals and antioxidants in normal physiological functions and human disease. Int J Biochem Cell Biol 39:44–84
4. Mao P (2013) Oxidative stress and its clinical applications in dementia. J Neurodegener Dis, Article ID 319898, 15 pages. http://dx.doi.org/10.1155/2013/319898
5. Wei A, Shibamoto T (2007) Antioxidant activities and volatile constituents of essential oils. J Agric Food Chem 54:1737–1742
6. Mayne ST (2003) Antioxidant nutrients and chronic disease: use of biomarkers of exposure and oxidative stress status in epidemiologic research. J Nutr 133:933S–940S
7. Floyd RA, Hensley K (2002) Oxidative stress in brain aging. Implications for therapeutics of neurodegenerative diseases. Neurobiol Aging 23:795–807
8. Halliwell B, Whiteman M (2004) Measuring reactive species and oxidative damage in vivo and in cell culture: how should you do it and what do the results mean? Br J Pharmacol 142:231–255
9. Ray PD, Huang BW, Tsuji Y (2012) Reactive oxygen species (ROS) homeostasis and redox regulation in cellular signaling. Cell Signal 4(5):981–990
10. Fujita M, Tsuruta R, Kasaoka S et al (2009) In vivo real-time measurement of superoxide anion radical with a novel electrochemical sensor. Free Radic Biol Med 47:1039–1048
11. Deshpande NN, Sorescu D, Seshiah P et al (2002) Mechanism of hydrogen peroxide-induced cell cycle arrest in vascular smooth muscle. Antioxid Redox Signal 4:845–854
12. Nguyen V, Bonds DV, Prokai L (2008) Measurement of hydroxyl-radical formation in the rat striatum by in vivo microdialysis and GC-MS. Chromatographia 68:57–62
13. Kanner J, German JB, Kinsella JE (1987) Initiation of lipid peroxidation in biological systems. Crit Rev Food Sci Nutr 25:317–364
14. Willcox JK, Ash SL, Catignani GL (2004) Antioxidants and prevention of chronic disease. Crit Rev Food Sci Nutr 44:275–295
15. Dalle-Donne I, Rossi R, Colombo R et al (2006) Biomarkers of oxidative damage in human disease. Clin Chem 52:601–623
16. Valko M, Izakovic M, Mazur M et al (2004) Role of oxygen radicals in DNA damage and cancer incidence. Mol Cell Biochem 266:37–56

17. Rao KS (2009) Free radical induced oxidative damage to DNA: relation to brain aging and neurological disorders. Indian J Biochem Biophys 46:9–15
18. Niki E, Yoshida Y, Saito Y, Noguchi N (2005) Lipid peroxidation: mechanisms, inhibition, and biological effects. Biochem Biophys Res Commun 338:668–676
19. Niki E (2012) Do antioxidants impair signaling by reactive oxygen species and lipid oxidation products? FEBS Lett 586:3767–3770
20. Belonogov RN, Titova NM, Lapeshin PV et al (2009) Changes in the content of protein and lipid oxidative modification products in tumor tissue at different stages of lung cancer. Bull Exp Biol Med 147:630–631
21. Monboisse JC, Gardes-Albert M, Randoux A (1988) Collagen degradation by superoxide anion in pulse and gamma radiolysis. Bioch Biophys Acta 965:29–35
22. Halliwell B (1996) Antioxidants in human health and disease. Annu Rev Nutr 16:33–50
23. Chance B, Sies H, Boveris A (1979) Hydroperoxide metabolism in mammalian organs. Physiol Rev 59:527–605
24. Aebi H (1984) Catalase in vitro. Meth Enzymol 105:121–126
25. McCord JM (2002) Superoxide dismutase in aging and disesase: an overview. Methods Enzymol 349:331–341
26. Fattman CL, Schaefer LM, Oury TD (2003) Extracellular superoxide dismutase in biology and medicine. Free Radic Biol Med 35:236–256
27. Halliwell B, Gutteridge JM (2007) Free radicals in biology and medicine. Oxford University Press, Oxford; New York
28. Tandogan B, Ulusu NN (2006) Kinetic mechanism and molecular properties of glutathione. J Pharm Sci 31:230–237
29. Wang JY, Wen LL, Huang YN et al (2006) Dual effects of antioxidants in neurodegeneration: direct neuroprotection against oxidative stress and indirect protection via suppression of glia-mediated inflammation. Curr Pharm Des 12:3521–3533
30. Clark JF, Sharp FR (2006) Bilirubin oxidation products (BOXes) and their role in cerebral vasospasm after subarachnoid hemorrhage. J Cereb Blood Flow Metab 26:1223–1233
31. Young AJ, Lowe GM (2001) Antioxidant and prooxidant properties of carotenoids. Arch Biochem Biophys 385:20–27
32. Takebe G, Yarimizu J, Saito Y et al (2002) A comparative study on the hydroperoxide and thiol specificity of the glutathione peroxidase family and selenoprotein P. J Biol Chem 277:41254–41258
33. Masella R, Di Benedetto R, Vari R et al (2005) Novel mechanisms of natural antioxidant compounds in biological systems: involvement of glutathione and glutathione-related enzymes. J Nutr Biochem 16:577–586
34. Nieto FJ, Iribarren C, Gross MD et al (2000) Uric acid and serum antioxidant capacity: a reaction to atherosclerosis? Atherosclerosis 148:131–139
35. Zadák Z, Hyspler R, Tichá A et al (2009) Antioxidants and vitamins in clinical conditions. Physiol Res 58:13–17
36. Choi J, Rees HD, Weintraub ST et al (2005) Oxidative modifications and aggregation of Cu, Zn-superoxide dismutase associated with Alzheimer and Parkinson diseases. J Biol Chem 280:11648–11655
37. Sayre LM, Perry G, Smith MA (2008) Oxidative stress and neurotoxicity. Chem Res Toxicol 21:172–188
38. Klaunig JE, Kamendulis LM, Hocevar BA (2010) Oxidative stress and oxidative damage in carcinogenesis. Toxicol Pathol 38:96–109
39. Sosa V, Moliné T, Somoza R et al (2013) Oxidative stress and cancer: an overview. Ageing Res Rev 12:376–390
40. Dhalla NS, Temsah RM, Netticadan T (2000) Role of oxidative stress in cardiovascular diseases. J Hypertens 18:655–673
41. Schnabel R, Blankenberg S (2007) Oxidative stress in cardiovascular disease: successful translation from bench to bedside? Circulation 116:1338–1340

42. Melo A, Monteiro L, Lima RM et al (2011) Oxidative stress in neurodegenerative diseases: mechanisms and therapeutic perspectives. Oxid Med Cell Longev 2011:467180. https://doi.org/10.1155/2011/467180
43. Walker JB, Sytsma KJ, Treutlein J et al (2004) *Salvia* (Lamiaceae) is not monophyletic: implications for the systematics, radiation, and ecological specializations of *Salvia* and tribe *Mentheae*. Am J Bot 91:1115–1125
44. Topcu G (2006) Bioactive triterpenoids from *Salvia* species. J Nat Prod 69:482–487
45. Karamanos AJ (2000) The cultivation of sage. Cultivation and breeding. In: Kintzios SE (ed) Sage: the genus *Salvia*. Medicinal and Aromatic Plants, vol 14. Harwood Academic Publishers, Amsterdam, p 93–108
46. Sáez L (2010) *Salvia* L. Vol. XII: Verbenaceae-Labiatae-Callitrichaceae. In: Morales R, Quintanar A, Cabezas F et al (eds) Flora ibérica, Real Jardín Botánico de Madrid (CSIC), Madrid, pp 298–326
47. Zarzuelo A, Gamez JM, Utrilla P et al (1995) Luteolin 5-rutinoside from *Salvia lavandulifolia* ssp. *oxyodon*. Phytochemistry 40:1321–1322
48. Palacio S, Milla R, Albuixech J et al (2008) Seasonal variability of dry matter content and its relationship with shoot growth and nonstructural carbohydrates. New Phytol 180:133–142
49. Hamidpour M, Hamidpour R, Hamidpour S et al (2014) Chemistry, pharmacology, and medicinal property of Sage (*Salvia*) to prevent and cure illnesses such as obesity, diabetes, depression, dementia, lupus, autism. heart disease, and cancer. J Tradit Complement Med 4:82–88
50. Blumenthal M, Goldberg A, Brinckmann J (2000) Herbal medicine expanded commission E monographs. American Botanical Council, Texas, pp 330–334
51. Ramos AA, Azqueta A, Pereira-Wilson C, Collins AR (2010) Polyphenolic compounds from *Salvia* species protect cellular DNA from oxidation and stimulate DNA repair in cultured human cells. J Agric Food Chem 58:7465–7471
52. Loizzo MR, Menichini F, Tundis R et al (2009) In vitro biological activity of *Salvia leriifolia* Benth. essential oil relevant to the treatment of Alzheimer's disease. J Oleo Sci 58:443–446
53. Porres-Martínez M, Gómez-Serranillos MP, Carretero ME (2012) *Salvia lavandulifolia* Vahl.: características botánicas, composición química y actividad biológica. Revista de Fitoterapia 12:53–63
54. Boukhary R, Raafat K, Ghoneim AI et al (2016) Anti-inflammatory and antioxidant activities of *Salvia fruticosa*: an HPLC determination of phenolic contents. Evid Based Complement Alternat Med ID 7178105:6
55. Ibrahim TA (2012) Chemical composition and biological activity of extracts from *Salvia bicolor* Desf. growing in Egypt. Molecules 17:11315–11334
56. Salimikia I, Monsef-Esfahani HR, Gohari AR et al (2016) Phytochemical analysis and antioxidant activity of *Salvia chloroleuca* aerial extracts. Iran Red Crescent Med J 18
57. Posadino AM, Porcu MC, Marongiu B (2012) Antioxidant activity of supercritical carbon dioxide extracts of *Salvia desoleana* on two human endothelial cell models. Food Res Int 46:354–359
58. Díaz JL (2013) *Salvia divinorum*: a psychopharmacological riddle and a mind-body prospect. Curr Drug Abuse Rev 6:43–53
59. Casselman I, Nock CJ, Wohlmuth H et al (2014) From local to global—Fifty years of research on *Salvia divinorum*. J Ethnopharmacol 151:768–783
60. Gali-Muhtasib H, Hilan C, Khater C (2000) Traditional uses of *Salvia libanotica* (East Mediterranean sage) and the effects of its essential oils. J Ethnopharmacol 71:513–520
61. Fu Z, Wang H, Hu X et al (2013) The pharmacological properties of *Salvia* essential oils. J App Pharm Sci 3:122–127
62. Miliauskas G, Venskutonis PR, Van Beek TA (2004) Screening of radical scavenging activity of some medicinal and aromatic plant extracts. Food Chem 85:231–237
63. Reyes-Caudillo E, Tecante A, Valdivia-López MA (2008) Dietary fibre content and antioxidant activity of phenolic compounds present in Mexican chia (*Salvia hispanica* L.) seeds. Food Chem 107:656–663

64. da Silva Marineli R, Moraes ÉA, Lenquiste SA et al (2014) Chemical characterization and antioxidant potential of Chilean chia seeds and oil (*Salvia hispanica* L.). LWT-Food Sci Technol 59:1304–1310
65. Tenore GC, Ciampaglia R, Arnold NA et al (2011) Antimicrobial and antioxidant properties of the essential oil of *Salvia lanigera* from Cyprus. Food Chem Toxicol 49:238–243
66. Perry NS, Bollen C, Perry EK, Ballard C (2003) *Salvia* for dementia therapy: review of pharmacological activity and pilot tolerability clinical trial. Pharmacol Biochem Behav 75:651–659
67. Loizzo MR, Abouali M, Salehi P et al (2014) In vitro antioxidant and antiproliferative activities of nine *Salvia* species. Nat Prod Res 28:2278–2285
68. Attaran Dowom S, Abrishamchi P, Asili J (2016) Essential oil (EO) composition and antioxidant activity of two *Salvia leriifolia* Benth.(Lamiaceae) populations from Iran. Nova Biol Reper 3
69. Tusi SK, Khodagholi F (2014) *Salvia macilenta* exhibits antiglycating activity and protects PC12 cells against H_2O_2-induced apoptosis. Cytotechnology 66:169–179
70. Tao S, Zheng Y, Lau A et al (2013) Tanshinone I activates the Nrf2-dependent antioxidant response and protects against As (III)-induced lung inflammation in vitro and in vivo. Antioxid Redox Signal 19:1647–1661
71. Chun-Yan SU, Qian-Liang MING, Rahman K et al (2015) *Salvia miltiorrhiza*: Traditional medicinal uses, chemistry, and pharmacology. Chin J Nat Med 13:163–182
72. Liu JJ, Wu HH, Chen TH et al (2015) 15,16-Dihydrotanshinone I from the functional food *Salvia miltiorrhiza* exhibits anticancer activity in human HL-60 Leukemia Cells: in vitro and in vivo studies. Int J Mol Sci 16:19387–19400
73. Asadi S, Khodagholi F, Esmaeili MA et al (2011) Chemical composition analysis, antioxidant, antiglycating activities and neuroprotective effects of *S. choloroleuca*, *S. mirzayanii* and *S. santolinifolia* from Iran. Am J Chin Med 39:615–638
74. Zarshenas MM, Krenn L (2015) Phytochemical and pharmacological aspects of *Salvia mirzayanii* Rech. f. & Esfand. J Evid-Based Complementary Altern Med 20:65–72
75. Baricevic D, Bartol T (2000) The Biological/Pharmacological Activity of the *Salvia* genus Pharmacology. In: Kintzios SE (ed). The Genus *Salvia*, Harwood Academic Publishers, Abingdon, Marston 143–184
76. Vladimir-Knežević S, Blažeković B, Kindl M et al (2014) Acetylcholinesterase inhibitory, antioxidant and phytochemical properties of selected medicinal plants of the Lamiaceae family. Molecules 19:767–782
77. Jantová S, Hudec R, Sekretár S et al (2014) *Salvia officinalis* L. extract and its new food antioxidant formulations induce apoptosis through mitochondrial/caspase pathway in leukemia L1210 cells. Interdiscip Toxicol 7:146–153
78. Hasanein P, Felehgari Z, Emamjomeh A (2016) Preventive effects of *Salvia officinalis* L. against learning and memory deficit induced by diabetes in rats: Possible hypoglycaemic and antioxidant mechanisms. Neurosci Lett 622:72–77
79. Al-Qudah MA, Al-Jaber HI, Zarga MHA, Orabi STA (2014) Flavonoid and phenolic compounds from *Salvia palaestina* L. growing wild in Jordan and their antioxidant activities. Phytochemistry 99:115–120
80. Zhang BB, He BQ, Sun JB et al (2015) Diterpenoids from *Salvia plebeia* R. Br. and their antioxidant and anti-inflammatory activities. Molecules 20:14879–14888
81. Jafari E, Andalib S, Abed A et al (2015) Neuroprotective, antimicrobial, antioxidant, chemotherapeutic, and antidiabetic properties of *Salvia reuterana*: a mini review. Avicenna J Phytomed 5:10–16
82. Esmaeili MA, Kanani MR, Sonboli A (2010) *Salvia reuterana* extract prevents formation of advanced glycation end products: an in vitro study. IJPS 6:33–50
83. Farhat MB, Landoulsi A, Chaouch-Hamada R et al (2013) Phytochemical composition and in vitro antioxidant activity of by-products of *Salvia verbenaca* L. growing wild in different habitats. Ind Crops Prod 49:373–379

84. Khlifi S, El Hachimi Y, Khalil A et al (2006) In vitro antioxidant properties of *Salvia verbenaca* L. hydromethanolic extract. Indian J Pharmacol 38:276
85. Farhat MB, Chaouch-Hamada R, Sotomayor JA et al (2015) Antioxidant properties and evaluation of phytochemical composition of *Salvia verbenaca* L. extracts at different developmental stages. Plant Foods Hum Nutr 70:15–20
86. Tepe B, Eminagaoglu O, Akpulat HA et al (2007) Antioxidant potentials and rosmarinic acid levels of the methanolic extracts of *Salvia verticillata* (L.) subsp. *verticillata* and *S. verticillata* (L.) subsp. *amasiaca* (Freyn & Bornm.) Bornm. Food Chem 100:985–989
87. Nickavar B, Rezaee J, Nickavar A (2016) Effect-directed analysis for the antioxidant compound in *Salvia verticillata*. Iran J Pharm Res 2016 15:241–246
88. Lu Y, Foo LY (2002) Polyphenolics of *Salvia*—a review. Phytochemistry 59(2):117–140
89. Sepahvand R, Delfan B, Ghanbarzadeh S et al (2014) Chemical composition, antioxidant activity and antibacterial effect of essential oil of the aerial parts of *Salvia sclareoides*. Asian Pac J Trop Med 7:S491–S496
90. Zupkó I, Hohmann J, Rédei D et al (2001) Antioxidant activity of leaves of *Salvia* species in enzyme-dependent and enzyme-independent systems of lipid peroxidation and their phenolic constituents. Planta Med 67:366–368
91. Asadi S, Ahmadiani A, Esmaeili MA et al (2010) In vitro antioxidant activities and an investigation of neuroprotection by six *Salvia* species from Iran: a comparative study. Food Chem Toxicol 48:1341–1349
92. Firuzi O, Miri R, Asadollahi M et al (2013) Cytotoxic, antioxidant and antimicrobial activities and phenolic contents of eleven *Salvia* species from Iran. Iran J Pharm Res 12:801–810
93. Perry N, Howes MJ, Houghton P, Perry E (2000) 14. Why sage may be a wise remedy: effects of *Salvia* on the nervous system. The Genus *Salvia* 207–223
94. Panagiotopoulos E, Skapeti M, Kapetanos C (2000) Production of secondary metabolites using liquid culture of *Salvia* plants: up to-date reports and scale-up potential. *Sage:* the Genus *Salvia*. Amsterdam: Harwood Academic Publishers 251–262
95. Ulubelen A (2000) III. Chemical constituents 4. Terpenoids in the genus *Salvia*. The Genus *Salvia*, 55
96. Cortese K, Daga A, Monticone M et al (2016) Carnosic acid induces proteasomal degradation of Cyclin B1, RB and SOX2 along with cell growth arrest and apoptosis in GBM cells. Phytomedicine 23:679–685
97. Fischedick JT, Standiford M, Johnson DA, Johnson JA (2013) Structure activity relationship of phenolic diterpenes from *Salvia officinalis* as activators of the nuclear factor E2-related factor 2 pathway. Bioorg Med Chem 21:2618–2622
98. Schwager J, Richard N, Fowler A et al (2016) Carnosol and related substances modulate chemokine and cytokine production in macrophages and chondrocytes. Molecules 21:465
99. Li J, He C, Tong W et al (2015) Tanshinone IIA blocks dexamethasone-induced apoptosis in osteoblasts through inhibiting Nox4-derived ROS production. Int J Clin Exp Pathol 8 (10):13695–13706
100. Ma S, Zhang D, Lou H et al (2016) Evaluation of the anti-inflammatory activities of tanshinones isolated from *Salvia miltiorrhiza* var. *alba* roots in THP-1 macrophages. J Ethnopharmacol 188:193–199
101. de Oliveira MR, Schuck PF, Dal Bosco SM (2017) Tanshinone I induces mitochondrial protection through an Nrf2-dependent mechanism in paraquat-treated human neuroblastoma SH-SY5Y cells. Mol Neurobiol 54(6):4597–4608
102. Shu M, Hu XR, Hung ZA et al (2016) Effects of tanshinone IIA on fibrosis in a rat model of cirrhosis through heme oxygenase-1, inflammation, oxidative stress and apoptosis. Mol Med Rep 13:3036–3042
103. Weng YS, Wang HF, Pai PY et al (2015) Tanshinone IIA prevents Leu27IGF-II-induced cardiomyocyte hypertrophy mediated by estrogen receptor and subsequent Akt activation. Am J Chin Med 43:1567–1591

104. Chen X, Li Q, He Y et al (2015) 15,16-dihydrotanshinone I induces apoptosis and inhibits the proliferation, migration of human osteosarcoma cell line 143B in vitro. Anticancer Agents Med Chem Oct 18
105. Kelen M, Tepe B (2008) Chemical composition, antioxidant and antimicrobial properties of the essential oils of three *Salvia* species from Turkish flora. Bioresource Technol 99:4096–4104
106. Giannouli AL, Kintzios SE (2000) 5 essential oils of *Salvia* spp.: examples of intraspecific and seasonal variation. The Genus *Salvia*, 69–79
107. Scalzo J, Politi A, Pellegrini N et al (2005) Plant genotype affects total antioxidant capacity and phenolic contents in fruit. Nutrition 21:207–213
108. Bozin B, Mimica-Dukic N, Samojlik I, Jovin E (2007) Antimicrobial and antioxidant properties of rosemary and sage (*Rosmarinus officinalis* L. and *Salvia officinalis* L., Lamiaceae) essential oils. J Agric Food Chem 55:7879–7885
109. Porres-Martínez M, González-Burgos E, Carretero ME, Gómez-Serranillos MP (2015) Major selected monoterpenes α-pinene and 1, 8-cineole found in *Salvia lavandulifolia* (Spanish sage) essential oil as regulators of cellular redox balance. Pharm Biol 53:921–929
110. Kozłowska M, Laudy AE, Przybył J et al (2015) Chemical composition and antibacterial activity of some medicinal plants from *Lamiaceae* family. Acta Pol Pharm 72:757–767
111. Chen CY, Li H, Yuan YN (2013) Antioxidant activity and components of a traditional chinese medicine formula consisting of *Crataegus pinnatifida* and *Salvia miltiorrhiza*. BMC Complement Altern Med 13:1–6
112. Ramu G, Dhanabal SP (2015) Antioxidant studies on ethanol extracts from two selected genera of Indian *Lamiaceae*. Indian J Pharm Sci 77:780–782
113. Guimarães AG, Oliveira GF, Melo MS et al (2010) Bioassay-guided evaluation of antioxidant and antinociceptive activities of carvacrol. Basic Clin Pharmacol Toxicol 107:949–957
114. López-Lázaro M (2009) Distribution and biological activities of the flavonoid luteolin. Mini Rev Med Chem 9:31–59
115. Hao W, Guo H, Zhang J (2014) Hydrogen peroxide is involved in salicylic acid-elicited rosmarinic acid production in *Salvia miltiorrhiza* cell cultures. Sci World J 2014:843764. https://doi.org/10.1155/2014/843764
116. Imanshahidi M, Hosseinzadeh H (2006) The pharmacological effects of *Salvia* species on the central nervous system. Phytother Res 20:427–437
117. Zhao GR, Zhang HM, Ye TX et al (2008) Characterization of the radical scavenging and antioxidant activities of danshensu and salvianolic acid B. Food Chem Toxicol 46:73–81
118. Guo H, Dang X, Dong J (2014) Hydrogen peroxide and nitric oxide are involved in salicylic acid-induced salvianolic acid B production in *Salvia miltiorrhiza* cell cultures. Molecules 19:5913–5924
119. Lin M, Zhai X, Wang G et al (2015) Salvianolic acid B protects against acetaminophen hepatotoxicity by inducing Nrf2 and phase II detoxification gene expression via activation of the PI3 K and PKC signaling pathways. J Pharmacol Sci 127:203–210
120. Ravipati AS, Zhang L, Koyyalamudi SR et al (2012) Antioxidant and anti-inflammatory activities of selected Chinese medicinal plants and their relation with antioxidant content. BMC Complement Altern Med 12:173
121. Farjam MH, Rustaiyan A, Ezzatzadeh E, Jassbi AR (2013) Labdane-type diterpene and two flavones from *Salvia sharifii* Rech. f. and Esfan. and their biological activities. Iran J Pharm Res 12:395–399
122. Grzegorczyk I, Matkowski A, Wysokińska H (2007) Antioxidant activity of extracts from in vitro cultures of *Salvia officinalis* L. Food Chem 104:536–541
123. Porres-Martínez M, González-Burgos E, Accame MEC, Gómez-Serranillos MP (2013) Phytochemical composition, antioxidant and cytoprotective activities of essential oil of *Salvia lavandulifolia* Vahl. Food Res Int 54:523–531

124. Porres-Martínez M, González-Burgos E, Carretero ME, Gómez-Serranillos MP (2014) Influence of phenological stage on chemical composition and antioxidant activity of *Salvia lavandulifolia* Vahl. essential oils. Ind Crops Prod 53:71–77
125. Orhan I, Kartal M, Naz Q et al (2007) Antioxidant and anticholinesterase evaluation of selected Turkish *Salvia* species. Food Chem 103:1247–1254
126. Porres-Martínez M, González-Burgos E, Carretero ME, Gómez-Serranillos MP (2015) Protective properties of *Salvia lavandulifolia* Vahl. essential oil against oxidative stress-induced neuronal injury. Food Chem Toxicol 80:154–162
127. Villa-Ruano N, Zurita-Vásquez GG, Pacheco-Hernández Y et al (2013) Anti-Iipase and antioxidant properties of 30 medicinal plants used in Oaxaca, México. Biol Res 46:153–160
128. Fei AH, Cao Q, Chen SY et al (2013) Salvianolate inhibits reactive oxygen species production in H_2O_2-treated mouse cardiomyocytes in vitro via the TGFβ pathway. Acta Pharmacol Sin 34:496–500
129. Garcia CS, Menti C, Lambert APF, Barcellos T (2016) Pharmacological perspectives from Brazilian *Salvia officinalis* (Lamiaceae): antioxidant, and antitumor in mammalian cells. An Acad Bras Ciên 88:281–292
130. Kasrati A, Jamali CA, Bekkouche K et al (2015) Comparative evaluation of antioxidant and insecticidal properties of essential oils from five Moroccan aromatic herbs. J Food Sci Tech 52:2312–2319
131. Kamatou GP, Viljoen AM, Gono-Bwalya AB et al (2005) The in vitro pharmacological activities and a chemical investigation of three South African *Salvia* species. J Ethnopharmacol 102:382–390

Anticancer Activity of *Salvia miltiorrhiza* and Its Secondary Metabolites

Ching-Fen Wu and Thomas Efferth

Abstract *Salvia miltiorrhiza* Bunge (*danshen*) is a well-known traditional Chinese medicinal herb. The root part of the plant contains most of the bioactive ingredients, which mainly belong to two groups of compounds: hydrophilic phenolic acids and hydrophobic tanshinones. These bioactive compounds refer to as secondary metabolites and exert multiple therapeutic activities, such as anti-oxidative stress, anti-neurodegenerative, anti-inflammatory, and anti-hypertensive effects without showing serious side effects. Besides these pharmacological activities, anticancer effects have also attracted scientific attention. This review will provide an updated summary of the anticancer effects and the proposed mechanisms of the major bioactive compounds isolated from *S. miltiorrhiza*.

Keywords *Salvia miltiorrhiza* · Phenolic acid · Tanshinone · Multidrug resistance Anticancer

List of Abbreviations

ABC	ATP-binding cassette
AMPK	AMP-activated protein kinase
APL	Acute promyelocytic leukemia
ATRA	All-trans retinoic acid
BCNU	1,3-bis-(2-chloroethyl)-1-nitosourea
BCRP	Breast cancer resistance protein
COX-2	Cyclooxygenase-2
CPT	Cryptotanshinone
DHT	Dihydrotestosterone
DMBA	7,12-dimethylbenz(a)anthracene
DMH	1,2-dimethylhydrazine
EGFR	Epidermal growth factor receptor

C.-F. Wu · T. Efferth (✉)
Department of Pharmaceutical Biology, Institute of Pharmacy and Biochemistry, Johannes Gutenberg University, Staudinger Weg 5, 55128 Mainz, Germany
e-mail: efferth@uni-mainz.de

V. Georgiev and A. Pavlov (eds.), *Salvia Biotechnology*,
https://doi.org/10.1007/978-3-319-73900-7_5

eIF4F	Eukaryotic initiation factor 4F
ER	Endoplasmic reticulum
GBD	Global burden of disease
H3K9	Histone H3 lysine 9
HIF-1α	Hypoxia-inducible factor 1 alpha
HNSCC	Head and neck squamous cell cancer
HUVECs	Human umbilical vein endothelial cells
LSD1	Lysine-specific demethylase 1
MDR	Reducing multidrug resistance
MMP	Mitochondrial membrane potential
MMP9	Matrix metalloproteinase 9
MRP1	Multidrug resistance associated protein 1
NQO1	NAD(P)H: quinone oxidoreductase 1
Nrf2	Nuclear factor erythroid-related factor 2
NSCLC	Non-small cell lung cancer
PGE2	Prostaglandin E2
PRMT1	Protein arginine methyltransferase
RA	Rosmarinic acid
ROS	Reactive oxygen species
Sal A	Salvianolic acid A
Sal B	Salvianolic acid B
SCC	Squamous cell carcinoma
STAT3	Signal transducer and activator of transcription 3
TCM	Traditional Chinese medicine
TIMP	Tissue inhibitor of matrix metalloproteinase protein
TRAIL	Tumor necrosis factor-related apoptosis-inducing ligand
VEGFs	Vascular endothelial growth factors
WHO	World Health Organization

1 Introduction

Cancer belongs to one of the most important public health issues. According to the report of the World Health Organization (WHO) on the global burden of disease (GBD), it is estimated that 14.9 million new cancer cases are diagnosed and 8.2 million deaths occur worldwide. In 2013, cancer ranked as the second leading cause of death behind cardiovascular diseases [1]. By 2030, the continuously ageing population will lead to a further increase of new cancer cases, which are expected to exceed 20 million every year [2]. Cancer refers to a multistep process, and genetic alterations lead to progressive transformation of healthy into malignant cells [3]. Hanahan and Weinberg have described the hallmarks of cancer: evading cell death, self-sufficiency in growth signals, insensitivity to anti-growth signals, activating invasion and metastasis, unrestricted replicative potential, sustained angiogenesis, impaired immune surveillance and reprogramming of energy metabolism [3].

Chemotherapy is still a mainstay in cancer treatment. The general function of classical chemotherapeutic agents is to interrupt cell division by targeting DNA, RNA, or specific proteins [4, 5]. Chemotherapeutic agents target fast-dividing cells, one of the main properties of cancer cells. Therefore, they also affect other fast-proliferating cells in normal tissues, e.g. cells in mucosa, bone marrow and hair follicles. Major side effects, which many patients under treatments of chemotherapy may encounter, involve hair loss, immunosuppression, anemia, gastrointestinal distress, infertility and organ damages (e.g. hepatotoxicity and nephrotoxicity).

In comparison with synthesized chemical drugs, natural products frequently show favorable profiles with low toxicity. Natural products are secondary metabolites, which serve for survival of organisms. These small molecules display molecular structures, named "privileged scaffolds", which are constructed with high-affinity to bind multiple targets (e.g. proteins) [6]. Natural products have been historically considered as important resources for anticancer drug discovery, and a variety of natural products have been identified as potent anticancer drugs [7, 8]. Most cancers reveal deregulated multiple signaling pathways causing either primary inherent resistance or secondary acquired resistance. Combination of targeted agents has been proposed to combat resistance development. Nevertheless, unexpected side effects may occur even with targeted drugs. In addition, the enormous costs of target therapies should be taken into consideration [9]. These drawbacks of targeted therapies revitalized research interests in natural products. Instead of monospecific chemotherapeutic drugs, which are prone to frequent resistance development, natural products with multifactorial specificity of action may be of advantage to overcome or bypass drug resistance [10].

Traditional Chinese medicine (TCM) is one of the most remarkable resources for natural products. Recently, the developments in modern cellular and molecular biology considerably contributed to the explanation of anticancer effects of TCM-derived drugs. TCM plays an important role in inhibiting cancer progression by inducing apoptosis, modulating the immune system, reducing multidrug resistance (MDR), etc. *Salvia miltiorrhiza* Bunge (*Lamiaceae*, also named *danshen* in Chinese, Fig. 1) is a well-known Chinese medicinal herb with more than 1000 years of clinical application. According to the record in Shen-Nung's *Pen-Ts'ao* [11], an ancient TCM book, *S. miltiorrhiza* has been classified as a "top-grade" herb, which means the herbal drug contains high therapeutic value with low side effects or toxicity in long-term use. In China, *S. miltiorrhiza* has been used not only as medicine, but also as dietary supplement or culinary ingredient for health improvement [12]. The dried root of *S. miltiorrhiza* is traditionally used as a hemorrheologic agent to promote circulation and remove blood stasis (a Chinese medicinal term named *huoxue huayu*) for the treatment of cardiovascular and cerebrovascular diseases [13]. In TCM, cancer result of blood stasis or *qi* stagnation [14, 15]. Herbal plants, which are able to promote circulation and remove blood stasis, are traditionally prescribed for cancer patients [16, 17]. Therefore, besides

Fig. 1 Pictures of plant *S. miltiorrhiza*. **a** Leaf and stem part of *S. miltiorrhiza*. **b** Root part of *S. miltiorrhiza*

the clinical use as hemorrheologic agent, *S. miltiorrhiza* has also been investigated for cancer treatment [18].

During the past 50 years, *danshen* received remarkable attention by modern scientists and more than 70 compounds were isolated. The root of the plant contains most of the bioactive ingredients, which mainly belong to two groups of compounds: hydrophilic phenolic acids (e.g. rosmarinic acid, salvianolic acid A, salvianolic acid B, danshensu, etc.) and hydrophobic tanshinones (e.g. tanshinone I, tanshinone IIA, miltirone, cryptotanshinone, etc.) [19–21]. These bioactive compounds refer to as secondary metabolites and exert multiple therapeutic activities, such as anti-oxidative stress, anti-neurodegenerative, anti-inflammatory, anti-hypertensive effects [22–24]. Accumulated evidences have revealed that these lipophobic and lipophilic compounds also showed significant anticancer activities. In this chapter, we summarize the current progress in research on secondary metabolites of *S. miltiorrhiza* for cancer therapy.

2 *S. miltiorrhiza* Against Multidrug Resistance

A crucial problem in cancer therapy is the occurrence of drug resistance. Cancers can develop drug resistance against chemotherapy via inherent cell heterogeneity, drug efflux, increased drug inactivation, drug target alteration, cell death inhibition, stimulated repair of DNA damage caused by chemotherapeutic agents, epigenetic alterations and epithelial-mesenchymal transition [25]. These mechanisms promote direct or indirect drug resistance independently or in combination through diverse molecular pathways.

2.1 P-glycoprotein

Multidrug resistance (MDR) represents a phenomenon of inherent or acquired resistance of cancer cells to structurally diverse chemotherapeutic agents, which target different cellular molecules [26]. Cellular or genetic alterations acquired during development of drug resistance involve increased efflux of cytotoxic drugs mediated by overexpression of the ATP-binding cassette (ABC) transporter. P-glycoprotein (P-gp, *ABCB1/MDR1*) is the best studied mediator in this family [27]. The overexpression of this transporter is responsible for drug efflux and resistance to several unrelated hydrophobic drugs used in cancer chemotherapy, e.g. anthracyclines, *Vinca* alkaloids, taxanes, epipodophyllotoxines, and probably hundreds of other compounds [28]. Most P-gp inhibitors revealed unfavorable properties in clinical trials, e.g. limited efficiency, increased toxicity, unrelated pharmacological effects, pharmacokinetic interaction with other drugs and lack of potency for specific mechanisms of resistance [29, 30]. To search for new P-gp inhibitors, a "triple e" strategies emerged: (1) engage: co-administration of P-gp inhibitors and cytotoxic agents; (2) evade: the use of cytotoxic agents that bypass P-gp-mediated efflux; (3) exploit: approach takes advantage of the collateral sensitivity of MDR cells [30].

S. miltiorrhiza root extract was cytotoxic towards acute lymphoblastic leukemia sensitive CCRF-CEM and its P-gp overexpression resistant CEM/ADR5000 cell lines with IC_{50} values of 1.9 ± 0.23 and 3.2 ± 0.83 μg/ml, respectively [31]. Doxorubicin is a substrate of P-gp and was used as control drug. It revealed IC_{50} values of 0.005 ± 0.0004 μM in sensitive and 16.6 ± 3.48 μM in resistant cells [32]. As indicated by the degrees of cross-resistance, doxorubicin was remarkably more active on sensitive cells than on resistant ones [32], while *S. miltiorrhiza* showed equal sensitive to both resistant and sensitive cells with a degree of resistance value of 1.68. Thus, the results implied that *S. miltiorrhiza* might bypass P-gp and can be used to treat P-gp-overexpressing multidrug-resistant cancer cells with similar efficacy as sensitive ones.

2.2 Tumor Suppressor p53

The tumor suppressor p53, encoded by *TP53* gene, plays an important role in maintaining genome stability and tumor prevention [33]. As a transcription factor, p53 mainly exerts its tumor suppressive function through transcriptional regulation of its target genes. A variety of stress signals, including DNA damage, nutrient deprivation, hypoxia and oncogene activation, which are all related in many ways to carcinogenesis, release p53 from negative regulators-mediated inhibition. This increases p53 protein levels and activity. Once activated, p53 regulates a group of genes targeting cell cycle arrest (e.g. p21, Gadd45, cdc25c and 14-3-3σ) [34], DNA repair (e.g. p48XPE) [35], apoptosis (e.g. Puma, Bax, Noxa) [36] or senescence (e.g. p21) [37]. Somatic p53 is frequently mutated in almost every type of human tumors [38]. In those tumors with low p53 mutation rates, p53 is often inactivated by alternative mechanisms. Mutations in the *TP53* tumor suppressor gene silence the function of the encoded p53 protein, which promotes oncogenic transformation and drug resistance [39, 40].

The IC_{50} values were 5.5 ± 0.83 and 4.3 ± 0.47 μg/ml for *S. miltiorrhiza* root extract against HCT116 ($p53^{+/+}$) and HCT116 ($p53^{-/-}$) cells [31]. *S. miltiorrhiza* exerted higher cytotoxicity towards HCT116 ($p53^{-/-}$) cells, implying that *S. miltiorrhiza* is more active on p53-deficient resistant cancer cells and might induce p53-independent cancer cell death.

2.3 Epidermal Growth Factor Receptor

The epidermal growth factor receptor (EGFR), a well-characterized oncogene also referred to ErbB1 or HER1, is a 170 kDa transmembrane protein belonging to the erB family of tyrosine kinase receptors [41–43]. Active EGFR dimers undergo auto-phosphorylation of tyrosine residues in the cytoplasmic tail of the receptor, subsequently activating multiple signal transduction pathways, including the MAPK signaling cascade, Src, STAT3/5, the phosphoinositide-3-kinase (PI3K) pathway, and the phospholipase Cγ pathway, which leads to protein kinase C (PKC) activation [43, 44]. Activation of these downstream signaling pathways results in oncogenesis, angiogenesis, cell cycle progression and differentiation in a variety of cancers. The overexpression of EGFR is presumably caused by multiple epigenetic mechanisms, gene amplification, and oncogenic viruses [45].

S. miltiorrhiza root extract was also active against U87.MG and U87.MGΔEGFR cells with IC_{50} values of 13.8 ± 0.7 and 6.3 ± 2.48 μg/ml, respectively [31]. It is remarkable that *S. miltiorrhiza* root extract was significantly more cytotoxic towards U87.MGΔEGFR cells than their parental wild-type cells, as indicated by degrees of resistance below 1.0. This kind of hypersensitivity is termed collateral sensitivity, meaning that *S. miltiorrhiza* might be able to preferentially kill EGFR-overexpressing cancer cells.

To sum up, these results have proved multifactorial properties of *S. miltiorrhiza* root extract to overcome drug resistance in different resistant cancer cell lines. The underlying mechanisms of how *S. miltiorrhiza* exerts effects against diverse drug resistances warrant further investigation.

3 Anticancer Effects of the Lipophilic Compounds

Tanshinones are the main lipid-soluble compounds of *S. miltiorrhiza*. They belong to a group of abietane diterpenes identified since 1930s [46]. The chemical structure of tanshinones is composed by four rings: naphthalene or tetrahydronaphthalene

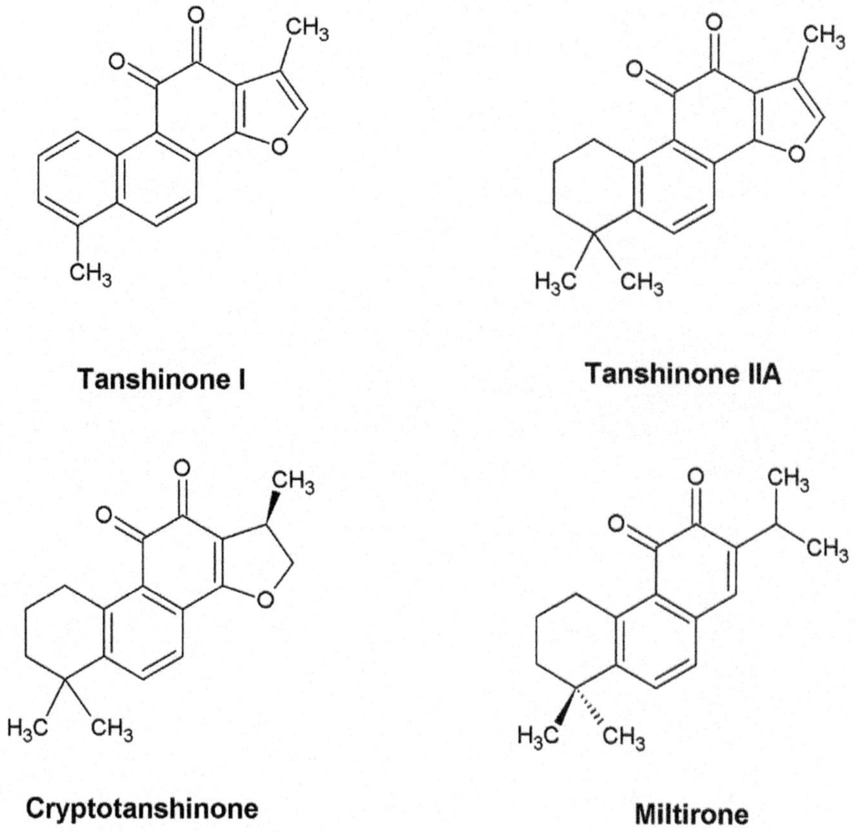

Fig. 2 Molecular structure of main lipophilic compounds isolated from *S. miltiorrhiza*

ring A and B, *ortho*- or *para*-quinone or lactone ring C, and a furan or dihydrofuran ring D [47]. The cytotoxic effects might be determined by the structure of ring A and C [48]. The most common tanshinones are the *ortho*-quinones such as tanshinone IIA, tanshinone I and cryptotanshinone. Among these tanshinones, tanshinone IIA are the major constituent and the most investigated ingredient for the pharmacological activities, including anticancer effects. The other tanshinones also gain attention in recent years (Fig. 2).

In general, *ortho*-quinones derivatives are redox-active species that may undergo electron reduction steps to reform the highly unstable catechol and/or semiquinones. Taking tanshinone IIA as an example, it is reduced to form catechol by NAD(P)H: quinone oxidoreductase 1 (NQO1), a cytosolic flavoenzyme, which catalyzes quinone reduction [49]. Subsequently, autooxidation of catechol futiles redox to generate multiple times of reactive oxygen species (ROS). ROS caused oxidative DNA damage, which in turn down-regulated expression of p53-dependent or -independent antiapoptotic Bcl-2. Consequently, Bcl-2-inhibited

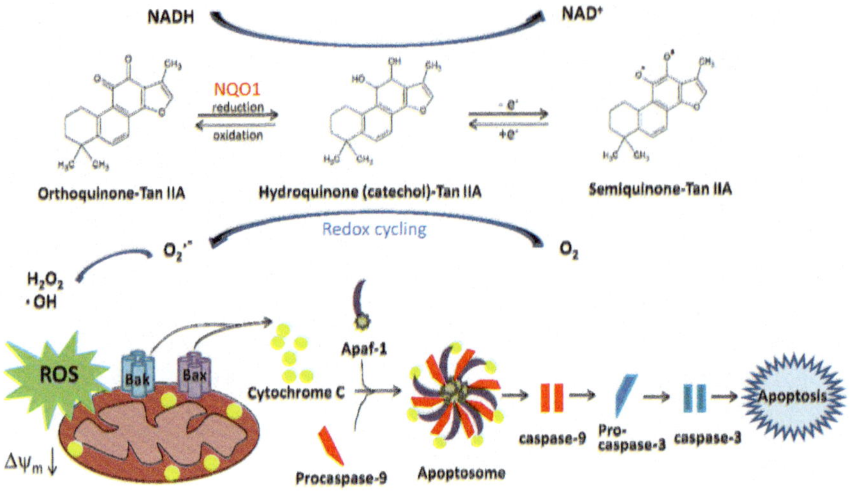

Fig. 3 A proposed scheme of tanshinone IIA (Tan IIA) initiating redox cycle and intrinsic apoptotic pathway. Images modified from [49]

Bax/Bak expression is elevated, which contributes to loss of mitochondrial membrane potential (MMP) and release of cytochrome C, leading to caspase-dependent apoptosis (Fig. 3).

3.1 Cryptotanshinone

Cryptotanshinone (CPT) exerted strong cytotoxicity towards CCRF-CEM and P-gp-overexpressing CEM/ADR5000 cells with similar efficacy [32]. CPT also decreased P-gp mRNA and protein levels, and inhibited ATPase activity of P-gp in colon cancer cells [50]. The compound suppressed doxorubicin efflux in P-gp-overexpressing HepG2 cells [51].

Microarray-based mRNA expression profiling of CCRF-CEM cells revealed that CPT predominantly modulated molecular functions such as cell death and survival, cell proliferation, ROS production, cell cycle arrest, DNA damage and cellular movement [32]. CPT stimulated the mitochondria-mediated intrinsic apoptotic pathway through reactive oxygen species (ROS) production, caused loss of mitochondrial membrane potential (MMP) and DNA damage, and activated caspase 3/7 and 9. Down-regulation of p65 translocation might result from a direct binding of CPT to IKKβ, providing supportive evidence for CPT-induced apoptosis. In addition, CPT inhibited cell adhesion to fibronectin. Based on the rapid accumulation of ROS by CPT, it is suggested that CPT-caused apoptosis, inhibition of adhesion and blocking of NFκB signaling may originate from ROS production [32]. Chen et al. have shown that CPT induced G2/M cell cycle arrest with upregulation of p53, p21CIP1/WAF1 and inhibition of CDK1, CDC25C and cyclin B1 [52].

Angiogenesis initiates cancer metastasis and promotes tumor growth. Vascular endothelial growth factors (VEGFs) and their receptors are crucial controllers accounting for development of angiogenesis. According to a reverse pharmacophore-based screening, angiogenesis-related signaling is one of the main pathways that CPT regulates cancer growth [53]. CPT inhibited tube formation of lymphatic endothelial cells by downregulating VEGF receptor 3 and phosphorylation of ERK1/2 [54]. Also, CPT significantly reduced cellular levels of VEGF and disturbed tube formation of human umbilical vein endothelial cells (HUVECs) [55]. CPT-inhibited tube formation of HUVECs might result from suppression of STAT3 and NFκB [56].

MAPK signaling pathway plays an important role in CPT-induced anticancer effects. CPT suppressed the expression of the apoptosis inhibitory protein, Bcl-2, and augmented the sensitivity of the death receptor Fas in prostate DU145 cancer cells. JNK and p38 MAPK act as regulators of Bcl-2 expression [57]. CPT stimulated ROS-mediated caspase-8 and p38 MAPK expression, which sensitized TNFα-induced apoptosis in chronic myeloid leukemia KBM-5 cells [58]. In another report, CPT-induced ROS production activated p38 MAPK and JNK, as well as inhibited ERK 1/2, leading to caspase-independent apoptosis in cancer cells [59]. CPT down-regulated phosphorylation of JNK, ERK1/2 and p38 MAPK, supporting the effect of CPT in targeting nuclear factor erythroid-related factor 2 (Nrf2), which mediated cytoprotective genes positively correlated to drug resistance [60].

Androgen and androgen receptor (AR) play a vital role in the development of prostate cancer. Therefore, they are therapeutic targets for anticancer agents. CPT has a similar chemical structure as dihydrotestosterone (DHT), a typical steroid which binds to AR and drives growth of prostate cells. CPT effectively inhibited DHT-induced AR transactivation and prostate cancer cell growth [61]. CPT is a novel inhibitor for AR signaling, and its inhibitory effects are due to disruption of AR dimerization and AR-co-regulation complex formation [62]. The mechanisms of CPT-affected AR signaling also include disruption of lysine-specific demethylase 1 (LSD1)-mediated demethylation of histone H3 lysine 9(H3K9) and represses the transcriptional activity of AR [63]. The anti-AR effects of CPT inhibited invasion of prostate cancer cells via downregulation of matrix metalloproteinase 9 (MMP9) [64].

Signal transducer and activator of transcription 3 (STAT3) is an important target for anticancer therapy, because it is constitutively activated in most human solid tumors and is involved in proliferation, angiogenesis and anti-apoptosis of cancer cells. CPT has been identified as potential STAT3 inhibitor. CPT inhibited phosphorylation of STAT3 at Tyr705 [65–68]. CPT might direct target STAT3, because it inhibited the formation of STAT3 dimers [65]. Inhibition of STAT3 decreased downstream target protein expression of cyclin D1 and survivin [65–69], Bcl-xL [65], and Bcl-2 [69, 70]. CPT could be a good apoptosis inducer via modification of STAT3 signaling.

The endoplasmic reticulum (ER) stress-induced apoptotic pathway has been suggested to be least affected by drug resistance, becoming a potential target for chemotherapy [71]. CPT generated ROS-mediated ER stress, leading to apoptosis of HepG2 and MCF7 cells, and sensitized the anticancer effects of 5-FU, cisplatin and arsenic through inducing ER stress [72, 73]. Upregulation of PPP1R15A/GADD34 (a growth arrest and DNA-damage inducible factor), and ER-stress-regulated CHOP/GADD153 suggested that the misfolded protein-stimulated PERK-eIF2α-ATF4 pathway was activated, and that ER stress-mediated apoptosis was triggered by CPT [74]. Tumor necrosis factor-related apoptosis-inducing ligand (TRAIL) induces apoptosis in cancer cells, but not in normal tissue cells. TRAIL resistance due to low TRAIL receptor expression is widely found in cancer cells. CPT-induced ROS production stimulated upregulation of CHOP and death receptor (DR5, TRAIL receptor 2), which consequently converted TRAIL-resistant lung A549 cancer cells into sensitive ones [75].

The PI3K/mTOR signaling pathway, a major hyperactivated signaling pathway in cancer development, controls the assembly of eukaryotic initiation factor 4F (eIF4F), which of eIF4A is a specific component with helicase activity in regulating protein translation initiation [76]. Using a computational approach, CPT was found to bind to PI3K/mTOR with a hydrogen bond at hinge residue Val882. In addition, CPT bound to eIF4A, in the same ATP binding region as pateamine A (a novel inhibitor for eIF4A), suggesting the role of CPT in interfering with protein translation initiation-associated cytotoxic effects [74]. CPT induced cytotoxicity in chronic myeloid leukemia K562/ADM cells through reducing eIF4E expression. Overexpression of eIF4E mediated resistance to CPT [77]. The *Bcr/ABL* gene is frequently overeypressed in leukemic cells with acquired resistance to imatinib. CPT increased the cytotoxic effects of imatinib by decreasing BCR/ABL expression, which may be attributed to deregulation of STAT3, eIF4E and mTOR [78].

In addition to regulation of apoptosis-related molecules, CPT potentially induces other types of cell death in apoptosis-resistant cancer cells, such as autophagy, an essential cellular degradation process for the clearance of damaged proteins and organelles. CPT-triggered autophagy is induced by ROS-mediated JNK signaling [79]. CPT induced less apoptosis in P-gp-overexpressing resistant colon cancer cells than sensitive one, but more autophagy in resistant cells than sensitive ones. In P-gp-overexpressing colon cancer cells, CPT-induced autophagy was p53-independent [80]. AMP-activated protein kinase (AMPK) signaling, which controls cellular energy homeostasis, was activated by CPT to induce autophagy on HepG2 cells [81].

3.2 *Tanshinone I*

Tanshinone I (Tan I) effectively in inhibited the growth of CL1-0 lung cancer cells via suppressing the expression of VEGF, Cyclin A, and Cyclin B proteins [82]. Furthermore, Tan I suppressed Aurora A expression, reduced final H1299 lung tumor weight associated by inhibiting proliferation and inducing apoptosis in an in vivo mice model. Aurora A, a member of a novel oncogenic family of mitotic serine/threonine kinases involved in spindle formation and G2/M transition, may be an important target for Tan I to suppress cancer progression [83]. Tan I-mediated Aurora A inhibition upregulated p53 in colorectal cancer cells [84]. ERK1/2 activation by Tan I contributes to cyclin D1 proteasomal degradation, which inhibited cell growth of human colorectal cancer cells [85]. Tan I induced mitochondrial-mediated apoptosis by increasing expression of p53, p21, bax and caspase-3 in colon cancer cells [86]. Tan I-induced growth-inhibition and apoptosis-induction in breast cancer cells [87] and myeloid leukemia cells [88] were mediated by suppression of the PI3K/Akt signaling pathway. In addition, Tan I may also serve as potent sensitizer of TRAIL to enhance apoptosis towards prostate cancer by upregulation of DR5 [89]. Notably, Tan I not only induced gastric cancer apoptosis, but also pro-survival autophagy by inducing expression of complex Beclin-1/VPS34. The combination with autophagy inhibitors enhanced apoptosis [90].

Tan I can reduce cancer metastasis and angiogenesis. Tan I suppressed the expression of angiogenic factor IL-8 through NF-κB and AP-1 pathway in CL1-5 lung cancer cells [91]. Tan I reduced the level of HIF-1α and phosphorylation of Stat3 at Tyr705 and decreased the mRNA level and protein secretion of VEGF in hypoxic MCF-7 breast cancer cells, suggesting a role of Tan I as angiogenesis inhibitor [92]. Tan I attenuated adhesion molecules, such as ICAM-1 and VCAM-1. Thereby, it prevented adhesion of MDA-MB-231 breast cancer cells to HUVECs stimulated with TNFα. Tan I also inhibited TNFα-induced VEGF production in MDA-MB-231 cells and migration of MDA-MB-231 cells through extracellular matrix. Additionally, a reduction of tumor mass volume and decrease of metastasis incidence by Tan I was observed in vivo [93].

3.3 Tanshinone IIA

Tanshinone IIA (Tan IIA) induced caspase-dependent apoptosis in TOV-21G, SKOV3, and OVCAR3 ovarian carcinoma cells without affecting Bcl-2, Bcl-xL and Bax. Tan IIA dose-dependently downregulated the expression of survivin via p38 MAPK activation [94]. The compound induced cytotoxicity in human lung cancer A549 cells possibly by p53-dependent, mitochondrial-mediated apoptosis with higher ratios of Bax/Bcl-2 [95]. Another report illustrated that Tan IIA-triggered p53 might result from altered DNA structure and RNA polymerase II (RNAPII) degradation [96]. Tan-IIA may inhibit tumor growth in a J5 xenograft animal model by increasing Bax and caspase 3 and decreasing CD31 expression [97]. Tan IIA generated G2/M [98, 99] or S phase [100] cell cycle arrest. The mitochondrial-dependent and caspase-mediated apoptotic pathway stimulated by Tan IIA might be related to activation of JNK [101, 102]. Tan IIA downregulated the expression of HPV E6 and E7 genes in HPV-positive CaSki cells, and caused S phase arrest. It also induced accumulation of p53 and altered expression of p53-dependent targets and caused p53-mediated apoptosis by modulating Bcl2, Bax, and caspase-3, expression and PARP cleavage [103]. Tan IIA induced apoptosis in human glioma stem cells in vitro and in vivo. Furthermore, Tan IIA increased the expression of differentiation and decreased expression of stem cell markers such as CD133 and nestin [104]. The inhibitory effects of Tan IIA on cancer stem cells might relate to interference of the IL6/STAT3 signaling pathway [104, 105].

Non-small cell lung cancer (NSCLC) cells showed resistance to TRAIL-mediated cell death. However, Tan IIA and TRAIL synergistically increased apoptosis in TRAIL-resistant NSCLC cells. Tan IIA stimulated DR5 and CHOP by activating the ER stress-initiated PERK-ATF4-CHOP pathway [106]. In another case, Tan IIA sensitized human ovarian carcinoma cells to TRAIL-induced extrinsic apoptosis. In addition, Tan IIA promoted JNK-mediated signaling to up-regulated CHOP and thereby induced DR5 expression [107]. Other results also supported that Tan IIA induced ER-stress-mediated CHOP upregulation, which led to cell cycle arrest [108, 109] and apoptosis of cancer cells via phosphorylation of JNK, ERK and p38 MAPK [110, 111].

Tan IIA-induced autophagy in various cancer cells such as prostate (PC-3), multiple myeloma (U266), lung cancer (NCI-H460), leukemia (kbm-5) and breast cancer (MDA-MB-231) cells. Tan IIA significantly increased the expression of microtubule-associated protein light chain 3 (LC3) II as hallmark of autophagy. Tan IIA augmented the phosphorylation of AMPK and attenuated the phosphorylation of mammalian target of rapamycin (mTOR) and p70 S6K. Also, Tan IIA activated the ERK signaling pathway including Raf, ERK and p90 RSK [112]. Tan IIA-generated ROS production simultaneously induced apoptosis and autophagy on PC-3 cancer cells. Autophagy inhibitors enhanced Tan IIA-induced cell apoptosis, while inhibition of apoptosis failed to stimulate more autophagic death, meaning that Tan IIA-induced autophagy was independent of apoptosis [113].

Tan IIA was a potent agent to regulate expression of VEGF and thereby inhibit cancer angiogenesis. In vascular endothelial cells, Tan IIA bound to the VEGFR2 kinase domain, blocking the downstream pathways of VEGF/VEGFR [114]. Tan IIA inhibited tumor size and angiogenesis through downregulating expression of cyclooxygenase-2 (COX-2) and VEGF in human colorectal cancer [115]. Tan IIA repressed HIF-1α, leading to the inhibition of VEGF in human breast cancer cells, in which the mTOR/p70S6K/4E-BP1 pathway was involved [116]. Tan IIA reduced cancer cell adhesion, migration and invasion [117, 118]. It inhibited invasion and metastasis by reducing levels of MMP-2, MMP-9 [119, 120] and urokinase plasminogen activator (uPA), and by increasing levels of tissue inhibitor of matrix metalloproteinase protein (TIMP)-1 and TIMP-2 [120].

In cancer cells, activation of NF-κB favors carcinogenic processes, including activation of anti-apoptotic genes, proliferation, angiogenesis, metastasis as well as resistance to anti-cancer drugs [121]. Tan IIA also suppressed NF-κB signaling [120, 122]. Tan IIA induced cell death in colon cancer cells and sensitized colon cancer cells to 5-FU therapy by inhibiting NF-κB activation. Tan IIA treatment decreased the level of the phosphorylated p65 and the NF-κB-regulated gene expression levels of VEGF, COX-2, c-Myc and Bcl-2 [123].

Clinically relevant mutations in AR convert antiandrogens into AR agonists, thereby promoting prostate tumor growth. Tan IIA is a potent antagonist of mutated

ARs. Tan IIA suppressed AR expression and induced apoptosis in LNC [124]. AR was essential to induce G1 arrest by tanshinone IIA in LNCa [125].

3.4 Miltirone

Miltirone was first identified in the 1970s [126], however, the pharmacological effects of miltirone was not explored until recently. Miltirone was cytotoxic against A549 and HCT-116 cells [127]. Miltirone stimulated ROS generation, disrupted MMP and caused DNA damage leading to the activation of caspases 3/7 and 9. Hence, miltirone induced mitochondria-driven intrinsic apoptosis. Inhibition of nuclear p65 translocation results from direct binding to IKKβ. Miltirone-induced DNA damage causes G2/M arrest by deregulating *CDC2*, *CCNB1* and *CDKN1A* expression. The expression of *TP53* (tumor suppressor p53), which regulates G2/M cell cycle arrest, did not change. Furthermore, miltirone inhibited cell adhesion and NF-κB signaling. Miltirone bound to the hinge region of IKK-β by forming a hydrogen bond with Cys99, and downregulated translocation of p65 into the nucleus [128]. Miltirone was cytotoxic towards both sensitive CCRF-CEM and multidrug-resistant CEM/ADR5000 cells. In addition, it is worthy to highlight that the cytotoxic effects of miltirone were more than that of other tanshinones [128, 129]. According to a report from Hu et al., miltirone had no inhibitory effects on P-gp function [50]. By contrast, a recent study illustrated that miltirone indeed interacted with P-gp and inhibited P-gp in doxorubicin-resistant HepG2 cells [130]. Miltirone triggered ROS-mediated MAPK signaling, which might also contribute to cancer cell apoptosis [130]. In colon cancer cells, miltirone stimulated the ROS- and p53-dependent proapoptotic pathway associated with accumulation of intracellular Ca^{+} and disruption of mitochondrial membrane potential [131]. Miltirone triggered ER stress through upregulation of phosphorylated PERK, eIF2a, GRP78, GRP94 and caspase 12. In addition, miltirone inhibited tumor growth and induced apoptosis in U937 xenograft model with low systemic toxicity [132].

4 Anticancer Effects of the Hydrophilic Compounds

Plant phenolic acids have been shown both preventive and treatment potential to inhibit the initiation and progression of cancers in vitro and in vivo. Although the cellular mechanisms are not fully elucidated, the anticancer effects of phenolic acids have been supported by plenty of evidences including scavenging free radicals, induction of enzymes involved in the metabolism of xenobiotics, regulation of gene expression, modulation of cellular signaling pathways involved in DNA damage repair, cell survival, apoptosis or programmed cell death, angiogenesis and invasion [133, 134]. The main water-soluble components isolated from *S. miltiorrhiza* are

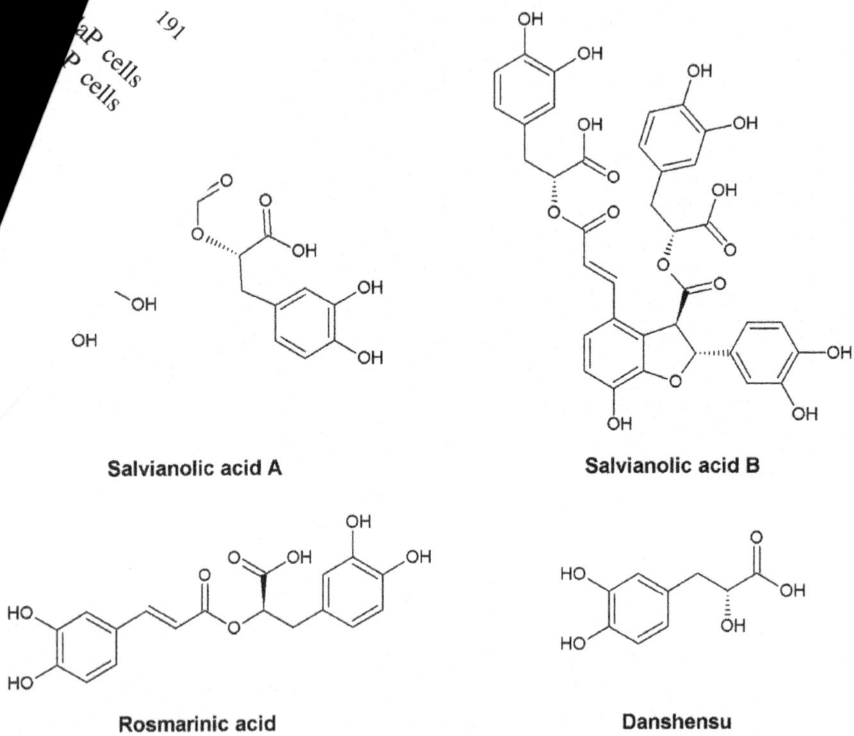

Fig. 4 Molecular structure of main hydrophilic compounds isolated from *S. miltiorrhiza*

phenolic acids containing salvianolic acids A–K, danshensu, rosmarinic acid, and so on (Fig. 4). In addition to anticancer properties, phenolic acid ingredients of *S. miltiorrhiza* reveal considerable pharmacological activities, such as cardiovascular protection [135], antidiabetic effect [136, 137], neuroprotection [138, 139], and anti-microbial effects [140, 141]. Remarkably, the combination of phenolic acids, with standard chemotherapeutic agents enhanced anticancer effects at lower toxicity and overcame drug resistance in tumor cells.

4.1 Rosmarinic Acid

Rosmarinic acid (RA), namely 3,4-dihydroxyphenyllactic acid, is an ester derivative of caffeic acid and found in numerous medicinal and culinary plants, such as *S. miltiorrhiza*, sweet basil, oregano and many others [142]. As a dietary polyphenolic agent, RA has been widely explored their cancer chemopreventive properties to arrest or reverse the process of carcinogenesis. One important aspect of carcinogenesis is recognized to be the involvement of inflammation. For instance, the overexpression of COX-2, a key enzyme responsible for the synthesis of inflammatory prostaglandin from arachidonic acid, is associated with colon, lung and prostate carcinogenesis [133]. RA showed inhibitory effects on COX-2 activity and proliferation of human lung cancer cell line A549, and inhibited colon cancer cell proliferation by repressing 12-*O*-tetradecanoylphorbol-13-acetate (TPA)-induced binding of AP-1, c-Jun and c-Fos to COX-2 promotor and antagonizing ERK1/2 activation [143, 144]. TPA-induced skin carcinogenesis leading to inflammatory activity, neutrophil infiltration and recruitment, ROS production, and COX-2 induction was significantly reduced by RA [145]. In addition, RA inhibited pro-inflammatory cytokines and miR-155 related to inflammation, suggesting that RA may suppress the Warburg effect via an inflammatory pathway, such as the IL-6/STAT3 pathway [146]. In the mouse model of cecal ligation and puncture (CLP)-mediated release of HMGB1, a key cytokine which mediates response to injury and inflammation, RA downregulated CLP-mediated HMGB1 release, the expression of HMGB1 receptor (TLR4), and HMGB1-mediated barrier disruption by increasing barrier integrity. RA also inhibited the expressions of cell adhesion molecules, which in turn reduced monocyte adhesion and migration in HUVECs [147]. In the 1,2-dimethylhydrazine (DMH)-induced colon carcinogenic model, Supplementation with RA to DMH treated rats significantly decreased tumor formation, reversed oxidative stress, and stimulated markers of antioxidant status. In addition, RA also modulated activity of xenobiotic enzymes, induced apoptosis of tumor cells and prevented DMH-induced DNA damage [148–150]. In 7,12-dimethylbenz(a)anthracene (DMBA)-induced oral carcinogenesis in hamsters, RA showed greater differentiation and significantly reduced the incidences and aggression of the tumors [151, 152]. Oral administration of RA inhibited UV-induced oxidative stress by regulating tyrosinase activity. Meanwhile, RA induced melanin production by activating the CRE-binding protein via protein kinase A, demonstrating that RA is a potential photo-protective agent [153, 154]. Taken together, RA exerted DNA protection and anti-carcinogenic effects via two independent mechanisms: inhibition of the inflammatory response and scavenging of reactive oxygen radicals. However, it is noteworthy that RA protected colon carcinoma CO115 cells from DNA damage induced by the genotoxic alkylating agent 1,3-bis-(2-chloroethyl)-1-nitosourea (BCNU) without markedly change DNA repair protein expression, implying RA may interfere with efficacy of alkylating agents used in cancer therapy [155, 156].

Anticancer effects of RA have been studied in different types of cancer cells. RA inhibited migration, adhesion and invasion of human colon carcinoma cells Ls174-T cells through downregulating the levels of ROS and expression of MMP-2 and 9 [157]. In the MDA-MB-231BO human bone-homing breast cancer cell model, RA repressed cell migration via the pathway of the receptor activator of NFκB ligand (RANKL)/RANK/osteoprotegerin (OPG) and by inhibiting the level of IL-8 [158]. Furthermore, RA induced HCT15 human colon cancer cell apoptosis by modulating phosphorylation of ERK [159]. In H22 hepatocarcinoma-bearing model, RA has been proved to suppress overexpression of angiogenic cytokines (e.g. VEGF, TGF-β, IL-6, and TNF-α) and pro-inflammatory factors (e.g. IL-1β,

IL-6, TNF-α and NFκB p65) in the xenograft microenvironment [160]. In another study, RA inhibited in vitro angiogenic processes involving HUVEC endothelial cell proliferation, migration, adhesion and tube formation. The underlying mechanism may relate to ROS-associated VEGF expression and IL-8 release [161]. RA induced apoptosis of HL-60 promyelocytic leukemia cells, exerted strong radical-scavenging effects and diminished cell cycle progression via significant inhibition of DNA synthesis [162]. RA induced necrosis and apoptosis through DNA damage and disruption of mitochondrial membrane potential (MMP) in a ROS- and caspase-independent manner in CCRF-CEM leukemia cells. Additionally, RA also inhibits NFκB signaling and cell adhesion [163]. Inhibition of TNFα-induced NFκB signaling and ROS formation by RA contributed to human U937 leukemia cells, sensitizing cell apoptosis and eliminating TNFα-induced pro-inflammatory effects [164]. RA also potentiated all-trans retinoic acid (ATRA)-induced macrophage differentiation in acute promyelocytic leukemia (APL) cells and stimulated macrophage functions including ROS production, phagocytic activity, expression of the chemokine receptors and the adhesion molecule ICAM-1 [165].

The therapeutic effects of RA against MDR has been demonstrated. RA revealed preferential cytotoxic effects on both sensitive CCRF-CEM and adriamycin-resistant CEM/ADR5000 leukemia cells. Notably, normal lymphocytes were able to tolerate RA, but not doxorubicin at doses necessary to kill MDR cells [163]. RA also reversed the MDR of human gastric cancer cell SGC-7901/ADR and breast cancer cell MCF7/ADR with the upregulation of intracellular concentration of adriamycin and downregulation of *MDR1* gene transcription and P-gp expression [166, 167]. RA enhanced the efflux activity of P-gp and multidrug resistance-associated protein 2 (MRP2) accompanied by marked increase in the intracellular ATP levels through the translocation of nuclear factor E2-related factor-2 (Nrf2) from cytoplasm to the nucleus in HepG2 cells [168]. RA was also identified as a potent inhibitor of EGFR targeting C-terminal hinge region of tyrosine kinase [169].

4.2 Salvianolic Acid A (Sal A)

Salvianolic acid (Sal A) induced apoptosis on A549 human lung cancer cells. It regulated the PI3K/AKT signaling pathway by mediating PTEN protein stability [170]. Sal A reversed paclitaxel resistance by mechanisms involving PI3 K/AKT pathway to suppress proliferation of paclitaxel resistant MCF-7 cells. In addition, Sal A inhibited upregulation of ABC transporter including P-gp, multidrug resistance associated protein 1 (MRP1), and breast cancer resistance protein (BCRP) [171]. The combination of Sal A and paclitaxel inhibited tumor migration and invasion, and suppressed the expression of transgelin 2 [172]. MCF-7/ADR breast cancer cells were collateral sensitive to Sal A. It arrested the cell cycle in the S phase, induced mitochondrial-mediated apoptosis (disruption of mitochondrial membrane potential, caspase-3 activation, Bcl-2 down-regulation, and Bax up-regulation) and modulated P-gp expression. In MDR xenograft tumors, Sal A showed better antitumor activity and less loss of body weight in contrast to doxorubicin [173]. Sal A also inhibited protein arginine methyltransferase 1 (PRMT1), a key molecule required for maintenance of P-gp (*MDR1*) overexpression. Sal A decreased *MDR1* activity promoter and enhanced the antitumor effect of doxorubicin in nude mice bearing drug-resistant tumors [174]. Therefore, Sal A may be a promising candidate to overcome MDR.

4.3 Salvianolic Acid B (Sal B)

The chemopreventive salvianolic acid B (Sal B) exerts antioxidant and anti-inflammatory properties [175]. It serves as a vascular protector, decreasing urethane-induced abnormal angiogenesis and lung carcinogenesis [176]. Sal B inhibits growth of head and neck squamous cell cancer (HNSCC) via anti-inflammation mechanisms. Aside from induction of caspase-dependent apoptosis, Sal B attenuated COX-2 expression in permanently cultured as well as primary in HNSCC cells isolated from xenograft tumors. Sal B also inhibited production of prostaglandin E2 (PGE2) with or without lipopolysaccharide stimulation [177]. Combining Sal B with other agents might also be an effective chemopreventive strategy, preventing HNSCC. Compared to high-dose Sal B or celecoxib (a selective COX-2 inhibitor) alone, Sal B in combination with low-dose celecoxib suppressed cell proliferation and enhanced apoptosis of HNSCC cells in vitro and in vivo through sequential inhibition of the COX-2/PGE2/EGFR pathway. Furthermore, the cardiotoxicity induced by celecoxib was reduced at lower doses [178]. Sal B inhibited malignant transformation of DMBA-induced oral precancerous lesion in hamsters. The reduced incidence of squamous cell carcinoma (SCC) was associated with inhibition of angiogenesis in dysplasia and SCC and concomitantly a downregulation of hypoxia-inducible factor 1 alpha (HIF-1α) and VEGF was observed [179]. Sal B attenuated or normalized genetic and biochemical alternations of inflammation and angiogenesis in terms of elevation of glycolysis, decreased cholesterol and myoinositol metabolism in the DMBA-induced tumor model [180].

Sal B inhibited oral squamous cancer cell growth. The anticancer effects were attributed to deregulated angiogenesis-associated gene expression, including HIF-1α, TNFα, MMP9 and THBS2 [181]. In addition, Sal B induced apoptosis of human U87 glioma cells through generating ROS production and phosphorylation of p38 MAPK and p53 [182]. Combination of Sal B and arsenic trioxide enhanced anticancer effects in comparison to arsenic trioxide alone and significantly improved arsenic trioxide-induced loss of cardiac function [183]. Sal B may be a potential inhibitor targeting Ku86, an initiator of DNA double strand break repair by non-homologous end joining (NHEJ). This suggests the use of Sal B in combination with radiotherapy against cancer [184].

4.4 Danshensu

Danshensu inhibited B16F10 melanoma cell invasion and migration in vitro and in vivo, which was attributed to the downregulation of MMP-2, -9 and VEGF [185]. Danshensu might be effective in cancer prevention and treatment by targeting COX-2 activity. In computational analysis, danshensu bound with high affinities to Arg120 and Tyr355, which are two main residues located in the active site of COX-2. Danshensu suppressed the proliferation of A549 cells and enhanced the activity of celecoxib in reducing tumor metastasis in vivo. Meanwhile, COX-2/PGE2/ERK signaling was involved in the anticancer activity of danshensu [144].

Danshensu significantly blocked the activity of monoamine oxidase B (MAOB), an enzyme localized in lungs. It catalyzed the oxidation of monoamine to produce hydrogen peroxide. In addition, danshensu attenuated NF-κB activation, leading to increased ionizing radiation (IR)-induced apoptosis, reduce cancer-associated inflammation and metastasis in NSCLC cells. In vivo danshensu elicited radiosensitization of NSCLC cells [186].

Danshensu derivatives also exerted significant anticancer activities. A synthetic *danshensu* derivative (D006) enhanced apoptosis and ameliorated cardiotoxicity of doxorubicin by mediating activation of HO-1 and mitochondrial biogenesis [187]. DT-010, a conjugate of danshensu and tetramethylpyrazine, inhibited mitochondrial complex II, which promoted ROS generation and disrupted mitochondrial function in breast cancer cells [188].

5 Conclusions and Perspectives

Accumulated evidence indicates that lipophilic and hydrophilic components isolated from *S. miltiorrhiza* are potential candidates for anticancer agents, especially tanshinones. These compounds exert diverse activities in inhibiting cancer development in vitro and in vivo with low side effects or toxicity. These compounds inhibit cancer cell growth and proliferation, induce cell cycle arrest and apoptosis, and suppress invasion and angiogenesis. It is also worth mentioning that they overcome drug resistance and synergistically enhance the anticancer effects, if combined with other standard chemotherapeutic agents.

Pharmacokinetic characteristics, such as poor water solubility of lipophilic compounds and poor stability and permeability of hydrophilic phenolic acids have limited their clinical use for cancer therapy [189]. Attempts have been made to overcome their poor bioavailability. For instance, structure modifications have improved the water solubility. Acetyltanshinone IIA exhibited increased water solubility and induced significant ROS production and mitochondrial-mediated apoptosis [190]. Nanotechnological or lipid-based drug delivery formulation have been investigated to enhance delivery and bioavailability of the compounds. The permeability of Sal B has been increased by phospholipid complexes [191].

Diverse drug delivery systems improved the solubility and dissolution rates and prolonged the circulating times of tanshinones. Solid dispersion and nanoparticles were promising formulations [192]. Unfortunately, most formulations lack tissue specificity. Therefore, it is necessary to develop drug delivery systems that specifically target cancer tissues to enhance therapeutic effects of these compounds.

Acknowledgements We are grateful to the German Academic Exchange Serve (DAAD) for a Ph.D. stipend to C.F.W. We Chunlan Hong for supporting literature survey and Dr. Maen Zeino for critical reading of the manuscript.

Conflict of Interest The authors declare that there is no conflict of interest.

References

1. Fitzmaurice C, Dicker D, Pain A et al (2015) Global burden of disease cancer collaboration. The global burden of cancer 2013. JAMA Oncol 1:505–527. https://doi.org/10.1001/jamaoncol.2015.0735
2. Bray F (2014) Transitions in human development and the global cancer burden. In: Wild CP, Stewart B (eds) World cancer report. International Agency for Research on Cancer, Lyon, pp 54–68
3. Hanahan D, Weinberg RA (2000) The hallmarks of cancer. Cell 100:57–70. https://doi.org/10.1016/S0092-8674(00)81683-9
4. Malhotra V, Perry MC (2003) Classical chemotherapy: mechanisms, toxicities and the therapeutic window. Cancer Biol Ther 2:S2–S4. https://doi.org/10.4161/cbt.199
5. Siddik ZH (2002) Mechanisms of action of cancer chemotherapeutic agents: DNA-interactive alkylating agents and antitumour platinum-based drugs. In: Alison MR (ed) The cancer handbook. Nature Publishing Group, London, pp 1295–1313
6. Evans BE, Rittle KE, Bock MG et al (1988) Methods for drug discovery: development of potent, selective, orally effective cholecystokinin antagonists. J Med Chem 31:2235–2246. https://doi.org/10.1021/jm00120a002
7. Swinney DC, Anthony J (2011) How were new medicines discovered? Nat Rev Drug Discov 10:507–519. https://doi.org/10.1038/nrd3480
8. Newman DJ, Cragg GM (2012) Natural products as sources of new drugs over the 30 years from 1981 to 2010. J Nat Prod 75:311–335. https://doi.org/10.1021/np200906s
9. Basmadjian C, Zhao Q, Bentouhami E et al (2014) Cancer wars: natural products strike back. Front Chem 2:20. https://doi.org/10.3389/fchem.2014.00020
10. Luo F, Gu J, Chen L et al (2014) Systems pharmacology strategies for anticancer drug discovery based on natural products. Mol BioSyst 10:1912–1917. https://doi.org/10.1039/c4mb00105b
11. Wang X, Morris-Natschke SL, Lee KH (2007) New developments in the chemistry and biology of the bioactive constituents of Tanshen. Med Res Rev 27:133–148. https://doi.org/10.1002/med.20077
12. Sung B, Chung HS, Kim M et al (2015) Cytotoxic effects of solvent-extracted active components of *Salvia miltiorrhiza* Bunge on human cancer cell lines. Exp Ther Med 9:1421–1428. https://doi.org/10.3892/etm.2015.2252
13. Wang BQ (2010) *Salvia miltiorrhiza*: chemical and pharmacological review of a medicinal plant. J Med Plants Res 4:2813–2820. ISSN 1996-0875

14. Zhang YH, Qin X, Xu J (2012) Analysis of Chinese medical syndrome features of patients with primary liver cancer before and after transcatheter arterial chemo-embolization. Zhongguo Zhong Xi I Jie He Za Zhi 32:111–1174. Print ISSN 1003-5370
15. Hu B, Wang SS, Du Q (2015) Traditional Chinese medicine for prevention and treatment of hepatocarcinoma: from bench to bedside. World J Hepatol 7:1209–1232. https://doi.org/10.4254/wjh.v7.i9.1209
16. Lahans T (2007) General pathophysiology and treatment: conventional and Chinese medicine. In: Lahans T (ed) Intergrating conventional and Chinese medicine in cancer care: a clinical guide. Elsevier Health Sciences, USA, pp 1–34
17. Chen X, Guo J, Bao J et al (2014) The anticancer properties of *Salvia miltiorrhiza* Bunge (Danshen): a systematic review. Med Res Rev 34:768–794. https://doi.org/10.1002/med.21304
18. Du G, Zhang J (2014) Overview of modern research on Danshen. In: Yan X (ed) Danshen (*Salvia miltiorrhiza*) in medicine. Springer, Dordrecht, pp 3–17
19. Zhou L, Zuo Z, Chow MS (2005) Danshen: an overview of its chemistry, pharmacology, pharmacokinetics, and clinical use. J Clin Pharmacol 45:1345–1359. https://doi.org/10.1177/0091270005282630
20. Liu AH, Li L, Xu M et al (2006) Simultaneous quantification of six major phenolic acids in the roots of *Salvia miltiorrhiza* and four related traditional Chinese medicinal preparations by HPLC–DAD method. J Pharm Biomed Anal 41:48–56. https://doi.org/10.1016/j.jpba.2005.10.021
21. Li HB, Chen F (2001) Preparative isolation and purification of six diterpenoids from the Chinese medicinal plant *Salvia miltiorrhiza* by high-speed counter-current chromatography. J Chromatogr A 925:109–114. https://doi.org/10.1016/s0021-9673(01)01026-3
22. Lin TH, Hsieh CL (2010) Pharmacological effects of *Salvia miltiorrhiza* (Danshen) on cerebral infarction. Chin Med 5:22. https://doi.org/10.1186/1749-8546-5-22
23. Zhang XZ, Qian SS, Zhang YJ et al (2016) *Salvia miltiorrhiza*: a source for anti-Alzheimer's disease drugs. Pharm Biol 54:18–24. https://doi.org/10.3109/13880209.2015.1027408
24. Cheng TO (2007) Cardiovascular effects of Danshen. Int J Cardiol 121:9–22. https://doi.org/10.1016/j.ijcard.2007.01.004
25. Housman G, Byler S, Heerboth S et al (2014) Drug resistance in cancer: an overview. Cancers (Basel) 6:1769–1792. https://doi.org/10.3390/cancers6031769
26. Gottesman MM (2002) Mechanisms of cancer drug resistance. Annu Rev Med 53:615–627. https://doi.org/10.1146/annurev.med.53.082901.103929
27. Gillet JP, Efferth T, Remacle J (2007) Chemotherapy-induced resistance by ATP-binding cassette transporter genes. Biochim Biophys Acta 1775:237–262. https://doi.org/10.1016/j.bbcam.2007.05.002
28. Scala S, Akhmed N, Rao US et al (1997) P-glycoprotein substrates and antagonists cluster into two distinct groups. Mol Pharmacol 51:1024–1033. https://doi.org/10.1124/mol.51.6.1024
29. Yague E, Armesilla AL, Harrison G et al (2003) P-glycoprotein (MDR1) expression in leukemic cells is regulated at two distinct steps, mRNA stabilization and translational initiation. J Biol Chem 278:10344–10352. https://doi.org/10.1074/jbc.m211093200
30. Szakács G, Paterson JK, Ludwig JA et al (2006) Targeting multidrug resistance in cancer. Nat Rev Drug Discov 5:219–234. https://doi.org/10.1038/nrd1984
31. Wu CF, Bohnert S, Thines E et al (2016) Cytotoxicity of the root extract of *Salvia miltiorrhiza* against multidrug-resistance cancer cells. Am J Chin Med 44:871–894. https://doi.org/10.1142/s0192415x16500488
32. Wu CF, Klauck SM, Efferth T (2016) Cytotoxicity of cryptotanshinone towards acute lymphoblastic leukemia cells. Arch Toxicol 90:2275–2286. https://doi.org/10.1007/s00204-015-1616-4
33. Liu J, Zhang C, Hu W et al (2015) Tumor suppressor p53 and its mutants in cancer metabolism. Cancer Lett 356:197–203. https://doi.org/10.1016/j.canlet.2013.12.025

34. Giono LE, Manfredi JJ (2006) The p53 tumor suppressor participates in multiple cell cycle checkpoints. J Cell Physiol 209:13–20. https://doi.org/10.1002/jcp.20689
35. Smith ML, Seo YR (2002) p53 regulation of DNA excision repair pathways. Mutagenesis 17:149–156. https://doi.org/10.1093/mutage/17.2.149
36. Amaral JD, Xavier JM, Steer CJ et al (2010) The role of p 53 in apoptosis. Discov Med 9:145–152. ISSN 1539-6509
37. Ben-Porath I, Weinberg RA (2005) The signals and pathways activating cellular senescence. Int J Biochem Cell Biol 37:961–976. https://doi.org/10.1016/j.biocel.2004.10.013
38. Levine AJ, Oren M (2009) The first 30 years of p53: growing ever more complex. Nat Rev Cancer 9:749–758. https://doi.org/10.1038/nrc2723
39. McCurrach ME, Connor TMF, Knudson CM et al (1997) Bax-deficiency promotes drug resistance and oncogenic transformation by attenuating p53-dependent apoptosis. Proc Natl Acad Sci U S A 94:2345–2349. https://doi.org/10.1073/pnas.94.6.2345
40. Giménez-Bonafé P, Tortosa A, Pérez-Tomás R (2009) Overcoming drug resistance by enhancing apoptosis of tumor cells. Curr Cancer Drug Targets 9:320–340. https://doi.org/10.2174/156800909788166600
41. Lynch TJ, Bell DW, Sordella R et al (2004) Activating mutations in the epidermal growth factor receptor underlying responsiveness of non-small-cell lung cancer to gefitinib. N Engl J Med 350:2129–2139. https://doi.org/10.1056/nejmoa040938
42. Amann J, Kalyankrishna S, Massion PP et al (2005) Aberrant epidermal growth factor receptor signaling and enhanced sensitivity to EGFR inhibitors in lung cancer. Cancer Res 65:226–235. Print ISSN 0008-5472
43. Tomas A, Futter CE, Eden ER (2014) EGF receptor trafficking: consequences for signaling and cancer. Trends Cell Biol 24:26–34. https://doi.org/10.1016/j.tcb.2013.11.002
44. Lee J, Moon C (2011) Current status of experimental therapeutics for head and neck cancer. Exp Biol Med 236:375–389. https://doi.org/10.1258/ebm.2010.010354
45. Zhang Z, Stiegler AL, Boggon TJ et al (2010) EGFR-mutated lung cancer: a paradigm of molecular oncology. Oncotarget 1:497–514. https://doi.org/10.18632/oncotarget.186
46. Dong Y, Morris-Natschke SL, Lee KH (2011) Biosynthesis, total synthesis, and antitumor activity of tanshinones and their analogs as potential therapeutic agents. Nat Prod Rep 28:529–542. https://doi.org/10.1039/c0np00035c
47. Tian XH, Wu JH (2013) Tanshinone derivatives: a patent review (January 2006–September 2012). Expert Opin Ther Pat 23:19–29. https://doi.org/10.1517/13543776.2013.736494
48. Li H, Zhang Q, Chu T et al (2012) Growth-inhibitory and apoptosis-inducing effects of tanshinones on hematological malignancy cells and their structure-activity relationship. Anticancer Drugs 23:846–855. https://doi.org/10.1097/cad.0b013e328351f896
49. Liu F, Yu G, Wang G et al (2012) An NQO1-initiated and p53-independent apoptotic pathway determines the anti-tumor effect of tanshinone IIA against non-small cell lung cancer. PLoS One 7:e42138. https://doi.org/10.1371/journal.pone.0042138
50. Hu T, To KK, Wang L et al (2014) Reversal of P-glycoprotein (P-gp) mediated multidrug resistance in colon cancer cells bycryptotanshinone and dihydrotanshinone of *Salvia miltiorrhiza*. Phytomedicine 21:1264–1272. https://doi.org/10.1016/j.phymed.2014.06.013
51. Lee WY, Cheung CC, Liu KW et al (2010) Cytotoxic effects of tanshinones from *Salvia miltiorrhiza* on doxorubicin-resistant human liver cancer cells. J Nat Prod 73:854–859. https://doi.org/10.1021/np900792p
52. Chen L, Wang HJ, Xie W et al (2014) Cryptotanshinone inhibits lung tumorigenesis and induces apoptosis in cancer cells in vitro and in vivo. Mol Med Rep 9:2447–2452. https://doi.org/10.3892/mmr.2014.2093
53. Yuan DP, Long J, Lu Y et al (2014) The forecast of anticancer targets of cryptotanshinone based on reverse pharmacophore-based screening technology. Chin J Nat Med 12:443–448. https://doi.org/10.1016/s1875-5364(14)60069-8
54. Luo Y, Chen W, Zhou H et al (2011) Cryptotanshinone inhibits lymphatic endothelial cell tube formation by suppressing VEGFR-3/ERK and small GTPase pathways. Cancer Prev Res (Phila) 4:2083–2091. https://doi.org/10.1158/1940-6207.capr-11-0319

55. Lee HJ, Jung DB, Sohn EJ et al (2012) Inhibition of hypoxia inducible factor alpha and astrocyte-elevated gene-1 mediates cryptotanshinone exerted antitumor activity in hypoxic PC-3 cells. Evid Based Complement Alternat Med 2012:390957. https://doi.org/10.1155/2012/390957
56. Zhu Z, Zhao Y, Li J et al (2016) Cryptotanshinone, a novel tumor angiogenesis inhibitor, destabilizes tumor necrosis factor-α mRNA via decreasing nuclear-cytoplasmic translocation of RNA-binding protein HuR. Mol Carcinog 55:1399–1410. https://doi.org/10.1002/mc.22383
57. Park IJ, Kim MJ, Park OJ et al (2010) Cryptotanshinone sensitizes DU145 prostate cancer cells to Fas (APO1/CD95)-mediated apoptosis through Bcl-2 and MAPK regulation. Cancer Lett 298:88–98. https://doi.org/10.1016/j.canlet.2010.06.006
58. Kim JH, Jeong SJ, Kwon TR et al (2011) Cryptotanshinone enhances TNF-α-induced apoptosis in chronic myeloid leukemia KBM-5 cells. Apoptosis 16:696–707. https://doi.org/10.1007/s10495-011-0605-1
59. Chen W, Liu L, Luo Y et al (2012) Cryptotanshinone activates p38/JNK and inhibits Erk1/2 leading to caspase-independent cell death in tumor cells. Cancer Prev Res (Phila) 5:778–787. https://doi.org/10.1158/1940-6207.capr-11-0551
60. Xia C, Bai X, Hou X et al (2015) Cryptotanshinone reverses cisplatin resistance of human lung carcinoma A549 cells through down-regulating Nrf2 pathway. Cell Physiol Biochem 37:816–824. https://doi.org/10.1159/000430398
61. Zhang Y, Won SH, Jiang C et al (2012) Tanshinones from Chinese medicinal herb Danshen (*Salvia miltiorrhiza* Bunge) suppress prostatecancer growth and androgen receptor signaling. Pharma Res 29:1595–1608. https://doi.org/10.1007/s11095-012-0670-3
62. Xu D, Lin TH, Li S et al (2012) Cryptotanshinone suppresses androgen receptor-mediated growth in androgen dependent and castration resistant prostate cancer cells. Cancer Lett 316:11–22. https://doi.org/10.1016/j.canlet.2011.10.006
63. Wu CY, Hsieh CY, Huang KE et al (2012) Cryptotanshinone down-regulates androgen receptor signaling by modulating lysine-specific demethylase 1 function. Int J Cancer 131:1423–1434. https://doi.org/10.1002/ijc.27343
64. Lin TH, Lee SO, Niu Y et al (2013) Differential androgen deprivation therapies with anti-androgens casodex/bicalutamide or MDV3100/enzalutamide versus anti-androgen receptor ASC-J9(R) lead to promotion versus suppression of prostate cancer metastasis. J Biol Chem 288:19359–19369. https://doi.org/10.1074/jbc.m113.477216
65. Shin DS, Kim HN, Shin KD et al (2009) Cryptotanshinone inhibits constitutive signal transducer and activator of transcription 3 function through blocking the dimerization in DU145 prostate cancer cells. Cancer Res 69:193–202. https://doi.org/10.1158/0008-5472.can-08-2575
66. Lu L, Li C, Li D et al (2013) Cryptotanshinone inhibits human glioma cell proliferation by suppressing STAT3 signaling. Mol Cell Biochem 381:273–282. https://doi.org/10.1007/s11010-013-1711-x
67. Ge Y, Yang B, Chen Z et al (2015) Cryptotanshinone suppresses the proliferation and induces the apoptosis of pancreatic cancercells via the STAT3 signaling pathway. Mol Med Rep 12:7782–7788. https://doi.org/10.3892/mmr.2015.4379
68. Yu HJ, Park C, Kim SJ et al (2014) Signal transducer and activators of transcription 3 regulates cryptotanshinone-induced apoptosis in human mucoepidermoid carcinoma cells. Pharmacogn Mag 10:S622–S629. https://doi.org/10.4103/0973-1296.139802
69. Li W, Saud SM, Young MR et al (2015) Cryptotanshinone, a Stat3 inhibitor, suppresses colorectal cancer proliferation and growth in vitro. Mol Cell Biochem 406:63–73. https://doi.org/10.1007/s11010-015-2424-0
70. Yan H, Guo BY, Zhang S (2016) Cancer-associated fibroblasts attenuate Cisplatin-induced apoptosis in ovarian cancer cells by promoting STAT3 signaling. Biochem Biophys Res Commun 470:947–954. https://doi.org/10.1016/j.bbrc.2016.01.131
71. Oyadomari S, Mori M (2004) Roles of CHOP/GADD153 in endoplasmic reticulum stress. Cell Death Differ 11:381–389. https://doi.org/10.1038/sj.cdd.4401373

72. Park IJ, Kim MJ, Park OJ et al (2012) Cryptotanshinone induces ER stress-mediated apoptosis in HepG2 and MCF7 cells. Apoptosis 17:248–257. https://doi.org/10.1007/s10495-011-0680-3
73. Zhang YF, Zhang M, Huang XL et al (2015) The combination of arsenic and cryptotanshinone induces apoptosis through induction of endoplasmic reticulum stress-reactive oxygen species in breast cancer cells. Metallomics 7:165–173. https://doi.org/10.1039/c4mt00263f
74. Wu CF, Seo EJ, Klauck SM et al (2016) Cryptotanshinone deregulates unfolded protein response and eukaryotic initiation factor signaling in acute lymphoblastic leukemia cells. Phytomedicine 23:174–180. https://doi.org/10.1016/j.phymed.2015.12.011
75. Tse AK, Chow KY, Cao HH et al (2013) The herbal compounnd cryptotanshinone restores sensitivity in cancer cells that are resistant to the tumor necrosis factor-related apoptosis-inducing ligand. J Biol Chem 288:29923–29933. https://doi.org/10.1074/jbc.m113.483909
76. Chu J, Pelletier J (2015) Targeting the eIF4A RNA helicase as an anti-neoplastic approach. Biochim Biophys Acta 1849:781–791. https://doi.org/10.1016/j.bbagrm.2014.09.006
77. Ge Y, Cheng R, Zhou Y et al (2012) Cryptotanshinone induces cell cycle arrest and apoptosis of multidrug resistant human chronic myeloid leukemia cells by inhibiting the activity of eukaryotic initiation factor 4E. Mol Cell Biochem 368:17–25. https://doi.org/10.1007/s11010-012-1338-3
78. Ge Y, Yang B, Xu X et al (2015) Cryptotanshinone acts synergistically with imatinib to induce apoptosis of human chronic myeloid leukemia cells. Leuk Lymphoma 56:730–738. https://doi.org/10.3109/10428194.2014.928934
79. Hao W, Zhang X, Zhao W et al (2016) Cryptotanshinone induces pro-death autophagy through JNK signaling mediated by reactive oxygen species generation in lung cancer cells. Anticancer Agents Med Chem 16:593–600. https://doi.org/10.2174/1871520615666150907093036
80. Hu T, Wang L, Zhang L et al (2015) Sensitivity of apoptosis-resistant colon cancer cells to tanshinones is mediated by autophagic cell death and p53-independent cytotoxicity. Phytomedicine 22:536–544. https://doi.org/10.1016/j.phymed.2015.03.010
81. Park IJ, Yang WK, Nam SH et al (2014) Cryptotanshinone induces G1 cell cycle arrest and autophagic cell death by activating the AMP-activated protein kinase signal pathway in HepG2 hepatoma. Apoptosis 19:615–628. https://doi.org/10.1007/s10495-013-0929-0
82. Tung YT, Chen HL, Lee CY et al (2013) Active component of Danshen (*Salvia miltiorrhiza* Bunge), tanshinone I, attenuates lung tumorigenesis via inhibitions of VEGF, cyclin A, and cyclin B expressions. Evid Based Complement Alternat Med 2013:319247. https://doi.org/10.1155/2013/319247
83. Li Y, Gong Y, Li L et al (2013) Bioactive tanshinone I inhibits the growth of lung cancer in part via downregulation of Aurora A function. Mol Carcinog 52:535–543. https://doi.org/10.1002/mc.21888
84. Lu M, Wang C, Wang J (2016) Tanshinone I induces human colorectal cancer cell apoptosis: the potential roles of Aurora A-p53 and survivin-mediated signaling pathways. Int J Oncol 49:603–610. https://doi.org/10.3892/ijo.2016.3565
85. Kim MK, Park GH, Eo HJ et al (2015) Tanshinone I induces cyclin D1 proteasomal degradation in an ERK1/2 dependent way in human colorectal cancer cells. Fitoterapia 101:162–168. https://doi.org/10.1016/j.fitote.2015.01.010
86. Su CC, Chen GW, Lin JG (2008) Growth inhibition and apoptosis induction by tanshinone I in human colon cancer Colo 205 cells. Int J Mol Med 22:613–618. https://doi.org/10.3892/ijmm_00000063
87. Wang L, Wu J, Lu J et al (2015) Regulation of the cell cycle and PI3K/Akt/mTOR signaling pathway by tanshinone I in human breastcancer cell lines. Mol Med Rep 11:931–939. https://doi.org/10.3892/mmr.2014.2819
88. Liu JJ, Liu WD, Yang HZ et al (2010) Inactivation of PI3k/Akt signaling pathway and activation of caspase-3 are involved in tanshinone I-induced apoptosis in myeloid leukemia cells in vitro. Ann Hematol 89:1089–1097. https://doi.org/10.1007/s00277-010-0996-z

89. Shin EA, Sohn EJ, Won G et al (2014) Upregulation of microRNA135a-3p and death receptor 5 plays a critical role in Tanshinone I sensitized prostate cancer cells to TRAIL induced apoptosis. Oncotarget 5:5624–5636. https://doi.org/10.18632/oncotarget.2152
90. Jing X, Xu Y, Cheng W et al (2016) Tanshinone I induces apoptosis and pro-survival autophagy in gastric cancers. Cancer Chemother Pharmacol 77:1171–1181. https://doi.org/10.1007/s00280-016-3034-6
91. Lee CY, Sher HF, Chen HW et al (2008) Anticancer effects of tanshinone I in human non-small cell lung cancer. Mol Cancer Ther 7:3527–3538. https://doi.org/10.1158/1535-7163.MCT-07-2288
92. Wang Y, Li JX, Wang YQ et al (2015) Tanshinone I inhibits tumor angiogenesis by reducing Stat3 phosphorylation at Tyr705 and hypoxia-induced HIF-1α accumulation in both endothelial and tumor cells. Oncotarget 6:16031–16042. https://doi.org/10.18632/oncotarget.3648
93. Nizamutdinova IT, Lee GW, Lee JS et al (2008) Tanshinone I suppresses growth and invasion of human breast cancer cells, MDA-MB-231, through regulation of adhesion molecules. Carcinogenesis 29:1885–1892. https://doi.org/10.1093/carcin/bgn151
94. Lin JY, Ke YM, Lai JS et al (2015) Tanshinone IIA enhances the effects of TRAIL by downregulating survivin in human ovarian carcinoma cells. Phytomedicine 22:929–938. https://doi.org/10.1016/j.phymed.2015.06.012
95. Chiu TL, Su CC (2010) Tanshinone IIA induces apoptosis in human lung cancer A549 cells through the induction of reactive oxygen species and decreasing the mitochondrial membrane potential. Int J Mol Med 25:231–236. https://doi.org/10.3892/ijmm_00000335
96. Zhang Z, Gao J, Wang Y et al (2009) Tanshinone IIA triggers p53 responses and apoptosis by RNA polymerase II upon DNA minor groove binding. Biochem Pharmacol 78:1316–1322. https://doi.org/10.1016/j.bcp.2009.06.110
97. Chien SY, Kuo SJ, Chen YL et al (2012) Tanshinone IIA inhibits human hepatocellular carcinoma J5 cell growth by increasing Bax and caspase 3 and decreasing CD31 expression in vivo. Mol Med Rep 5:282–286. https://doi.org/10.3892/mmr.2011.631
98. Chen J, Shi DY, Liu SL et al (2012) Tanshinone IIA induces growth inhibition and apoptosis in gastric cancer in vitro and in vivo. Oncol Rep 27:523–528. https://doi.org/10.3892/or.2011.1524F
99. Wang JF, Feng JG, Han J et al (2014) The molecular mechanisms of Tanshinone IIA on the apoptosis and arrest of human esophageal carcinoma cells. Biomed Res Int 2014:582730. https://doi.org/10.1155/2014/582730
100. Wei X, Zhou L, Hu L et al (2012) Tanshinone IIA arrests cell cycle and induces apoptosis in 786-O human renal cell carcinoma cells. Oncol Lett 3:1144–1148. https://doi.org/10.3892/ol.2012.626
101. Yun SM, Jeong SJ, Kim JH et al (2013) Activation of c-Jun N-terminal kinase mediates tanshinone IIA-induced apoptosis in KBM-5 chronic myeloid leukemia cells. Biol Pharm Bull 36:208–214. https://doi.org/10.1248/bpb.b12-00537
102. Zhang J, Wang J, Jiang JY et al (2014) Tanshinone IIA induces cytochrome c-mediated caspase cascade apoptosis in A549 human lungcancer cells via the JNK pathway. Int J Oncol 45:683–690. https://doi.org/10.3892/ijo.2014.2471
103. Munagala R, Aqil F, Jeyabalan J et al (2015) Tanshinone IIA inhibits viral oncogene expression leading to apoptosis and inhibition of cervicalcancer. Cancer Lett 356:536–546. https://doi.org/10.1016/j.canlet.2014.09.037
104. Yang L, Guo H, Dong L et al (2014) Tanshinone IIA inhibits the growth, attenuates the stemness and induces the apoptosis of human glioma stem cells. Oncol Rep 32:1303–1311. https://doi.org/10.3892/or.2014.3293
105. Lin C, Wang L, Wang H et al (2013) Tanshinone IIA inhibits breast cancer stem cells growth in vitro and in vivo through attenuation of IL-6/STAT3/NF-kB signaling pathways. J Cell Biochem 114:2061–2070. https://doi.org/10.1002/jcb.24553

106. Kim EO, Kang SE, Im CR et al (2016) Tanshinone IIA induces TRAIL sensitization of human lung cancer cells through selective ER stress induction. Int J Oncol 48:2205–2212. https://doi.org/10.3892/ijo.2016.3441
107. Chang CC, Kuan CP, Lin JY et al (2015) Tanshinone IIA facilitates TRAIL sensitization by up-regulating DR5 through the ROS-JNK-CHOP signaling axis in human ovarian carcinoma cell lines. Chem Res Toxicol 28:1574–1583. https://doi.org/10.1021/acs.chemrestox.5b00150
108. Cheng CY, Su CC (2010) Tanshinone IIA inhibits Hep-J5 cells by increasing calreticulin, caspase 12 and GADD153 protein expression. Int J Mol Med 26:379–385. https://doi.org/10.3892/ijmm_00000476
109. Chiu SC, Huang SY, Chen SP et al (2013) Tanshinone IIA inhibits human prostate cancer cells growth by induction of endoplasmic reticulum stress in vitro and in vivo. Prostate Cancer Prostatic Dis 16:315–322. https://doi.org/10.1038/pcan.2013.38
110. Yan MY, Chien SY, Kuo SJ et al (2012) Tanshinone IIA inhibits BT-20 human breast cancer cell proliferation through increasing caspase 12, GADD153 and phospho-p38 protein expression. Int J Mol Med 29:855–863. https://doi.org/10.3892/ijmm.2012.908
111. Pan TL, Wang PW, Hung YC et al (2013) Proteomic analysis reveals tanshinone IIA enhances apoptosis of advanced cervix carcinoma CaSki cells through mitochondria intrinsic and endoplasmic reticulum stress pathways. Proteomics 13:3411–3423. https://doi.org/10.1002/pmic.201300274
112. Yun SM, Jung JH, Jeong SJ et al (2014) Tanshinone IIA induces autophagic cell death via activation of AMPK and ERK and inhibition of mTOR and p70 S6K in KBM-5 leukemia cells. Phytother Res 28:458–464. https://doi.org/10.1002/ptr.5015
113. Li C, Han X, Zhang H et al (2016) The interplay between autophagy and apoptosis induced by tanshinone IIA in prostate cancercells. Tumor Biol 37:7667–7674. https://doi.org/10.1007/s13277-015-4602-9
114. Xie J, Liu J, Liu H et al (2015) The antitumor effect of tanshinone IIA on anti-proliferation and decreasing VEGF/VEGFR2 expression on the human non-small cell lung cancer A549 cell line. Acta Pharm Sin B 5:554–563. https://doi.org/10.1016/j.apsb.2015.07.008
115. Zhou LH, Hu Q, Sui H et al (2012) Tanshinone II–a inhibits angiogenesis through down regulation of COX-2 in human colorectalcancer. Asian Pac J Cancer Prev 13:4453–4458. https://doi.org/10.7314/apjcp.2012.13.9.4453
116. Li G, Shan C, Liu L et al (2015) Tanshinone IIA inhibits HIF-1α and VEGF expression in breast cancer cells via mTOR/p70S6K/RPS6/4E-BP1 signaling pathway. PLoS ONE 10: e0117440. https://doi.org/10.1371/journal.pone.0117440
117. Liu JJ, Lin DJ, Liu PQ et al (2006) Induction of apoptosis and inhibition of cell adhesive and invasive effects by tanshinone IIA in acute promyelocytic leukemia cells in vitro. J Biomed Sci 13:813–823. https://doi.org/10.1007/s11373-006-9110-x
118. Chiu SC, Huang SY, Chang SF et al (2014) Potential therapeutic roles of tanshinone IIA in human bladder cancer cells. Int J Mol Sci 15:15622–15637. https://doi.org/10.3390/ijms150915622
119. Zhang Y, Wei RX, Zhu XB et al (2012) Tanshinone IIA induces apoptosis and inhibits the proliferation, migration, and invasion of the osteosarcoma MG-63 cell line in vitro. Anticancer Drugs 23:212–219. https://doi.org/10.1097/cad.0b013e32834e5592
120. Shan YF, Shen X, Xie YK et al (2009) Inhibitory effects of tanshinone II-A on invasion and metastasis of human colon carcinoma cells. Acta Pharmacol Sin 30:1537–1542. https://doi.org/10.1038/aps.2009.139
121. Hoesel B, Schmid JA (2013) The complexity of NF-κB signaling in inflammation and cancer. Mol Cancer 12:86. https://doi.org/10.1186/1476-4598-12-86
122. Su CC, Chien SY, Kuo SJ et al (2012) Tanshinone IIA inhibits human breast cancer MDA-MB-231 cells by decreasing LC3-II, Erb-B2 and NF-κBp65. Mol Med Rep 5: 1019–1022. https://doi.org/10.3892/mmr.2012.756

123. Bai Y, Zhang L, Fang X et al (2016) Tanshinone IIA enhances chemosensitivity of colon cancer cells by suppressing nuclear factor-κB. Exp Ther Med 11:1085–1089. https://doi.org/10.3892/etm.2016.2984
124. Liu W, Zhou J, Geng G et al (2012) Antiandrogenic, maspin induction, and antiprostate cancer activities of tanshinone IIA and its novel derivatives with modification in ring A. J Med Chem 55:971–975. https://doi.org/10.1021/jm2015292
125. Won SH, Lee HJ, Jeong SJ et al (2012) Activation of p53 signaling and inhibition of androgen receptor mediate tanshinone IIA induced G1 arrest in LNCaP prostate cancer cells. Phytother Res 26:669–674. https://doi.org/10.1002/ptr.3616
126. Hayashi T, Kakisawa H, Hsu HY et al (1970) The structure of miltirone, a new diterpenoid quinone. J Chem Soc D 5:299. https://doi.org/10.1039/c2970000299a
127. Huang W, Li J, Zhang W et al (2006) Synthesis of miltirone analogues as inhibitors of Cdc25 phosphatases. Bioorg Med Chem Lett 16:1905–1908. https://doi.org/10.1016/j.bmcl.2005.12.080
128. Wu CF, Efferth T (2015) Miltirone induces G2/M cell cycle arrest and apoptosis in CCRF-CEM acute lymphoblastic leukemia cells. J Nat Prod 78:1339–1347. https://doi.org/10.1021/acs.jnatprod.5b00158
129. Efferth T, Kahl S, Paulus K et al (2008) Phytochemistry and pharmacogenomics of natural products derived from traditional Chinese medicine and Chinese materia medica with activity against tumor cells. Mol Cancer Ther 7:152–161. https://doi.org/10.1158/1535-7163.MCT-07-0073
130. Zhou X, Wang Y, Lee WY et al (2015) Miltirone is a dual inhibitor of p-glycoprotein and cell growth in doxorubicin-resistant HepG2 cells. J Nat Prod 78:2266–2275. https://doi.org/10.1021/acs.jnatprod.5b00516
131. Wang L, Hu T, Shen J et al (2016) Miltirone induced mitochondrial dysfunction and ROS-dependent apoptosis in colon cancer cells. Life Sci 151:224–234. https://doi.org/10.1016/j.lfs.2016.02.083
132. Zhou L, Jiang L, Xu M et al (2016) Miltirone exhibits antileukemic activity by ROS-mediated endoplasmic reticulum stress and mitochondrial dysfunction pathways. Sci Rep 6:20585. https://doi.org/10.1038/srep20585
133. Rosa LS, Silva NJA, Soares NCP et al (2016) Anticancer properties of phenolic acids in colon cancer—a review. J Nutr Food Sci 6:468. https://doi.org/10.4172/2155-9600.1000468
134. Wahle KW, Brown I, Rotondo D et al (2010) Plant phenolics in the prevention and treatment of cancer. Adv Exp Med Biol 698:36–51. https://doi.org/10.1007/978-1-4419-7347-4-4
135. Ho JH, Hong CY (2011) Salvianolic acids: small compounds with multiple mechanisms for cardiovascular protection. J Biomed Sci 18:30. https://doi.org/10.1186/1423-0127-18-30
136. Wu P, Yan Y, Ma LL et al (2016) Effects of the Nrf2 modulator salvianolic acid A alone or combined with metformin on diabetes-associated microvascular and renal injury. J Biol Chem 291:22288–22301. https://doi.org/10.1074/jbc.m115.712703
137. Qiang G, Yang X, Shi L et al (2015) Antidiabetic effect of salvianolic acid A on diabetic animal models via AMPK activation and mitochondrial regulation. Cell Physiol Biochem 36:395–408. https://doi.org/10.1159/000430258
138. Hamaguchi T, Ono K, Murase A et al (2009) Phenolic compounds prevent Alzheimer's pathology through different effects on the amyloid-beta aggregation pathway. Am J Pathol 175:2557–2565. https://doi.org/10.2353/ajpath.2009.090417
139. Hasanein P, Mahtaj AK (2015) Ameliorative effect of rosmarinic acid on scopolamine-induced memory impairment in rats. Neurosci Lett 585:23–27. https://doi.org/10.1016/j.neulet.2014.11.027
140. Hooker CW, Lott WB, Harrich D (2001) Inhibitors of human immunodeficiency virus type 1 reverse transcriptase target distinct phases of early reverse transcription. J Virol 75:3095–3104. https://doi.org/10.1128/jvi.75.7.3095-3104.2001
141. Chung YC, Hsieh FC, Lin YJ et al (2015) Magnesium lithospermate B and rosmarinic acid, two compounds present in *Salvia miltiorrhiza*, have potent antiviral activity against

enterovirus 71 infections. Eur J Pharmacol 755:127–133. https://doi.org/10.1016/j.ejphar.2015.02.046

142. Petersen M, Simmonds MS (2003) Rosmarinic acid. Phytochemistry 62:121–125. https://doi.org/10.1016/s0031-9422(02)00513-7
143. Scheckel KA, Degner SC, Romagnolo DF (2008) Rosmarinic acid antagonizes activator protein-1-dependent activation of cyclooxygenase-2 expression in human cancer and nonmalignant cell lines. J Nutr 138:2098–2105. https://doi.org/10.3945/jn.108.090431
144. Tao L, Wang S, Zhao Y et al (2014) Phenolcarboxylic acids from medicinal herbs exert anticancer effects through disruption of COX-2 activity. Phytomedicine 21:1473–1482. https://doi.org/10.1016/j.phymed.2014.05.001
145. Osakabe N, Yasuda A, Natsume M et al (2004) Rosmarinic acid inhibits epidermal inflammatory responses: anti-carcinogenetic effects of *Perilla frutescens* extract in the murine two-stage skin mode. Carcinogenesis 25:549–557. https://doi.org/10.1093/carcin/bgh034
146. Han S, Yang S, Cai Z et al (2015) Anti-Warburg effect of rosmarinic acid via miR-155 in gastric cancer cells. Drug Des Dev Ther 9:2695–2703. https://doi.org/10.2147/dddt.s82342
147. Yang EJ, Ku SK, Lee W et al (2013) Barrier protective effects of rosmarinic acid on HMGB1-induced inflammatory responses in vitro and in vivo. J Cell Physiol 228:975–982. https://doi.org/10.1002/jcp.24243
148. Venkatachalam K, Gunasekaran S, Jesudoss VA et al (2013) The effect of rosmarinic acid on 1,2-dimethylhydrazine induced colon carcinogenesis. Exp Toxicol Pathol 65:409–418. https://doi.org/10.1016/j.etp.2011.12.005
149. Venkatachalam K, Gunasekaran S, Namasivayam N (2016) Biochemical and molecular mechanisms underlying the chemopreventive efficacy of rosmarinic acid in a rat colon cancer. Eur J Pharmacol 791:37–50. https://doi.org/10.1016/j.ejphar.2016.07.051
150. Furtado RA, Oliveira BR, Silva LR et al (2015) Chemopreventive effects of rosmarinic acid on rat colon carcinogenesis. Eur J Cancer Prev 24:106–112. https://doi.org/10.1097/cej.0000000000000055
151. Baldasquin-Caceres B, Gomez-Garcia FJ, López-Jornet P et al (2014) Chemopreventive potential of phenolic compounds in oral carcinogenesis. Arch Oral Biol 59:1101–1107. https://doi.org/10.1016/j.archoralbio.2014.06.007
152. Anusuya C, Manoharan S (2011) Antitumor initiating potential of rosmarinic acid in 7,12-dimethylbenz(a)anthracene-induced hamster buccal pouch carcinogenesis. J Environ Pathol Toxicol Oncol 30:199–211. https://doi.org/10.1615/jenvironpatholtoxicoloncol.v30.i3.30
153. Lee J, Kim YS, Park D et al (2007) Rosmarinic acid induces melanogenesis through protein kinase A activation signaling. Biochem Pharmacol 74:960–968
154. Sánchez-Campillo M, Gabaldon JA, Castillo J et al (2009) Rosmarinic acid, a photo-protective agent against UV and other ionizing radiations. Food Chem Toxicol 47:386–392. https://doi.org/10.1016/j.fct.2008.11.026
155. Ramos AA, Pedro D, Collins AR et al (2012) Protection by *Salvia* extracts against oxidative and alkylation damage to DNA in human HCT15 and CO115 cells. J Toxicol Environ Health A 75:765–775. https://doi.org/10.1080/15287394.2012.689804
156. Lozano-Baena MD, Tasset I, Muñoz-Serrano A et al (2016) Cancer prevention and health benefices of traditionally consumed *Borago officinalis* plants. Nutrients 8: pii: E48. https://doi.org/10.3390/nu8010048
157. Xu Y, Xu G, Liu L et al (2010) Anti-invasion effect of rosmarinic acid via the extracellular signal-regulated kinase and oxidation-reduction pathway in Ls174-T cells. J Cell Biochem 111:370–379. https://doi.org/10.1002/jcb.22708
158. Xu Y, Jiang Z, Ji G et al (2010) Inhibition of bone metastasis from breast carcinoma by rosmarinic acid. Planta Med 76:956–962. https://doi.org/10.1055/s-0029-1240893
159. Xavier CP, Lima CF, Fernandes-Ferreira M et al (2009) *Salvia fruticosa*, *Salvia officinalis*, and rosmarinic acid induce apoptosis and inhibit proliferation of human colorectal cell lines:

the role in MAPK/ERK pathway. Nutr Cancer 61:564–571. https://doi.org/10.1080/01635580802710733
160. Cao W, Hu C, Wu L et al (2016) Rosmarinic acid inhibits inflammation and angiogenesis of hepatocellular carcinoma by suppression of NF-κB signaling in H22 tumor-bearing mice. J Pharmacol Sci 132:131–137. https://doi.org/10.1016/j.jphs.2016.09.003
161. Huang SS, Zheng RL (2006) Rosmarinic acid inhibits angiogenesis and its mechanism of action in vitro. Cancer Lett 239:271–280. https://doi.org/10.1016/j.canlet.2005.08.025
162. Saiko P, Steinmann MT, Schuster H et al (2015) Epigallocatechin gallate, ellagic acid, and rosmarinic acid perturb dNTP pools and inhibit de novo DNA synthesis and proliferation of human HL-60 promyelocytic leukemia cells: synergism with arabinofuranosylcytosine. Phytomedicine 22:213–222. https://doi.org/10.1016/j.phymed.2014.11.017
163. Wu CF, Hong C, Klauck SM et al (2015) Molecular mechanisms of rosmarinic acid from *Salvia miltiorrhiza* in acute lymphoblastic leukemia cells. J Ethnopharmacol 176:55–68. https://doi.org/10.1016/j.jep.2015.10.020
164. Moon DO, Kim MO, Lee JD et al (2010) Rosmarinic acid sensitizes cell death through suppression of TNF-alpha-induced NF-kappaB activation and ROS generation in human leukemia U937 cells. Cancer Lett 288:183–191. https://doi.org/10.1016/j.canlet.2009.06.033
165. Heo SK, Noh EK, Yoon DJ et al (2015) Rosmarinic acid potentiates ATRA-induced macrophage differentiation in acute promyelocytic leukemia NB4 cells. Eur J Pharmacol 747:36–44. https://doi.org/10.1016/j.ejphar.2014.10.064
166. Li FR, Fu YY, Jiang DH et al (2013) Reversal effect of rosmarinic acid on multidrug resistance in SGC7901/Adr cell. J Asian Nat Prod Res 15:276–285. https://doi.org/10.1080/10286020.2012.762910
167. Berdowska I, Zieliński B, Fecka I et al (2013) Cytotoxic impact of phenolics from *Lamiaceae* species on human breast cancer cells. Food Chem 141:1313–1321. https://doi.org/10.1016/j.foodchem.2013.03.090
168. Wu J, Zhu Y, Li F et al (2016) Spica prunellae and its marker compound rosmarinic acid induced the expression of efflux transporters through activation of Nrf2-mediated signaling pathway in HepG2 cells. J Ethnopharmacol 193:1–11. https://doi.org/10.1016/j.jep.2016.07.021
169. Hsu KC, Sung TY, Lin CT et al (2015) Anchor-based classification and type-C inhibitors for tyrosine kinases. Sci Rep 5:10938. https://doi.org/10.1038/srep10938
170. Bi L, Chen J, Yuan X et al (2013) Salvianolic acid A positively regulates PTEN protein level and inhibits growth of A549 lung cancercells. Biomed Rep 1:213–217. https://doi.org/10.3892/br.2012.33
171. Cai J, Chen S, Zhang W et al (2014) Salvianolic acid A reverses paclitaxel resistance in human breast cancer MCF-7 cells via targeting the expression of transgelin 2 and attenuating PI3K/Akt pathway. Phytomedicine 21:1725–1732. https://doi.org/10.1016/j.phymed.2014.08.007
172. Zheng X, Chen S, Yang Q et al (2015) Salvianolic acid A reverses the paclitaxel resistance and inhibits the migration and invasion abilities of human breast cancer cells by inactivating transgelin 2. Cancer Biol Ther 16:1407–1414. https://doi.org/10.1080/15384047.2015.1070990
173. Wang X, Wang C, Zhang L et al (2015) Salvianolic acid A shows selective cytotoxicity against multidrug-resistant MCF-7 breast cancer cells. Anticancer Drugs 26:210–223. https://doi.org/10.1097/cad.0000000000000184
174. Li T, Kong AT, Ma Z et al (2016) Protein arginine methyltransferase 1 may be involved in pregnane x receptor-activated overexpression of multidrug resistance 1 gene during acquired multidrug resistant. Oncotarget 7:20236–20248. https://doi.org/10.18632/oncotarget.7752
175. Zhao Y, Guo Y, Gu X (2011) Salvianolic Acid B, a potential chemopreventive agent, for head and neck squamous cell cancer. J Oncol 2011:534548. https://doi.org/10.1155/2011/534548

176. Li GG, Guo ZZ, Ma XF et al (2016) The M2 macrophages induce autophagic vascular disorder and promote mouse sensitivity to urethane-related lung carcinogenesis. Dev Comp Immunol 59:89–98. https://doi.org/10.1016/j.dci.2016.01.010
177. Hao Y, Xie T, Korotcov A et al (2009) Salvianolic acid B inhibits growth of head and neck squamous cell carcinoma in vitro and in vivo via cyclooxygenase-2 and apoptotic pathways. Int J Cancer 124:2200–2209. https://doi.org/10.1002/ijc.24160
178. Zhao Y, Hao Y, Ji H et al (2010) Combination effects of salvianolic acid B with low-dose celecoxib on inhibition of head and neck squamous cell carcinoma growth in vitro and in vivo. Cancer Prev Res (Phila) 3:787–796. https://doi.org/10.1158/1940-6207.CAPR-09-0243
179. Zhou ZT, Yang Y, Ge JP (2006) The preventive effect of salvianolic acid B on malignant transformation of DMBA-induced oral premalignant lesion in hamsters. Carcinogenesis 27:826–832. https://doi.org/10.1093/carcin/bgi271
180. Wei J, Xie G, Ge S et al (2012) Metabolic transformation of DMBA-induced carcinogenesis and inhibitory effect of salvianolic acid b and breviscapine treatment. J Proteome Res 11:1302–1316. https://doi.org/10.1021/pr2009725
181. Yang Y, Ge PJ, Jiang L et al (2011) Modulation of growth and angiogenic potential of oral squamous carcinoma cells in vitro using salvianolic acid B. BMC Complement Altern Med 11:54. https://doi.org/10.1186/1472-6882-11-54
182. Wang ZS, Luo P, Dai SH et al (2013) Salvianolic acid B induces apoptosis in human glioma U87 cells through p38-mediated ROS generation. Cell Mol Neurobiol 33:921–928. https://doi.org/10.1007/s10571-013-9958-z
183. Wang M, Sun G, Wu P et al (2013) Salvianolic Acid B prevents arsenic trioxide-induced cardiotoxicity in vivo and enhances its anticancer activity in vitro. Evid Based Complement Alternat Med 2013:759483. https://doi.org/10.1155/2013/759483
184. Sun MF, Chang TT, Chang KW et al (2011) Blocking the DNA repair system by traditional Chinese medicine? J Biomol Struct Dyn 28:895–906. https://doi.org/10.1080/07391102.2011.10508616
185. Zhang LJ, Chen L, Lu Y et al (2010) Danshensu has anti-tumor activity in B16F10 melanoma by inhibiting angiogenesis and tumor cell invasion. Eur J Pharmacol 643:195–201. https://doi.org/10.1016/j.ejphar.2010.06.045
186. Son B, Jun SY, Seo H et al (2016) Inhibitory effect of traditional oriental medicine-derived monoamine oxidase B inhibitor on radioresistance of non-small cell lung cancer. Sci Rep 6:21986. https://doi.org/10.1038/srep21986
187. Wang L, Zhang X, Chan JY et al (2016) A novel danshensu derivative prevents cardiac dysfunction and improves the chemotherapeutic efficacy of doxorubicin in breast cancer cells. J Cell Biochem 117:94–105. https://doi.org/10.1002/jcb.25253
188. Wang L, Zhang X, Cui G et al (2016) A novel agent exerts antitumor activity in breast cancer cells by targeting mitochondrial complex II. Oncotarget 7:32054–32064. https://doi.org/10.18632/oncotarget.8410
189. Bi X, Liu X, Di L et al (2016) Improved oral bioavailability using a solid self-microemulsifying drug delivery system containing a multicomponent mixture extracted from *Salvia miltiorrhiza*. Molecules 21:456. https://doi.org/10.3390/molecules21040456
190. Tian HL, Yu T, Xu NN et al (2010) A novel compound modified from tanshinone inhibits tumor growth in vivo via activation of the intrinsic apoptotic pathway. Cancer Lett 297:18–30. https://doi.org/10.1016/j.canlet.2010.04.020
191. Li J, Liu P, Liu JP et al (2013) Bioavailability and foam cells permeability enhancement of Salvianolic acid B pellets based on drug-phospholipids complex technique. Eur J Pharm Biopharm 83:76–86. https://doi.org/10.1016/j.ejpb.2012.09.021
192. Cai Y, Zhang W, Chen Z et al (2016) Recent insights into the biological activities and drug delivery systems of tanshinones. Int J Nanomed 11:121–130. https://doi.org/10.2147/ijn.S84035

Strategies for Optimization of the Production of Rosmarinic Acid in *Salvia officinalis* L. and *Salvia dolomitica* Codd Biomass with Several Biotechnological Approaches

Marco Savona, Sara Barberini, Laura Bassolino, Eric Mozzanini, Laura Pistelli, Luisa Pistelli and Barbara Ruffoni

Abstract Rosmarinic acid (RA) is a caffeic acid ester widely used by pharmaceutical and food industry. Its presence in several medicinal plant species and herb has been correlated with their biological activities and health beneficial effects. RA is widely used as additive to preserve foods, and because the chemical synthesis, even possible, remains time consuming and cost inefficient, there is a growing interest in development of biotechnological strategies to produce this compound. RA is abundant within the Lamiaceae family, particularly in plants belonging to genus *Salvia*. The use of plant tissues and cell cultures may represent an alternative

---

M. Savona · S. Barberini · E. Mozzanini · B. Ruffoni (✉)
CREA Research Centre for Vegetable and Ornamental Crops, Sanremo, Italy
e-mail: ruffoni.barbara@crea.gov.it

M. Savona
e-mail: marco.savona@crea.gov.it

S. Barberini
e-mail: sara-barberini@libero.it

E. Mozzanini
e-mail: eric.bussi.mozzanini@gmail.com

L. Bassolino
CREA Research, Centre for Genomics and Bioinformatics, Montanaso Lombardo, Italy
e-mail: laura.bassolino@crea.gov.it

L. Pistelli
Department of Agrarian, Food and Agro-environmental Sciences, University of Pisa, Pisa, Italy
e-mail: laura.pistelli@unipi.it

L. Pistelli
Department of Pharmacy, University of Pisa, Pisa, Italy
e-mail: luisa.pistelli@unipi.it

V. Georgiev and A. Pavlov (eds.), *Salvia Biotechnology*,
https://doi.org/10.1007/978-3-319-73900-7_6

strategy for the production of highly valuable plant metabolites and could be even adopted by pharmaceutical industries to develop new drugs and formulations. Protocols for shoots and calli regeneration, as well as cell cultures have been already established for *Salvia officinalis* and *Salvia dolomitica* with the aim to evaluate the content of RA in various tissues at different growth conditions. In particular, *S. dolomitica* Codd has been shown to be a good model for tissue culture studies due to its flexible propagation under in vitro conditions. The *hydroxyphenylpyruvate reductase*, *SoHPPR*, gene encoding a hydroxyphenylpyruvate reductase, a key biosynthetic enzyme of RA pathway, has been characterized in cell cultures of *S. officinalis*, representing a good target for metabolic engineering strategies.

Keywords *Hydroxyphenylpyruvate reductase* (*HPPR*) · *Agrobacterium tumefaciens* · Tissue culture · HPLC

List of Abbreviations

2,4-D	2,4-dichlorophenoxyacetic acid
4CL	4-coumarate CoA ligase
BA	6-benzyladenine
C4H	Cinnamic acid 4-hydroxylase
CA	Carnosic acid
CAR	Carnosol
DW	Dry weight
EO	Essential oils
FW	Fresh weight
HPPR	Hydroxyphenylpyruvate reductase
HS	Headspace
IAA	Indol acetic acid
Kin	Kinetin
MeJA	Methyl jasmonate
PAL	Phenylalanine ammonia lyase
PGR	Plant growth regulators
RA	Rosmarinic acid
RAS	Rosmarinic acid synthase
SA	Salicylic acid
SM	Secondary metabolites
SoHPPR	*Salvia officinalis* hydroxyphenylpyruvate reductase gene
TAT	Tyrosine aminotransferase
VOCs	Volatile organic compounds
ZEA	Zeatin

1 Introduction

The genus *Salvia* (Lamiaceae) encompasses about 900 species worldwide, 26 of which found in Southern Africa, like *S. dolomitica* Codd, whereas others are native to the Mediterranean basin like *S. officinalis* L. [1]. The genus name is derived from the Latin verb "Salvare" which means 'to save the life of someone else or to be safe' with a clear reference to the phytochemical and medicinal properties of the active compounds derived from these species. In addition to their well-recognized ornamental value, *Salvia* species have been widely accepted as medicinal plants and used in traditional Chinese medicine for treatment or prevention of various pathological conditions because of their health promoting effects [2–4]. Moreover, they are also used as ingredients in cooking and as a source of natural antioxidants in food industry [5].

RA ($C_{18}H_{16}O_8$) is a caffeic acid ester that exhibits several bioactivities, including antiviral, antibacterial, anti-inflammatory, anti-allergic, and anti-oxidant properties [6]. Due to its use as an additive to preserve foods, there is a growing interest in development of biotechnological strategies to produce this compound. Chemical synthesis was proposed, but is commercially inefficient because of the high cost and labour time. RA is abundant in medicinal plants within the Lamiaceae family, especially in members of *Salvia* genus. It is well known that the use of plant tissues and cell cultures may serve as an alternative platform for the production of high-valuable plant metabolites, and even could be used in pharmaceutical industry in development of new drugs and formulations [7]. In vitro technology has been considered suitable for biodiversity preservation and may represent an alternative production method to satisfy the increasing demand for both volatile (e.g. Essential Oils—EO) and non-volatile bioactive secondary metabolites. Moreover, since the production of secondary metabolites in plants has always been considered as defensive response to biotic and abiotic stress, several physical and chemical elicitors can be applied to stimulate plants to increase concentrations of desired compound or a group of compounds. Genetic transformation and metabolic engineering are other powerful tools for increasing the accumulation of specific plant metabolite. The biosynthetic pathway leading to RA, is a branch of the more general phenylpropanoid pathway, and has been elucidated for the first time in *Coleus blumei* [8]. The genes encoding for the enzymes involved in this process have been also identified. The *hydroxyphenylpyruvate reductase gene* (*SoHPPR*), encoding a *hydroxyphenylpyruvate reductase*, a key biosynthetic enzyme of RA pathway, was isolated from *S. officinalis* [9] and its induction has been correlated with an increase in RA accumulation in *S. miltiorrhiza* hairy roots [10], as well as in cell suspension culture of *S. officinalis* [9]. *SoHPPR* represents a good target for metabolic engineering strategies. Micropropagated *S. dolomitica* plantlets and callus were assessed for their RA contents in comparison with in vivo grown plants. It was showed that the addition of hormones like 6-benzyladenine (BA) and 2,4-dichlorophenoxyacetic acid (2,4-D) into the growth medium for shoots and calli development, can induce a remarkable increase in accumulation of this phenolic compound, compared to the mother plant [11]. Moreover, leaves derived from in vivo grown plants accumulate higher amount of RA in respect to the other aerial tissues. By using a transgenic approach, two weeks old calli from *S. officinalis* were transformed via *A. tumefaciens* carrying a binary vector with the *SoHPPR* gene under the control of 35S promoter. This chapter presents an overview of RA accumulation by plants of Lamiaceae family, and particularly, the possibility to obtain this compound by using several tissue cultures of *S. officinalis* and *S. dolomitica* species. Moreover, as an example among *Salvia* species, having high in vitro performance, *S. dolomitica* was proposed as suitable model plant for conducting physiological and molecular studies.

2 *Salvia officinalis* and *Salvia dolomitica:* Botanical Description and Main Active Metabolites

Salvia officinalis L. is an evergreen herbaceous aromatic plant that usually grows on dry hillsides, on clayish soils, in sunny places, up to 300 m above sea level (Fig. 6.1). The plant is generally 80 cm tall, stems are woody, quadrangular at the base. Leaves are opposite, rough, finely toothed, the lower ones petioled, the higher ones sessile, whitish beneath and grayish-green above. Flowers are blue-purple, 10–30 cm long, with corolla 17–25 mm long [12].

Salvia dolomitica Codd is an aromatic perennial shrub originated from the north-east province of Transvaal (Fig. 6.2). Its scientific name "dolomitica" is related to its common growth environment which is characterized by heavy soils and dolomitic rock between 1000 and 1500 m in altitude. Moreover, *S. dolomitica* is drought resistant [13] and highly scented, thus is mainly commercialized for ornamental purposes. Furthermore, this sage grows well in dry substrates and is cultivated in pots and used as groundcover, shrub border or edging [14]. Sage mainly accumulate phenolic compounds and terpenoids in specialized glands, trichomes, whose morphology, abundance and functions can vary among species belonging to Lamiaceae family [11, 15, 16].

The aerial parts of *Salvia* species are the principal source of flavonoids, triterpenoids and volatile compounds (e.g. monoterpenoids), while in the radical parts diterpenoids are found [17].

Terpenoids with low molecular weight (mono- and sesqui-terpenes and their oxygenated derivatives) represents the main components of the of the sage EO: interestingly, *S. officinalis* is considered to have the highest EO yield among *Salvia* species, together with a higher total ketone content and a lower total alcohol content [17]. The smell of aromatic plants is caused by the green leaf volatile compounds, mainly terpenes, a subset of Volatile Organic Compounds (VOCs). EO profile and volatile substance composition vary within each species, depending on geno/chemo-type but also on season and environmental conditions such as temperature, day length and light intensity [18–20]. In *S. officinalis* the major constituents are

Fig. 6.1 Particular of the flower of *S. officinalis*

Fig. 6.2 Particular of the flower of *S. dolomitica*

α- and ß-thujones (35–50%, mainly α); other terpenoids include 1,8 cineole, borneol, camphor, caryophyllene and linalyl acetate. The quantity of the EO is influenced more by the environment than the genotype, viceversa is true for the qualitative characteristics. In 1991, Bernath et al. [21] found that maximum dry mass and EO production occurred when cultured on the field, recording minimum levels under cold treatment and low light regime. Under cold climatic conditions, additionally, there was a significant reduction of the content of α-pinene and 1,8 cineole, compared to the plants grown in the warm. Different accessions—chemotypes—could also be classified as having high, medium, or low total thujone contents [22, 23]: flowering parts of *S. officinalis* had higher oil contents (1.6 vs. 1.1%) and β-pinene levels (27 vs. 10%), and lower thujone levels (16 vs. 31%) than leaves.

Terpenoids were found also in oleoresin sage, i.e. in *Salvia officinalis* [24]: six major compounds were purified and identified by IR, MS, and ^1H NMR spectrometry as carnosol, carnosic acid, rosmadial, rosmanol, epirosmanol, and methyl carnosate.

S. officinalis is also a natural source of flavonoids and polyphenolic compounds, having strong antioxidant, radical scavenging and antibacterial activities [25, 26]. The main antioxidant activity can be attributed to phenolic diterpenes carnosic acid (CA) and carnosol (CAR), and to the caffeic derivative rosmarinic acid (RA). Usually, these compounds are extracted from sage with water-alcoholic mixtures, although the types of solvent and the extraction method strongly influenced the quality and quantity of polyphenol extracts. When different extracts, obtained by various extraction methods, such as supercritical-CO_2, acetone and methanol/water, where compared, it was found that the acetone extracts possessed the highest antioxidant activity [23]. Conventional extraction methods are usually based on selection of the proper solvent and heating the solution to increase the solubility of the compounds. Other "green" methods comprising as example, the microwave-assisted extraction, have been used to shortening the extraction time without influencing the phenolic composition of the extract.

Traditional "herbal tea" of sage has been used over the last century with proved positive effect on gastrointestinal tract [27]. During that period, many papers report modifications of different extraction methods by using water or other aqueous solutions. Ollanketo et al. [28] reported that the extraction of sage antioxidants with pressurized hot water was more effective than ultrasonication-assisted methanol extraction, hydrodistillation and maceration with 70% ethanol. Recently Martins et al. [29] evaluated the antifungal and antioxidant properties of different aqueous extracts of *S. officinalis*, obtained by infusion and decoction processes or by methanol/water extraction. In their study, decoction and methanol/water extracts showed the highest antioxidant and antifungal properties, due to the higher concentration of polyphenols.

However, independently of the used solvent, the high antioxidant power of *Salvia* extracts was due to the presence of phenolic components belonging to the group of caffeic acid derivatives.

The EO characterization from fresh aerial parts of *S. dolomitica* Codd grown in open fields in South Africa was reported by Fisher [30] and Kamatou et al. [16]. The major constituents were geraniol, lynalyl acetate and linalool which are usually rare in *Salvia*. These components made the difference among the other South African indigenous species [31]. It was noted that species such as *S. verbenaca*, *S. radula* and *S. dolomitica,* which displayed good antimicrobial activity (MIC: 0.10 mg/mL), also exhibited good antibacterial activity. Infections caused by *Staphylococcus aureus* are among the most difficult to treat with standard clinical antibacterial agents [32]. The growth of *S. aureus* was inhibited by all extracts especially by *S. radula, S. chamelaeagnea, S. aurita* and *S. dolomitica* (MIC $\leq$ 0.06 mg/mL). Thus, these plants could yield compounds that may improve the treatment of infections caused by *S. aureus* in particular. It was reported that many species of *Salvia* are used in decoctions in South Africa by Sotho people against various bacterial infections such as throat inflammation, colds and coughs and to treat pulmonary diseases such as tuberculosis [2]. Kamatou et al. [16] identified some major compounds in the solvent extracts of *Salvia dolomitica*: caffeic acid, rosmarinic acid, oleanolic/ursolic acid and betulafolientriol oxide. Moreover, the authors investigated the in vitro antioxidant and anti-inflammatory activities of *S. dolomitica* extract, obtained via methanol: chloroform (1:1) extraction. Interestingly, mainly due to the lack of CA and derivatives, the extracts of *S. dolomitica* exhibited poor antioxidant activity even though RA is one of the major determinants [4, 33].

The major compounds in *S. dolomitica* are α-pinene (7.1%), δ-3 carene (7.5%), limonene (9.7%), 1,8-cineole (17.6%), β-caryophyllene (17.4%) and borneol (8.5%) [6].

Both methanol and acetone extracts of *S. dolomitica* demonstrated substantial activity against *Bacillus cereus* and *S. aureus* whilst the EO showed slight activity only against *B. cereus*. *S. dolomitica*. Acetone extract showed minimal activity against *E. coli*. When the agents were tested at one-half the MIC, only 1,8-cineole significantly reduced the tolerance of *S. aureus* to sodium chloride. The antimicrobial activity of linalool and 1,8-cineole has been demonstrated separately and it has also been observed that the antimicrobial activity is increased when they are used in combination [34].

The EO of *S. dolomitica* exhibited good antiplasmodial and anti-inflammatory activities [4, 33, 35] and also some degree of toxicity (IC_{50} value is 8 $\pm$ 0.3 μg/mL) compared to the anticancer drug 5-fluo-rouracil with IC_{50} value equal to 136 $\pm$ 16.6 μg/mL. The presence of the above compounds, although present in small concentrations, may synergistically increase the toxicity of the EO. Since the essential oils exhibited various degrees of toxicity, they should be used with caution [16] and further biological tests are necessary to assess their potential use for human end-consumers.

Moreover Gram-positive restricted antimycobacterial activity was described by Fisher [30], whereas anticancer properties were reported by Kamatou et al. [4]. Due to the characteristics of its oil, *S. dolomitica* is a good candidate to produce highly valuable plant metabolites that could be further exploited by pharmaceutical industries to develop new drugs and formulations.

However, research for sage-derived compounds and bioprospecting lead to a high pressure on wild over-utilized populations thus destroying indigenous biodiversity.

3 Rosmarinic Acid: Biosynthetic Pathway and Biological Activity

RA was firstly isolated from *Rosmarinus officinalis* and detected in several plant species from Lamiaceae family (sage, thyme and peppermint) which act as food preserves. The RA content in medicinal plant species and herbs can vary across species and generally correlates with their biological and health beneficial properties. Indeed, it's widely accepted that this phenolic compound exhibits several properties including antiviral, antibacterial, anti-inflammatory, anti-allergic, and anti-oxidant properties [6]. As others secondary metabolites, RA is always found at low amounts in plants of Lamiaceae family. Its production by chemical synthesis is possible, but economically ineffective. Nowadays, due to the high medicinal and economic values of this secondary metabolite, there is a growing interest to find out biotechnological strategies to enhance RA production. Before clarifying the chemical structure, RA was considered a compound similar to tannins. It was described as a catechine condensed, or belonging to the family of ellagitannins, probably containing caffeic acid isolated from *Melissa officinalis* [67]. Shortly after, in 1958, Scarpati and Oriente [36] isolated the pure acidic compound from the rosemary for the first time, and decided to call it RA in agreement with the plant from which it was extracted, the common rosemary (*Rosmarinus officinalis*). The chemical structure was clarified by the two authors as an ester of caffeic acid and 3,4-hydroxyphenyllactic acid [36].

Caffeic acid and RA have been extensively reported in *S. officinalis* [37–39], with the latter one considered as a promising molecule due to multitude of biological activities combined with a very low toxicity (LD_{50} in mice of 561 mg/kg) [40]. Hence the strong antioxidant activity connected with the low toxicity makes RA also interesting in food industry. RA has been analysed in various herbs and spices using different analytical methods: *Salvia officinalis* leaf extracts are not an exception. HPLC with spectrometric detectors has been most widely used for the separation of RA in various plant extracts, in addition, there is a tendency to use a combination of more than one separation/detection/measurement method in order to increase the effectiveness of analysis, as reported by Bandoniene et al. [41]. Since there is no standardization in the scientific literature regarding the units in which phenolic compounds are presented, it is difficult to establish the best extraction method for RA. However, HPLC can resolve the peak of the RA in all the type of tissue extracted: fresh or dry tissues, herbal teas, suspension cells [42–46].

Studies on biogenesis of the compound began in 1970 when Ellis and Towers [47] inserted in the culture medium of common mint (*Mentha officinalis*) radiolabelled amino acids showing that the amino acid phenylalanine and tyrosine were incorporated into RA. The portion formed by caffeic acid is only derived by phenylalanine while part of 3,4-dihydroxyphenyl lactic acid comes from tyrosine. These results were then compared with a similar experiment in cell cultures of *Coleus blumei* conducted by Razzaque and Ellis [8].

The enzymes involved in the biosynthesis of RA (Fig. 6.3) from amino acids have been studied in *Anchusa officinalis* (Fam. Boraginaceae) by Kelley et al. [49] and in *Coleus blumei* (Fam. Labiateae) by Petersen and co-authors [50, 51].

Summarising, RA biosynthesis follows two pathways: a phenylpropanoid branch, which drives to the caffeic acid moiety, and a tyrosine-derived branch that leads to the dihydroxyphenyl lactic acid. In the phenylpropanoid pathway, L-phenylalanine is transformed to 4-coumaroyl-CoA by coupling the reactions catalyzed by phenylalanine ammonia lyase (PAL), cinnamic acid 4-hydroxylase (C4H), and 4-coumarate CoA ligase (4CL). For the tyrosine-derived moiety, the first step is the transamination of L-tyrosine by 2-oxoglutarate as a second substrate to form 4-hydroxyphenylpyruvate by tyrosine aminotransferase (TAT). Then 4-hydroxyphenylpyruvate is transformed by HPPR to 4-hydroxyphenyllactate. This product is then coupled with 4-coumaroyl CoA to form an intermediate by the enzyme rosmarinic acid synthase (RAS), whereby the aromatic 3- and 3′-hydroxyl positions are hydroxylated via cytochrome P450 to form RA molecule.

As regards the distribution of RA in the Plant Kingdom, it was seen that its presence is reflected in the family of Boraginaceae, in some species of the family of Blechnaceae [52], monocotyledon plants in the family Zosteraceae [53], while in the Lamiaceae family RA presence is restricted to the subfamily of Nepetoideae, part of which is *S. officinalis* [54]. This distribution shows that rosmarinic acid can not be used as chemo-taxonomic marker [26]. Some examples of distribution of RA in the plant kingdom are summarized in Table 6.1.

Numerous pharmacological and biological activities of RA have been described. However its high antioxidant capacity is the most noticeable, principally due to the high redox potential and to the hydroxyl groups presented in the molecule [64, 65].

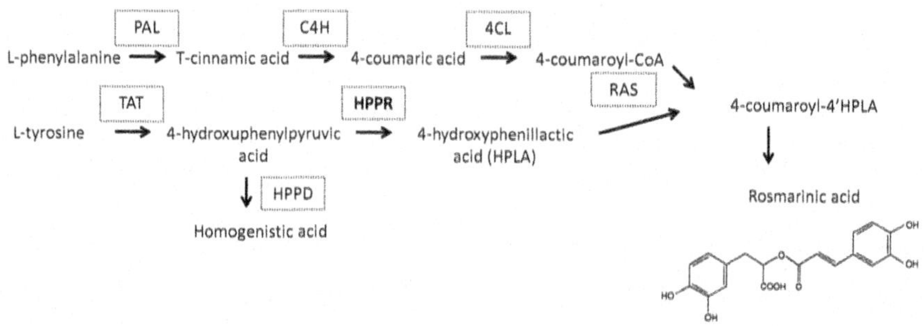

Fig. 6.3 A simplified diagram of enzymes and products involved in rosmarinic acid biosynthesis. PAL phenylalanine ammonia lyase; CAH cinnamic acid 4-hydroxylase; 4CL 4-coumarate CoA ligase; TAT tyrosine aminotransferase; HPPD hydroxyphenylpyruvate dioxygenase; HPPR (bold) hydroxyphenilpyruvate reductase; RAS rosmarinic acid synthase

Table 6.1 Some examples of RA distribution in the Plant Kingdom (modified from [48])

Family	Reference
Dicotyledonae	[36]
Lamiaceae	[54]
Boraginaceae	[49]
Apiaceae	[55]
Araliaceae	[56]
Cucurbitaceae	[57]
Rubiaceae	[58]
Plantaginaceae	[59]
Sterculiaceae	[60]
Tiliaceae	[61]
Monocotyledonae	
Cannaceae	[62]
Zosteraceae	[53]
Pteridopsida	
Blechnaceae	[52]
Anthoceropsida	
Anthocerotaceae	[63]

RA action has been associated with membrane stabilization and further hindrance of radical propagation that leads to the protection of the membranes against oxidative stress. Furthermore, RA is positively involved in liposomes stability, and also in reduction of damages produced by UVB in human keratinocytes [66]. Petersen and Simmonds [26] reported other biological activities as antiviral, antibacterial and anti-inflammatory: these are confirmed by studies on *Human immunodeficiency virus* (HIV-1) and *Herpes simplex virus* (HSV-1) or by antimicrobial essay against *Helicobacter pylori* and plants associated pathogens [67–69]. Moreover, important novel effects for RA regarding cognitive performance were reported: the administration of RA prevents the development of Alzheimer's pathology [70] and protects neurons from degenerative diseases [71]. Finally, RA possesses antiangiogenic activity and antifibrotic effects [72, 73]. One of the last challenges in pharmacological study is to demonstrate the possible use of RA to treat or prevent cancer disease: recent studies correlated RA with the inhibition of some TNFs (Tumour Necrosis Factors) [66].

4 Biotechnological Platform for in Vitro Manipulation of Sage Tissue

The micropropagation is reported for several *Salvia* species, among them the economically important *S. officinalis* [74] and *S. miltiorrhiza* [75]. Reports are also available for a number of ornamental species [76] (Fig. 6.4). A recent review about the use of in vitro cell culture for the production of secondary metabolites has been reported by Marchev et al. [1].

Fig. 6.4 *Salvia officinalis* L. in vitro cultures. **a** Leaves were used as starting material for *in vitro* callus induction. **b** Callus cultures on agarized "T8" medium [76] after 30 days. **c** Cell synchronization through the passage into a metal colander, before cell suspension cultures in Erlenmayer flasks. **d** Green callus explants (50 mg each) ready to be transformed. **e** The pRCS2-*SoHHPR* vector, containing the *SoHPPR* gene, used for callus agro-transformation. **f** Callus co-transformed with *35S::uidA* and *2X35S::SoHPPR*

Biotechnological techniques significantly facilitate plant propagation and production of some important bioactive compounds from the genus *Salvia* [77]. In *S. officinalis*, both callus, cell suspension, immobilized cell and hairy root cultures were established [78–80] with the aim to analyse secondary metabolites, with particular attention to RA.

Santos-Gomes et al. [81] confirmed that in vitro cultures of *S. officinalis* would afford a very good and efficient system model for studying the accumulation of antioxidant compounds, under controlled environmental and nutritional conditions.

Grzegorczyk et al. [82] gave a strong effort on the study of in vitro culture of *Salvia officinalis*, publishing many contributions from 2004. They confirmed that the biotechnological methods based on in vitro culture can produce standardized material; they also focused their studies on the concept that the production of secondary metabolites from in vitro cultured plants can be increased by genetic transformation with the soil-borne pathogen *Agrobacterium rhizogenes*.

Kintzios et al. [77] induced callus from leaf explants of *Salvia officinalis*, where donor plantlets were obtained from in vivo germinated seeds. The successful medium was MS [83] solidified with agar-agar and supplemented with 3% sucrose, 10 mg/L ascorbic acid, 1 mg/L 2,4-D and 1 mg/L Kin. After 3 weeks of induction, callus pieces were used for RA content determination, at weekly intervals. The procedure was: 70% ethanol extraction, separation and purification by TLC and finally recording by spectrophotometrically quantification of RA concentration.

Equimolar auxin and cytokinin concentrations, after many PGRs factorial experiments, ensured higher callus induction rates. As expected, *S. officinalis* callus followed a typical growth curve, with a very slow growth rate: up to the 4th and the 5th week after callus initiation, the callus fresh weight reached a maximum value, and then a decreasing was observed, possibly due to cell desiccation and/or cell death. At the same time with fresh weight measurements, calli formed on 1 mg/L 2,4-D and 1 mg/L Kin were assessed for RA accumulation over the same period of 6 weeks.

RA production started with callus culture establishment and increased progressively with callus growth, but RA accumulation declined rapidly during the 3rd and 4th week (i.e. during maximum callus growth) and increased again after the 5th week. Kintzios et al. [77] concluded that RA accumulation in *S. officinalis* was inversely related to callus growth, since the maximum RA production was observed between the 2nd and the 3rd week, as well as between the 5th and the 6th week, i.e. when callus growth was stabilized. When the authors [77] compared the obtained in vitro results to those ones recorded in vivo, they concluded that the accumulation was lower (ca. 35 μmol/g leaf fresh weight). This aspect was already well stated for several secondary metabolites produced in vitro [84]. An outstanding potential for RA production in vitro is clear and the observed growth and production patterns of *Salvia officinalis* callus cultures offer new perspectives for studying the relationship between cell behaviour in culture and in vitro secondary metabolism. Furthermore, the authors set up similar experiments by using suspension cultures rather than callus, in order to evaluate and accurate estimate of the in vitro RA production, since a better determination could rise on dry weight basis. In fact, preliminary indications reported by Hippolyte et al. [79] referred to a production of 6.4 g/L RA from *S. officinalis* suspension cultures. Santos-Gomes et al. [81] induced calli in *Salvia officinalis* starting from shoot internode explants and they evaluated the effect of different hormonal supplementation. In order to obtain suspended cells, from each gelled medium, they transferred the induced calli into liquid medium containing the same PGRs supplementation chosen for the growth and proliferation phase.

The proper evaluation of the callus induction medium is required since many authors reported that calli became gradually brownish and showed low growth rate. It has been demonstrated [81] that a decrease of the 2,4-D concentration (0.05 mg/L), keeping cytokinin influence, in particular BA (1.5 mg/L), resulted in a satisfactory growth green micro-calli, or even, dark green. The phenolic extract obtained from these calli permitted to identify 12 compounds and they were grouped into three classes of phenolic compounds: (1) phenolic acids: gallic acid, 3-*O*-caffeoylquinic acid, 5-*O*-caffeoylquinic acid, caffeic acid and rosmarinic acid; (2) flavonoids: hesperetin, hispidulin and genkwanin; (3) phenolic diterpenes: epirosmanol, carnosol, carnosic acid, and methyl carnosate. RA has previously been identified in nature-growing sage plants from the genotypes used in the establishment of the calli and suspension cultures here reported [85].

Several years after that, Ruffoni et al. [76] obtained *S. officinalis* cells from leaf tissue and perform the growth in bioreactor in batch culture (*Close System*) for 12 days: they defined a trend with a typical sigmoid growth curve showing a logarithmic phase between 4th and 8th day reaching a plateau at senescence (Fig. 6.5). These data are useful for determination of FW/DW rate. Furthermore the evaluation of the antioxidant activity during cell culture was reported [76], showing a maximum of activity in the line SOF2 between day 2 to day 7 corresponding to a minimum level of EC_{50}, indirectly related to the antioxidant capacity.

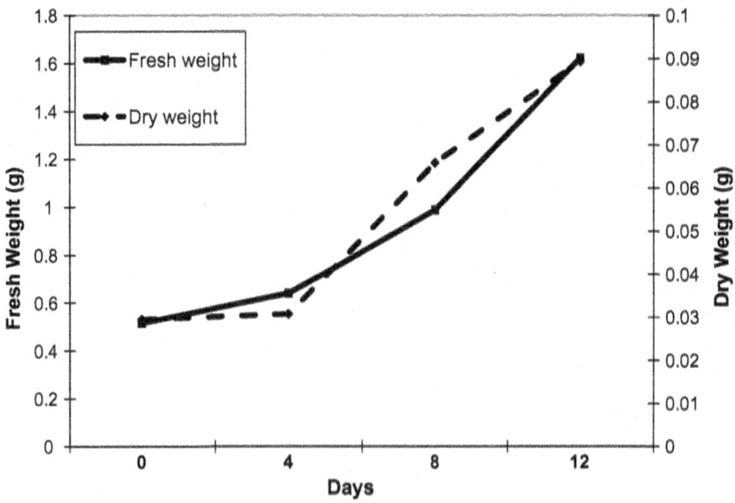

Fig. 6.5 Growth curve of *Salvia officinalis* cells in batch culture. The typical sigmoid curve showed a logarithmic phase between the 4th and 10th day [7]

The presence of RA in sage extracts, associated with other phenolic acids and/or flavonoids or their derivatives, has also been reported [37, 86–89]. From molecular point of view, the expression of some genes, responsible for the synthesis of certain secondary metabolites, is dependent upon a type of organ or tissue specific differentiation which switches on the respective promoter regions [90]. Another reason of this differential presence of compounds could be retrieved in differential level: more differentiated the sage cultures are, more phenolic diversity is present.

Santos-Gomes et al. [81] found that the highest specific content of total phenolic compounds was recorded for calli, followed by suspended cells grown with 0.05 mg/L 2,4-D and 0.5 mg/L Kin. The content of total phenolics accumulated by these suspension cultures, 15 days after the beginning of the subculture, was really high, compared to others PGRs treatments. Also in this study, the accumulation of total phenolic compounds is inversely correlated with the growth of the cultures. According to the authors, this could be a consequence of a greater demand of carbon and energy and promoted metabolic flux for alternative biosynthetic products, namely those of primary metabolism, which take part of the dry biomass.

It was also observed that the highest increase (221%) in dry biomass of calli was recorded at the end of the growth cycle (33 days) on agarized substrate supplemented with 0.05 mg/L 2,4-D and 1.5 mg/L ZEA [81]. It was demonstrated that the suspensions culture, originated from the calli induced on solid medium of the same type, in liquid MS medium supplemented with 0.05 mg/L 2,4-D and 1.5 mg/L BA, grew in the form of compact microcalli. It seems that the combination of 0.05 mg/L 2,4-D with 1.5 mg/L ZEA was the one that provide the highest growth of sage suspensions. However, suspensions from all three hormonal variants, as well as, the green calli, grown on 0.05 mg/L 2,4-D and 1.5 mg/L BA, were analysed for the accumulation of antioxidative phenolics, and twelve compounds have been identified [81].

The addition of different cytokinins to the suspension cultures had a crucial role on the accumulation of RA; for example, the use of ZEA may prevent the production of RA by sage suspensions while KIN has been the chosen cytokinin to stimulate the production of RA in *S. officinalis* in vitro cultures [77, 79].

As already reported by Santos Gomes et al. [91], in vitro sage shoots accumulated RA 3- to 6-fold over that of a studied commercial sage plants (482 μg/g of biomass dry weight). The same type of comparison gives values of RA contents in calli, grown with 1.5 mg/L BA, and suspended cells, grown with 0.5 mg/L Kin, 7-fold over that of the referred commercial sage sample, and 2-fold over in suspensions grown with 1.5 mg/L Kin.

Grzegorczyk et al. [92] evaluated the antioxidant capacities and discovered many other compounds from in vitro culture of *S. officinalis*. Three different assays methods were applied (DPPH free radical scavenging activity, reduction of transition metal ions by phosphomolybdenum complex and inhibition of lipid peroxidation). The analysis were conducted on methanolic and acetone extracts of various in vitro cultures of sage (callus, cell suspension, shoots and hairy roots) as well as shoots and roots of in vitro propagated plants. All extracts were chemically characterized by HPLC method in order to find the relationship between their activity and chemical composition of extracts.

The activity of the extracts was compared with that of methanol and acetone extracts of sage plants obtained from seeds and grown in the field. It was found that extracts of sage roots exhibited strongest antioxidant activity. Moreover, due to different nature of antioxidant compounds presented in sage, the antioxidant activity of the extracts was strongly dependent to the assay method and the method of sample extraction (e.g., the polarity of solvent used for sample extraction). For example, the antioxidant properties, measured by DPPH method were greater in polar fractions, i.e., methanol extracts, containing both diterpenoids and RA. The main finding of this work is the fact that extracts of sage organ cultures, especially hairy roots, obtained as a result of transformation through *A. rhizogenes* possessed higher antioxidant activity, comparable to, or even higher than that of extracts of shoots and roots of field-collected *S. officinalis* plants. With respect to this, it is possible to consider the culture as a potential source of natural antioxidants [76].

The advantages of transformed roots are: their biochemical and genetic stability and relatively rapid growth in the absence of exogenous growth regulators. As expected, the extracts of undifferentiated callus and cell cultures of sage proved to be less active in all methods tested. It is accepted that the dedifferentiation of plant tissues during establishment of callus and cell cultures is often connected with reduction of content of secondary metabolites [116].

Another research, conducted by the same group, focused the attention on the production of compounds with antioxidant activity in shoot cultures of *S. officinalis* grown in laboratory-scale sprinkle nutrient bioreactors [45]. A stock of shoots was maintained on MS agarized medium supplemented with IAA (0.1 mg/L) and BAP (0.45 mg/L). The growth of roots and shoots in the bioreactor was measured in terms of fresh and dry weights. Additionally, for shoot culture, the total number of shoots produced in the bioreactor over a three week period (multiplication rate) was counted. The cultivation of sage shoots in the bioreactor resulted in biomass accumulation of 25.2 $\pm$ 2.5 g/L FW and 2.25 $\pm$ 0.18 g/L DW on the 21st day. There was roughly a 42–45 fold increase over the initial inoculation. Differences in multiplication rate (estimated by the number of shoots produced after 3 weeks) were observed between cultures in the bioreactor (4.4 $\pm$ 0.31 shoots per explant) and Magenta vessels (3 $\pm$ 0.19 shoots/explant). The biggest problem was that, all shoots formed in bioreactor had hyperhydrated leaves (translucent, curled and thickened) and stems. The reason was probably the high water potential of sage leaves. The bioreactor is not suitable for sage shoot multiplication because of hyperhydricity problems. Further studies are needed to optimize culture conditions such as regulating the time that the shoots are in contact with the liquid medium to overcome the hyperhydricity problem and to enhance the production of diterpenoids in sage shoots, grown in the bioreactor. Problem of hyperhydricity is quite common, unfortunately, when explants are cultivated in liquid medium in fully-immersion system. RA was produced by sage shoot cultured in the bioreactor. Its content was 26.2 $\pm$ 0.48 mg/g DW after three weeks of culture. The value was similar to that detected in sage shoots grown in Magenta vessels (24.89 $\pm$ 0.59 mg/g DW) and over two times higher than RA level in the leaves of *S. officinalis* plants growing in the field (12.2 ± 0.17 mg/g DW) [93].

Grzegorczyk and Wysokińska [94] reported the effect of three treatment with methyl jasmonate (MeJA) (20, 50 and 100 μM) on the shoots of *S. officinalis* growing in MS liquid medium supplemented with IAA (0.1 mg/L) and BAP (0.45 mg/L), with the aim to increase the production of compounds with antioxidant activity, especially RA. The MeJA signalling pathway induces a wide variety of plant secondary products such as terpenoids, flavonoids, alkaloids and phenylpropanoids. The increase in metabolite production depends on MeJA concentration, the period of exposure to elicitor and the type of elicited compound. The highest amount of RA (about 41 mg/g DW) was achieved with 50 or 100 μM of MeJA on the 5th day after elicitation. It was almost 2-fold higher compared to the control (cultures treated with only ethanol). Elicitation is a well-known method which aims to enhance the production of secondary metabolites and has been used in many plant species. The stimulus is received by receptors and results in the activation of secondary messengers. These transmit the signals into the cell leading to genes expression and biochemical changes. These results are in agreement with the fact that changes in plant genes expression are induced by micro molar concentrations of MeJA and that the time required to reach the maximum level of genes expression is different for each biosynthetic pathway and even for each compound gene type. In sage shoot culture, MeJA rapidly stimulated secondary metabolites production. It is also noteworthy that MeJA did not have a negative effect on the growth of sage shoots when the elicitor was added on 14th day of cultivation period.

Kračun-Kolarević et al. [95] evaluated the influence of sodium salicylate on RA accumulation by *S.officinalis* L. shoots grown in vitro. Salicylic Acid (SA) is a well-known elicitor and phytohormone [96] that controls plant growth, thermogenesis, flowering, ion uptake, stomatal movement, photosynthetic rate, and production of SMs, and thus participates in endogenous signalling pathways leading to defence against pathogens [96]. In many plants SA increases the amount of alkaloids [97] and stimulates the biosynthesis of RA [98]. In *Salvia* species SA application determines EO composition [99].

Kračun-Kolarević et al. [95] started an aseptic culture from sterilized seeds (surface-sterilized with commercial sodium hypochlorite: 2% active chlorine) germinated onto basal medium containing MS salts, 0.1 g/L myo-inositol, 30 g/L sucrose and 7 g/L agar. For the multiplication phase, one node stem segments of 4 weeks old seedlings were propagated into the suggested medium, but supplemented with 6-benzylaminopurine (BAP) 0.2 mg/L and indole-3-butyric acid (IBA) 0.1 mg/L. Different NaSA concentrations, ranging from 30 to 500 μM, were tested.

The results indicated a dose-dependent NaSA effect on *S. officinalis* shoot growth, which was obvious 3 weeks after the treatment. It was shown that SA negatively affect plant growth, particularly when applied in high doses [96], and during the intensive growth period after the transfer of explants on fresh medium [97]. However, accumulation of RA was not affected by the NaSA treatment.

Significant progress has been made in the use of tissue culture and genetic transformation to modify pathways for the biosynthesis of target metabolites. Hairy roots are unique in their genetic and biosynthetic stability and their fast growth offers an additional advantage to their use as continuous source for the production of valuable secondary metabolites. The *Agrobacterium rhizogenes* oncogenes effects have been only recently clarified and, as suggested by Pistelli et al. [100], permitted a new use of the *A. rhizogenes* genes potential. Instead of transformation with the bacterium wild strains, it is possible to use strains with selected pool of *Rol* genes with more driven results in order to optimize metabolite production.

5 *Salvia dolomitica:* An Example of a Versatile Sage for Different Biotechnological Approaches

Salvia dolomitica has been shown to be a good model for tissue culture studies due to its flexibility in propagation under in vitro conditions [11].

Genetic transformation procedures and metabolic engineering strategies allows selective redirection of metabolic fluxes towards the production of a specific plant secondary metabolite. However, set up stable genetic transformation in aromatic plants is still challenging mainly due to difficulties in plant regeneration and the presence of endogenous secondary metabolites which play an important role as defensive chemicals against microbial pathogens including *Agrobacteria* strains and thus strongly interfering with the transformation process [101].

In a recent work, Bassolino et al. [11] evaluated the composition of essential oils in micropropagated plants of *S. dolomitica* compared to the in vivo derived plant biomass. Moreover, micropropagated plants have been exposed to a physical elicitor (e.g. high light) with the aim to understand whether the aromatic profile and composition of EO was modulated by the treatment.

Kračun-Kolarević et al. [95] established in vitro manipulation protocols of selected *S. dolomitica* plants for fast propagation, biomass production, genetic improvement support, and biotechnological strategy for extraction of secondary metabolites.

In vitro culture and micropropagated plants were established starting from a *S. dolomitica* Codd mother plant selected for its intense fragrance. The mother plants have been grown in an unheated greenhouse at the CREA Research Centre for Vegetable and Ornamental Crops located in Sanremo (IM), Italy, at latitude 43L490 050 0N and longitude 07L450 300 0E (Fig. 6.6).

Apical shoots and nodes from the mother plant were cut and treated as follows: (1) washed with tap water, (2) dipped in 70% ethanol for 30 s, (3) sterilized in 1.25% of active chlorine sodium hypochlorite solution for 15 min, (4) rinsed twice with sterile distilled water for 10 min each. After sterilization the explants were cultured on MS [83] base medium composed of macro and microelements, vitamins and 30 mg/L sucrose, supplemented with BA (0.3 mg/L) as reported by Mascarello et al. [102]. The cultivation was performed at the following environmental conditions:

Fig. 6.6 **a** In vitro production of *S. dolomitica* biomass. Nodal explants grown on MS supplemented with BA. **b** Callus plated on MS supplemented with 2,4-D and Kinetin. **c** Cell suspension culture of *S. dolomitica* grown in flasks containing MS + 2,4-D and NAA. **d** FDA assay to assess cell vitality: visible light (left) and UV light (right), 400×. **e** In vivo production of *S. dolomitica* biomass. Cutting preparation, development of root apparatus, inflorescence, and en plein air *S. dolomitica* plants after 20 days. **f** Callus development of *S. dolomitica* plated on MS + 2,4-D and increase of callus fresh weight over time (average ± standard error)

23 ± 1 °C, 16 h lighting photoperiod at 30 μE/m^2 s. Massive micropropagation was then performed by 28-day subcultures in the same medium and at the same thermo-photoperiodic conditions. Rooting was achieved from shoots after transfer to MS medium. Complete plantlets were then acclimatized in the greenhouse. The multiplication rate was on average 2.5 in several subcultures. Micropropagated shoots rooted easily in absence of hormones and were acclimatized in the greenhouse successfully. A protocol for callus induction was also developed. Leaves of in vitro material were cut along the main rib and the fragments were placed on agarized (8 g/L technical agar) substrate composed of MS added with 2,4-D (0.5 mg/L), Kin (0.5 mg/L) at pH 5.7. Thermo-photoperiodic conditions were as indicated above. The material was sub-cultured every 28 days. Callus induction from leaf explants started after 15 days in the presence of 0.5 mg/L 2,4-D and 0.5 mg/L Kin. After 28 days 100% of fragments showed friable callus development. This callus, transferred onto the same medium, showed massive and constant proliferation over a period of 24 days with a weight growth index of 3.4. Once established, the callus culture can be shifted to liquid media to obtain cell suspension culture of *S. dolomitica* that might be used to scale up the production of specific metabolites by following a molecular farming approach. It is also possible to obtain

in vivo fresh biomass of *S. dolomitica* via in vivo cutting propagation. In detail, cuttings were performed in June starting from the same mother plant described above. Sixty apical tips with at least two nodes, 10 cm long, were used as cutting. The plant material was planted in plastic alveolar containers (with 60 holes) filled with a mixture of commercial potting soil (Klasmann) for sowing and perlite (Agrilit), 30:70 v/v and grown during July 2013 in an unheated glasshouse with an average temperature of 27.8 °C (with a wide range between 20.4 and 35.2 °C). In August 2013 the rooted cuttings were transplanted in 1.3 L plastic pots and moved to open field. New potting was necessary after a further 2–2.5 months to allow plants a suitable growth. After 20 days of culture in the greenhouse, 70% of cuttings developed complete vegetative tissues and root apparatus. Rooted plants were then transferred to 10 cm diameter pots for growing and to produce fresh biomass for EO evaluation. The good rooting capacity of *S. dolomitica* cuttings leaded to high propagation performance, uniform and quick growth and a good ratio of plant height and width in agreement with data reported by Cervelli et al. [14].

The obtained in vivo and in vitro plant materials were used to perform a phytochemical screening, focused on the analysis of the EOs. Static headspaces (HS) of *S. dolomitica* in vivo plants in comparison with in vitro plantlets, grown in normal conditions and under high light treatment, have been performed in order to define their volatile fingerprints [11].

Interestingly, in total 83, 88 and 90 compounds were identified in samples by in vivo plants (a), micropropagated plants (b) and high light elicited in vitro plants (c) respectively. The yields of EOs were 0.10% w/v for sample **a**, and 0.17% w/v for sample **b**, while surprisingly sample **c**, stressed by high light condition, gave 0.30% w/v. These data are in agreement with the study of Fernandes et al. [103] in *Ocimum gratissimum*, supporting the evidence that the EO yield/plant increased with increasing light intensity. Moreover, this was not related to variation in trichomes morphology or distribution, since in *S. dolomitica* the distribution of glandular trichomes was not affected by high light treatment. The EO composition of the in vivo grown plants (sample **a**) was characterized by a high percentage of sesquiterpenes (85.15%), shared between hydrocarbons (71.54%) and oxygenated sesquiterpenes, OS (13.64%). The main constituent was b-caryophyllene (25.58%) followed by a-guaiene (8.93%), d-cadinene (7.49%) and valencene (5.27%). Caryophyllene oxide was the more abundant oxygenated sesquiterpene (6.16%) in this sample. The chemical compositions of the EOs from micropropagated plants (samples **b** and **c**) were very similar to each other. Both showed very high percentages of total monoterpenes (59.27% in sample **b** and 69.91% in sample **c**), while the total sesquiterpenes ranged from 39.93% in sample **b** to 29.73% for sample **c**. The average percentage of total hydrogenated compounds represented about 60% in both sample **b** and sample **c** where the monoterpene hydrocarbons, MH (33.54%, **b**; 39.06%, **c**) and sesquiterpene hydrocarbons, SH (29.81%, **b**; 18.62%, **c**) were more abundant in comparison with the corresponding oxygenated classes of compounds (25.73%, **b** and 30.85%, **c** for OM; 10.11%, **b** and 11.11%, **c** for OS). Regarding the pure constituents present in these EOs, a-pinene (7.24%, **b**; 8.36%, **c**), d-3-carene (4.94%, **b**; 4.91%, **c**), b-phellandrene (6.50%, **b**; 7.66%, **c**), 1,8-cineole

(13.28%, **b**; 16.62%, **c**), borneol (10.45%, **b**; 11.73%, **c**) and b-caryophyllene (14.58%, **b**; 10.75%, **c**) showed the highest percentages, ranging from about 5 to 16.62%. Arikat et al. [104] reported that the higher amount of borneol in the microshoots may be due to the presence of BA in the medium in *S. fruticosa*. It was demonstrated, that in vitro conditions influenced the qualitative and quantitative compositions of EO by *S. dolomitica* in comparison with the in vivo grown mother plants. In fact the percentage of total sesquiterpenes decreased strongly in in vitro grown plantlets and high light exposed plants [11]. On the contrary, the total monoterpenes abundance was higher these plants. This different behaviour could be associated with the juvenile phase, maintained in in vitro conditions [105], or to be due to the artificial growth conditions in in vitro environment. EO of in vitro cultured plants was predominantly composed of hydrogenated and oxygenated monoterpenes in comparison with in vivo grown plants. Variations in EO composition of micropropagated materials, in respect to the propagated one, could be the consequence of the different ontological stage. This hypothesis is supported by the fact that in vitro cultured plants are considered, by definition, at juvenile stage and it is well-known that accumulation of the monoterpenes is restricted to young leaf tissues due to the highest capacity of biosynthesis [20, 105, 106]. It's remarkable that the EO profile of the in vivo sample does not correspond with that reported in the literature [16, 33] for a *S. dolomitica* grown in South Africa, which was dominated by the presence of oxygenated monoterpenes (72%), with high amounts of geraniol, linalyl acetate (in similar amount of about 20%) and linalool (17%). Interestingly, both EOs, the one analysed in Kamatou's study [6] and the ones analyzed by Bassolino et al. [11] (sample **a**, **b** and **c**), do not contain camphor, which might be of interest for application of this species as food additive. The head space analysis (SPME) demonstrated that limonene was the main constituent of in vivo samples, even though at half concentration (11.93%), if compared with in vitro samples (24.83%, **b**; 24.44%, **c**, respectively). Other differences were found regarding myrcene, which was present in very low concentrations in sample **a** (3.41%), compared to the in vitro samples (12.69%, sample **b** and 17.37%, sample **c**). Furthermore, sesquiterpenes have been found to characterize the aroma of the in vivo plants (sample **a**). Among them, germacrene d, b-caryophyllene, a-guaiene and c-patchoulene were presented in very high amounts. However, the same compounds were absent or presented in traces in in vitro plant materials. It is interesting to note that sesquiterpenes were the predominant class of constituents in EO of in vivo plants (85.18%), while their percentage decreased in its HS composition (59.34%), whereas monoterpenes became more abundant (from 12.17% in EO to 40.24%) in HS. In conclusion, it is possible to underline that micropropagation and light treatment caused an increase in monoterpene amount while sesquiterpene content decreased. Putting the HS and EO data together it is possible to speculate that light selectively modulates the branches of terpenoids biosynthetic pathway, leading to monoterpenes and sesquiterpenes accumulation, by interfering at a transcriptional level with the activity of specific enzymes like monoterpene and sesquiterpene synthetases. Moreover, a-pinene, b-phellandrene and borneol accumulate predominantly in micropropagated plants suggesting that in vitro manipulation of *S. dolomitica* could be exploited to increase the relative amount of specific compounds.

As per EO, the possibility to modulate in vitro the metabolic profile by physical elicitation opens the door to establish in vitro protocols for hyperproduction of target metabolites stating from species with high performances, high flexibility and high manipulation possibilities as *S. dolomitica* for example.

6 Molecular Cloning of *HPPR* and Metabolic Engineering

Phytochemicals have been extracted from various natural plant resources since the beginning of human mankind. However, during the last century the supply of phytochemicals became difficult due to overexploitation of plant field collections, exhausting the natural environment. Apart the chemical synthesis, not always possible due to the complexity of the chemical molecules, plant biotechnology can provide an alternative system for plant secondary metabolites production.

Particularly for production and accumulation of metabolites of interest, cell cultures provide an excellent model to study biosynthetic pathways. Once the genes functions are known, cell suspensions offer a rapid way to manipulate their expression level by modulation of the micro-environmental conditions, and to push-up the production of target metabolite. In the 90s a new technique, called metabolic engineering, emerged allowing the analysing the metabolic pathway of a microorganism to determine the key factors responsible in biosynthesis of desired compounds. Then the technique has been applied to other superior organisms, including plants. Plants metabolic engineering is generally defined as the redirection of one or more enzymatic reactions to produce new compounds, to improve the production of existing compounds, or mediate the degradation of compounds [107].

First requirements for the development of metabolic engineering of a particular molecule are the use of the basic molecular biology technologies such as cloning, promoter analysis, protein targeting, plant transformation, biochemical genetics, and other areas of plant biology.

The molecular structure of the RA is known since many years [36], as well as the biosynthetic precursors and the specific enzymes involved in its biosynthesis [108–110].

The research on the RA biosynthesis has greatly profited from the fact that in vitro cultures of suitable plant species, usually synthesize and accumulate RA in acceptable amounts. Among these species, *S. officinalis* displayed good capacity to accumulate RA in plant organs and cell suspensions. Apart from production and accumulation of metabolites of interest, cell cultures provide an excellent model to study biosynthetic pathways. Once the genes are known, cell suspensions offer a rapid matrix to study their expression level after modulation of micro-environmental conditions, or to push-up the production of target metabolites. Hence, empirical strategies combined with metabolic engineering tools can increase RA production in plant cell platforms in a more rational way.

Despite the botanical and pharmacological interest of *S. officinalis*, and the potential possibility of obtaining RA from both cell suspensions and hairy roots, only few reports are available regarding the elucidation of RA biosynthesis in *S. officinalis* [9].

According to the hypothesis that a key enzyme for pushing-up the RA biosynthesis in cell suspension could be in the final part of the biosynthetic pathway, Barberini et al. [9] focused the attention on a peculiar enzyme, HPPR, (Hydroxyphenylpyruvate Reductase), involved in the biosynthesis of the phenylpropanoid moiety of RA [9]. The hydroxyphenylpyruvate reductase (HPPR; E.C. 1.1.1.237) is considered the first specific key-enzyme responsible for the metabolic biosynthesis of RA [109, 111]. This enzyme, which reduces hydroxyphenylpyruvate to the corresponding hydroxyphenyllactate in dependence of NAD(P)H, was first characterized in cell-free extracts obtained from suspension-cultured cells of *C. blumei* [112, 113] and was then purified and sequenced [109]. Recently, *HPPR* content was characterized in *S. miltiorrhiza* by Solexa deep RNA- sequencing [114]. This novel technology enabled the whole transcriptome of a plant species to be evaluated in every phase of its life cycle. It appears that the *HPPR* gene is ubiquitously expressed in plants, although the expression pattern showed maximal expression levels during seed germination and in leaf explants in the reproductive stage. However, to get better insight into the metabolic way of RA formation, it is necessary to activate or suppress the key enzymes involved in its production.

Deep investigation on *HPPR* gene expression in *Salvia officinalis* (Table 6.2), in order to correlate gene expression levels and RA accumulation, have been performed by monitoring *HPPR* expression in cell suspension cultures and performing over expression on common sage calli [114]. Previously, Ruffoni et al. [76] set up a protocol to obtain a high biomass from sage cell suspension, starting from a particular cell line, rich in antioxidant power. The authors started from the same cell line (SOFF1) and monitored the antioxidant capacity during the cell suspension cycle.

In this study, starting from leaves of common sage, molecular cloning of *SoHPPR* in the pGEMT-easy vector was carried out and, after sequenced, the sequence was registered in GenBank with the accession n° JX566894.1. This sequence showed high levels of orthology with other plant *HPPR,* in particular with the Chinese *Salvia* Dan-Shen (*S. miltiorrhiza*), which recently has been deeply investigated in its secondary metabolism with the strategy of RNA-seq.

Table 6.2 *SoHPPR* identity percentage among *HPPR* found in literature

Sequence	Identity (%)
Salvia miltiorrhiza HPPR mRNA putative (GenBank DQ099741.1)	93
Salvia miltiorrhiza HPPR CDS (GenBank DQ266514.1)	93
Coleus blumei HPPR mRNA (GenBank AJ507733.2)	89
Salvia miltiorrhiza HPPR gene (GenBank EF458148.1)	95
Perilla frutescens HPPR gene (GenBank HM 152567.1)	92

Alignment was performed using BLAST® software

By following the *SoHHPR* expression pattern during the production of RA in cell suspension cultures, Barberini et al. [9] were able to positively correlate the gene expression level with the *de novo* production of RA in cells. However, since the RA production persists over the whole culture period, the mRNA level decreased after few days, suggesting cooperation with other genes in pushing the RA biosynthesis on. Considering that nor DNA neither EST databases are available for *S. officinalis*, molecular cloning of single genes for this species is still an important tool to facilitate full understanding of some secondary metabolite pathways. Taken together, the findings confirm the involvement of *HPPR* in RA biosynthesis pathway, and suggest it possible role as marker gene for secondary metabolism mapping

Genetic transformation is another powerful tool to study gene function in plants. For this technique, adaptation of transformation protocols and plant regeneration methods are the first step when dealing with a new species. Usually native strains of *A. rhizogenes* or *A. tumefaciens* are used to induce hairy roots (HR) or other type of tumours, respectively. However, despite that RA producing HR of *S. officinalis* have been obtained years ago [45], there are no attempts to insert foreign DNA in cells of this species. Regarding organs or the whole plant transformation, only few reports are available for the genus *Salvia*, most of them relating to *S. miltiorrhiza* [115–117]. It was reported that, the problem of achieving stable transgene integration in sage is related to its high antibacterial activity against a wide range of microorganisms [118]. It was found that active ingredients, accumulated in *S. officinalis,* had a negative effect on bacterial growth and development, and thus, represent the major obstacle for conducting an effective sage transformation by using *A. tumefaciens* strain LBA4404 [118].

Contrary to the work of Luwańska et al. [118], which failed to develop a useful method for sage shoots transformation, in 2014 Barberini [119] made preliminary attempts to transform calli. Since RA has been found in all plant tissues, it was proposed that undifferentiated cells could be a suitable model for experiments to study RA genes overexpression. Moreover, they could be cultivated in a form of cell suspension cultures that can be easily manipulated. Barberini [119] reported a methodology to transform callus of *S. officinalis* with two different strains of *A. tumefaciens* (strains AGLO and EHA105), both carrying the binary plasmid pKIWI105 with *uidA* gene (GUS) and the *nptII* gene(neomycin phosphotransferase—NPTII). The results with transient expression of the marker gene protein GUS indicated a good percentage of transformation, which is a good base for further investigation in metabolic pathway. Subsequently, an experiment for increase the RA production by the over expression of *SoHPPR* was performed. Two binary vectors (pKIWI 105 carrying GUS and pRCS2 carrying *SoHPPR*) have been used in *A. tumefaciens* AGLO to achieve stable transformation. In this study, the optimal combination of bacterial concentration and co-cultivation time was determined experimentally.

The overexpression of *SoHPPR* in calli resulted in pale explants with reduced growth capacity. However, as a result, a suitable protocol for cells manipulation in order to understand the genetic basis of RA biosynthesis in *S. officinalis* has been developed [119]. After 2 months of cultivation in growth chamber (16 h light photoperiod, 24 °C), most of the transformed cells were still viable, even the author noticed a slow growth rate and a different phenotype (pale colour instead of bright green). The green non-transformed calli were able to double their biomass (measured as fresh weight/piece of callus) after one month of cultivation, while pale *Agrobacterium*-transformed explants showed low or null biomass increasing.

The plausible hypothesis is that the transformation events occurred in the cell line may have changed the metabolic profile, affecting viability and photosynthetic capacity. When overexpressing *HPPR* in *C. blumei* hairy roots, Hücherig and Petersen [120] found a high copy number of this gene, even not toxic for cells. Barberini et al. [9] were not able to positively correlate the copy number of *SoHPPR* with RA production due to the low amount of callus obtained. Therefore, further experiments with other binary vectors, carrying the cloned gene, have to be done.

7 Closing Remarks and Future Perspectives

The biotechnological platform established for the two species reviewed in this chapter reveals a wide field of possible developments in *Salvia* metabolic engineering research. The in vitro grown tissues of *S. dolomitica* and *S. officinalis* could be used as perspective model plant systems from Labiatae family in order to produce secondary metabolites or to study the accumulation of VOCs and EO composition in response to chemical and physical elicitors. The recent elucidation of RA metabolic pathway could be used as a base for future projects on evaluation of feasibility to increase metabolite production by *Salvia* plant tissues, cultivated under controlled environment. The *SoHPPR* has been cloned and sequenced, opening the way for further genetic engineering procedures. Despite that no clear evidences for overproduction of RA in transformed calli could be provided yet, the effective transformation protocols have been developed. To the best of our knowledge, this is the first transformation protocol applied to *Salvia* with positive outcomes, which could be used as a good starting point for future research. To date, this is the first report for *Salvia officinalis* callus transformation, and this protocol is suitable for developing further experiments with cell suspensions with rapid *in planta* detection of secondary metabolites.

References

1. Marchev A, Haas C, Schulz S, Georgiev V, Steingroewer J, Bley T, Pavlov A (2014) Sage in vitro culture: a promising tool for the production of bioactive terpenes and phenolic substances. Biotech Lett 36:211–221
2. Watt JM, Breyer-Brandwijk MG (1962) The medicinal and poisonous plants of southern and eastern Africa. E.&S. Livingstone Ltd, Edinburgh, vol XII, 1457 pp

3. Shoemaker M, Hamilton B, Dairkee SH, Cohen I, Campbell MJ (2005) In vitro anticancer activity of twelve Chinese medicinal herbs. Phytotherapy Res 19(7):649–651
4. Kamatou GPP, van Zyl RL, Davids H, van Heerden FR, Lourens ACU, Viljoen AM (2008) Antimalarial ad anticancer activities of selected South African *Salvia* species and isolated compounds from *S. radula*. S Afr J Bot 74:238–243
5. Thorsen MA, Hildebrandt KS (2003) Quantitative determination of phenolic diterpenes in rosemary extracts: aspects of accurate quantification. J Chromatogr A 995:119–125
6. Kamatou GPP, Chen W, Viljoen AM (2012) Quantification of Rosmarinic acid in *Salvia* species indigenous to South Africa by HPTLC. J Planar Chromatogr 25(5):403–408
7. Ruffoni B, La Pistelli, Bertoli A, Pistelli Lu (2010) Plant cell cultures: bioreactors for industrial production. Adv Exp Med Biol 698:203–218
8. Razzaque A, Ellis BE (1977) Rosmarinic acid production in *Coleus* cell cultures. Planta 137 (3):287–291. https://doi.org/10.1007/BF00388164
9. Barberini S, Savona M, Raffi D, Leonardi M, La Pistelli, Stochmal A, Vainstein A, Pistelli Lu, Ruffoni B (2013) Molecular cloning of *SoHPPR* encoding a *hydroxyphenylpyruvate reductase*, and its expression in cell suspension cultures of *Salvia officinalis*. Plant Cell Tissue Organ Cult 112(3):131–138
10. Xiao Y, Zhang L, Gao S, Saechao S, Di P, Chen J, Chen W (2011) The c4h, tat, hppr and hppd genes prompted engineering of rosmarinic acid biosynthetic pathway in *Salvia miltiorrhiza* hairy root cultures. PLoS ONE 6(12):e29713. https://doi.org/10.1371/journal.pone.0029713
11. Bassolino L, Giacomelli E, Giovanelli S, Pistelli L, Cassetti A, Bisio A, Damonte G, Ruffoni B (2014) Tissue culture and aromatic profile in *Salvia dolomitica* Codd. Plant Cell Tissue Organ Cult 121:83–95
12. Cervelli C (2011) Salvie. Caratteristiche, usi e coltivazione. Schede monografiche. Ed. Ace2, pp 179–189
13. Caser M, Ruffoni B, Scariot V (2012) Screening for drought tollerance in *Salvia* spp. and *Helichrysium petiolare*: a way to select low maintenance ornamental plants. Act Hort 953:240–246
14. Cervelli C, Capponi A, Mascarello B, Ruffoni B, Del Gaudio C (2013) New species and cultivars of *Salvia* as ornamental pot plants. Acta Hort 1000:35–41
15. Araújo C, Sousa MJ, Ferreira MF, Leão C (2003) Activity of essential oils from Mediterranean Lamiaceae species against food spoilage yeasts. J Food Prot 66:625–632
16. Kamatou GPP, Viljoen AM, Figueiredo AC, Tilney PM, Van Zyl RL, Barroso JG, Pedro LG, Van Vuuren SF (2007) Trichomes, essential oil composition and biological activities of *Salvia albicaulis* Benth. and *S. dolomitica* Codd, two species from the Cape region of South Africa. S Afr J Bot 73:102–108
17. Ulubelen A, Öksüz S, Kolak U, Bozok-Johansson C, Çelik C, Voelter W (2000) Antibacterial diterpenes from the roots of *Salvia viridis*. Planta Med 66(05):458–462
18. Burbott AJ, Loomis WD (1967) Effects of light and temperature on the monoterpenes of peppermint. Plant Physiol 42:20–28
19. Santos-Gomes PC, Fernandes-Ferreira M (2001) Organ- and season-dependent variation in the essential oil composition of *Salvia officinalis* L. cultivated at two different sites. J Agric Food Chem 49(6):2908–2916
20. Grausgruber-Gröger S, Schmiderer C, Steinborn R, Novak J (2012) Seasonal influence on gene expression of monoterpene synthases in *Salvia officinalis* (*Lamiaceae*). J Plant Physiol 169:353–359
21. Bernath J, Danos B, Hethelyi E (1991) Variation in essential oil spectrum on *Salvia* species affected by environment. Herba Hung 30(1–2):35–48
22. Boelens MH, Boelens H (1997) Chemical and sensory evaluation of three sage oils. Perfum Flavor 22:19–40
23. Perry NB, Anderson RE, Brennan NJ, Douglas MH, Heaney AJ, McGimpsey JA, Smallfield BM (1999) Essential oils from Dalmatian sage (*Salvia officinalis* L.): variations among individuals, plant parts, seasons, and sites. J Agric Food Chem 47(5):2048–2054

24. Cuvelier M-E, Berset C, Richard H (1994) Antioxidant constituents in sage (*Salvia officinalis*). J Agric Food Chem 42:665–669
25. Lu Y, Foo LY (2001) Antioxidant activities of polyphenols from sage (*Salvia officinalis*). Food Chem 75(2):197–202
26. Petersen M, Simmonds MSJ (2003) Molecules of interest: rosmarinic acid. Phytochemistry 62:121–125
27. Büyükbalci A, El SN (2008) Determination of in vitro antidiabetic effects, antioxidant activities and phenol contents of some herbal teas. Plant Foods Hum Nutr 63(1):27–33
28. Ollanketo M, Peltoketo A, Hartonen K, Hiltunen R, Riekkola ML (2002) Extraction of sage (*Salvia officinalis* L.) by pressurized hot water and conventional methods: antioxidant activity of the extracts. Eur Food Res Technol 215(2):158–163
29. Martins N, Barros L, Santos-Buelga C, Henriques M, Silva S, Ferreira IC (2015) Evaluation of bioactive properties and phenolic compounds in different extracts prepared from *Salvia officinalis* L. Food Chem 170:378–385
30. Fisher VL (2005) Indigenous *Salvia* species—an investigation of the antimicrobial activity, anti-oxidant activity and chemical composition of leaf extracts. M.Sc. dissertation, University of the Witwatersrand, Johannesburg, South Africa
31. Kamatou GPP, Viljoen AM, Gono-Bwalya AB, van Zyl RL, van Vuuren SF, Lourens ACU, Başer KHC, Demirci B, Lindsey KL, van Staden J, Steenkamp P (2005) The in vitro pharmacological activities and a chemical investigation of three South African *Salvia* species. J Ethnopharmacol 102(3):382–390
32. Tomás-Barberán FA, Tomas-Lorente F, Rumbero A (1990) Anti-fungal phloroglucinol derivatives and lipophilic flavonoids from *Helichrysum decumbens*. Phytochemistry 28 (6):1613–1615
33. Kamatou PP, Viljoen A, Steenkamp P (2010) Antioxidant, anti-inflammatory activities and HPLC analysis of South African *Salvia* species. Food Chem 119:684–688
34. Faleiro ML, Miguel MG, Ladeiro F, Venancio F, Tavares R, J Brito C, Figueiredo AC, Barroso JG, Pedro LG (2003) Antimicrobial activity of essential oils isolated from Portuguese endemic species of *Thymus*. Lett Appl Microbiol 36:35–40
35. Kamatou GPP, van Vuuren SF, van Heerden FR, Seaman T, Viljoen AM (2007) Antibacterial ad antimycobacterial activities of South African *Salvia* species and isolated compounds from *S. chamelaegnea*. S Afr J Bot 73:552–557
36. Scarpati ML, Oriente G (1958) Isolamento e costituzione dell'acido rosmarinico dal (*Rosmarinus off.*). Ric Sci 28:2329–2333
37. Lu Y, Foo LY (1999) Rosmarinic acid derivatives from *Salvia officinalis*. Phytochemistry 51:91–94
38. Hossain MB, Barry-Ryan C, Martin-Diana AB, Brunton NP (2010) Effect of drying method on the antioxidant capacity of six Lamiaceae herbs. Food Chem 123(1):85–91
39. Zimmermann BF, Walch SG, Tinzoh LN, Stühlinger W, Lachenmeier DW (2011) Rapid UHPLC determination of polyphenols in aqueous infusions of *Salvia officinalis* L. (sage tea). J Chromatogr B 879(24):2459–2464
40. Parnham MJ, Kesselring K (1985) Rosmarinic acid. Drugs Future 10(9):756–757
41. Bandoniene D, Murkovic M, Venskutonis PR (2005) Determination of rosmarinic acid in sage and borage leaves by high-performance liquid chromatography with different detection methods. J Chromatogr Sci 43(7):372–376
42. Zheng W, Wang SY (2001) Antioxidant activity and phenolic compounds in selected herbs. J Agric Food Chem 49(11):5165–5170
43. Shan B, Cai YZ, Sun M, Corke H (2005) Antioxidant capacity of 26 spice extracts and characterization of their phenolic constituents. J Agric Food Chem 53(20):7749–7759
44. Fecka I, Turek S (2007) Determination of water-soluble polyphenolic compounds in commercial herbal teas from Lamiaceae: peppermint, melissa, and sage. J Agric Food Chem 55(26):10908–10917

45. Grzegorczyk I, Wysokińska H (2010) Antioxidant compounds in *Salvia officinalis* L. shoot and hairy root cultures in the nutrient sprinkle bioreactor. Acta Soc Botanicorum Pol 79 (1):7–10
46. Roby MHH, Sarhan MA, Selim KAH, Khalel KI (2013) Evaluation of antioxidant activity, total phenols and phenolic compounds in thyme (*Thymus vulgaris* L.), sage (*Salvia officinalis* L.), and marjoram (*Origanum majorana* L.) extracts. Ind Crops Prod 43:827–831
47. Ellis BE, Towers GHN (1970) Biogenesis of rosmarinic acid in *Mentha*. Biochem J 118:291–297
48. Petersen M, Abdulla Y, Benner J, Eberle D, Gehlen K, Hücherig S, Janiak V, Kim KH, Sander M, Weitzel C, Wolters S (2009) Evolution of rosmarinic acid biosynthesis. Phytochemistry 70(15–16):1663–1679
49. Kelley CJ, Mahajan JR, Brooks LC, Neubert LA, Breneman WR, Carmack M (1975) Polyphenolic acids of *Lithospermum ruderale* (Boraginaceae). I. Isolation and structure determination of lithospermic acid. J Org Chem 40(12):1804–1815
50. Petersen M, Hausler E, Karwatzki B, Mainhard J (1993) Proposed biosynthetic pathway for rosmarinic acid in cell culture of *Coleus blumei*. Plant Cell Tissue Organ Cult 38:171–179
51. Petersen M (1997) Cytochrome p450-dependent hydroxylation in the biosynthesis of rosmarinic acid in *Coleus*. Phytochemistry 45:1165–1172
52. Häusler E, Petersen M, Alfermann A (1992) Rosmarinsäure in Blechnum-spezies. In: Botanikertagung H, Haschke P, Schnarrenberger C (eds) Akademie-Verlag, Berlin, p 507
53. Ravn H, Pedersen MF, Borum J, Andary C, Anthoni U, Christophersen C, Nielsen PH (1994) Seasonal variation and distribution of two phenolic compounds, rosmarinic acid and caffeic acid, in leaves and roots-rhizomes of eelgrass (*Zostera marina* L.). Ophelia 40:51–61
54. Litvinenko VI, Popova TP, Simonjan AV, Zoz IG, Sokolov VS (1975) "Gerbstoffe" und oxyzimtsäureabkömmlinge in Labiaten. Planta Med 27:372–380
55. Hiller K (1965) Zur Kenntnis der Inhaltsstoffe einiger Saniculoidae. 1. Mitteilung: *Sanicula europaea* L.—Isolierung und quantitative Erfas-sung von Chlorogen-und Rosmarinsäure. Pharmazie 20:574–579
56. Trute A, Nahrstedt A (1996) Separation of rosmarinic acid enantiomers by three different chromatographic methods (HPLC, CE, GC) and the determination of Rosmarinic Acid in *Hedera helix* L.
57. De Tommasi N, De Simone F, De Feo V, Pizza C (1991) Phenylpropanoid glycosides and rosmarinic acid from *Momordica balsamina*. Planta Med 57:201
58. Aquino R, Ciavatta ML, De Simone F, Pizza C (1990) A flavanone glycoside from *Hamelia patens*. Phytochemistry 29:2358–2360
59. Holzmannova V (1995) Kyselina rosmarinová její biologická aktivita. Chem Listy 90:486–496
60. Satake T, Kamiya K, Saiki Y, Hama T, Fujimoto Y, Kitanaka S, Kimura Y, Uzawa J, Endang H, Umar M (1999) Studies on the constituents of fruits of *Helicteres isora* L. Chem Pharm Bull 47:1444–1447
61. Lasure A, Van Poel B, Pieters L, Claeys M, Gupta M, Vanden Berghe D, Vlietinck AJ (1994) Complement-inhibiting properties of *Apeiba tibourbou*. Planta Med 60:276–277
62. Simmonds MSJ, Stevenson PC, Hanson FE (submitted) Rosmarinic acid in *Canna generalis* (Cannaceae) activates the medial deterrent chemosensory neurone and deters feeding in the tobacco hornworm, *Manduca sexta* (Lepidoptera). Physiol Entomol
63. Takeda R, Hasegawa J, Sinozaki M (1990) The first isolation of lignans, megacerotonic acid and anthocerotonic acid, from non-vascular plants, Anthocerotae (hornworts). Tetrahedron Lett 31:4159–4162
64. Chen JH, Ho C-T (1997) antioxidant activities of caffeic acid and its related hydroxycinnamic acid compounds. J Agric Food Chem 45(7):2374–2378
65. Furtado C, Moraes F, Carvalho ADM (2008) Geometric phases in graphitic cones. Phys Lett A 372:5368

66. Bulgakov VP, Shkryl YN, Veremeichik GN, Gorpenchenko TY, Inyushkina YV (2011) Application of *Agrobacterium* rol genes in plant biotechnology: a natural phenomenon of secondary metabolism regulation. In: Prof. Alvarez M (ed) *Genetic transformation*, vol 13, pp 261–270
67. Sanchez-Medina A, Etheridge CJ, Hawkes GE, Hylands PJ, Pendry BA, Hughes MJ, Corcoran O (2007) Comparison of rosmarinic acid content in commercial tinctures produced from fresh and dried lemon balm (*Melissa officinalis*). J Pharm Pharmaceut Sci 10:455–463
68. Chun SS, Vattem DA, Lin YT, Shetty K (2005) Phenolic antioxidants from clonaloregano (*Origanum vulgare*) with antimicrobial activity against *Helicobacter pylori*. Process Biochem 40:809–816
69. Szabo E, Thelen A, Petersen M (1999) Fungal elicitor preparations and methyl jasmonate enhance rosmarinic acid accumulation in suspension cultures of *Coleus blumei*. Plant Cell Rep 18(6):485–489
70. Hamaguchi T, Ono K, Murase A, Yamada M (2009) Phenolic compounds prevent Alzheimer's pathology through different effects on the amyloid-beta aggregation pathway. Am J Pathol Dec 175(6):2557–2565
71. Fallarini S, Miglio G, Paoletti T, Minassi A, Amoruso A, Bardelli C, Brunelleschi S, Lombardi G (2009) Clovamide and rosmarinic acid induce neuroprotective effects in vitro models of neuronal death. Br J Pharmacol 157(6):1072–1084
72. Kim OT, Bang KH, Kim YC, Hyun DY, Kim MY, Cha SW (2009) Upregulation of ginsenoside and gene expression related to triterpene biosynthesis in ginseng hairy root cultures elicited by methyl jasmonate. Plant Cell Tiss Organ Cult 98:25–33
73. Li G-S, Jiang W-L, Tian J-W, Qu G-W, Zhu H-B, Fu F-H (2010) In vitro and in vivo antifibrotic effects of rosmarinic acid on experimental liver fibrosis. Phytomedicine 17(3–4):282–288
74. Avato P, Fortunato IM, Ruta C, D'Elia R (2005) Glandular hairs and essential oils in micropropagated plants of *Salvia officinalis* L. Plant Sci 169:29–36
75. Wang JW, Wu JY (2010) Tanshinone biosynthesis in *Salvia miltiorrhiza* and production in plant tissue culture. Appl Microbiol Biotechnol 88(2):437–449
76. Ruffoni B, Raffi D, Rizzo A, Oleszek W, Giardi MT, Bertoli A, Pistelli L (2009) Establishment of in vitro *Salvia* cell biomass for the controlled production of antioxidant metabolites. Acta Hort 829:423–427
77. Kintzios S, Nikolaou A, Skoula M (1999) Somatic embryogenesis and in vitro rosmarinic acid accumulation in *Salvia officinalis* and *S. fruticose* leaf callus cultures. Plant Cell Rep 18:462–466
78. Funk C, Koepp AE, Croteau R (1992) Introduction and characterization of a cytochrome P-450-dependent camphor hydroxylase in tissue culture of common sage (*Salvia officinalis*). Arch Biochem Biophys 294:306–313
79. Hippolyte I, Marin B, Baccou JC, Jonard R (1992) Growth and rosmarinic production in cell suspension cultures of *Salvia officinalis* L. Plant Cell Rep 11:109–112
80. Tawfic AA, Read PE, Cuppert SL (1992) Stimulation of growth and monoterpene production of sage (*Salvia officinalis*) by benzylandenine in vitro. Plant Growth Regul Soc Am Q 20:200–206
81. Santos-Gomes PC, Seabra RM, Andrade PB, Fernandes-Ferreira M (2003) Determination of phenolic antioxidant compounds produced by calli and cell suspensions of sage (*Salvia officinalis* L.). J Plant Physiol 160:1025–1032
82. Grzegorczyk I, Bilichowski I, Mikiciuk-Olasik E, Wysokińska H (2004) In vitro cultures of *Salvia officinalis* L. as a source of antioxidant compounds. Acta Soc Botanicorum Pol 74. https://doi.org/10.5586/asbp.2005.003
83. Murashige T, Skoog F (1962) A revised method for rapid growth and biassays with tobacco tissue cultures. Physiol Plant 15:472–497
84. Van Der Plas LHW, Eijkelboom C, Hagendoorn MJM (1995) Relation between primary and secondary metabolism in plant cell suspensions. Plant Cell Tissue Organ Cult 43:111–116

85. Areias F, Valentão P, Andrade PB, Ferreres F, Seabra RM (2000) Flavonoids and phenolic acids of sage: influence of some agricultural factors. J Agric Food Chem 48:6081–6084
86. Hohmann J, Zupkó I, Rédei D, Csányi M, Falkay G, Máthé I, Janicsák G (1999) Protective effects of the aerial parts of *Salvia officinalis*, *Melissa officinalis* and *Lavandula angustifolia* and their constituents against enzyme-dependent and enzyme-independent lipid peroxidation. Planta Med 65:576–578
87. Lamaison JL, Petitjean-Freytet C, Carnat A (1990) Teneures en acide rosmarinique, en dérivés hydroxycinnamiques totaux et activité antioxydant chez les Apiacées, les Borraginacées et les Lamiacées médicinales. Ass Pharm Fr 48:103–108
88. Lamaison JL, Petitjean-Freytet C, Carnat A (1991) Lamiacées médicinales à propriétés antioxydantes, sources potentielles d'acid rosmarinic. Pharm Acta Helv 66:185–188
89. Wang M, Li J, Rangarajan M, Shao Y, LaVoie EJ, Huang T-C, Ho C-T (1998) Antioxidative phenolic compounds from sage (*Salvia officinalis*). J Agric Food Chem 46:4869–4873
90. Lindsey K, Jones MGK (1995) Plant biotechnology in agriculture. Wiley, Chichester
91. Santos-Gomes PC, Seabra RM, Andrade PB, Fernandes-Ferreira M (2002) Phenolic antioxidant compounds produced by in vitro shoots of sage (*Salvia officinalis* L.). Plant Sci 162:981–987
92. Grzegorczyk I, Matkowski A, Wysokińska H (2007) Antioxidant activity of extracts from in vitro cultures of *Salvia officinalis* L. Food Chem 104:536–541
93. Grzegorczyk I, Wysokińska H (2008) Liquid shoot cultures of *Salvia officinalis* L. for micropopagration and production of antioxidant compounds; effect of triacontanol. Acta Soc Pol 77:99–104
94. Grzegorczyk I, Wysokińska H (2009) The effect of methyl jasmonate on production of antioxidant compounds in shoot cultures of *Salvia officinalis* L. Herba Pol 55(3)
95. Kračun-Kolarević M, Dmitrović S, Filipović B, Perić M, Mišić D, Simonović A, Todorović S (2015) Influence of sodium salicylate on rosmarinic acid, carnosol and carnosic acid accumulation by *Salvia officinalis* L. shoots grown in vitro. Biotech Lett 37:1693–1701
96. Hayat S, Ali B, Ahmad A (2007) Salicylic acid: biosynthesis, metabolism and physiological roles in plants. In: Hayat S, Ahmad A (eds) Salicylic acid: a plant hormone. Springer, Dordrecht, pp 1–14 (ISBN-10 1-4020-5184-0)
97. Kang S-M, Jung H-Y, Kang Y-M, Yun D-J, Bahk J-D, Yang J-K, Choi M-S (2004) Effects of methyl jasmonate and salicylic acid on the production of tropane alkaloids and the expression of PMT and H6H in adventitious root cultures of *Scopolina parviflora*. Plant Sci 166:745–751
98. Li W, Koike K, Asada Y, Yoshikawa T, Nokaido T (2005) Rosmarinic acid production by *Coleus forskohlii* hairy root cultures. Plant Cell Tissue Organ Cult 80:151–155
99. Rowshan V, Khosh Khoi M, Javidnia K (2010) Effects of salicylic acid on quality and quantity of essential oil components in *Salvia macrosiphon*. J Biol Environ Sci 4:77–82
100. La Pistelli, Giovannini A, Ruffoni B, Bertoli A, Pistelli Lu (2010) Hairy root cultures for secondary metabolites production. Adv Exp Med Biol 698:167–184
101. Bennett RN, Wallsgrove RM (1994) Secondary metabolites in plant defence mechanisms. New Phytol 127:617–633
102. Mascarello C, Mantovani E, Ruffoni B (2006) In vitro culture of several ornamental and medicinal *Salvia* species. Acta Hortic 723:375–380
103. Fernandes VF, Bezerra L, da S. Feijó EVR, Silva D, Oliveira R, Mielke MS, do B. Costa LC (2013) Light intensity on growth, leaf micromorphology and essential oil production of *Ocimum gratissimum*. Rev Bras Farmacogn [online] 23(3):419–424
104. Arikat NA, Jawad FM, Karam NS, Shibli RA (2003) Micropropagation and accumulation of essential oils in wide sage (*Salvia fruticosa* Mill.). Sci Hortic 100:193–202
105. Azam A, Qian J, Zhang B, Xu C, Chen K (2013) Citrus leaf volatiles as affected by developmental stageand genetic type. Int J Mol Sci 14(9):17744–17766
106. Croteau R, Felton M, Karp F, Kjonaas R (1981) Relationship of camphor biosynthesis to leaf development in sage (*Salvia officinalis*). Plant Physiol 67:820–824

107. Kumar A (2015) Metabolic engineering in plants. In: Bahadur B, Venkat Rajam M, Sahijram L, Krishnamurthy K (eds) Plant biology and biotechnology. Springer, New Delhi
108. Petersen M, Hausler E, Meinhard J, Karwatzki B, Gertlowski C (1994) The biosynthesis of rosmarinic acid in suspension cultures of *Coleus blumei*. Plant Cell Tiss Org Cult 38:171–179
109. Kim KH, Janiak V, Petersen M (2004) Purification, cloning and functional expression of hydroxyphenylpyruvate reductase involved in rosmarinic acid biosynthesis in cell cultures of *Coleus blumei*. Plant Mol Biol 54:311–332
110. Eberle D, Ullmann P, Werk-Reichhart D, Petersen M (2009) CDNA cloning and functional characterization of CYP98A14 and NADPH:cytochrome P450 reductase from *Coleus blumei* involved in Rosmarinic acid biosynthesis. Plant Mol Biol 69:239–253
111. Janiak V, Petersen M, Zentgraf M, Klebe G, Heine A (2010) Structure and substrate docking of a hydroxy(phenyl)pyruvate reductase from the higher plant *Coleus blumei* Benth. Acta Crystallogr D Biol Crystallogr 66:593–603
112. Petersen M, Alfermann AW (1988) Two new enzymes of rosmarinic acid biosynthesis from cell cultures of *Coleus blumei*: hydroxyphenylpyruvate reductase and rosmarinic acid synthase. Z Naturforsch C: Biosci 43:501–504
113. Häusler E, Petersen M, Alfermann AW (1991) Hydroxyphenylpyruvate reductase from cell suspension cultures of *Coleus blumei* Benth. Z Naturforsch 46(c):371–376
114. Wenping H, Yuan Z, Jie S, Lijun Z (2011) De novo transcriptome sequencing in *Salvia miltiorrhiza* to identify genes involved in the biosynthesis of active ingredients. Genomics 98:272–279
115. Chen H, Chen F, Zhang Y-L, Song J-Y (1999) Production of lithospermic acid B and rosmarinic acid in hairy root cultures of *Salvia miltiorrhiza*. J Ind Microbiol Biotechnol 22(3):133–138
116. Zhang Y, Song J, Qi J, Lu G (1997) The plant regeneration of *Salvia miltiorrhiza* Bge. Transformed by Agrobacterium. Zhongguo Zhong Yao Za Zhi 22(5):274–275 (in Chinese)
117. Yan Y, Wang Z (2007) Genetic transformation of the medicinal plant *Salvia miltiorrhiza* by *Agrobacterium tumefaciens*-mediated method. Plant Cell Tiss Organ Cult 88:175
118. Luwańska A, Wielgus K, Seidler-Łożykowska K, Lipiński D, Słomski R (2017) Evaluation of *Agrobacterium tumefaciens* usefulness for the transformation of sage (*Salvia officinalis* L.). In: Transgenesis and secondary metabolism, pp 153–176
119. Barberini S (2014) Biotechnological approach in Lamiaceae species for the production of antioxidant and antibacterial compounds. PhD thesis
120. Hücherig S, Petersen M (2012) RNAi suppression and overexpression studies of hydroxyphenylpyruvate reductase (*HPPR*) and rosmarinic acid synthase (*RAS*) genes related to rosmarinic acid biosynthesis in hairy roots cultures of *Coleus blumei*. Plant Cell Tiss Org Cult. https://doi.org/10.1007/s11240-012-0277-8

Biosynthesis of *Salvia* Specialized Metabolites and Biotechnological Approaches to Increase Their Production

Vincenzo D'Amelia, Alessandra Ruggiero, Valentina Tranchida-Lombardo, Antonietta Leone, Marina Tucci and Teresa Docimo

Abstract Aromatic *Salvia* species are particularly valuable for providing several bioactive compounds used as food additives, pigments, cosmetics, perfumes and fine chemicals. Within the Lamiaceae family, the *Salvia* genus, with more than 900 species, biosynthesizes a plethora of beneficial metabolites including terpenes, steroids and polyphenols. The whole plant can be considered a factory of bioactive compounds, but plant cell and tissue cultures are also an attractive sustainable alternative to cultivation. *Salvia* cell cultures can readily be initiated from different explants, including leaves, roots, stems, petioles, anthers and seedlings; however high metabolites accumulation in plant tissue and cell culture is a prerequisite for massive production of these bioactive compounds. In this chapter, the occurrence and tissue distribution of specialized metabolites in several *Salvia* species, especially flavonoids and diterpenoids, will be reviewed along with recent advances in the understanding of biosynthetic pathways as well as regulatory mechanisms leading to their biosynthesis. We will focus on the recent biotechnological approaches aimed at enhancing the final biomass and metabolite accumulation in *Salvia* cell and tissue cultures. Advances in metabolic engineering strategies will be also summarized, reporting relevant and successful results and potential pitfalls, in order to provide valuable perspectives for developing cell and tissue cultures as a reliable and standardized biomass platform for the extraction of *Salvia* bioactive metabolites.

Keywords *Salvia* · Bioactive specialized metabolites · Biosynthetic enzymes Biotechnology · Metabolic engineering

A. Ruggiero · V. Tranchida-Lombardo · M. Tucci · T. Docimo (✉)
Institute of Biosciences and Bioresources, Research (CNR-IBBR),
National Research Council of Italy, Portici, NA, Italy
e-mail: teresa.docimo@ibbr.cnr.it; teresdocimo@gmail.com

V. D'Amelia
Department of Agricultural Sciences, University of Naples "Federico II",
Portici, NA, Italy

A. Leone
Department of Pharmacy, University of Salerno, Fisciano, SA, Italy

V. Georgiev and A. Pavlov (eds.), *Salvia Biotechnology*,
https://doi.org/10.1007/978-3-319-73900-7_7

Abbreviations

4C-DHPL	4-coumaroyl-3′4′-dihydroxyphenyllactic acid
4CL	4-coumarate-CoA ligase
AtPAP1	Arabidopsis Production of Anthocyanin Pigment
AtPIF5	Phytochrome-interacting factor 5
BA	6-benzylaminopurine
BABA	β-Aminobutyricacid
C4H	Cinnamic acid 4-hydroxylase
CcCls	*Cistus creticus* copal-8-ol diphosphate synthase
CGA	Chlorogenic acid
CPS	Copalyl diphosphate synthase
CT	cryptotanshinone
CYP	Cytochrome P450
DHPL	3,4-dihydroxyphenyllactic acid
diTPS	Diterpene synthases
DMAPP	Dimethylallyl diphosphate
DW	Dry weight
FPP	(E,E)-farnesyl diphosphate
GGPP	(E,E,E)-geranylgeranyl diphosphate
GPP	geranyl diphosphate
HPPR	Hydroxyphenylpyruvate reductase
IAA	Indole-3-acetic acid
IBA	Indole-3-butyric acid
IPP	Isopentenyl diphosphate
KSL	Kaurene synthase-like
KSs	Ent-kaurene synthases
MEP	2-C-methyl-D-erythritol-4-phosphate
MJ	Methyl jasmonate
MVA	Mevalonate pathway
NAA	Naphthaleneacetic acid
PAL	Phenylalanine ammonia-lyase
RA	Rosmarinic acid
RAS	Rosmarinic acid synthase
SsSS	*S. sclarea* sclareol synthase
TAT	Tyrosine aminotransferase
TDZ	Thidiazuron
TF	Transcriptional factors
WHO	World Health Organization
YE	Yeast extract

1 Introduction

Plants produce an extraordinary plethora of diverse specialized metabolites, whose complexity is even enriched by their extraordinary stereo-chemical variation [1]. These specialized metabolites, also referred to as natural products, have been named as secondary because they are synthesized after modification of primary ones and they are in low abundance in plants [2]. In spite of their exclusion from primary functions, it has been largely proved that several specialized metabolites play a relevant role in fundamental plant functions, e.g. gibberellins and abscisic acid acting as phytohormones [3], flavonoids involved in auxin transport or acting as developmental regulators in Arabidopsis [4–6], or lignins involved in cell wall strengthening and chemical defense [7, 8]. However, these metabolites are able also to exert different pharmacological or toxicological effects in humans and other animals, as largely evidenced by several studies on plant-based remedies used in folk and traditional medicine. The beneficial action of these metabolites either in food or as disease remedies have been experienced by "trial and error" for thousands of years, testing edibility or health effectiveness, thus contributing to our present knowledge of natural products and of their beneficial functions for human health. This ethno-botanical knowledge has promoted further investigations on the various plant constituents and their different properties, as well as on their efficacy as potential drugs. According to the World Health Organization (WHO), 80% of people still relies on plant-based traditional medicines for primary health care and 80% of the 122 plant-derived drugs are related to their original ethno-pharmacological use [9, 10]. In addition, the majority of anti-cancer drugs used in modern medicine are plant-derived (e.g. paclitaxel, vincristine, vinblastine, camptothecin, irinotecan, topotecan, teniposide and others) [11].

Distinct ecological roles of some specialized metabolites have been suggested by analyzing their occurrence within taxonomically restricted species. Taxa are often characterized by distinct groups of secondary metabolites along with several minor compounds [12]. For example, the tropane alkaloids and calystegines are taxonomic markers for the Solanaceae [13], glucosinolates for Brassicaceae [14, 15] and quinolizidine alkaloids for Leguminosae [16]. Within the Lamiaceae, *Salvia* is a rich source of structurally diverse terpenoids and caffeic acids derivatives. The biosynthesis and accumulation of some *Salvia* metabolites is often tissue-specific, since different metabolites are synthesized in leaves and roots, and some of them are localized in different subcellular compartments, such as gland trichomes, suggesting specialized functions in plants [12].

Salvia has been used for centuries in traditional medicine, for example in China as longevity herb or in Roman ceremonies as a sacred herb. Sage seeds were used in the Middle Ages for clearing vision, from which derives its popular name 'clary' or 'clear eye'. Sage leaves have been traditionally used for tea preparation to treat digestive and circulation disorders and reported as a remedy for calming muscular pain [17]. More recently, herbal extracts and essential oils from different *Salvia* species have been proved to be beneficial remedies for treating disorders of the nervous system, such as depression and cerebral ischemia as well as the Alzheimer's disease [18, 19]. Currently, in several Asian countries the treatment of cerebrovascular diseases is largely based on the use of *S. miltiorrhiza* dried roots, known as Danshen or Tanshen. Currently, more than seven hundred pharmaceutical companies produce Danshen-based preparations, which is the first Chinese herbal medicine approved by the *Food and Drug Administration* for clinical tests in the United States [20].

Despite the numerous therapeutic and health-promoting effects of *Salvia* bioactive metabolites, they are synthesized at very low level in different parts of the plant, preventing in many cases their thorough pharmacological characterization and, ultimately, their industrial exploitation. This pitfall have prompted several efforts toward the development of biotechnological approaches based on cell cultures or hairy roots of *Salvia* species as a source for the extraction of secondary metabolites alternative to plant cultivation, to ensure a constant, stable and standardized biomass as well as to provide a feasible system for boosting their production.

In this chapter, the occurrence and tissue distribution of specialized metabolites in *Salvia* species is reviewed, reporting also the recent advances in the understanding of their biosynthetic pathways. We will review also the recent biotechnological approaches aimed at producing a high biomass and promoting metabolite accumulation, by elicitation or metabolic engineering. Relevant and successful advances and potential pitfalls will be discussed, in order to provide valuable perspectives for designing and developing cell and tissue cultures as a reliable and standardized biomass platform for the extraction of *Salvia* bioactive metabolites.

2 Evolution and Chemotaxonomy in *Salvia* Species

The *Salvia* L. genus is the more species-rich of the Lamiaceae family, with more than 900 species, characterized by a high diversity in plant morphology and pollination biology. This genus underwent remarkable evolutionary divergence in three regions of the world: Central and South America (500 species), central Asia/Mediterranean (250 species), and eastern Asia (90 species) and is separated from other members of the tribe Menthae, as revealed by its unusual staminal structure. Whilst most of the Menthae species has four stamens, *Salvia* species have only two stamens, and the theca on each stamen is separated by an elongate connective. In addition, the staminal structure in *Salvia* species is often associated with an unusual pollination mechanism which makes *Salvia* regarded as a monophyletic genus [21]. The tremendous diversity of staminal structure, floral morphology, and vegetative habit found across the *Salvia* species has led to considerable taxonomic misunderstanding, especially regarding intra-generic delimitations.

Metabolic profiling has revealed both intra- and inter-species variation, which can be used to define chemotaxonomic markers [22] and to distinguish *Salvia* species [23, 24]. This chemical diversity is also recognized by traditional ethno-botanical studies which support the use of different parts of *Salvia* plants for the treatment of specific diseases, on the basis of their peculiar chemical composition [17].

In this regard, the endemic species of Central and South America are particularly rich of diterpenoids, but they do not synthesize alkaloids [17]. Particularly interesting is the Mexican *S. divinorum*, which produces the psychoactive cloredane diterpene salvinorin A. Several abietane diterpenoids are characteristic of Californian species. Stereoisomers and derivatives of carnosic acid and ferruginol (i.e. 16-hydroxycarnosic acid, 10-acetylferruginol, methyl carnosoate) are synthesized in *Salvia* species, such as *S. melissodora* and *S. mellifera* [20]. A similar abundance and variability of terpenes have been identified also in Central Asian or Mediterranean *Salvia* species, and specific terpenoids are also present in African species. To the terpenoid complexity of *Salvia* species contribute also sesquiterpenes and monoterpenes as major constituents of essential oils, together with low-molecular-weight acids, aldehydes, esters and polyphenols [25–27]. Polyphenols, such as rosmarinic acid and flavonoids, are particularly abundant in African indigenous *Salvia* species, which are, in most cases, responsible for the pharmacological properties of these plants [28]. For example, *S. runcinata* is particularly rich in bisabolol and might be used as a natural source of this commercially important sesquiterpene [22, 28]. Noticeably, bioactive abietane diterpenes and caffeic acid derivatives are present in the East Asian *Salvia* species [29], such as *S. miltiorrhiza* roots, which are a valuable source of this class of beneficial compounds.

Such a large chemical variability points out the importance of an in-depth understanding of *Salvia* chemotaxonomy and of the biosynthetic machinery for these specialized metabolites especially when a biotechnological approach is planned to boost the production of desired bioactive compound.

2.1 Bioactive Metabolites in Salvia *Species*

2.1.1 Recent Advances in Understanding Metabolic Pathways

Thanks to their pharmacological effects and therapeutic uses, extensive chemical and pharmacological studies have been conducted in different *Salvia* species to identify bioactive compounds and to uncover their mechanisms of action [30, 31]. In a recent review, [20] reports that, of the 791 identified constituents in *Salvia* species, 545 were diterpenes and the remaining were mostly caffeic acid derivatives. Here we summarize recent advances toward elucidating the biosynthetic pathways of these major representatives of *Salvia* bioactive compounds.

Rosmarinic Acid

Polyphenols, mainly phenolic acids and flavonoids, are the main metabolites found in a water-soluble extracts of several *Salvia* species [20]. Flavonoids are found mostly in the aerial parts of *Salvia* plants, especially in the flowers and leaves, whereas phenolic acids are mainly present in the roots. *S. miltiorrhiza*, *S. chinensis* and *S. yunnanensis* are the species in which these compounds were mainly identified [32]. *Salvia* species, as well as other species of the Lamiaceae family, produce Rosmarinic Acid (RA), an ester of caffeic and 3,4-dihydroxyphenyllactic acid, distributed in all the tissues of *Salvia* plants, as result of either in situ synthesis and/or transport from leaves [33]. Beside this monomer, unique caffeic acid oligomers, called depsides or salvianolic acids, are usually synthesized in different *Salvia* species [20, 32].

RA is a strong antioxidant, with antibacterial and antiviral activity and it has been also proved to be effective in preventing the Alzheimer's disease, among other biological properties [34]. Recently, RA have been also tested as feedstocks for lignocellulosic biofuels, since this hydroxycinnamic ester can be used as monolignol substitutive, thus permitting a more efficient delignification and enzymatic hydrolysis of the plant cell walls [35]. This wide range of uses of RA has prompted several efforts to elucidate its biosynthetic pathway and to uncover potential regulatory circuits in order to design appropriate biotechnological strategies aimed at enhancing its accumulation. RA biosynthesis has been characterized in *Coleus blumei* [36], in *S. miltiorrhiza* and *Lithospermum erythrorhizon* [37, 38]. Two biosynthetic pathways participate in the biosynthesis of this phenolic acid: the phenylpropanoid pathway and the tyrosine-derived pathway. In the phenylpropanoid pathway, phenylalanine is transformed into 4-coumaroyl-CoA by three subsequent enzymatic steps, catalyzed by the phenylalanine ammonia-lyase (PAL), the cinnamic acid 4-hydroxylase (C4H) and the 4-coumarate: CoA ligase (4CL). In the tyrosine-derived pathway, tyrosine is converted to 4-hydroxyphenyllactate by the tyrosine aminotransferase (TAT) and the 4-hydroxyphenylpyruvate reductase (HPPR), leading to the formation of the 3,4-dihydroxyphenyllactic acid (DHPL, or Salvianic acid A). Then, 4-coumaroyl-CoA and DHPL are further coupled by the BAHD enzyme rosmarinic acid synthase (RAS) to form 4-coumaroyl-3′4′-dihydroxyphenyllactic acid (4C-DHPL). 4C-DHPL is oxidized by the CYP98A14, which introduces a 3-hydroxyl group to form RA (Fig. 1). Genome and transcriptional analyses in *S. miltiorrhiza* have indicated that of the different members belonging to the PAL family, *SmPAL1* and *SmPAL3* expression paralleled the RA high content found in roots, suggesting also the roots as the active site for its biosynthesis [39]. Recently, in *S. miltiorrhiza* hairy roots most of the genes encoding the enzymes of the tyrosine pathway were found to be up-regulated in response to different elicitors, which triggered also a significant increase in the content of phenolic acids and suggesting a possible limiting role of this pathway in the biosynthesis of RA [40].

Shared biosynthetic enzymes and correlated production of RA with chlorogenic acid (CGA) [32] led to hypothesize a common evolution of the biosynthetic pathway of these two compounds [36]. This hypothesis has been further supported by a study in which the overexpression of a MYB transcription factor enhanced accumulation of RA in *S. sclarea* hairy roots and CGA in tobacco plants, thus highlighting a possible co-regulation of phenylpropanoid pathway [41] (Fig. 1).

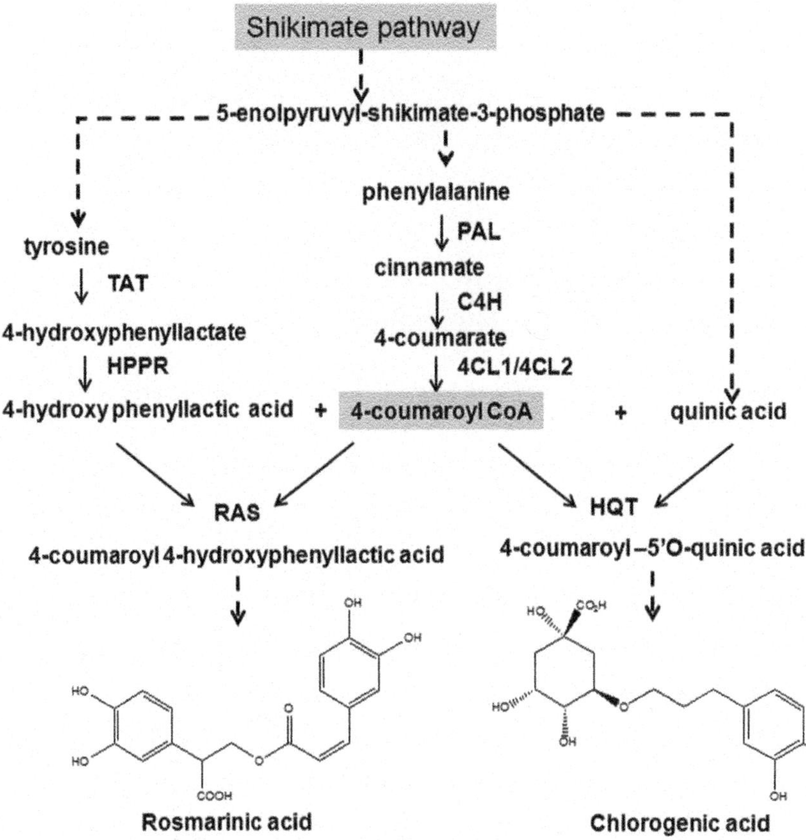

Fig. 1 Pathway to Rosmarinic Acid and Chlorogenic acid (modified from [41])

In addition, it has been proposed that a R2R3MYB TF, homologous to the *At*PAP1, which is a key regulator of the anthocyanin pathway in *A. thaliana*, might control the phenolic metabolism in *Salvia miltiorrhiza* [42].

Sclareol

Essential oils are peculiar constituents of aromatic plants belonging to the *Lamiaceae* family. In *Salvia* species these compounds gained much attention for the wide spectrum of uses as fragrances in perfumery industry, as drug additives in pharmaceutical preparations, or as spices in culinary tradition. Among essential oil constituents, the diterpene sclareol, mainly produced and extracted from cultivated *S. sclarea* plants, is the sole natural precursor for the synthesis and commercial production of ambroxide, highly priced in the fragrance industry. Sclareol has been also reported to possess a promising pharmacological activity as antimicrobial agent against a wide spectrum of bacteria [43]. *S. sclarea* oils are also rich in the monoterpenes linalool and linalyl acetate (Fig. 2).

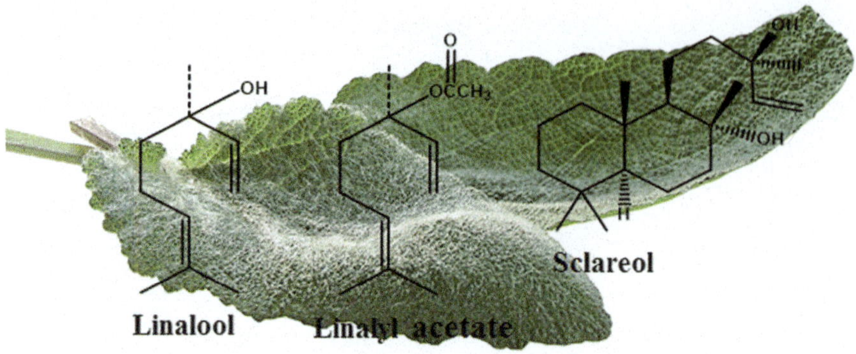

Fig. 2 Major oil constituents of *S. sclarea* leaves: Linalool, Linalyl acetate and Sclareol

The ecological role of essential oils has been widely studied in many plant species [44, 45]. Consistently with their function as attractants to pollinators [46] or as repellant toward herbivores [47], these molecules accumulate mostly in leaves and flowers. However, the variable accumulation and distribution of sclareol within oil gland types, as reported by [48], pointed out that the exact biosynthetic site of the sclareol is still unclear.

Generally, terpene biosynthesis proceeds in plants thorough two pathways: the mevalonate pathway (MVA) in the cytosol and the 2-C-methyl-D-erythritol-4-phosphate (MEP) pathway in the plastids [49, 50]. The main building block of terpenes is an isoprene unit derived from isopentenyl diphosphate (IPP) and dimethylallyl diphosphate (DMAPP). By the action of prenyltransferases on IPP and DMAPP, the up-stream building blocks of terpenes are generated: geranyl diphosphate (GPP) for monoterpene synthesis, (E,E)-farnesyl diphosphate (FPP) for sesquiterpenes and (E,E,E)-geranylgeranyl diphosphate (GGPP) for diterpene synthesis (Fig. 3).

Diterpenes are formed through the cyclization of geranylgeranyl diphosphate (GGPP), catalyzed by diterpene synthases (diTPS). The large diversity of diterpenes is due to two types of distinct cyclization mechanisms, operated by two types of diTPS: the class I and class II diTPS [51]. Sclareol follows the general biosynthesis of bicyclic labdane-type diterpenes, which proceeds via a stepwise ionization and cycloisomerization of *(E,E,E)*-geranylgeranyl diphosphate (GGPP) [20]. In angiosperms, such an enzymatic cascade requires the consecutive activity of two monofunctional diTPSs. A class II diTPS, defined as copalyl diphosphate synthase (CPS), catalyzes the protonation-dependent cyclization of GGPP to form a bicyclic copalyl diphosphate, an intermediate of variable stereochemistry and hydroxylation [52]. Subsequently, class I diTPSs facilitate the ionization of the diphosphate group and often the additional cyclization and rearrangement reactions. The class I diterpene cyclases are named ent-kaurene synthase-like and produce miltiradiene, which is the precursor of labdane type diterpenoids (Fig. 4).

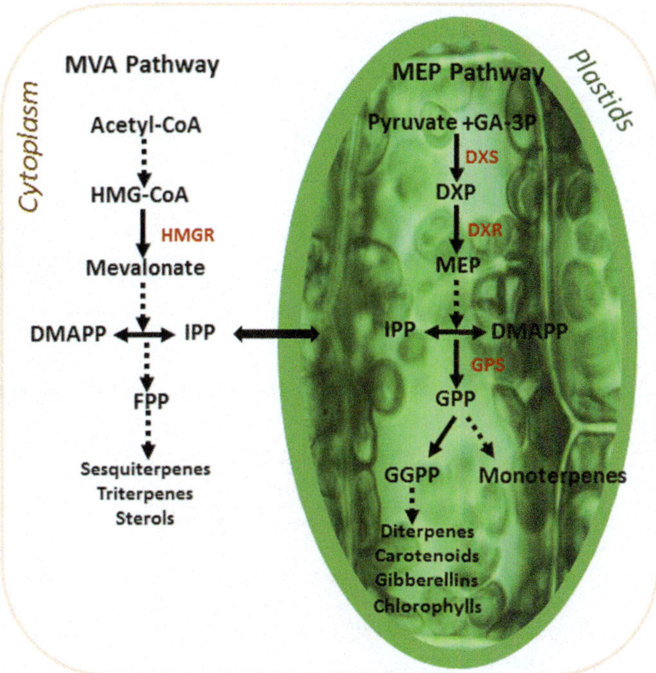

Fig. 3 Isoprenoid biosynthesis in plants. Abbreviations: MVA—mevalonic acid; HMG-CoA—hydroxymethylglutaryl CoA; HMGR—HMG-CoA reductase; DMAPP—dimethylallyl diphosphate; IPP—isopentenyl diphosphate; FPP—farnesyl diphosphate; MEP—methylerythritol 4-phosphate; GA-3P glyceraldehyde 3-phosphate; DXS—d-xylulose 5-phosphate synthase; DXP—deoxyxylulose-5-phosphate; DXR—DXP reductoisomerase; MEP—methylerythritol 4-phosphate; GPS—geranyl phosphate synthase; GPP—geranyl diphosphate; GGPP—geranylgeranyl diphosphate

Due to the commercial relevance of sclareol, the elucidation of its biosynthesis has received increasing interest [53, 54] and was recently elucidated through the transcriptomic analysis of the *S. sclarea* flower calyx, which has led to the identification of two diterpene synthases responsible for sclareol biosynthesis [54, 55]. Sclareol is synthesized in a two-step enzymatic reaction catalyzed by these two monofunctional diTPSs (*Ss*LPPS and S*s*SS), thus following, as expected, the usual pathway for diterpene formation in angiosperm. Moreover the plastidial localization of these enzymes has suggested that sclareol biosynthesis might occur in the plastids of flower calyces. Most likely sclareol is synthesized in capitates trichomes and then secreted on the surface of the inflorescences [54, 55].

Carnosic Acid

Carnosic acid is another prominent labdane-related diterpenoid, used in food and cosmetic industry, which has also relevant antioxidant, anti-inflammatory and anticancer properties [56–58]. Carnosic acid was extracted for the first time about

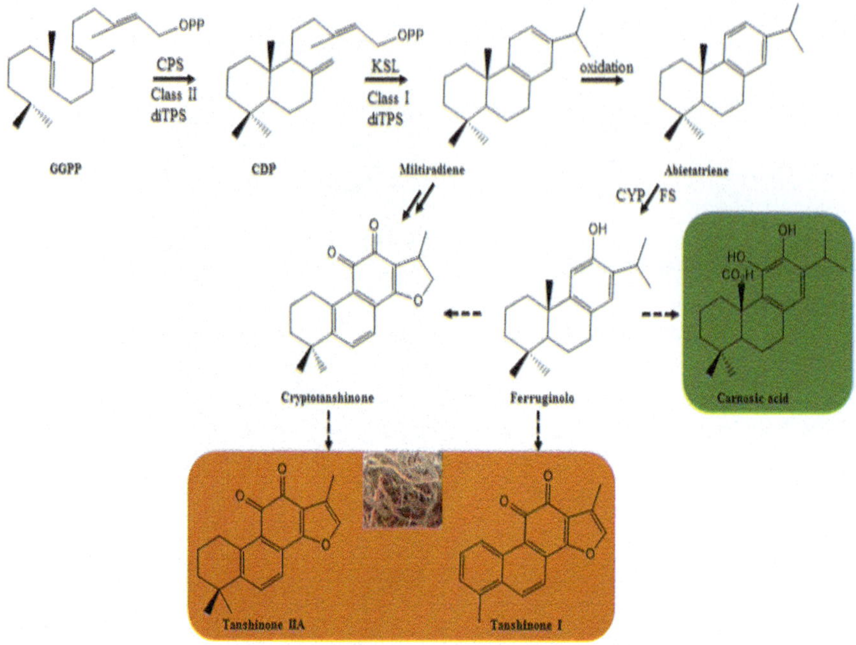

Fig. 4 Schematic representation of labdane diterpene pathway leading to the biosynthesis of carnosic acid and tanshinones

70 years ago from different *Salvia* species and its chemical structure was resolved about 20 years later. This diterpene is synthesized in plants of seven genera of the Mentheae tribe, including *Salvia*, *Rosmarinus*, *Lepechinia*, *Oreganum*, *Thymus* and *Hyssopus*, and in *Ocimum* of the Ocimeae tribe [59]. This wide spectrum of biological properties has prompted several studies aimed at elucidating the carnosic acid biosynthetic pathway in Lamiaceae. Carnosic acid is mostly present in the aerial parts of the plant, and its biosynthesis in leaves has been thoroughly studied in *S. fruticosa* and *R. officinalis* [59–61]. It has been proposed that the glandular trichomes are the main organs where the first or all steps of the carnosic acid biosynthesis occur, from which a precursor, or carnosic acid itself, seems to be transported to the leaves [62]. Recently, by analyzing the transcriptome of glandular trichomes of *S. fruticosa* and *R. officinalis*, key biosynthetic genes of carnosic acid pathway have been identified, including two diterpene synthases and a Cytochrome P450 (CYP) gene involved in abietatriene hydroxylation (Fig. 4). The two diterpene syntheses, namely *SfCPS* and *SfKSL* from *S. fruticosa*, are highly similar to genes encoding the copalyl diphosphate synthase (CPS) and kaurene synthase-like (KSL). The CYP gene, named *SfFS*, is presumably involved in catalyzing ferruginol formation [62]. Interestingly, as previously described for *S. sclarea* sclareol synthase (*Ss*SS) in sclareol biosynthesis, *Sf*KSL along with *Cistus creticus* copal-8-ol diphosphate synthase (*Cc*Cls) and 8-hydroxy-copalyl diphosphate

synthase from *Nicotiana tabacum* (*Nt*CPS2) [63, 64], are monofunctional class II diTPS lacking the usual 'internal' gamma domain, which plays a role in class II cycloisomerization reaction. As well described [65], these genes are examples of gene duplication and neo-functionalization, where new genes, arisen from ancestral diTPS result to be no longer necessary for gibberellins biosynthesis and participate in the biosynthesis of other specialized secondary metabolites. In this regard, the distinctive hydroxylation in the *ortho*-position of abietatriene to form carnosic acid might strongly contribute to its antioxidant properties, and, along with its organelle distribution, suggests a protective role of phenolic diterpenoid from oxidative stress in chloroplasts and various organelle membranes.

Tanshinone

Tanshinone I, tanshinone IIA, cryptotanshinone and dihydrotanshinone I are the main diterpene quinones constituents of *S. miltiorrhiza* roots [66], although they are minor compounds in other *Salvia* species e.g. *S. trijuga*, *S. nipponica* and *S. glutinosa* [20]. These bioactive molecules exhibit powerful biological activities, such as heart-protective, anti-ischemic, antioxidant, antibacterial, and antitumor properties [67–69]. Similarly, tanshinone-related molecules, i.e. aethiopinone and salvipisone, occurring mostly in *S. sclarea* and *S. aethiopis* roots [70, 71], are receiving great interest, although their biosynthesis is not fully resolved.

The increasing clinical demand of tanshinones in Asian countries makes the sole cultivation of *S. miltiorrhiza* plants an insufficient source of these compounds and therefore, several biotechnological approaches have been proposed to obtain higher amounts of this class of compounds (see paragraphs 3 and 4). Several recent transcriptomic, metabolomics and genetic studies have been focused on elucidating the tanshinone pathway, in order to identify biosynthetic and regulatory genes controlling the synthesis of this class of diterpenes. As reported for carnosic acid, tanshinone biosynthesis requires a pair of cyclization to produce the abietane miltiradiene, reaction catalyzed by a KSL homologous to the ent-kaurene synthases (KSs) involved in GA biosynthesis [72] (Fig. 4). However, the hypothesis that tanshinone biosynthesis might proceed through carnosic acid still needs to be proven.

An interesting recent work, based on combined biochemical and genetic studies, has demonstrated that a functional divergence exists among the five isolated *SmCPSs* in *S. miltiorrhiza*. While *SmCPS3* appears to be inactive, the other *SmCPSs* seem to take part in separate diterpenoid pathways [73]. In particular, the presence of distinct *SmCPS1* and *SmCPS2* suggest possible distinct roles in the roots versus the aerial tissues [73]. *SmCPS1* and *SmKSL* are abundantly expressed in roots, and RNAi silencing of *SmCPS1* caused a decrease in the tanshinones content in the roots, whereas their production was unaffected in the leaves. *SmCPS2* is predominantly expressed in leaves, indicating further that *SmCPS2* mediates tanshinone biosynthesis in the aerial organs of *S. miltiorrhiza* plants. Rather than reflecting redundancy, these findings suggest the existence of similar biochemical activity of SmCPS1and SmCPS2 in different plant organs, where the labdane-related diterpenoids may play different roles and have diverse function in plant development and adaptation [73]. This is another example of neo-functionalization of CPS enzymes in Lamiaceae, by gene duplication, leading to *SmCPS1* and *SmCPS2*. Further localization and up-regulation studies of these two genes in *S. miltiorrhiza* enabled to establish that tanshinone biosynthesis occur in root periderm cells [74]. These findings represent a breakthrough for future genetic engineering strategies aimed at boosting the tanshinone production.

3 Tissue Cultures for Secondary Metabolites Production

As mentioned before, since *Salvia* plants are an important source of food additives as well as of pharmacologically interesting metabolites, recently the demand of these products is strongly increasing. The supply of biomass from cultivation might be strongly limited by unpredictable biotic and abiotic constraints and, in this regard, plant cell and tissues cultures are a valuable alternative source for a controlled and constant extraction and production of these molecules.

There are several different advantages in extracting valuable secondary metabolites from plant cell and tissue cultures, rather than from cultivated plants. These include: (a) production can be more reliable, simpler, and more predictable; (b) isolation of the phytochemical can be rapid and efficient, as compared to extraction from complex plant matrices; (c) quite often the metabolic profile of plant cells mirror that of the whole plant; (d) interfering compounds that occur in the field-grown plants can be avoided in cell and tissue cultures; (e) tissue and cell cultures can yield defined standard phytochemicals in large volumes; (f) cell and tissue culture can be generated on a continuous year-round basis without variation in quality and quantity of metabolites of interest due to seasonal constraints; (g) tissue and cell cultures are also suitable systems to study and elucidate factors controlling metabolite biosynthesis.

3.1 Cell and Hairy Root Cultures as Source for the Production of Salvia *Bioactive Metabolites*

Due to the high demand of *Salvia* bioactive metabolites, cell and hairy roots cultures have been widely exploited as source of their extraction and production. However, several studies report that *Salvia* cell suspensions often failed to produce high amount of secondary metabolites and were characterized by slow growth, low biomass density, high tendency to form aggregates and higher degree of genetic and metabolic instability during sub-cultivation cycles [75]. The low metabolite production has been mostly associated to the absence of cell differentiation and, hence, of specific tissues, necessary for biosynthesis or sequestration [76, 77]. This drawback may, in most cases, be circumvented by using molecules, known as

Table 1 *Salvia* species and metabolites produced in several tissue culture types

Salvia species	Tissue culture type	Metabolites	References
Salvia officinalis	Cell culture Hairy roots	*Rosmarinic acid*	[78, 80]
Salvia fruticosa	Callus, cell and root cultures	*Rosmarinic acid*	[79]
Salvia officinalis	Shoot cultures	*Carnosic acid*	[80]
Salvia miltiorrhiza	Cell cultures	*Criptotanshinone* *Ferruginol*	[88, 89]
	Roots culture	*Tanshinone*	[90]
	Ti/transformed cell line	*Tanshinone*	[91]
Salvia miltiorrhiza	Hairy roots	*Tanshinone* *Criptotanshinone* *Rosmarinic acid*	[97]
	Hairy roots	*Tanshinone* *Ferruginol*	[98]
Salvia sclarea	Hairy roots	*salvipisone, ferruginol, aethiopinone* *1-oxoaethiopinone*	[93]
Salvia officinalis	Hairy roots	*Rosmarinic acid*	[99]

elicitors, which mimic external stimuli and induce transcriptional activation of biosynthetic and regulatory genes of a specific pathway. Thus, the use of elicitors has become one of the most common approaches to induce the accumulation of a bioactive secondary metabolite in plant cells and tissue cultures. In the next paragraphs, we report and discuss the results thus far achieved for the production of the main class of secondary metabolites of *Salvia* species in cell suspensions and hairy root cultures (Table 1).

Phenolic Acids

Successful results have been reported for RA production in *Salvia* cell cultures. Since RA biosynthesis occurs in all plant tissues, it does not require cell differentiation and the accumulation in *Salvia* cell cultures may result even higher than in the whole plant. [78] reported an extremely high level of RA production in *S. officinalis* cell culture (reaching 6.4 mg/l of suspension culture), though this amount has not been obtained in other similar studies. More recently, RA accumulation in *S. fruticosa* callus, cell suspension and root cultures was studied by [79]. The largest callus (0.79 g) was obtained from leaves cultured on MS medium containing 6.9 mM thidiazuron (TDZ) and 3 mM indole-3-acetic acid (IAA). Callus culture for 5 weeks resulted in maximum callus growth and RA yield (2.12 mg/100 mg dw). Roots cultured with 4% (w/v) sucrose exhibited the highest fresh (3.06 g) and dry (0.28 g) weight and produced the highest amount of RA (2.62 mg/100 mg dry weight). In a comparative study, using different in vitro systems with different levels of differentiation, it was found that *S. officinalis* hairy roots and cell suspensions ensured the highest RA production, thus highlighting that differentiation was not necessary for biosynthesis of this specialized metabolite [80] (Table 1).

Diterpenes

Opposite to RA, organogenesis is necessary for the induction of volatile substances in various tissue cultures of plants of species belonging to the Laminaceae. Tissue differentiation is also critical in diterpene synthesis [81]. No volatiles were accumulated in undifferentiated callus cultures of *Origanum vulgare, Rosmarinus officinalis* and *S. officinalis* [82]. Similarly, no monoterpenes could be detected in undifferentiated tissue cultures of *Lavandula angustifolia* and *R. officinalis* [83, 84]. Regarding production of carnosic acid in *S. officinalis*, [80] demonstrated that only shoot cultures were able to produce this diterpene. This finding is in accordance with previous results, where the differentiation level did not affect the production of RA, whereas the biosynthesis of carnosic acid and carnosol was negatively influenced by the lack of morphological differentiation [85].

So far, several studies have been conducted using calli and/or cell suspension cultures of Danshen (*S. miltiorrhiza* Bunge) to produce diterpenes [75]. Micropropagation of *S. miltiorrhiza* plantlets was first achieved two decades ago through in vitro organogenesis [86]. Calluses were induced from young leaf and petiole explants on Murashige and Skoog medium, with appropriate concentrations of phytohormones. In particular, 4 mg/L of 6-benzylaminopurine (BA) and 3 mg/L of naphthaleneacetic acid (NAA) were the best concentration for callus induction [87]. In an early study, *S. miltiorrhiza* seedlings were used to induce a cell line capable of producing significant amounts of cryptotanshinone (CT) and ferruginol [88]. Furthermore, medium composition was proved to exert a profound effect on the production of CT and ferruginol, since medium optimization led to a much higher accumulation of CT (110 vs. 87.4 mg/L) and slightly lower ferruginol (53.2 vs. 69.3 mg/L). In addition, the ferruginol content in cells cultured under light was only 56% of that cultured in the dark [89]. Similarly, medium composition was found to be crucial also for higher terpenes production in *S. miltiorrhiza* root cultures: roots, subcultured in liquid Gamborg B5 medium supplemented with 0.5 mg/L indole-3-butyric acid (IBA) for eight weeks, had the highest tanshinone content, about 80 mg/g dry weight (dw), six times higher than the amount produced in the roots of the parent plant [90]. Another way to produce higher content of tanshinone was based on infecting *S. miltiorrhiza* plantlets with the *A. tumefaciens* strain C58 and culturing the Ti-transformed cell lines, which accumulated ten times more total tanshinones than field-grown Danshen roots [91].

Since many bioactive metabolites are synthesized and/or accumulated in the roots of several *Salvia* plants it is worth to notice that the establishment of hairy root culture has been widely used as an alternative source for their extraction. Hairy roots are induced by infection of wounded plant tissues with the soil bacterium *Agrobacterium rhizogenes* [92]. Beside the ability to synthesize desired compounds, hairy roots cultures have the advantages of being genetically and biochemically stable and of showing rapid and hormone-independent growth rates. For these reasons, metabolites production may be significantly higher in hairy roots than the content in roots of the field-growing plants or in cell suspensions [93]. In vitro grown young *Salvia* plants seemed to be the best sources of explants for transformation with *A. rhizogenes* [94] in comparison with leaves from in vivo grown plants [95], likely because the transformation of in vivo *Salvia* leaves is accompanied by an extensive release of phenolic compounds, which negatively affects transformation [96]. In addition, the efficiency of hairy root induction also depends on the *Agrobacterium* strain [95]. The first reported hairy root culture of *S. miltiorrhiza* was obtained by infecting plant explants with different *A. rhizogenes* strains [97]. Hairy roots have been mostly induced by infection with *A. rhizogenes* ATCC 15834, which showed high transformation efficiency in comparison with other strains, but did not result in higher metabolites production. For example, *A. rhizogenes* ATCC 15834 transformed *S. miltiorrhiza* hairy roots produced a small amount of tanshinone and 0.04 mg/L of CT, but a higher level of RA (1.24% w/w). It is widely proved that composition of the culture medium strongly affected metabolites production. When the hairy roots were cultured in liquid MS medium without ammonium nitrate, seven major tanshinone and ferruginol compounds were detected in the hairy roots and the culture medium. The dry weight of hairy roots increased by 22-fold in 20 days, while the total tanshinone content of roots reached 43 mg/g dw (20 mg/g dw as CT) [98]. Similarly, the total diterpenoid content of *S. sclarea* was nine-fold higher in hairy roots than in the roots of field-grown plants [93] and *S. officinalis* hairy roots cultures were found to accumulate higher amount of rosmarinic acid (31 mg/g dw) than mature plant roots [99] (Table 1).

Further increase in metabolite production has been reported for Salvia hairy roots treated with elicitors, as reported below.

3.2 *Enhanced Production of* Salvia *Bioactive Metabolites by Elicitation*

Secondary metabolites contribute partially to the plant defence against microbial pathogens and environmental stress, and therefore biotic and abiotic elicitors can stimulate their production in plant tissue cultures. Elicitors refer to agents inducing plant defence responses, such as the accumulation of phytoalexins (antimicrobial secondary metabolites) in plant cells [100].

In *S. miltiorrhiza* hairy root cultures, the tanshinones are normally produced at low yields, and different well-documented elicitation strategies have been applied to boost the synthesis of this interesting class of compounds, such as the treatments with biotic and abiotic elicitors. For example, *S. miltiorrhiza* hairy roots elicited with Yeast extract (YE), accumulated a reduced constitutive level of RA, whereas the level of CT was enhanced greatly, although the increase was associated to a concomitant decrease in biomass accumulation, which indicated that elicitation partially switched the cells from primary to secondary metabolism [101]. YE showed a dose-dependent stimulating effect on tanshinone production in *S. miltiorrhiza* hairy root cultures, most significant when added at 100 mg/L, with a tanshinone yield about 3.8-fold higher than the amount found in unelicited hairy roots [102]. The non-protein amino acid β-aminobutyricacid (BABA), another well-known inducer of plant defence responses, could strongly potentiate or enhance YE-induced tanshinone production in *S. miltiorrhiza* hairy root cultures [103]. The pre-treatment of hairy roots with BABA for 3 days before the addition of YE resulted in a tanshinone yield much higher than YE treatment alone, possibly because the BABA pre-treatment primed root cells to be more sensitive and responsive to elicitation.

Inorganic ions or abiotic stressors, including heavy metals and hyperosmotic stress, have been also effective in stimulating tanshinone production in *S. miltiorrhiza* hairy root cultures. [104] supplied silver ion Ag^{+} (30 μM) to *S. miltiorrhiza* hairy root cultures 2–3 days before the stationary growth phase, increasing the total tanshinone content of roots by two fold over the control but suppressing root growth. The pre-treatment of hairy roots with Ag^{+} for 24–48 h also potentiated the YE-induced diterpene production (mainly CT) [105]. Another heavy metal ion, La^{3+}, applied to *S. miltiorrhiza* hairy root cultures, improved tanshinones production by about 50% (up to 27 mg tanshinones/g) [106]. Hyperosmotic stress is another effective way to elicit secondary metabolism in hairy root cultures. Sorbitol added to the culture medium of *S. miltiorrhiza* hairy roots at 50–70 g/L (1.5–2 times osmolarity of the control MS medium) increased the total tanshinone content of roots by 4.5-fold vs the control untreated roots [107]. A particularly interesting compound for inducing hyperosmotic stress is sucrose, because it can induce osmotic stress and simultaneously serve as the main carbon source for the cultured plant cells [108].

Although microbial elicitors in the form of dead cells fragments have been used since early studies, more recently treatment optimization with biological agents has greatly improved tanshinone production in *S. miltiorrhiza* [109, 110]. Inoculation at the beginning of the cultivation process of *S. miltiorrhiza* hairy roots, with various concentrations of *B. cereus* live bacteria, had a very strong positive effect on tanshinone accumulation (12- to18-fold increase), but inhibited significantly root growth (up to 50% less biomass), so that the overall yield remained low [109]. After optimization of the inoculation time, a 12-fold increase in the tanshinones yield (from 0.20 to 2.67 mg/g dw) was achieved by adding a suspension of *B. cereus* live bacteria (2.5% v/v) to an 18-day-old *S. miltiorrhiza* hairy root culture [110]. [91] used autoclaved mycelial liquid of the plant pathogenic fungus *Armillaria mellea* as elicitor to crown gall cell cultures of *S. miltiorrhiza*, achieving a maximum tanshinone yield of 147 mg/L, 9-fold over the control.

Since elicitation can negatively influence growth [111], optimization of elicitor concentration as well as time point of addition might be crucial [107]. [93] reported that concentration and exposure time to the elicitor were indeed factors that strongly affected the diterpenoid production of *Salvia sclarea* hairy roots. The highest diterpenoid accumulation was achieved, by applying 125 μM methyl jasmonate

(MJ) to 23-day-old hairy roots for seven days in a bioreactor culture. Interestingly, these studies highlighted that MJ was able to efficiently stimulate the bioactive diterpenoids accumulation in shake flask as well as in sprinkle bioreactor cultures. Anyway, MJ elicitation of *S. sclarea* hairy roots in a bioreactor was more advantageous probably due to a higher root biomass and, consequently, exposing more receptors to the elicitor. Recently, also [112] used MJ as elicitor to examine the regulation of the tanshinone biosynthesis pathway in *S. miltiorrhiza* hairy roots. In this integrative study on gene transcriptional activation and tanshinone biosynthesis, tanshinone accumulation resulted dramatically stimulated by MJ and all the tested genes encoding related enzymes involved in the early steps of tanshinone biosynthesis were coordinately induced by MJ. Nevertheless, in the same study was evaluated also the effect of YE or Ag^{+}, which either alone or in combination determined the induction of tanshinone biosynthesis, probably due to the induction of late unidentified specific tanshinone biosynthetic genes. Therefore, a better understanding the whole tanshinone biosynthetic pathway could provide useful information to promote its production in *S. miltiorrhiza* hairy root culture by a synergic combination of biotic and abiotic elicitors.

The elicitation studies in different *Salvia* species coupled with transcriptome analysis have revealed that the boosting effects of different elicitors activity is based on their ability to up-regulate the expression of biosynthetic genes or regulatory genes. Therefore, great efforts have been aimed at improving the biosynthetic potential of *Salvia* in cell and tissue cultures by over-expressing, up-regulating, silencing or introducing new genes that encode rate limiting enzymes or key transcription factors responsible for secondary metabolite formation [113].

4 Metabolic Engineering of *Salvia* Secondary Metabolites

4.1 Molecular Approaches for Metabolites Production in Salvia

The molecular machine underpinning the biosynthesis of secondary metabolites is spatio-temporally orchestrated by regulatory and biosynthetic genes [114]. This may limit the possibility to exploit the *Salvia* biochemical biodiversity using cell culture methodology. Strategies of metabolic engineering are, in many cases, the main tool to enhance the synthesis of target compounds in cell cultures. Hence, the identification and characterization of the biosynthetic and regulatory genes involved in the production of a compound as well as of its specific isoforms are fundamental for any metabolic engineering applications [115].

In the following paragraphs we will report and discuss the most recent results obtained on the biosynthetic and regulatory genes that can be used in biotechnological strategies to boost the synthesis of the main class of compounds that characterizes *Salvia* ssp. An important contribution has been provided by the recent data available from the sequencing of the *S. miltiorrhiza* genome [116].

4.2 Candidate and Engineered Genes to Increase Polyphenols Biosynthesis in Salvia Species

The RA biosynthetic pathway is under strict developmental and environmental control, which may influence the accumulation of RA and its derivatives [117]. As mentioned, the biosynthesis of RA and its derivatives is quite complex since it requires the contribution of both the phenylpropanoid pathway (through hydroxycinnamic acid branches) and the tyrosine-derived pathway. The phenylpropanoid pathway is highly controlled at the transcriptional level [118], and several transcription factors have been identified in many species that control expression of the genes encoding PAL, C4H, and 4CL, which are essential to produce phenolic acids, or other fundamental cell components as lignin. Different transcriptional factors (TF) and other protein regulators have been characterized and may be used to enhance the precursor-flow of the bioactive *Salvia* phenolic acids (Table 2).

Table 2 Candidate and engineered biosynthetic genes isolated from different species (gene source) and used to increase polyphenol biosynthesis in different *Salvia* species

Structural genes	Gene source	Target compounds	References
C4H	*Salvia miltiorrhiza*	RA and SalB	[117]
TAT			
HPPR			
HPPD			
CCR	*Salvia miltiorrhiza*	RA and SalB	[122]
COMT			
Regulatory genes			
AmDELILA	*Antirrhinum majus*	Flavonoids/phenolic acids	[119]
AmROSEA1			
AtPAP1	*Arabidopsis thaliana*	Flavonoids/phenolic acids	[122]
OsMyb4	*Oryza sativa*	Phenolic acids	[41]
SmMYB1	*Salvia miltiorrhiza*	Phenolic acids	[123]
SmMYB13			
SmMYB16			
SmMYB35			
SmMYB62			
SmMYB63			
SmMYB79			
Sm008	*Salvia miltiorrhiza*	Phenolic acids	[124]
Sm166			

The snapdragon *AmDELILA* (Am*DEL*) and *AmROSEA1* (*AmROS1*), belonging to R2R3 MYB and bHLH family, respectively, have been largely used to activate flavonols as well as the general phenylpropanoid pathway in different plant species [119–121] constitutively co-expressed *AmDEL* and *AmROS1* in *S. miltiorrhiza*, enriching the plant with both RA or Salvianolic acid B (*Sal B*) and total flavonoids. Similarly to *AmDEL* and *AmROS1* activity, the overexpression of Arabidopsis R2R3 MYB *Arabidopsis Production of Anthocyanin Pigment 1* (*AtPAP1*), coupled with the silencing of the lignin key enzymes *SmCCR* and *SmCOMT*, induced the accumulation of both RA and *Sal B* [122]. In this combined genetic manipulation strategy the metabolic flux was completely redirected to produce phenolic acids, since the competitive lignin branches did not consume the increased metabolic precursors obtained through *AtPAP1* overexpression [122]. This system can be successful applicable to cell cultures since reduced content of lignin is potentially not detrimental for in vitro cultivated cells.

Several *cis* regulators of *Salvia* phenolic acids have been identified by whole genome sequencing of *Salvia* spp. Genome-wide comparative analysis allowed to identify seven R2R3 MYB TFs (*Sm*MYB1, *Sm*MYB13, *Sm*MYB16, *Sm*MYB35, *Sm*MYB62, *Sm*MYB63 and *Sm*MYB79) which were identified as putative regulators of phenolic acids production in *S. miltiorrhiza* [123]. These TFs candidates may prove useful, alone or in combination, to re-direct metabolic flux towards the desired phenylpropanoid branches. In a similar study, [124] identified two TFs, *Sm*008 and *Sm*166, belonging to the AP2/ERF family, which may theoretically bind the *SmHPPR*, and *SmRAS* promoters of the RA pathway. Overexpression of heterologous rice MYB TF was also proved to be effective in boosting RA accumulation. When *Os*MYB4 TF was overexpressed in *S. sclarea* hairy roots and tobacco plants a higher amount of RA and CGA, respectively, was obtained. The enhanced accumulation of these species-specific phenolic acids suggested that several genes belonging to the phenylpropanoid pathway were coordinately up-regulated by this MYBTF [41], thus confirmed an evolutionary conserved regulatory role of this TF in phenylpropanoid biosynthesis [125].

Successful technological results for increasing the amount of phenolic acids have been obtained by up-regulating key genes leading to RA production [117] (Table 1). In *S. miltiorrhiza* hairy roots a strong increment of RA and *SalB* production was obtained through overexpression of both *SmTAT* and *SmHPPR*. This increment was higher than the increase obtained in MJ-elicited wild type cells, suggesting that it can be an efficient system for large-scale production of phenolic acids RA and *SalB* [117].

4.3 *Candidate and Engineered Genes to Increase terpenoids Biosynthesis in* Salvia *Species*

All *Salvia* terpenoids have demonstrated a wide range of bioactivities for human health [30, 126]. However, recently the genome sequencing of *S. miltiorrhiza*, which is particularly rich in these compounds, has contributed to elucidate the molecular regulation of diterpene biosynthesis [98, 116].

As mentioned previously, diterpenes are mainly derived from the plastidial MEP pathway [127]. Several limiting steps of the MEP pathway can negatively affect the production of precursors used by diterpene synthases (diTPS). To overcome these limiting steps, the key genes encoding the1-deoxy-D-xylulose 5-phosphate synthase (*DXS*), 3-hydroxy-3-methylglutaryl coenzyme A reductase (*HMGR*), deoxyxylulose phosphate reductoisomerase (*DXR*) and geranylgeranyl diphosphate synthase (*GGPPS*) were over-expressed in *S. miltiorrhiza* in *S. sclarea* hairy roots [128–130] (Table 2). Despite the increase in the final yield in tanshinone or other abietane diterpenes in both *Salvia* species by overexpressing different biosynthetic genes alone or in combination, the boosting effect is not sufficient for a commercial scale production [128]. This is particularly evident when MEP genes are overexpressed. Competitive hormonal and photosynthetic pigment pathways can cause a drop in diterpenes production because many isoprene intermediates are also used in photosynthesis and phytohormone-mediated pathway [127, 131]. Particularly limiting in this respect is geranylgeranyl diphosphate (GGPP), the substrate used by diTPSs, which may also be incorporated in photosynthetic pigments or used to feed the gibberellin pathway [127].

Engineering strategies target specifically the diTPS genes may exert less effects on competitive pathways. Of the five *SmCPSs* reported by [132], only *SmCPS1* and *SmCPS2* were associated to tanshinone biosynthesis [72, 73, 132] and consequently are the best candidates for a metabolic engineering approach. Interestingly, while *SmCPS1* is mainly involved in the biosynthesis of this class of compound in root, *SmCPS2* is associated to tanshinone synthesis in aerial tissues, [73]. This can be very useful since leaves can be sources of tanshinones throughout the plant lifespan, contrarily to the roots, whose harvesting destroys the entire plant [73]. In this regard, suppression of the tanshinone biosynthesis in the roots, by silencing of root-specific endogenous biosynthetic gene *SmCPS1* by the current silencing approaches, including also the novel CRISPR-CAS9 technology, is expected to promote tanshinone production in leaves. Parallely, since it has been reported that the overexpression of *S. fruticosa CPSs* and *KSLs* as diterpene synthases induces the accumulation of miltiradiene as ferruginol precursor [61, 62] it has been suggested that it is possible to increase CA production, by overexpressing these genes along with the *R. officinalis* ferruginol synthase [62].

Although transcription factors can be particularly attractive for engineering the biosynthesis of terpenoids, so far few engineering approaches based on the over-expression of TFs have been reported in *Salvia* species (Table 3).

Table 3 Candidate and engineered genes to increase *Salvia* terpenoids biosynthesis

Structural genes	Gene source	Target compounds	References
DXS	*Arabidopsis thaliana*	Aethiopinone	[129]
HMGR			
GGPPS			
DXS	*Arabidopsis thaliana*	Tanshinones	[128]
HMGR			
GGPPS			
CPS1	*Salvia miltiorrhiza*	Tanshinones	[132]
CPS2			
CPS	*Salvia fruticosa*	Carnosic acid	[62]
KSL			
RoFS1	*Rosmarinus officinalis*		
RoFS2			
Regulatory genes			
C1	*Zea mays*	Tanshinones	[133]
ORCA2 *ORCA3*	*Catharantus roseus*	Terpenoid indole alkaloid/tanshinones	[136, 138]
SmbHLH37 *SmbHLH74*	*Salvia miltiorrhiza*	Tanshinones	[140]
PIF5	*Arabidopsis thaliana*	Chlorophylls, carotenoids and isopentenyl diphosphate	[131]

Overexpression of the heterologous maize R2R3 MYB TF *C1* was successful in obtaining high tanshinone biosynthesis in *S. miltiorrhiza* hairy roots [133]. Though *C1* is well-known to be involved in flavonoid biosynthesis [134, 135], its constitutive expression switched the cellular metabolism from flavonoid to terpenes, influencing also *SmMDC* and *SmPMK* expression [133].

Transcriptome analysis in *S. miltiorrhiza* identified putative transcription factors belonging to different families [124, 136, 137]. Transcripts of the AP2/ERF and GRAS TF families were particularly abundant in *S. miltiorrhiza* roots, suggesting some potential regulatory proteins for tanshinone biosynthesis belonging to these two classes [124, 136, 137]. Among AP2/ERF proteins, ORCA2 and ORCA3, which are well-known to regulate the terpenoid indole alkaloid metabolism in *Catharanthus roseus* [136, 138, 139], were also suggested to be involved in tanshinone biosynthesis in *S. miltiorrhiza* roots [136].

Other in silico studies confirmed that all these TFs could potentially bind *cis*-element of MEP and specific diterpene synthases genes [124]. In spite of these promising results, it has been widely reported that many of these TFs play simultaneous roles in plant growth and development other than controlling exclusively secondary metabolism [136]. Consequently, engineering strategy of these genes should take into account that their overexpression may imply undesired pleiotropic effects on cell development and, ultimately, cause yield penalty.

Upon MJ-elicitation of *S. miltiorrhiza* hairy roots, the induction of bHLHs TFs paralleled tanshinone production, thus supporting their regulatory role in tanshinone biosynthesis. In this regard, particularly promising are the *SmbHLH37* and *SmbHLH74* genes [140]. Previous studies have demonstrated that *AtPIF5* (*Phytochrome-interacting factor 5*) positively regulates the MEP pathway in Arabidopsis T87 cell cultures. In general, *AtPIF5* overexpression increased also the accumulation of chlorophylls, carotenoids an disoprenoids [131]. The combined action of *AtPIF5* with specific diTPSs in *Salvia* could be an interesting approach to fuel the produced isoprenoid precursors toward diterpene production.

5 Conclusions and Future Prospects

In this chapter, we have reported and discussed the most recent advancements on specialized metabolites biosynthesis in *Salvia* species, along with the most successful biotechnological approaches to improve the production of these bioactive molecules. The release of the *S. miltiorrhiza* draft genome and the availability of transcriptomic data from other *Salvia* species are strongly contributing to fill the gaps in the identification and characterization of key biosynthesis genes of the main specialized secondary metabolites of *Salvia* species. The identification of key-limiting biosynthetic steps and genes along with the elucidation of the regulatory mechanisms under-pinning the biosynthesis of plant specialized metabolites is crucial for developing and optimizing any biotechnological strategy for increasing their content in plant cells and tissues. However, the forefront objective in the metabolic engineering of plant secondary metabolites is the possibility to identify natural and/or generate, by genome editing, allelic variants of biosynthetic genes encoding enzymes with peculiar catalytic activity. This will contribute to overcome possible metabolic constraints limiting a higher production of plant bioactive metabolites as well as will provide extraordinary molecular tools to diversify *in planta* their functional groups and the opportunity to design new drugs or leads compounds with improved or novel pharmacological activities. Beside the biotechnological progresses towards the improvement of *Salvia* metabolites biosynthesis, innovative strategies of bioprocess engineering by the creation of ad hoc bioreactors for tissue cultures, will guarantee to scale up the production to large scale. Higher productivity of these biosynthetic processes along with costs reduction could envisage in a near future the commercial availability of *Salvia* plant tissue culture with a rich collection of bioactive molecules.

Acknowledgements The data reported in this chapter reflect the advances reported for *Salvia* species according to our current knowledge of the bibliographic data. We apologize if we have omitted any major work but, in spite of every effort by all the authors, it is impossible to cite every reference.

References

1. Stavrinides A, Tatsis EC, Caputi L et al (2016) Structural investigation of heteroyohimbine alkaloid synthesis reveals active site elements that control stereoselectivity. Nat Commun 7:12116. https://doi.org/10.1038/ncomms12116
2. Pichersky E, Gang DR (2000) Genetics and biochemistry of secondary metabolites in plants: an evolutionary perspective. Trends Plant Sci 5:439–445
3. Weiss D, Ori N (2007) Mechanisms of cross talk between gibberellin and other hormones. Plant Physiol 144:1240–1246. https://doi.org/10.1104/pp.107.100370
4. Buer CS, Muday GK, Carolina N (2004) The transparent testa4 mutation prevents flavonoid synthesis and alters auxin transport and the response of *arabidopsis* roots to gravity and light. Plant Cell 16:1191–1205. https://doi.org/10.1105/tpc.020313.tiles
5. Taylor LP, Grotewold E (2005) Flavonoids as developmental regulators. Curr Opin Plant Biol 8:317–323. https://doi.org/10.1016/j.pbi.2005.03.005
6. Peer WA, Murphy A (2007) Flavonoids and auxin transport: modulators or regulators? Trends Plant Sci 12:556–563. https://doi.org/10.1016/j.tplants.2007.10.003
7. Last RL (2015) Enzyme recruitment to specialized metabolism something old, something new: conserved enzymes and the evolution of novelty in plant specialized metabolism. Plant Physiol 3:1512–1523. https://doi.org/10.1104/pp.15.00994
8. Weng J, Chapple C (2010) The origin and evolution of lignin biosynthesis. New Phytol 187:273–285
9. Dias DA, Urban S, Roessner U (2012) A historical overview of natural products in drug discovery. Metabolites 2:303–336. https://doi.org/10.3390/metabo2020303
10. Harvey AL, Edrada-Ebel R, Quinn RJ (2015) The re-emergence of natural products for drug discovery in the genomics era. Nat Rev Drug Discovery 14:111–129. https://doi.org/10.1038/nrd4510
11. Cragg GM, Grothaus PG, Newman DJ (2009) Impact of natural products on developing new anti-cancer agents. Chem Rev 109:3012–3043. https://doi.org/10.1021/cr900019j
12. Wink M (2003) Evolution of secondary metabolites from an ecological and molecular phylogenetic perspective. Phytochemistry 64:3–19. https://doi.org/10.1016/s0031-9422(03)00300-5
13. Pigatto AGS, Blanco CC, Mentz LA, Soares GLG (2015) Tropane alkaloids and calystegines as chemotaxonomic markers in the Solanaceae. An Acad Bras Cienc 87:2139–2149. https://doi.org/10.1590/0001-3765201520140231
14. Yan X, Chen S (2007) Regulation of plant glucosinolate metabolism. Planta 226:1343–1352. https://doi.org/10.1007/s00425-007-0627-7
15. Zang YX, Kim HU, Kim JA et al (2009) Genome-wide identification of glucosinolate synthesis genes in *Brassica rapa*. FEBS J 276:3559–3574. https://doi.org/10.1111/j.1742-4658.2009.07076.x
16. Ohmiya S, Saito K, Murakoshi I (2000) Recent progress on the lupine alkaloids in leguminous plants growing mainly in Japan. Yakugaku Zasshi 120:923–934
17. Jenks AA, Kim SC (2013) Medicinal plant complexes of *Salvia* subgenus Calosphace: an ethnobotanical study of new world sages. J Ethnopharmacol 146:214–224. https://doi.org/10.1016/j.jep.2012.12.035
18. Akhondzadeh S, Noroozian M, Mohammadi M et al (2003) *Salvia officinalis* extract in the treatment of patients with mild to moderate Alzheimer' s disease: a double blind, randomized and placebo-controlled trial. J Clin Pharm Ther 28:53–59. https://doi.org/10.1046/j.1365-2710.2003.00463.x
19. Scholey AB, Tildesley NTJ, Ballard CG et al (2008) An extract of *Salvia* (sage) with anticholinesterase properties improves memory and attention in healthy older volunteers. Psychopharmacology 198:127–139. https://doi.org/10.1007/s00213-008-1101-3
20. Wu YB, Ni ZY, Shi QW et al (2012) Constituents from *salvia* species and their biological activities. Chem Rev 112:5967–6026. https://doi.org/10.1021/cr200058f

21. Walker JB, Sytsma KJ, Treutlein J, Wink M (2004) *Salvia* (Lamiaceae) is not monophyletic: implications for the systematics, radiation, and ecological specializations of *Salvia* and tribe Mentheae. Am J Bot 91:1115–1125. https://doi.org/10.3732/ajb.91.7.1115
22. Sandasi M, Kamatou GP, Viljoen AM (2012) An untargeted metabolomic approach in the chemotaxonomic assessment of two *Salvia* species as a potential source of alpha-bisabolol. Phytochemistry 84:94–101. https://doi.org/10.1016/j.phytochem.2012.08.009
23. Nakiboğlu M (2002) The classification of the *Salvia* L. (Labiatae) species distributed in West Anatolia according to phenolic compounds. Turk J Bot 26:103–108
24. Rzepa J, Wojtal L, Staszek D et al (2009) Fingerprint of selected *Salvia* species by HS-GC-MS analysis of their volatile fraction. J Chromatogr Sci 47:575–580
25. Máthé I, Máthé Á, Hohmann J, Janicsák G (2010) Volatile and some non-volatile chemical constituents of Mediterranean *Salvia* species beyond their native area. Isr J Plant Sci 58:273–277. https://doi.org/10.1560/ijps.58.3-4.273
26. Jassbi AR, Asadollahi M, Masroor M et al (2012) Chemical classification of the essential oils of the Iranian *Salvia* species in comparison with their botanical taxonomy. Chem Biodivers 9:1254–1271. https://doi.org/10.1002/cbdv.201100209
27. Russo A, Formisano C, Rigano D et al (2013) Chemical composition and anticancer activity of essential oils of Mediterranean sage (*Salvia officinalis* L.) grown in different environmental conditions. Food Chem Toxicol 55:42–47. https://doi.org/10.1016/j.fct.2012.12.036
28. Kamatou GPP, Makunga NP, Ramogola WPN, Viljoen AM (2008) South African *Salvia* species: a review of biological activities and phytochemistry. J Ethnopharmacol 119:664–672. https://doi.org/10.1016/j.jep.2008.06.030
29. Li M, Li Q, Zhang C et al (2013) An ethnopharmacological investigation of medicinal *Salvia* plants (Lamiaceae) in China. Acta Pharm Sin B 3:273–280. https://doi.org/10.1016/j.apsb.2013.06.001
30. Imanshahidi M, Hosseinzadeh H (2006) The pharmacological effects of *Salvia* species. Phyther Res 437:427–437. https://doi.org/10.1002/ptr
31. Kharazian N (2014) Flavonoid constituents in some species of *Salvia* L. (Lamiaceae) in Iran. J Sci I R Iran 25:219–227
32. Lu Y, Foo LY (2002) Polyphenolics of *Salvia*—a review. Phytochemistry 59:117–140
33. del Bano MJ, Lorente J, Castillo J et al (2003) Phenolic diterpenes, flavones, and rosmarinic acid distribution during the development of leaves, flowers, stems, and roots of *Rosmarinus officinalis*. Antioxidant activity. J Agric Food Chem 51:4247–4253. https://doi.org/10.1021/jf0300745
34. Khojasteh A, Mirjalili MH, Palazon J, Hidalgo D (2014) New trends in biotechnological production of rosmarinic acid. Biotechnol Lett 36:2393–2406. https://doi.org/10.1007/s10529-014-1640-0
35. Vega-Sanchez ME, Ronald PC (2010) Genetic and biotechnological approaches for biofuel crop improvement. Curr Opin Biotechnol 21:218–224. https://doi.org/10.1016/j.copbio.2010.02.002
36. Petersen M, Abdullah Y, Benner J et al (2009) Evolution of rosmarinic acid biosynthesis. Phytochemistry 70:1663–1679. https://doi.org/10.1016/j.phytochem.2009.05.010
37. Yamamura Y, Ogihara Y, Mizukami H (2001) Cinnamic acid 4-hydroxylase from Lithospermum erythrorhizon: cDNA cloning and gene expression. Plant Cell Rep 20:655–662
38. Matsuno M, Nagatsu A, Ogihara Y et al (2002) CYP98A6 from Lithospermum erythrorhizon encodes 4-coumaroyl-4′-hydroxyphenyllactic acid 3-hydroxylase involved in rosmarinic acid biosynthesis. FEBS Lett 514:219–224
39. Hou X, Shao F, Ma Y, Lu S (2013) The phenylalanine ammonia-lyase gene family in *Salvia miltiorrhiza*: genome-wide characterization, molecular cloning and expression analysis. Mol Biol Rep 40:4301–4310. https://doi.org/10.1007/s11033-013-2517-3
40. Zhang S, Yan Y, Wang B et al (2014) Selective responses of enzymes in the two parallel pathways of rosmarinic acid biosynthetic pathway to elicitors in *Salvia miltiorrhiza* hairy root cultures. J Biosci Bioeng 117:645–651. https://doi.org/10.1016/j.jbiosc.2013.10.013

41. Docimo T, Mattana M, Fasano R et al (2013) Ectopic expression of the *Osmyb4* rice gene enhances synthesis of hydroxycinnamic acid derivatives in tobacco and clary sage. Biol Plant 57:179–183
42. Ge Q, Zhang Y, Hua W et al (2015) Combination of transcriptomic and metabolomic analyses reveals a JAZ repressor in the jasmonate signaling pathway of *Salvia miltiorrhiza*. Sci Rep 5:1–14. https://doi.org/10.1038/srep14048
43. Ulubelen A, Topcu G, Eri C et al (1994) Terpenoids from *Salvia sclarea*. Phytochemistry 36:971–974. https://doi.org/10.1016/s0031-9422(00)90474-6
44. Muhlemann JK, Klempien A, Dudareva N (2014) Floral volatiles: from biosynthesis to function. Plant Cell Environ 37:1936–1949. https://doi.org/10.1111/pce.12314
45. Dudareva N, Cseke L, Blanc VM, Pichersky E (1996) Evolution of floral scent in Clarkia: novel patterns of S-Linalool synthase gene expression in the C. *breweri* Flower. Plant Cell Am Soc Plant Physiol 8:1137–1148. https://doi.org/10.1105/tpc.8.7.1137
46. Raguso RA, Pichersky E (1999) New perspectives in pollination biology: floral fragrances. A day in the life of a linalool molecule: chemical communication in a plant-pollinator system. Part 1: linalool biosynthesis in flowering plants. Plant Sci Biol 14:95–120
47. Pichersky E, Gershenzon J (2002) The formation and function of plant volatiles: perfumes for pollinator attraction and defense. Curr Opin Plant Biol 5:237–243
48. Schmiderer C, Grassi P, Novak J et al (2008) Diversity of essential oil glands of clary sage (*Salvia sclarea* L., Lamiaceae). Plant Biol 10:433–440. https://doi.org/10.1111/j.1438-8677.2008.00053.x
49. Aharoni A, Jongsma MA, Bouwmeester HJ (2005) Volatile science? Metabolic engineering of terpenoids in plants. Trends Plant Sci 10:594–602. https://doi.org/10.1016/j.tplants.2005.10.005
50. Bouwmeester HJ (2006) Engineering the essence of plants. Nat Biotechnol 24:1359–1361. https://doi.org/10.1038/nbt1106-1359
51. Zerbe P, Hamberger B, Yuen MMS et al (2013) Gene discovery of modular diterpene metabolism in nonmodel systems. Plant Physiol 162:1073–1091. https://doi.org/10.1104/pp.113.218347
52. Peters RJ (2010) Two rings in them all: the labdane-related diterpenoids. Nat Prod Rep 27:1521–1530. https://doi.org/10.1039/c0np00019a
53. Caissard JC, Olivier T, Delbecque C et al (2012) Extracellular localization of the diterpene sclareol in Clary Sage (*Salvia sclarea* L., Lamiaceae). PLoS ONE 7:1–8. https://doi.org/10.1371/journal.pone.0048253
54. Caniard A, Zerbe P, Legrand S et al (2012) Discovery and functional characterization of two diterpene synthases for sclareol biosynthesis in *Salvia sclarea* (L.) and their relevance for perfume manufacture. BMC Plant Biol 12:1. https://doi.org/10.1186/1471-2229-12-119
55. Schalk M, Pastore L, Mirata MA et al (2012) Toward a biosynthetic route to sclareol and amber odorants. J Am Chem Soc 134:18900–18903
56. Rau O, Wurglics M, Paulke A et al (2006) Carnosic acid and carnosol, phenolic diterpene compounds of the labiate herbs rosemary and sage, are activators of the human peroxisome proliferator-activated receptor gamma. Planta Med 72:881–887. https://doi.org/10.1055/s-2006-946680
57. Poeckel D, Greiner C, Verhoff M et al (2008) Carnosic acid and carnosol potently inhibit human 5-lipoxygenase and suppress pro-inflammatory responses of stimulated human polymorphonuclear leukocytes. Biochem Pharmacol 76:91–97. https://doi.org/10.1016/j.bcp.2008.04.013
58. Bauer J, Kuehnl S, Rollinger JM et al (2012) Carnosol and carnosic acids from *Salvia officinalis* inhibit microsomal prostaglandin E(2) synthase-1. J Pharmacol Exp Ther 342:169–176. https://doi.org/10.1124/jpet.112.193847
59. Birtić S, Dussort P, Pierre FX et al (2015) Carnosic acid. Phytochemistry 115:9–19. https://doi.org/10.1016/j.phytochem.2014.12.026

60. Abreu ME, Müller M, Alegre L, Munné-Bosch S (2008) Phenolic diterpene and α-tocopherol contents in leaf extracts of 60 *Salvia* species. J Sci Food Agric 88:2648–2653. https://doi.org/10.1002/jsfa.3384
61. Brückner K, Božić D, Manzano D et al (2014) Characterization of two genes for the biosynthesis of abietane-type diterpenes in rosemary (*Rosmarinus officinalis*) glandular trichomes. Phytochemistry 101:52–64. https://doi.org/10.1016/j.phytochem.2014.01.021
62. Božić D, Papaefthimiou D, Brückner K et al (2015) Towards elucidating carnosic acid biosynthesis in Lamiaceae: functional characterization of the three first steps of the pathway in *Salvia fruticosa* and *Rosmarinus officinalis*. PLoS ONE 10:1–28. https://doi.org/10.1371/journal.pone.0124106
63. Falara V, Pichersky E, Kanellis AK (2010) A copal-8-ol diphosphate synthase from the angiosperm *Cistus creticus* subsp. *creticus* is a putative key enzyme for the formation of pharmacologically active. Oxygen-containing labdane-type diterpenes. Plant Physiol 154:301–310. https://doi.org/10.1104/pp.110.159566
64. Sallaud C, Giacalone C, Töpfer R et al (2012) Characterization of two genes for the biosynthesis of the labdane diterpene Z-abienol in tobacco (*Nicotiana tabacum*) glandular trichomes. Plant J 72:1–17. https://doi.org/10.1111/j.1365-313x.2012.05068.x
65. Hillwig ML, Xu M, Toyomasu T et al (2011) Domain loss has independently occurred multiple times in plant terpene synthase evolution. Plant J 68:1051–1060. https://doi.org/10.1111/j.1365-313x.2011.04756.x
66. Zhong GX, Peng LI, Zeng LJ et al (2009) Chemical characteristics of *Salvia miltiorrhiza* (Danshen) collected from different locations in China. J Agric Food Chem 57:6879–6887. https://doi.org/10.1021/jf901462s
67. Chen X, Guo J, Bao J et al (2014) The anticancer properties of *Salvia miltiorrhiza* Bunge (Danshen): a systematic review. Med Res Rev 34:768–794. https://doi.org/10.1002/med.21304
68. Hong HJ, Liu JC, Chen PY et al (2012) Tanshinone IIA prevents doxorubicin-induced cardiomyocyte apoptosis through Akt-dependent pathway. Int J Cardiol 157:174–179. https://doi.org/10.1016/j.ijcard.2010.12.012
69. Liu JR, Chen GF, Shih HN, Kuo PC (2008) Enhanced antioxidant bioactivity of *Salvia miltiorrhiza* (Danshen) products prepared using nanotechnology. Phytomedicine 15:23–30. https://doi.org/10.1016/j.phymed.2007.11.012
70. Boya MT, Valverde S (1981) An orthoquinone isolated from *Salvia aethiopis*. Phytochemistry 20:1367–1368. https://doi.org/10.1016/0031-9422(81)80041-6
71. Walencka E, Rozalska S, Wysokinska H et al (2007) Salvipisone and aethiopinone from *Salvia sclarea* hairy roots modulate staphylococcal antibiotic resistance and express anti-biofilm activity. Planta Med 73:545–551. https://doi.org/10.1055/s-2007-967179
72. Gao W, Hillwig ML, Huango L et al (2009) A functional genomics approach to tanshinone biosynthesis provides stereochemical insights. Org Lett 11:5170–5173. https://doi.org/10.1021/ol902051v
73. Cui G, Duan L, Jin B et al (2015) Functional divergence of diterpene syntheses in the medicinal plant *Salvia miltiorrhiza*. Plant Physiol 169:1607–1618. https://doi.org/10.1104/pp.15.00695
74. Xu Z, Peters RJ, Weirather J et al (2015) Full-length transcriptome sequences and splice variants obtained by a combination of sequencing platforms applied to different root tissues of *Salvia miltiorrhiza* and tanshinone biosynthesis. Plant J 82:951–961. https://doi.org/10.1111/tpj.12865
75. Marchev A, Haas C, Schulz S, Pavlov A (2014) Sage in vitro cultures: a promising tool for the production of bioactive terpenes and phenolic substances. Biotechnol Lett 36:211–221. https://doi.org/10.1007/s10529-013-1350-z
76. Facchini PJ (2001) Alkaloids biosynthesis in plants: biochemistry, cell biology, molecular regulation, and metabolic engineering applications. Ann Rev Plant Physiol Plant Mol Biol 59:29–66

77. Ziegler J, Facchini PJ (2008) Alkaloid biosynthesis: metabolism and trafficking. Annu Rev Plant Biol 59:735–769. https://doi.org/10.1146/annurev.arplant.59.032607.092730
78. Hippolyte I, Marin B, Baccou JC et al (1992) Growth and rosmarinic acid production in cell suspension of *Salvia officinalis* L. Plant Cell Rep 11:109–112. https://doi.org/10.1007/bf00232160
79. Karam NS, Jawad FM, Arikat NA et al (2003) Growth and rosmarinic acid accumulation in callus, cell suspension, and root cultures of wild *Salvia fruticosa*. Plant Cell Tissue Organ Cult 73:117–121. https://doi.org/10.1023/a:1022806420209
80. Grzegorczyk I, Krolicka A, Wysokinska H (2006) Establishment of *Salvia officinalis* L. hairy root cultures for the production of rosmarinic acid. Z Naturforsch C 61:351–356
81. Mulder-Krieger TH, Verpoorte R, Svendsen AB et al (1988) Production of essential oils and flavours in plant cell and tissue cultures. A review. Plant Cell Tissue Organ Cult 13:85–154. https://doi.org/10.1007/bf00034451
82. Becker H (1970) Untersuchungen zur Frage der Bildung fliichtiger Stoffwechselprodukte in. Calluskulturen Biochem Physiol Pftanz 161:425–441
83. Katagi H, Honda K, Inui M et al (1983) Essential oil in adventitious shoots of cultured Lavandula vera cells. Nippon Nõgeikagaku Kaishi 57:771–773. https://doi.org/10.1271/nogeikagaku1924.57.771
84. Webb JK, Banthorpe DV, Watson DG (1984) Monoterpene synthesis in shoots regenerated from callus cultures. Phytochemistry 23:903–904. https://doi.org/10.1016/S0031-9422(00)85056-6
85. Kuhlmann A, Röhl C (2006) Phenolic antioxidant compounds produced by in vitro cultures of rosemary (*Rosmarinus officinalis*) and their anti-inflammatory effect on lipopolysaccharide-activated microglia. Pharm Biol 44:401–410. https://doi.org/10.1080/13880200600794063
86. Wang JY, Liu D (1987) Organogenesis of *Salvia miltiorrhiza*. Plant Physiol Commun 6:46
87. Lemraski MG, Eftekhari M, Faraji M, Zarrini SS (2014) Study of callus induction in common sage (*Salvia Officinalis* L.). Int J Agric Crop Sci 7:386–389
88. Miyasaka H, Nasu M, Yamamoto T et al (1986) Production of cryptotanshinone and ferruginol by immobilized cultured cells of *Salvia miltiorrhiza*. Phytochemistry 25:1621–1624. https://doi.org/10.1016/S0031-9422(00)81221-2
89. Miyasaka H, Nasu M, Yamamoto T et al (1987) Effect of nutritional factors on cryptotanshinone and ferruginol production by cell suspension cultures of *Salvia miltiorrhiza*. Phytochemistry 26:1421–1424. https://doi.org/10.1016/S0031-9422(00)81826-9
90. Shimomura K, Kitazawa T, Okamura N, Yagi A (1991) Tanshinone production in adventitious roots and regenerates of *Salvia miltiorrhiza*. J Nat Prod 54:1583–1587. https://doi.org/10.1021/np50078a014
91. Song JY, Qi JJ, Ren CL et al (2000) Dynamics of growth and total tanshinones accumulation in crown gall cultures of *salvia* miltiorrhiza. Acta Pharm Sin 35:929–931
92. Saito K (1993) Genetic engineering in tissue culture of medicinal plants. Plant Tissue Cult Lett 10:8. https://doi.org/10.5511/plantbiotechnology1984.10.1
93. Kuzma L, Bruchajzer E, Wysokinska H (2008) Diterpenoid production in hairy root culture of *Salvia sclarea* L. Z Naturforsch C 63:621–624
94. Hao G, Ji H, Li Y et al (2012) Exogenous ABA and polyamines enhanced salvianolic acids contents in hairy root cultures of *Salvia miltiorrhiza* Bge. f. alba. Plant Omics J 5:446–452
95. Gupta SK, Liu R, Liaw S, Chan H (2011) Enhanced tanshinone production in hairy roots of '*Salvia miltiorrhiza* Bunge' under the influence of plant growth regulators in liquid culture. Bot Stud 52:435–443
96. Marchev A, Georgiev V, Ivanov I et al (2011) Two-phase temporary immersion system for *Agrobacterium rhizogenes* genetic transformation of sage (*Salvia tomentosa* Mill.). Biotechnol Lett 33:1873–1878. https://doi.org/10.1007/s10529-011-0625-5
97. Zhi BH, Alfermann AW (1993) Diterpenoid production in hairy root cultures of *Salvia miltiorrhiza*. Phytochemistry 32:699–703. https://doi.org/10.1016/S0031-9422(00)95156-2

98. Wang JW, Wu JY (2010) Tanshinone biosynthesis in *Salvia miltiorrhiza* and production in plant tissue cultures. Appl Microbiol Biotechnol 88:437–449. https://doi.org/10.1007/s00253-010-2797-7
99. Chen H, Chen F, Chiu FCK, Lo CMY (2001) The effect of yeast elicitor on the growth and secondary metabolism of hairy root cultures of *Salvia miltiorrhiza*. Enzym Microb Technol 28:100–105
100. Zhao J, Davis LC, Verpoorte R (2005) Elicitor signal transduction leading to production of plant secondary metabolites. Biotechnol Adv 23:283–333. https://doi.org/10.1016/j.biotechadv.2005.01.003
101. Chen H, Chen F (2000) Effect of yeast elicitor on the secondary metabolism of Ti-transformed *Salvia miltiorrhiza* cell suspension cultures. Plant Cell Rep 19:710–717
102. Yan Q, Hu Z, Xiang R, Wu J (2005) Efficient production and recovery of diterpenoid tanshinones in *Salvia miltiorrhiza* hairy root cultures with in situ adsorption, elicitation and semi-continuous operation. J Biotechnol 119:416–424. https://doi.org/10.1016/j.jbiotec.2005.04.020
103. Ge X, Wu J (2005) Induction and potentiation of diterpenoid tanshinone accumulation in *Salvia miltiorrhiza* hairy roots by β-aminobutyric acid. Appl Microbiol Biotechnol 68:183–188. https://doi.org/10.1007/s00253-004-1873-2
104. Zhang C, Yan Q, Cheuk W, Wu J (2004) Enhancement of tanshinone production in *Salvia miltiorrhiza* hairy root culture by Ag^+ elicitation and nutrient feeding. Planta Med 70:147–151. https://doi.org/10.1055/s-2004-815492
105. Ge X, Wu J (2005) Tanshinone production and isoprenoid pathways in *Salvia miltiorrhiza* hairy roots induced by Ag+ and yeast elicitor. Plant Sci 168:487–491. https://doi.org/10.1016/j.plantsci.2004.09.012
106. Zhou J, Fang L, Wang X et al (2013) La Dramaticaly enhances the accumulation of tanshinones in *Salvia miltiorrhiza* hairy root cultures. Earth Sci Res J 2:187–192. https://doi.org/10.5539/esr.v2n1p187
107. Shi M, Kwok KW, Wu JY (2007) Enhancement of tanshinone production in *Salvia miltiorrhiza* Bunge (red or chinese sage) hairy-root culture by hyperosmotic stress and yeast elicitor. Biotechnol Appl Biochem 196:191–196. https://doi.org/10.1042/ba20060147
108. Wu J-Y, Shi M (2008) Ultrahigh diterpenoid tanshinone production through repeated osmotic stress and elicitor stimulation in fed-batch culture of *Salvia miltiorrhiza* hairy roots. Appl Microbiol Biotechnol 78:441–448. https://doi.org/10.1007/s00253-007-1332-y
109. Wu J-Y, Ng J, Shi M, Wu S-J (2007) Enhanced secondary metabolite (tanshinone) production of *Salvia miltiorrhiza* hairy roots in a novel root-bacteria coculture process. Appl Microbiol Biotechnol 77:543–550. https://doi.org/10.1007/s00253-007-1192-5
110. Zhao J, Zhou L, Wu J (2010) Effects of biotic and abiotic elicitors on cell growth and tanshinone accumulation in *Salvia miltiorrhiza* cell cultures. Appl Microbiol Biotechnol 137–144. https://doi.org/10.1007/s00253-010-2443-4
111. Zhao J, Zhou L, Wu J (2010) Promotion of *Salvia miltiorrhiza* hairy root growth and tanshinone production by polysaccharide—protein fractions of plant growth-promoting rhizobacterium *Bacillus cereus*. Process Biochem 45:1517–1522. https://doi.org/10.1016/j.procbio.2010.05.034
112. Kai G, Liao P, Xu H, Wang J (2012) Molecular mechanism of elicitor-induced tanshinone accumulation in *Salvia miltiorrhiza* hairy root cultures. Acta Physiol Plant 34:1421–1433. https://doi.org/10.1007/s11738-012-0940-z
113. Zhou M, Zhu X, Shao J (2011) Production and metabolic engineering of bioactive substances in plant hairy root culture. Appl Microbiol Biotechnol 90:1229–1239. https://doi.org/10.1007/s00253-011-3228-0
114. Patra B, Schluttenhofer C, Wu Y et al (2013) Transcriptional regulation of secondary metabolite biosynthesis in plants. Biochim Biophys Acta-Gene Regul Mech 1829:1236–1247. https://doi.org/10.1016/j.bbagrm.2013.09.006
115. Roberts SC (2007) Production and engineering of terpenoids in plant cell culture. Nat Chem Biol 3:387–395. https://doi.org/10.1038/nchembio.2007.8 [pii]\r

116. Xu H, Song J, Luo H et al (2016) Analysis of the genome sequence of the medicinal plant *Salvia miltiorrhiza*. Mol Plant 9:949–952. https://doi.org/10.1016/j.molp.2016.03.010
117. Xiao Y, Zhang L, Gao S et al (2011) The *c4h, tat, hppr* and *hppd* genes prompted engineering of rosmarinic acid biosynthetic pathway in *Salvia miltiorrhiza* hairy root cultures. PLoS ONE. https://doi.org/10.1371/journal.pone.0029713
118. Xu F, Ning Y, Zhang W et al (2014) An R2R3-MYB transcription factor as a negative regulator of the flavonoid biosynthesis pathway in *Ginkgo biloba*. Funct Integr Genomics 14:177–189. https://doi.org/10.1007/s10142-013-0352-1
119. Wang D, Song Y, Chen Y et al (2013) Metabolic pools of phenolic acids in *Salvia miltiorrhiza* are enhanced by co-expression of *Antirrhinum majus* Delila and Roseal transcription factors. Biochem Eng J 74:115–120. https://doi.org/10.1016/j.bej.2013.02.014
120. Schwinn K, Venail J, Shang Y et al (2006) A small family of MYB-regulatory genes controls floral pigmentation intensity and patterning in the genus *Antirrhinum*. Plant Cell 18:831–851. https://doi.org/10.1105/tpc.105.039255
121. Outchkourov NS, Carollo CA, Gomez-Roldan V et al (2014) Control of anthocyanin and non-flavonoid compounds by anthocyanin-regulating MYB and bHLH transcription factors in *Nicotiana benthamiana* leaves. Front Plant Sci 5:519. https://doi.org/10.3389/fpls.2014.00519
122. Zhang Y, Yan YP, Wu YC et al (2014) Pathway engineering for phenolic acid accumulations in *Salvia miltiorrhiza* by combinational genetic manipulation. Metab Eng 21:71–80. https://doi.org/10.1016/j.ymben.2013.10.009
123. Li C, Lu S (2014) Genome-wide characterization and comparative analysis of R2R3-MYB transcription factors shows the complexity of MYB-associated regulatory networks in *Salvia miltiorrhiza*. BMC Genomics 15:277. https://doi.org/10.1186/1471-2164-15-277
124. Ji AJ, Luo HM, Xu ZC et al (2016) Genome-wide identification of the AP2/ ERF gene family involved in active constituent biosynthesis in *Salvia miltiorrhiza*. Plant Genome 146:1–11. https://doi.org/10.3835/plantgenome2015.08.0077
125. Park SU, Uddin R, Xu H et al (2008) Biotechnological applications for rosmarinic acid production in plant. Afr J Biotechnol 7:4959–4965
126. Akaberi M, Mehri S, Iranshahi M (2015) Multiple pro-apoptotic targets of abietane diterpenoids from *Salvia* species. Fitoterapia 100:118–132. https://doi.org/10.1016/j.fitote.2014.11.008
127. Zi J, Mafu S, Peters RJ (2014) To gibberellins and beyond! Surveying the evolution of (Di) terpenoid metabolism. Ann Rev Plant Biol 65:259–286. https://doi.org/10.1146/annurev-arplant-050213-035705
128. Kai G, Xu H, Zhou C et al (2011) Metabolic engineering tanshinone biosynthetic pathway in *Salvia miltiorrhiza* hairy root cultures. Metab Eng 13:319–327. https://doi.org/10.1016/j.ymben.2011.02.003
129. Vaccaro M, Malafronte N, Alfieri M et al (2014) Enhanced biosynthesis of bioactive abietane diterpenes by overexpressing AtDXS or AtDXR genes in *Salvia sclarea* hairy roots. Plant Cell Tissue Organ Cult 119:65–77. https://doi.org/10.1007/s11240-014-0514-4
130. Ikram NKBK, Zhan X, Pan X et al (2015) Stable heterologous expression of biologically active terpenoids in green plant cells. Front Plant Sci 6:129. https://doi.org/10.3389/fpls.2015.00129
131. Mannen K, Matsumoto T, Takahashi S et al (2014) Biochemical and biophysical research communications coordinated transcriptional regulation of isopentenyl diphosphate biosynthetic pathway enzymes in plastids by phytochrome-interacting factor 5. Biochem Biophys Res Commun 443:768–774. https://doi.org/10.1016/j.bbrc.2013.12.040
132. Ma Y, Yuan L, Wu B et al (2012) Genome-wide identification and characterization of novel genes involved in terpenoid biosynthesis in *Salvia miltiorrhiza*. J Exp Bot 63:2809–2823. https://doi.org/10.1093/jxb/err466

133. Zhao S, Zhang J, Tan R et al (2015) Enhancing diterpenoid concentration in *Salvia miltiorrhiza* hairy roots through pathway engineering with maize C1 transcription factor. J Exp Bot 66:7211–7226. https://doi.org/10.1093/jxb/erv418
134. Grotewold E, Chamberlin M, Snook M et al (1998) Engineering secondary metabolism in maize cells by ectopic expression of transcription factors. Plant Cell 10:721–740. https://doi.org/10.1105/tpc.10.5.721
135. Bovy A, De Vos R, Kemper M et al (2002) High-flavonol tomatoes resulting from the heterologous expression of the maize transcription factor genes LC and C1. Plant Cell 14:2509–2526. https://doi.org/10.1105/tpc.004218.growth
136. Yang L, Ding G, Lin H et al (2013) Transcriptome analysis of medicinal plant *Salvia miltiorrhiza* and identification of genes related to tanshinone biosynthesis. PLoS ONE 8:1–13. https://doi.org/10.1371/journal.pone.0080464
137. Wenping H, Yuan Z, Jie S et al (2011) De novo transcriptome sequencing in *Salvia miltiorrhiza* to identify genes involved in the biosynthesis of active ingredients. Genomics 98:272–279. https://doi.org/10.1016/j.ygeno.2011.03.012
138. Van Der Fits L, Memelink J, Van Der Fits L, Memelink J (2001) The jasmonate-inducible AP2/ERF-domain transcription factor ORCA3 activates gene expression via interaction with a jasmonate-responsive promoter element. Plant J 25:43–53. https://doi.org/10.1111/j.1365-313x.2001.00932.x
139. Menke FLH, Champion A, Kijne JW, Memelink J (1999) A novel jasmonate- and elicitor-responsive element in the periwinkle secondary metabolite biosynthetic gene *Str* interacts with a jasmonate- and elicitor-inducible AP2-domain transcription factor, ORCA 2. EMBO J 18:4455–4463
140. Zhang X, Luo H, Xu Z et al (2015) Genome-wide characterisation and analysis of bHLH transcription factors related to tanshinone biosynthesis in *Salvia miltiorrhiza*. Sci Rep 5:11244. https://doi.org/10.1038/srep11244

Hairy Roots of *Salvia* Species for Bioactive Substances Production

Sibylle Kümmritz, Christiane Haas, Katja Winkler, Vasil Georgiev and Atanas Pavlov

Abstract The biosynthesis of biologically active secondary metabolites in plant tissues is frequently related to their differentiation. From this point of view, the hairy roots are a suitable plant in vitro system for producing bioactive substances due to their degree of differentiation and metabolic profiles identical to or similar to that of the root systems of the intact plants from which they are derived. In addition, they are grown in growth regulators-free nutrient media, which is essential for the subsequent application of their metabolites in variety of nutritional and pharmaceutical systems. They are characterized by high genetic and biochemical stability, making possible the scale-up of the cultivation processes in bioreactors. This provides complete control and management of the biosynthetic process and results in high yields of target metabolites for a short period of time, regardless of seasons. *Salvia* plant species are widely used in traditional medicine. The major biological activities they possessed are antioxidant, anti-inflammatory, anti-tumor activities and etc. These activities are consequence of biosynthesized biologically active substances: sterols, various polyphenol structures, di- and triterpenes, etc. Data about *Salvia* hairy roots in the scientific literature are limited. Nevertheless, it is clear that they biosynthesized mainly diterpenes and phenolic compounds. Current progress of investigation of *Salvia* hairy roots is limited to the phytochemical profiling and assessment of the influence of different independent variables (mainly elicitors) on the yields of the target bioactive substances and the development of commercial production process is not visible at the time. Therefore in the near

Sibylle Kümmritz and Christiane Haas are contributed equally to this work.

S. Kümmritz · C. Haas · K. Winkler
Technische Universität Dresden, Institute of Natural Materials Technology,
Bergstraße 120, 01069 Dresden, Germany

V. Georgiev · A. Pavlov (✉)
University of Food Technologies, 26 Maritza Blvd., 4002 Plovdiv, Bulgaria
e-mail: at_pavlov@yahoo.com

V. Georgiev · A. Pavlov
Laboratory of Applied Biotechnologies, Institute of Microbiology,
Bulgarian Academy of Sciences, 139 Ruski Blvd., 4000 Plovdiv, Bulgaria

V. Georgiev and A. Pavlov (eds.), *Salvia Biotechnology*,
https://doi.org/10.1007/978-3-319-73900-7_8

future it is necessary an integrated approach for optimizing production processes to be developed. In our opinion, this is the basis for successful achievement of the commercially important yields of bioactive substances produced by hairy roots of *Salvia* species.

Keywords *Agrobacterium rhizogenes* · Hairy roots · Bioactive substances Salvia · Secondary metabolites

1 Introduction

Transformed root cultures (hairy roots) are root-like forms developed after infection of plant tissues with soil bacterium *Agrobacterium rhizogenes*. Infected plant develops an intensive lateral branching root tissue [1].

Using of hairy roots in plant biotechnologies started during 1930s [2]. During 1980s and in the beginning of 1990s this type in vitro cultures attracted attention of the researchers to investigate them as an experimental matrix for production of valuable secondary metabolites [3–6].

Hairy roots are prospective alternative for producing bioactive substances possessing huge number of advantages as in vitro plant technology [7–10]. The advantages and the problems for their industrial implementation are summarized in Table 1.

Plants from *Salvia* species are widely used in traditional medicine. The major biological activities possessed by *Salvia* plants are antioxidant, anti-inflammatory, anti-tumor activities and etc. These activities are consequence of biosynthesized biologically active substances: sterols, various polyphenol structures, di- and

Table 1 Advantages and disadvantages of hairy roots for producing plant secondary metabolites

Advantages	Disadvantages
Genetic and biochemical stability	Availability in their genome of t-DNA of *Agrobacterium rhizogenes*
Growing in media free of growth regulators	Mainly dicotyledonous plants could be transformed with *Agrobacterium rhizogenes*
Synthesized metabolites are with comparable yields to the intact plants	Produce mainly metabolites that are synthesized in the roots of corresponded intact plants
Stable growth characteristics at the cultivation in bioreactors	Bioreactor design should be adopted to the morphological peculiarities of the hairy roots
Fast growth and short doubling time comparable to plant suspension cultures	
Missing of geotropism	
Possibility of manipulation of biosynthetic pathways	

triterpenes, etc. Based on their widespread ethno-application, some *Salvia* species (*S. sclarea*, *S. officinalis*, *S. fruticosa syn. triloba* and *S. lavandulaefolia*) are recognized as safe in the USA (FDA, 2012) and their medicinal products are officially presented in the European Pharmacopoeia VI edition. This determines the importance of the *Salvia* genus, as well as its application in the food, pharmaceutical and cosmetic industries [11].

The wide range of ethnopharmacological use of *Salvia* plants, as well as the huge diversity of biologically active secondary metabolites, found in this genus, have been the driving force in development of plant in vitro biotechnology for exploring their biosynthetic potential and commercial production of pharmaceuticals. However, at the time of writing this chapter, only several reports for *Agrobacterium rhizogenes* mediated transformation of limited *Salvia* species (*S. miltiorrhiza*, *S. officinalis*, *Salvia austriaca*, *Salvia runcinata*, *Salvia wagneriana* and *S. sclarea*) have been found [11]. Among them, the transformed hairy root cultures of *Salvia miltiorrhiza* have been the most well studied [12].

In this chapter we present an overview of the major advantages of hairy roots as in vitro systems for the production of biological secondary plant metabolites. We examine the main metabolic and molecular events during genetic transformation, the problems with the transformation of *Agrobacterium rhizogenes* of tissues from intact plants of *Salvia* genus and approaches to overcome them. A summary of the biologically active substances produced from hairy roots of *Salvia* plants, as well as our point view about developing economically significant processes for their production are presented as well.

2 *Agrobacterium rhizogenes*: A Natural Genetic Engineer

High interest in the genus *Agrobacterium* Conn 1942 is related to the investigation of apple plant diseases that caused enormous financial losses in the 1930s [13]. It has been found that the causative agents of the hairy roots and crown gall are *Agrobacterium rhizogenes* Conn 1942 and *Agrobacterium tumefaciens* Conn 1942, respectively, which together with the non-pathogenic bacterium *Agrobacterium radiobacter* Conn 1942 were separated from the genus *Phytomonas Donovan* 1909 into the genus *Agrobacterium* Conn 1942 [14]. Later, to this genus were included the *Agrobacterium rubi* Starr and Weiss 1943, which causes tumor formation on sugar cane; *Agrobacterium vitis* Ophel and Kerr 1990 causing tumor formation on the grape plants and *Agrobacterium larrymoorei* Bouzar and Jones 2001, causing airborne tumors on the fig plants [14–18].

Bacteria of the genus *Agrobacterium* are small, short non-spore forming rod-like Gram-negative bacteria. They are mobile and have 1–4 flagella. They are found primarily in the soil or as pathogens attacking root systems or causing plant stem hypertrophy [14, 19]. The systematic of the genus *Agrobacterium* has not yet been uniformed. Nowadays, some scientists [20] supported "taxonomic validity of the genus *Agrobacterium*" and other [14, 21, 22] its assignment to the genus *Rhizobium*

Frank 1889. It seems the dispute will continue. Valuable discussion on this topic is presented by Young and co-authors [22]. For the purposes of the plant biotechnology, another classification of phytopathogens of the genus *Agrobacterium* is often used. It has a practical application and it is based on the type of opines that are synthesized by the transformed tissue after infection. According to this classification, the existing strains of agrobacteria can be applied to the following types: octopine, chrysopine/succinamopine, nopaline, succinamopine, cucumopine/mikimopine, agropine, mikimopine, mannopine, agropine/mannopine, chrysopine/nopaline, octopine/cucumopine and cucumopine [23].

The ability of *Agrobacterium rhizogenes* to promote the formation of transformed roots is due to the presence of a relatively large plasmid (200–800 kb) named Ri (root-inducing) plasmid [24–27]. The transfer of a t-DNA segment from this plasmid to the plant cell is mediated by contact of agrobacteria with injured plant tissues [19, 28–30]. It has been established that injured plant tissues secrete into the environment phenolic components and sugars as a protective reaction. These compounds, mainly the acetosyringone (Fig. 1), activate the expression of virulent genes from the bacterial plasmid responsible for t-DNA transfer.

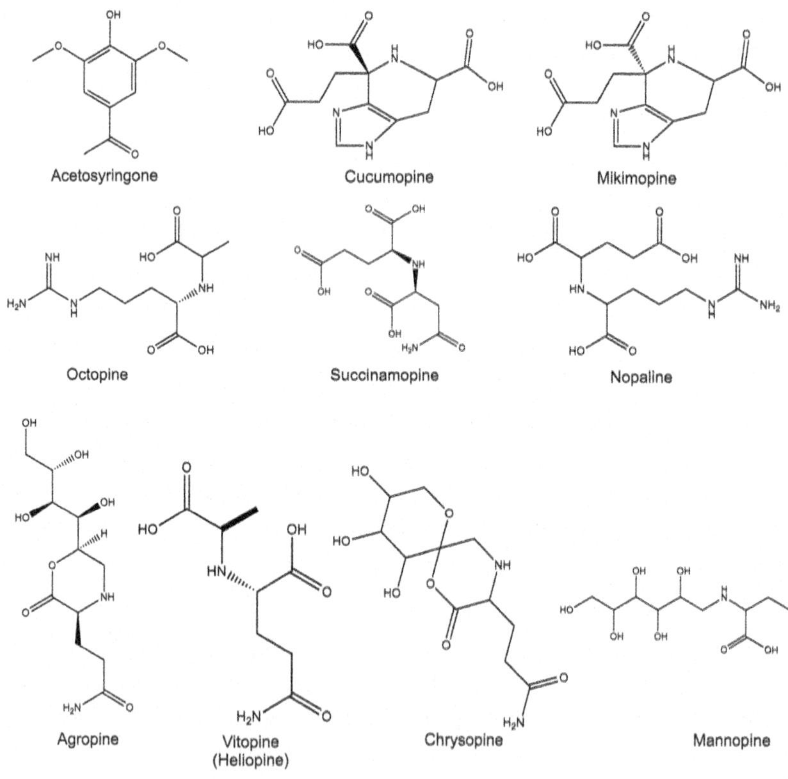

Fig. 1 Structures of acetosyringone and some opines

Transported into the plant cell, t-DNA is integrated permanently into the plant nucleus DNA [28, 31–33] and induces unnatural growth of plant tissue in the form of fine lateral branched roots [13, 34, 35]. Thus, agrobacteria perform "genetic colonization" of susceptible plant species by forcing them to synthesize and secrete into the environment non-specific compounds called opines [27, 36].

Different strains of *Agrobacteria* provoke the synthesis of various opines [37, 38]. The opines are compounds that are synthesized by condensation of α-keto or amino acid compounds with sugars (Fig. 1). They are utilized as a carbon source by the *Agrobacteria*, thus giving them an advantage over other soil microorganisms. These compounds are exclusively synthesized by transformed plant cells [36–38]. Each bacterial Ri-plasmid contains the information on the synthesis and the catabolism of one or several opines [23, 36, 39, 40].

The transferred t-DNA is a single chain short DNA part of the *Agrobacterium* Ri-plasmid, limited at both ends by right and left boundaries, consisting of 25nb "imperfect direct border repeats" [32, 41, 42]. It was established that t-DNA genes do not participate in the transport and integration processes, and any DNA fragment located between the right and left borders can be transferred and integrated into the appropriate host plant cell [32, 42–44]. Transfer of t-DNA is mediated by T-flagella, which are a virulent-specific type IV secretion (T4S) or so-called type IV transport in Gram-negative pathogens [40, 45–47]. The processes of transfer and integration of t-DNA, as well as the genes and proteins involved in the process, are well studied and described [13, 15, 28, 34, 41, 42, 48–50].

Most strains of agrobacteria transfer one DNA fragment, but some such as the agropine-type *A. rhizogenes* strains carry the transfer of two independent t-DNAs—TL-DNA and TR-DNA fragments [13, 51–53]. Both fragments—TR and TL-DNA are transferred and integrated into the plant genome independently of each other, but the transfer of TL-DNA is essential for the formation of hairy root phenotype [42].

By now, the molecular mechanism of root formation after *A. rhizogenes* transformation has not yet been fully elucidated. The abundant root formation in *A. rhizogenes* transformed plants is not a consequence of a change in the hormonal balance of plant cells, but most likely of their sensitivity to auxin action [13, 50]. A key role in this process plays the products of the four role genes—Rol A, Rol B, Rol C and Rol D. For example, the *rolB* gene product—the Rol B protein, with its tyrosine phosphatase activity, enhances the sensitivity and susceptibility of the cell to the auxins. This increased sensitivity to the auxins leads to formation of meristem cells. The newly formed meristems differentiate into tissues depending on the local phytohormonal balance, the physiological status of the plant and the general condition of the cell [27, 34]. This fact explains the occurrence of a root phenotype as far as the roots are usually the first differentiated organs of the plant organism [34].

In summary, the description of the natural transgenic system of *Agrobacterium* bacteria is a fundamental basis for the development of algorithms for the production of both high productive plant in vitro systems and new transgenic plants as well. Much of the mechanisms leading to the appearance of hairy roots are described at the molecular level, but important questions remain unclear and should be answered—

generally about the specific activities of some of the oncogene products. From a bioprocess engineering point of view, the use of hairy roots is preferable to other types of plant in vitro systems since they are genetically homogeneous and stable regarding biosynthetic and morphological characteristics.

3 Frequently Used Methods for *Agrobacterium rhizogenes* Mediated Transformation

There is no necessity for special equipment to obtain hairy roots [54, 55]. Practically, the procedure for *Agrobacterium* transformation involves co-cultivation in aseptic conditions of a sterile explant (injured parts of a plant) and a suspension of *Agrobacterium rhizogenes* [56]. Most frequently, two approaches are used—a co-cultivation method and a direct infection method (Fig. 2). The direct infection method is a simple inoculation of sterile plant explants with a suspension of *Agrobacterium rhizogenes.* The co-cultivation method involves a step of co-cultivating the injured plant explant and the bacterial suspension in a liquid culture medium over a period up to 6 h. This step aims advance activation of virulent bacterial genes by secreted signal molecules from the injured plant tissue and is used primarily for "more difficult" plant species [57]. The next step is the cultivation of infected plant explants on a solid medium to effect the transformation (about 72 h), removal of agrobacteria by transferring to media supplemented with antibiotic and isolation of lines from the resulting hairy roots (Fig. 2). The use of the appropriate antibiotics is essential as far as it is necessary to effectively remove the bacterial cells without killing the plant explant. The most commonly used antibiotics are cefotaxime [54] and penicillin derivatives [58]. The developed hairy roots could be used for separation of lines growing as a lateral branching root mass and they are cultivated on standard nutrient media without growth regulators [52]. After selection, these lines can be used as producers of secondary metabolites in vitro [56, 59, 60], for regeneration of transgenic plants [61], for the production of artificial seeds [62], for phytoremediation [63], and for expression of heterologous protein in molecular farming [64, 65]. Essential for successful transformation are the plant species, the type of *Agrobacterium* strain used, the age and degree of differentiation of the selected plant explant, the concentration and age of the bacteria, the time and the co-cultivation method, the ratio of the bacterial cells to the plant explants, the pH of the medium as well as the addition of transformation-enhancing substances such as acetosyringone, sugars and etc. [56, 57, 66]. By now, different methods have been developed to produce hairy roots from a number of "difficult" for transformation plants:

- **Ultrasonic treatment**—By this method are successfully transformed *Leucojum aestivum* L. [67], *Papaver somniferum* L. [68], *Phtheirospermum japonicum* [69]. The use of this approach allows the opening of micropores in the cell walls of the transformed plant cells, which facilitates the process of transformation.

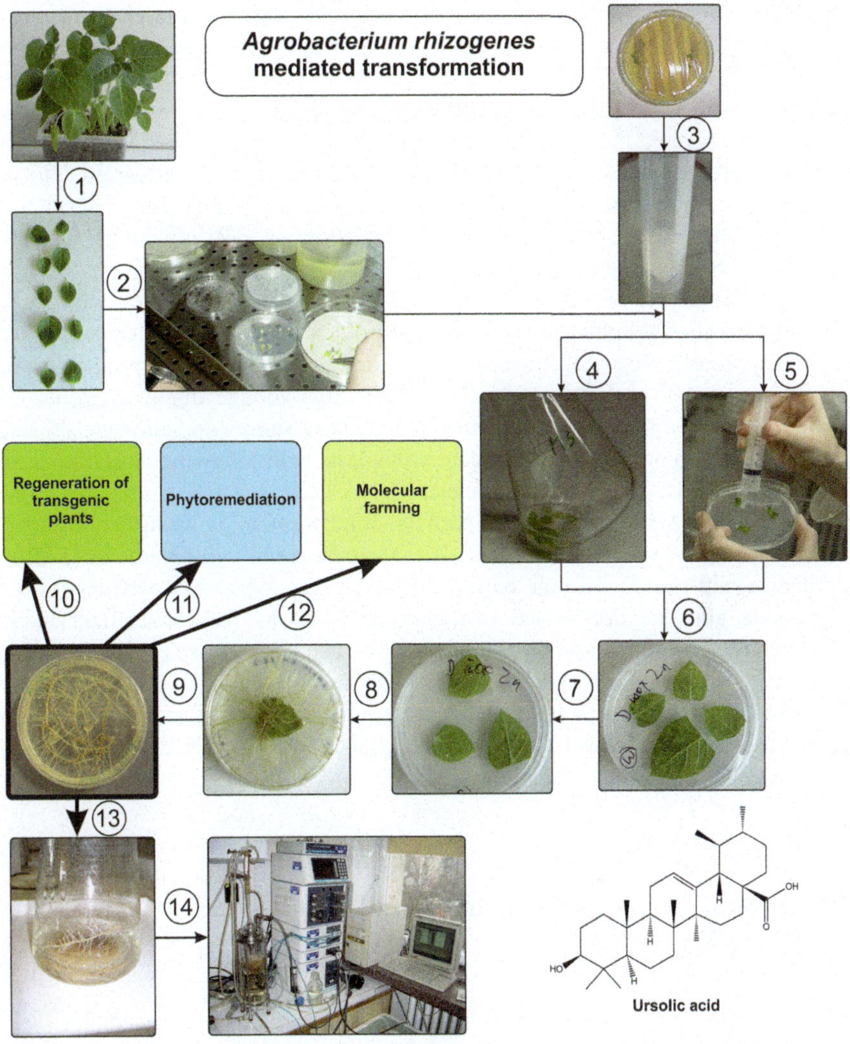

Fig. 2 General scheme for hairy roots obtaining and some of their applications [1]: 1 Selection of explant; 2 Sterilization; 3 Preparation of *Agrobacterium rhizogenes* liquid culture; 4 Co-cultivation of plant explant with bacteria in liquid medium for 2–3 h. on shaker; 5 Direct infection of plant explant with bacterial suspension; 6 Cultivation of infected explants on solid medium for 72 h; 7 Cultivation of transformed explants on solid medium with antibiotic for removal of bacteria; 8 Separation of individual transformed root lines; 9 Selection of high producing root lines; 10 Regeneration of transgenic plants; 11 Phytoremediation of waste water; 12 Expression of heterologous protein in molecular farming; 13 Adaptation to conditions of submerged cultivation; 14 Biotechnological production of valuable secondary metabolites

- **Under vacuum**—This is another approach that facilitates the crossing of the cell wall and cytoplasmic membrane by increasing the size of the membrane pores [70].
- **Treatment with different chemicals**—Plant protoplasts treated with polyethylene glycol, for example, are susceptible to passing the DNA from the environmental medium. Subsequently, this DNA is stably integrated into the plant cell chromosomal DNA [71]. Then, the protoplasts are cultured in cell wall regeneration media and the regenerated, transformed cells initiate the development of hairy roots [72].
- **Electroporation**—Electroporation usually is used in combination with protoplasts because the plant cell wall prevents the crossing macromolecules [73]. In this approach, electrical impulses are applied to a mixture of a protoplast suspension and a bacterial DNA. Short-term, high-voltage impulses induce the formation of micropores in cell membranes for a short time, allowing the bacterial DNA to pass through the plant protoplasts with following its stable steady integration into the plant cell's nucleus DNA [72].
- **Gene Gun Microprojectiles**—Particle bombardment is using to introduce bacterial DNA into plant cells [74–77]. Most frequently gold particles (1–2 μm in size) are coated with the bacterial DNA. Then, the coated particles using a particle gun, are accelerated to highspeed either by the electrostatic energy (highvoltage applied to a droplet of water) or using pressurized helium gas. The target are intact plant tissues, calli or plant cell suspensions. When the microprojectiles enter into the plant cells, bacterial DNA is released from the surface of the gold particles and subsequent stable incorporates into the nucleus DNA of the transforming plant cells [72, 78].

4 *Agrobacterium* Transformation of *Salvia* spp.: Problems and Decisions

It is well known that *Salvia* plants accumulate a wide range of metabolites with antibacterial activity [78–83]. Various plant extracts showed strong antibacterial activity against both gram-positive and gram-negative microorganisms, including *Agrobacteria* [84]. This fact is the basis of the observed difficulties in *Agrobacterium*-mediated transformations of various *Salvia* species. Grzegorczyk and co-workers [85] have reported that the success of *Agrobacterium rhizogenes* transformation directly depends on the type of explant, the *Agrobacterium rhizogenes* strain, and the presence of acetosyringone in the medium. The authors failed to obtain hairy roots when using leaves of intact plants of *Salvia officinalis* as explants. Probably the basis of this fact is the accumulation of large quantities of polyphenols that inhibit the growth of *Agrobacteria* and can also negatively affect the growth of newly formed young hairy roots. Based on this, Grzegorczyk and co-workers [85] achieved a successful transformation using in vitro grown shoots as explants known to accumulate significantly lower amounts of polyphenols than intact tissues. Later Luwańska and co-authors [84] concluded that the considerable difficulties in transforming sage with *Agrobacterium tumefaciens* are associated with the high antibacterial activity of the metabolic mix against a wide range of microorganisms, including *Agrobacteria*, as the negative effect of intact explants on the growth of *Agrobacterium tumefaciens* reduced the probability of the successful integration of T-DNA from the bacterial plasmid into the genome of transformed plant cells.

Similar problems have been observed also in our laboratory in the obtaining hairy roots of different Bulgarian species of the genus *Salvia* [80]. We have established, however, that beside the growth of *Agrobacterium rhizogenes*, polyphenol structures accumulated in intact plant tissues also affect the development of the young hairy roots [86]. To investigate and solve the problem, we used a model system for obtaining hairy roots of *Salvia tomentosa* after transformation with *Agrobacterium rhizogenes* ATCC 15834. Necrosis of the explants used from intact plants was observed, which began from the wounding points and rapidly spread over the entire explant. Obviously, this necrosis was due to the polyphenols secreted by plant cells in wounding points—leakages of brownsap from the wounding points were observed. To prevent necrosis of the explants and to prevent inhibition of the development of young hairy roots, the solution was to find a way to remove secreted polyphenols from the culture system. Addition of the adsorption resin Amberlite XAD-4 was the right decision. As the addition of Amberlite XAD-4 packages to the culture medium, mechanically destroyed the newly obtained hairy roots, we combined the addition of Amberlite XAD-4 with explant cultivation in a temporary immersion cultivation system. The design of this type of bioreactors allows the separation of the resin from the plant explants. Thus, developed new "Two-phase, temporary immersion cultivation system" resulted in removing 93% of the phenolic acids secreted in the medium and 100% of the secreted flavanoids. The developed new approach to *Agrobacterium rhizogenes* transformation solved the problem of the negative effect of secreted, from intact explants, polyphenol structures that possess antimicrobial and phytotoxic activities. In our opinion, the developed protocol is applicable both for transformations with *Agrobacterium rhizogenes* and for transformations with *Agrobacterium tumefaciens* of various plants that produce large quantities of polyphenol structures.

5 Bioactive Substances by Hairy Roots of *Salvia*

To the best our knowledge, the first successful transformation of tissues from a plant of the genus *Salvia* was carried out by Hu and Alfermann during 1993 [87]. They obtained hairy roots of *Salvia miltiorrhiza* by genetic transformation of in vitro plants with different *Agrobacterium rhizogenes* strains—LBA 9402, ATCC 15834, TR 105, R 1601, A 4 1027 and proved biosynthesis oftanshinone I, tanshinone II, tanshinone IIa, tanshinone IIb, tanshinone V, dihydrotanshinone I, criptotanshinone, tanshinone VI and ferrugiol.

Salvia miltiorrhiza is one of the most frequently used plants in traditional Chinese medicine for improving blood circulation, in treating neural and cardiovascular diseases [88]. On this basis, the publications on hairy roots of *Salvia miltiorrhiza* are the major part of the scientific information, currently available, presenting biosynthesis of biologically active secondary metabolites in these plant in vitro systems. The main metabolites accumulated in *Salvia miltiorrhiza* hairy roots can be summarized in two groups: (1) polar phenolic acids [88, 89] and (2) unipolar, mainly tanshinones [87]. Detailed information on the individual biologically active structures biosynthesized by *Salvia miltiorrhiza* hairy roots is presented in Table 2. Hairy roots from this plant were used for fundamental research to identify and structurally elucidate new structures [90]; for molecular biological studies of the biosynthetic pathways of the secondary metabolites [91–94]; to determine their biological activities [95], for investigation of the possibilities for increasing the yields of biologically active secondary metabolites and for bioprocess engineering investigations [88, 89, 96–102], for the regeneration (micropropagation) of intact plants [103], as well as for the development of new cultivation systems based on co-cultivation of hairy roots and bacteria [96].

Kuzma and co-authors [104, 105] reported on successfully obtained hairy roots from *Salvia austriaca*, which biosynthesize the unknown diterpene, a taxodione derivate—7-(2-oxohexyl)-11-hydroxy-6,12-dioxo-7,9(11),13-abietatriene [7-(2-oxohexyl)-taxodione] in parallel with known bioactive diterpenes—taxodione, royleanone and 15-deoxyfuerstione. In vitro studies showed that the identified new diterpene had a high cytotoxic activity—10 times higher than that of taxodione [105], as well as antibacterial/antibiofilm-forming activity [106].

Kuzma and co-workers [107, 108] investigated hairy roots cultures of *Salvia sclarea* for biosynthesis of diterpenes. They observed increased yields of these structures during the cultivation of the hairy roots in a sprinkle bioreactor compared to their cultivation in flasks on a shaker. After elicitation with methyl jasmonate, the increases in the yields of the investigated diterpenes were between 3.8- and 9-folds. These results were probably due to improved mass-exchange in the bioreactor system and are of practical significance to the further development of a commercial biosynthetic process [109].

Figlan and Makunda [110] investigated the genetic transformation of *Salvia runcinata* with two agropine strains *Agrobacterium rhizogenes* (A4T and LBA 9402). The authors found out that stable hairy root lines were obtained only after transformation with *Agrobacterium rhizogenes* A4T, although initial root formation was more intense after the transformation of leaf explants with *Agrobacterium rhizogenes* LBA 9402. The next GC/MS metabolic profiling showed that the obtained hairy roots are potential producers of rosmarinic, shikimik and caffeic acids, which are known for their antioxidant and antimicrobial activities.

Ruffoni et al. [111] induced hairy roots of *Salvia wagneriana* by transformation with *Agrobacterium rhizogenes* of leaf lamina and petiole fragments. Obtained hairy roots synthesized significant amounts of rosmarinic acid, carnosol and

Table 2 Bioactive substances biosynthesized from *Salvia* hairy roots

Hairy root in vitro system	Bioactive substances	References
Salvia miltiorrhiza	– Rosmarinic acid – Salvanolic acid A – Salvanolic acid B – Lithospermic acid B – t-cinamic acid – 4-coumaric acid – Flavonoids – Tanshinone I – Tanshinone II – Tanshinone IIa – Tanshinone IIb – Tanshinone V – Dihydrotanshinone I – Criptotanshinone – Tanshinone VI – Ferrugiol	[86, 87, 115]
Salvia austriaca	– Taxodione – Royleanone – 15-deoxyfuerstione – 7-(2-oxohexyl)-11-hydroxy-6, 12-dioxo-7,9 (11),13-abietatriene [=7-(2-oxohexyl)-taxodione]	[104, 105]
Salvia sclarea	– Salvipisone – Aethiopinone – 1-oxoaethiopine – Salvipisone – Ferrugiol	[106–108]
Salvia runcinata	– Caffeic acid – Rosmarinic acid – Shikimic acid	[110]
Salvia wagneriana	– Rosmarinic acid – Carnosol – Carnosic acid	[111]
Salvia broussonetii	– Brussonol – Iguestol	[116]
Salvia officinalis	– Rosmarinic acid – Carnosic acid – Carnosol – *p*-coumaric acid – ferulic acid – Gallic acid – Protocatechuic acid – Salicylic acid – Sinapic acid – Syringic acid – *trans*-cinnamic acid – Vanillic acid	[84, 112, 113]

carnosic acid. The biosynthesis of rosmarinic acid was affected by the addition of casein hydrolyzate to the cultivation system, but was not affected by elicitation with methyl jasmonate.

Grzegorczyk et al. [87] initially reported for rosmarinic acid biosynthesis of *Salvia officinalis* hairy roots and investigated also the antioxidant capacity of culture extracts [112] and explained this biological activity to biosynthesized polyphenols and diterpenes. They proved that extracts from transformed root cultures have a higher antioxidant activity compared to extracts from undifferentiated callus and suspension cultures. Grzegorczyk and Wysokinska [113] then investigated the antioxidant structures synthesized by this culture during its cultivation in a nutrient sprinkle bioreactor. The authors report significant yields of rosmarinic acid (34.7 mg/L), carnosic acid (1.7 mg/L) and carnosol (1.3 mg/L). The results obtained could be determined as a good basis for the next development of a commercial process for obtaining them.

Delenk et al. [114] investigated extracts from *Salvia officinalis* hairy root cultures for antifungal activity against cellulose degrading molds *Chaetomium globosum* and *Trichoderma viride*. The authors analyzed the patterns of the phenolic acids synthesized from the hairy roots and found that two of them, *trans*-cinnamic acid and salicylic acid, had antifungal activity against *Chaetomium globosum* comparable to that of cyproconazole. In addition, sinapic acid inhibited the development of both tested molds at lower concentrations, whereas cyproconazole, as a fungicidal control, inhibited only the development of *Trichoderma viride*.

6 Conclusion and Future Prospects

The reports in the scientific literature on hairy roots of genus *Salvia* are limited. The most *Salvia* hairy roots are obtained after *Agrobacterium rhizogenes* transformation of in vitro grown explants as far as intact plants are characterized with a high content of polyphenols. This affects the transformation negatively, suppressing the growth of the bacterial strain or causing explant necrosis. On this basis, it is necessary to optimize the conditions of the genetic transformation—the type and concentration of the *Agrobacteria*, the induction of the process by acetosyringone, etc., as well as to develop principle new experimental approaches to overcome this problem.

The *Salvia* hairy roots are producers of diterpenes and phenolic compounds. There are no reports on accumulation of triterpenes from them. Nowadays, the investigations on development of production processes, based on *Salvia* hairy roots, are still in early stage and many points still have to be addressed before the scale-up and the commercialization of the processes to take place. Investigations are concentrated on phytochemical profiling and simple assessment of influence of different independent variables (mainly elicitors) on the yields of the target bioactive substances. In our opinion, economically significant yields of secondary metabolites from hairy root in vitro systems could be obtained only on the basis of complex optimization of the individual components of the biological system (conditions of genetic transformation; nutrient medium—factorial analysis of the impact of the main food components and full factor experiment for the optimization of relationships in the biological system; bioreactor system—design and optimization of cultivation conditions; scale-up of the production process), as well as on the basis of detailed knowledge of the pathways of target metabolites and the mechanisms of their regulation. In recent years considerable progress is visible in the research of plant endophytes and in particular their use for the manipulation of the biosynthesis. When fungal medium filtrate of plant endophytes was applied on Salvia fruticose suspension culture the yield of triterpenes were increased [117]. A similar yield increasing effect for biologically active secondary metabolites produced by hairy roots was observed by [118]. Therefore this aspect should also be included as a component of the complex optimization of the biological system based on *Salvia* hairy roots. Only by using an integrated approach for optimizing production processes could be achieve economically significant yields and on this basis could be developed commercial processes.

References

1. Georgiev V (2008) Biosynthesis of betalain pigments by plant in vitro systems of Beta vulgaris. Ph.D. thesis, The Stefan Angeloff Institute of Microbiology, Bulgarian Academy of Sciences, in Bulgarian
2. White PR (1934) Potentially unlimited growth of excised tomato roots in culture. Plant Physiol 9:586–600
3. Flores HE, Hoy MW, Pickard JJ (1987) Secondary metabolites from root cultures. Trends Biotechnol 5(3):64–69
4. Charlwood BV, Charlwood KA, Molina-Torres J (1990) Accumulation of secondary compounds by organized plant cultures. In: Charlwood BV, Rhodes MJC (eds) Secondary products from plant tissue culture. Oxford University Press, New York, pp 167–200. ISBN 0198577176
5. Robins RJ (1998) The application of root cultures to problems of biological chemistry. Nat Prod Rep 15(6):549–570
6. Walton NJ, Alfermann AW, Rhodes MJC (1999) Production of secondary metabolites in cell and differentiated organ cultures. In: Wink M (ed) Functions of plant secondary metabolites and their exploitation in biotechnology. CRC Press, Boca Raton, pp 311–346. ISBN-13: 978-0849340864
7. Georgiev MI, Pavlov AI, Bley Th (2007) Hairy root type plant in vitro systems as sources of bioactive substances. Appl Microbiol Biotechnol 74:1175–1185
8. Chandra Sh, Chandra R (2011) Engineering secondary metabolite production in hairy roots. Phytochem Rev 10:371–396
9. Zhou M-L, Zhu X-M, Shao J-R, Tang Y-X, Wu Y-M (2011) Production and metabolic engineering of bioactive substances in plant hairy root culture. App Microbiol Biotechnol 90 (4):1229–1239
10. Steingroewer J, Bley Th, Georgiev V, Ivanov I, Lenk F, Marchev A, Pavlov A (2013) Bioprocessing of differentiated plant in vitro systems. Eng Life Sci 13(1):26–38

11. Marchev A (2014) Biologically active substances from rare Bulgarian Salvia species and their in vitro cultures. Ph.D. thesis, The Stefan Angeloff Institute of Microbiology, Bulgarian Academy of Sciences, in Bulgarian
12. Dreger M, Krajewska-Patan A, Górska-Paukszta M, Pieszak M, Buchwald W, Mikołajczak P (2010) Production of the secondary metabolites in *Salvia miltiorrhiza* in vitro cultures. Herba Polonica 56(4):78–90
13. Nilsson O, Olsson O (1997) Getting to the root: the role of the *Agrobacterium rhizogenes* rol genes in the formation of hairy roots. Physiol Plant 100(3):463–473
14. Young JM (2008) Agrobacterium—Taxonomy of plant-pathogenic rhizobium species. In: Tzfira T, Citovsky V (eds) Agrobacterium: from biology to biotechnology. Springer, New York, pp 183–220. ISBN 978-0-387-72289-4
15. Gelvin SB (2003) Agrobacterium-mediated plant transformation: the biology behind the "Gene-jockeying" Tool. Microbiol Mol Biol Rev 67(1):16–37
16. Escobar MA, Dandekar AM (2003) *Agrobacterium tumefaciens* as an agent of disease. Trends Plant Sci 8(8):380–386
17. Tzfira T, Citovsky V (2006) Agrobacterium—mediated genetic transformation of plants: biology and biotechnology. Curr Opin Biotechnol 17:147–154
18. Ozyigit II, Dogan I, Tarhan EA (2013) *Agrobacterium rhizogenes*-mediated transformation and its biotechnological applications in crops. In: Hakeem KR, Ahmad P, Ozturk M (eds) Crop improvement—new approaches and modern techniques. Springer, Berlin, pp 1–48. ISBN: 978-1-4614-7027-4
19. Matveeva TV, Sokornova SV (2016) *Agrobacterium rhizogenes*-mediated transformation of plants for improvement of yields of secondary metabolites. In: Pavlov A, Bley Th (eds) Bioprocessing of plant in vitro systems, Reference Series in Phytochemistry. https://doi.org/10.1007/978-3-319-32004-5_18-1. Springer International Publishing AG, Switzerland. (In press)
20. Farrand SK, Van Berkum PB, Oger P (2003) *Agrobacterium* is a definable genus of the family *Rhizobiaceae*. Int J Syst Evol Microbiol 53(Pt 5):1681–1687
21. Young JM, Kuykendall LD, Martinez-Romero E, Kerr A, Sawada H (2001) A revision of Rhizobium Frank 1889, with an emended description of the genus, and the inclusion of all species of Agrobacterium Conn 1942 and Allorhizobium Undicola de Lajudie et al. 1998 as new combinations: *Rhizobium radiobacter*, *R. rhizogenes*, *R. rubi*, *R. undicola* and *R. vitis*. Int J Syst Evol Microbiol 51(1):89–103
22. Young JM, Kuykendall LD, Martinez-Romero E, Kerr A, Sawada H (2003) Classification and nomenclature of Agrobacterium and Rhizobium—a reply to Farrand et al. (2003). Int J System Evol Microbiol 53(5):1689–1695
23. Otten L, Burr T, Szegedi E (2008) Agrobacterium: a disease-causing bacterium. In: Tzfira T, Citovsky V (eds) Agrobacterium: from biology to biotechnology. Springer, New York, pp 1–46. ISBN 978-0-387-72289-4
24. White FF, Nester EW (19980a) Relationship of plasmids responsible for hairy root and crown gall tumorigenicity. J Bacteriol 144(2):710–720
25. White FF, Nester EW (1980) Hairy root: plasmid encodes virulence traits in *Agrobacterium rhizogenes*. J Bacteriol 141(3):1134–1141
26. Gartland JS (1995) *Agrobacterium* virulence. In: Gartland KMA, Davey MR (eds) Agrobacterium protocols. Humana Press Inc, New Jersey, pp 15–28. ISBN 978-0-89603-302-3
27. Georgiev M, Georgiev V, Weber J, Bley Th, Ilieva M, Pavlov A (2008) *Agrobacterium rhizogenes*-mediated genetic transformations: a powerful tool for the production of metabolites. In: Wolf TV, Koch JP (eds) Genetically modified plants. Nova Science Publishers, Hauppauge, pp 99–126. ISBN 978-1-60456-696-3
28. Winans SC (1992) Two-way chemical signaling in Agrobacterium-plant interactions. Microbiol Rev 56(1):12–31

29. Veena Jiang H, Doerge RW, Gelvin SB (2003) Transfer of t-DNA and vir proteins to plant cells by *Agrobacterium tumefaciens* induces expression of host genes involved in mediating transformation and suppresses host defense gene expression. Plant J 35(2):219–236
30. Tzfira T, Citovsky V (2003) The Agrobacterium-plant cell interaction. Taking biology lessons from a bug. Plant Physiol 133(3):943–947
31. White FF, Garfinkel DJ, Huffman GA (1983) Sequences homologous to *Agrobacterium rhizogenes* t-DNA in the genomes of uninfected plants. Nature 301(5898):348–350
32. Caplan AB, Van Montagu M, Schell J (1985) Genetic analysis of integration mediated by single t-DNA borders. J Bacteriol 161(2):655–664
33. Tzfira T, Citovsky V (2000) From host recognition to t-DNA integration: the function of bacterial and plant genes in the Agrobacterium-plant cell interaction. Mol Plant Pathol 1 (4):201–212
34. Britton MT, Escobar MA, Dandekar AM (2008) The oncogenes of *Agrobacterium tumefaciens* and *Agrobacterium rhizogenes*. In: Tzfira T, Citovsky V (eds) Agrobacterium: from biology to biotechnology. Springer, New York, pp 523–563. ISBN 978-0-387-72289-4
35. Tzfira T, Lacroix B, Citovsky V (2013) Nuclear import of Agrobacterium t-DNA. Madame Curie Bioscience Database [Internet]
36. Saito K, Yamazaki M, Murakoshi I (1992) Transgenic medicinal plants: Agrobacterium-mediated foreign gene transfer and production of secondary metabolites. J Nat Prod 55(2):149–162
37. Dessaux Y, Petit A, Tempe J (1993) Chemistry and biochemistry of opines, chemical mediators of parasitism. Phytochem 34(1):31–38
38. Nester EW (2015) Agrobacterium: nature's genetic engineer. Front Plant Sci/Plant Biotechnol 5, Article 730:730
39. Cho K, Fuqua C, Winans SC (1997) Transcriptional regulation and locations of *Agrobacterium tumefaciens* genes required for complete catabolism of octopine. J Bacteriol 179(1):1–8
40. Llosa M, Zupan J, Baron C, Zambryski P (2000) The n- and c-terminal portions of the Agrobacterium virB1 protein independently enhance tumorigenesis. J Bacteriol 182 (12):3437–3445
41. Ziemienowicz A, Tzfira T, Hohn B (2008) Mechanisms of t-DNA integration. In: Tzfira T, Citovsky V (eds) Agrobacterium: from biology to biotechnology. Springer, New York, pp 395–440. ISBN 978-0-387-72289-4
42. Hillebrand H, Hell R (2011) Plant Biotechnology. In: Wink M (ed) An introduction to molecular biotechnology. Molecular fundamentals, methods and application in modern biotechnology. Wiley-VCH Verlag GmbH & Co. KGaA, Germany, pp 431–450. ISBN: 978-3-527-32637-2
43. Fritze K, Walden R (1995) Gene activation by t-DNA tagging. Methods Mol Biol (Clifton, N.J.) 44:281–294
44. Bako L, Umeda M, Tiburcio AF, Schell J, Koncz C (2003) The virD2 pilot protein of Agrobacterium-transferred DNA interacts with the tata box-binding protein and a nuclear protein kinase in plants. Proc Nat Acad Sci USA 100(17):10108–10113
45. Ward DV, Zambryski PC (2001) The six functions of Agrobacterium virE2. Proc Nat Acad Sci USA 98(2):385–386
46. Winans SC, Bassler BL (2002) Mob psychology. J Bacteriol 184(4):873–883
47. Atmakuri K, Christie PJ (2008) Translocation of oncogenic T-DNA and effector proteins to plant cells. In: Tzfira T, Citovsky V (eds) Agrobacterium: from biology to biotechnology. Springer, New York, pp 315–364. ISBN 978-0-387-72289-4
48. Hoekema A, Hooykaas PJ, Schilperoort RA (1984) Transfer of the octopine T-DNA segment to plant cells mediated by different types of Agrobacterium tumor- or root-inducing plasmids: generality of virulence systems. J Bacteriol 158(1):383–385
49. Gelvin SB (2000) Agrobacterium and plant genes involved in t-DNA transfer and integration. Ann Rev Plant Biol 51:223–256

50. Altamura MM (2004) Agrobacterium rhizogenes rolb and rold genes: regulation and involvement in plant development. Plant Cell Tissue Org Cult 77(1):89–101
51. Huffman GA, White FF, Gordon MP, Nester EW (1984) Hairy-root-inducing plasmid: physical map and homology to tumor-inducing plasmids. J Bacteriol 157(1):269–276
52. Rhodes MJC, Robins RJ, Hamill JD, Parr AJ, Hilton MG, Walton NJ (1990) Properties of transformed root cultures. In: Charlwood BV, Rhodesa MJC (eds) Secondary products from plant tissue culture. Oxford University Press, New York, pp 201–226. ISBN 0198577176
53. Lemcke K, Schmulling T (1998) Gain of function assays identify non-rol genes from *Agrobacterium rhizogenes* TL-DNA that alter plant morphogenesis or hormone sensitivity. Plant J 15(3):423–433
54. Hamill JD, Lidgett AJ (1997) Hairy root cultures opportunities and key protocols for studies metabolic engineering. In: Doran PM (ed) Hairy roots culture and applications. Amsterdam, Harwood Academic Publishers, pp 1–31. ISBN 10: 905702117X
55. Dandekar AM, Fisk HJ (2005) Plant transformation: Agrobacterium-mediated gene transfer. Methods Mol Biol (Clifton, N.J.) 286:35–46
56. Sevon N, Oksman-Caldentey KM (2002) *Agrobacterium rhizogenes*-mediated transformation: root cultures as a source of alkaloids. Planta Med 68(10):859–868
57. Soltani J, Paul G, van Heusden H, Hooykaas PJJ (2008) Agrobacterium-mediated transformation of non-plant organisms. In: Tzfira T, Citovsky V (eds) Agrobacterium: from biology to biotechnology. Springer, New York, pp 649–675. ISBN 978-0-387-72289-4
58. Ur Rahman L, Ikenaga T, Kitamura Y (2004) Penicillin derivatives induce chemical structure-dependent root development, and application for plant transformation. Plant Cell Rep 22(9):668–677
59. Kuzovkina IN (1992) Cultivation of genetically transformed plant roots: possibilities and prospects for use in plant physiology. Soviet Plant Physiol 39(6):810–814
60. Shanks JV, Morgan J (199) Plant 'hairy root' culture. Curr Opin Biotechnol 10(2):151–155
61. Christey MC, Sinclair BK, Braun RH, Wyke L (1997) Regeneration of transgenic vegetable Brassicas (*Brassica oleracea* and *B. campestris*) via Ri-mediated transformation. Plant Cell Rep 16:587–593
62. Uozumi N (2004) Large-scale production of hairy root. Adv Biochem Eng/Biotechnol 91:75–103
63. Guillon S, Tremouillaux-Guiller J, Pati PK, Rideau M, Gantet P (2006) Harnessing the potential of hairy roots: dawn of a new era. Trends Biotechnol 24(9):403–409
64. Xu J, Towler M, Weathers PJ (2016) Platforms for plant-based protein production. In: Pavlov A, Bley Th (eds) Bioprocessing of plant in vitro systems, Reference Series in Phytochemistry. Springer International Publishing AG, Switzerland. https://doi.org/10.1007/978-3-319-32004-5_14-1. (In press)
65. Georgiev V (2015) Mass propagation of plant cells – An emerging technology platform for sustainable production of biopharmaceuticals. Biochem Pharmacol (Los Angel) 4:e180
66. Lin YH, Binns AN, Lynn DG (2008) The initial steps in *Agrobacterium tumefaciens* pathogenesis: chemical biology of host recognition. In: Tzfira T, Citovsky V (eds) Agrobacterium: from biology to biotechnology. Springer, New York, pp 221–241. ISBN 978-0-387-72289-4
67. Diop MF, Hehn A, Ptak A, Chretien F, Doerper S, Gontier E, Bourgaud F, Henry M, Chapleur Y, Laurain-Mattar D (2007) Hairy root and tissue cultures of *Leucojum aestivum* L.—relationships to galanthamine content. Phytochem Rev 6(1):137–141
68. Le Flem-Bonhomme V, Laurain-Mattar D, Fliniaux MA (2004) Hairy root induction of *Papaver somniferum* var. *album*, a difficult-to-transform plant, by *A. rhizogenes* LBA 9402. Planta 218:890–893
69. Ishida JK, Yoshida S, Ito M, Namba S, Shirasu K (2011) *Agrobacterium rhizogenes*-mediated transformation of the parasitic plant *Phtheirospermum japonicum*. PLoS ONE 6 (10):e25802

70. Theboral J, Arun M, Manickvasagam M, Natesan S, Ganapathi A (2017) Sonification and vacuum infiltration enhanced *Agrobacterium rhizogenes* mediated transformation in soybean. Innovare J Agricult Sci 5(2):1–8
71. Mathur J, Koncz C. (1997) PEG-mediated protoplast transformation with naked DNA. Methods Mol Biol 82:267–276. ISSN 1064-3745
72. Narusaka Y, Narusaka M, Yamasaki S, Iwabuchi M (2012) Methods to transfer foreign genes to plants. In: YO Çiftçi (ed) Transgenic plants—advances and limitations. InTech, pp 173–188. ISBN: 978-953-51-0181-9
73. Bates GW (1999) Plant transformation via protoplast electroporation. Methods Mol Biol 111:359–366. ISSN 1064-3745
74. Birch RG, Franks T (1991) Development and optimization of microprojectile systems for plant genetic transformation. Aust J Plant Physiol 18:453–469
75. Christou P (1992) Genetic transformation of crop plants using microprojectile bombardment. Plant J 2:275–281
76. Christou P (1995) Strategies for variety-independent genetic transformation of important cereals, legumes and woody species utilizing particle bombardment. Euphytica 85:13–27
77. Gan C (1998) Gene gun accelerates DNA-coated particles to transform intact cells. Sci 3(18):25
78. Takahashi M, Nishihara M, Yamamura S, Nishizawa S, Irifune K, Morikawa H (1998) Stable transformation of *Eustoma grandiflorum* by particle bombardment. Plant Cell Rep 17(6–7):504–507
79. Delamare APL, Moschen-Pistorello IT, Artico L, Atti-Serafini L, Echeverrigaray S (2007) Antibacterial activity of the essential oils of *Salvia officinalis* L. and *Salvia triloba* L. cultivated in South Brazil. Food Chem 100:603–608
80. Marchev A, Haas C, Schulz S, Georgiev V, Steingroewer J, Bley T, Pavlov A (2014) Sage in vitro cultures: a promising tool for the production of bioactive terpenes and phenolic substances. Biotechnol Lett 36:211–221
81. Viuda-Martos M, Ruiz-Navajas Y, Fernández-López J, Pérez-Álvarez JA (2008) Antibacterial activity of different essential oils obtained from spices widely used in Mediterranean diet. Int J Food Sci Technol 43:526–531
82. Stanojevic D, Comic L, Stefanovic O, Solujic-Sukdolak S (2010) In vitro synergistic antibacterial activity of *Salvia officinalis* L. and some preservatives. Arch Biol Sci 62(1): 167–173
83. Ardalan A, Majid S (2012) Essential oil composition, phenolic content, antioxidant and antimicrobial activity in *Salvia officinalis* L. cultivated in Iran. Adv Environ Biol 6:221–226
84. Luwańska A, Wielgus K, Seidler-Łożykowska K, Lipiński D, Słomski R (2017) Evaluation of *Agrobacterium tumefaciens* usefulness for the transformation of sage (*Salvia officinalis* L.). In: Jha S (ed) Transgenesis and secondary metabolism. Springer, Switzerland, pp 153–176. ISBN 978-3-319-28668-6
85. Grzegorczyk I, Krolicka A, Wysokinska H (2006) Establishment of *Salvia officinalis* L. hairy root cultures for the production of rosmarinic acid. Z Naturforsch 61c:351–356
86. Marchev A, Georgiev V, Ivanov I, Badjakov I, Pavlov A (2011) Two-phase temporary immersion system for *Agrobacterium rhizogenes* genetic transformation of sage (*Salvia tomentosa* Mill.). Biotechnol Lett 33:1873–1878
87. Hu ZB, Alfermann AW (1993) Diterpenoid production in hairy root cultures of *Salvia miltiorrhiza*. Phytochem 32(3):699–703
88. Zhang S, Li H, Liang X, Yan Y, Xia P, Jia Y, Lianga Z (2015) Enhanced production of phenolic acids in *Salvia miltiorrhiza* hairy root cultures by combing the RNAi-mediated silencing of chalcone synthase gene with salicylic acid treatment. Biochem Eng J 103: 185–192
89. Yan Q, Shi M, Ng J, Wu JY (2006) Elicitor-induced rosmarinic acid accumulation and secondary metabolism enzyme activities in *Salvia miltiorrhiza* hairy roots. Plant Sci 170:853–858

90. Wang JW, Wu JY (2010) Tanshinone biosynthesis in *Salvia miltiorrhiza* and production in plant tissue cultures. Appl Microbiol Biotechnol 88:437–449
91. Kai G, Liao P, Xu H, Wang J, Zhou C, Zhou W, Qi Y, Xiao J, Wang Y, Zhang L (2012) Molecular mechanism of elicitor-induced tanshinone accumulation in *Salvia miltiorrhiza* hairy root cultures. Acta Physiol Plant 34:1421–1433
92. Zhao S, Zhang J, Tan R, Yang L, Zheng X (2015) Enhancing diterpenoid concentration in *Salvia miltiorrhiza* hairy roots through pathway engineering with maize C1 transcription factor. J Exp Bot 66(22):7211–7226
93. Ma X-H, Ma Y, Tang J-F, He Y-L, Liu Y-C, Ma X-J, Shen Y, Cui G-H, Lin H-X, Rong Q-X, Guo J, Huang L-Q (2015) The biosynthetic pathways of tanshinones and phenolic acids in *Salvia miltiorrhiza*. Molecules 20:16235–16254
94. Li B, Cui G, Shen G, Zhan Z, Huang L, Chen J, Qi X (2017) Targeted mutagenesis in the medicinal plant *Salvia miltiorrhiza*. Sci Rep 7:43320. https://doi.org/10.1038/srep43320
95. Lou J, Mao Z, Shana T, Wang Q, Zhou L (2014) Chemical composition, antibacterial and antioxidant properties of the essential oils from the roots and cultures of *Salvia miltiorrhiza*. J Essent Oil Bear Plants 17(3):380–384
96. Wu J-Y, Ng J, Shi M, Wu S-J (2007) Enhanced secondary metabolite (tanshinone) production of *Salvia miltiorrhiza* hairy roots in a novel root–bacteria coculture process. Appl Microbiol Biotechnol 77:543–550
97. Gupta SK, Liu R-B, Liaw S-Y, Chan H-S, Tsay H-S (2011) Enhanced tanshinone production in hairy roots of *Salvia miltiorrhiza* Bunge' under the influence of plant growth regulators in liquid culture. Bot Stud 52:435–443
98. Han M, Guo W, Liang Z, Yang D, Yan X, Zhu Y, Liu Y (2015) Effects of cerous nitrate on growth and tanshinone production in *Salvia miltiorrhiza* hairy roots. J Rare Earths 33 (11):1228–12035
99. Xing B, Yang D, Guo W, Liang Z, Yan X, Zhu Y, Liu Y (2015) Ag+ as a More effective elicitor for production of tanshinones than phenolic acids in *Salvia miltiorrhiza* hairy roots. Molecules 20:309–324
100. Wang Y, Shen Y, Shen Z, Zhao L, Ning D, Jiang C, Zhao R, Huang L (2016) Comparative proteomic analysis of the response to silver ions and yeast extract in *Salvia miltiorrhiza* hairy root cultures. Plant Physiol Biochem 107:364–373
101. Wang CH, Zheng LP, Tian H, Wang JW (2016) Synergistic effects of ultraviolet-B and methyl jasmonate on tanshinone biosynthesis in *Salvia miltiorrhiza* hairy roots. J Photochem Photobiol B: Biol 159:93–100
102. Yang D, Huang Z, Xing B, Jin W, Yan X, Guo Z, Liang Z (2016) Regulation of folic acid on phenolic acids production in *Salvia miltiorrhiza* hairy roots. Plant Cell Tissue Org Cult 127:175–185
103. Wang QJ, Zheng LP, Yuan HY, Wang JW (2013) Propagation of *Salvia miltiorrhiza* from hairy root explants via somatic embryogenesis and tanshinone content in obtained plants. Ind Crops Prod 50:648–653
104. Kuzma Ł, Kisiel W, Królicka A, Wysokinska H (2011) Genetic transformation of *Salvia austriaca* by *Agrobacterium rhizogenes* and diterpenoid isolation. Pharmazie 66:904–907
105. Kuźma L, Wysokińska H, Różalski M, Krajewska U, Kisiel W (2012) An unusual taxodione derivative from hairy roots of *Salvia austriaca*. Fitoterapia 83:770–773
106. Kuzma L, Wysokinska H, Rózilski M, Budzynska A, Wieckowska-Szakielc M, Sadowska B (2012) Antimicrobial and anti-biofilm properties of new taxodione derivative from hairy roots of *Salvia austriaca*. Phytomedicine 19:1285–1287
107. Kuzma Ł, Skrzypek Z, Wysokinska H (2006) Diterpenoids and triterpenoids in hairy roots of *Salvia sclarea*. Plant Cell Tissue Organ Cult 84:171–179
108. Kuzma L, Bruchajzer E, Wysokinska H (2008) Diterpenoid production in hairy root culture of *Salvia sclarea* L. Z Naturforsch 63c:621–624
109. Kuzma L, Bruchajzer E, Wysokinska H (2009) Methyl jasmonate effect on diterpenoid accumulation in *Salvia sclarea* hairy root culture in shake flasks and sprinkle bioreactor. Enzyme Microbiol Technol 44:406–410

110. Figlan S, Makunga N (2017) Genetic transformation of the medicinal plant *Salvia runcinata* L. f. using *Agrobacterium rhizogenes*. S Afr J Bot 112:193–202
111. Ruffoni B, Bertoli A, Pistelli L (2016) Micropropagation of *Salvia wagneriana* Polak and hairy root cultures with rosmarinic acid production. Nat Prod Res 4:1–7
112. Grzegorczyk I, Matkowski A, Wysokinska H (2007) Antioxidant activity of extracts from in vitro cultures of *Salvia officinalis* L. Food Chem 104:536–541
113. Grzegorczyk I, Wysokinska H (2010) Antioxidant compounds in *Salvia officinalis* L. shoots and hairy root cultures in the nutrient sprinkle bioreactor. Acta Soc Bot Pol 79:7–10
114. Delenk H, Haas C, Gantz S, Marchev A, Pavlov A, Steudler S, Unbehaun H, Steingroewer J, Bley T, Wagenführ A (2015) Influence of *Salvia officinalis* L. hairy roots derived phenolic acids on the growth of *Chaetomium globosum* and *Trichoderma viride*. PRO Ligno 11(4): 268–274
115. Chen H, Chen F, Zhang YL, Song JY (1999) Production of lithospermic acid B and rosmarinic acid in hairy root cultures of *Salvia miltiorrhiza*. J Ind Microbiol Biotechnol 22:133–138
116. Fraga BM, Diäaz CE, Guadaño A, Lez-Coloma AG (2005) Diterpenes from *Salvia broussonetii* transformed roots and their insecticidal activity. J Agricult Food Chem 53:5200–5206
117. Kümmritz S, Louis M, Haas C, Oehmichen F, Gantz S, Delenk H, Steudler S, Bley T, Steingroewer J (2016) Fungal elicitors combined with a sucrose feed significantly enhance triterpene production of a *Salvia fruticosa* cell suspension. Appl Microbiol Biotechnol 100(16):7071–7082
118. Yan Y, Zhang S, Zhang J, Ma P, Duan J, Liang Z (2014) Effect and mechanism of endophytic bacteria on growth and secondary metabolite synthesis in *Salvia miltiorrhiza* hairy roots. Acta Physiol Plant 36:1095–1105

Molecular Markers in *Salvia* L.: Past, Present and Future

Mehmet Karaca and Ayse Gul Ince

Abstract Sage (*Salvia* L.) is one of the largest genera in the family Lamiaceae. Species in the genus are used to preserve foods and a spice for flavoring, ornamental plants in parks and gardens. In addition, several species of this genus are utilized in medicine and fragrance industries. The utilization of molecular markers in *Salvia* lags behind many other plant species. Molecular markers are proteins/isozymes, secondary metabolites, deoxyribonucleic acid or ribonucleic acid (DNA/RNA) sequences that could be thought as signs or marks differentiation one locus or an individual from others. DNA markers (DMs) define the location of traits, genes or indicate differences within and between genomes of individuals. The earlier application of DMs was the fingerprinting studies and later on DMs were extensively used in plant genetic mapping and gene identification studies. Currently DMs are routinely used in plant variety identification, protection, conservation, genetic stock development, marker-assisted breeding, association and ecological studies. Genotyping by sequencing (GBS), high-throughput marker systems based on next generation sequencing (NGS), makes DMs more effective. Future technologies of molecular markers would not only detect and locate the genome-wide genetic differences at the single nucleotide level, but also will detect the epigenetic differences in the whole genome. DMs will be extensively used in genome mapping and genome selection research in next generation breeding studies. In this chapter, the technical aspects, types and principles of traditional and NGS based DMs used in plant research are briefly introduced. DMs are technically classified as traditional PCR-based, hybridization-based, PCR-restriction enzyme based, PCR-hybridization based and NGS-based GBS markers. DMs are also classified as genetic and genic (functional) markers. Furthermore, DMs could be classified as low-throughput and high-throughput markers. In some other resources, molecular markers are classified as the first, second, third, fourth and next generation marker systems. DMs could be classified as nuclear, plastid (chloroplast), mitochondrial markers, or *in silico* based

M. Karaca (✉)
Department of Field Crops, Akdeniz University, 07070 Antalya, Turkey
e-mail: mkaraca@akdeniz.edu.tr

A. G. Ince
Vocational School of Technical Sciences, Akdeniz University, 07070 Antalya, Turkey

V. Georgiev and A. Pavlov (eds.), *Salvia Biotechnology*,
https://doi.org/10.1007/978-3-319-73900-7_9

markers. In this chapter, DMs are reclassified based on their technical principles and will be revisited emphasizing on their use in *Salvia*. This chapter also deals with genome-wide genetic markers and molecular markers for secondary metabolites.

Keywords Classification of markers · Functional markers · Genotyping by sequencing · Genic and genetic markers · NGS technologies · Secondary metabolites

Abbreviations

2b-RAD-Seq	Type IIB restriction enzyme digestion restriction site associated DNA sequencing
2-D PAGE	Two-dimensional polyacrylamide gel electrophoresis
6-PGD	6-phosphogluconate dehydrogenase
AAT	Aspartate aminotransferase
ABI	Applied biosystem instruments
ABS	Amplicon based sequencing
AFLP	Amplified fragment length polymorphism
AMP	Adenosine monophosphate
AMP-PCR	Anchored microsatellite-primed PCR
AP-PCR	Arbitrarily primed PCR
APS	Adenosine-5′-phosphosulfate
ASB	Allele-specific blocker
ASP	Allele-specific primer
AS-PCR	Allele-specific PCR
ASSRs	Anchored simple sequence repeats
ATP	Adenosine triphosphate
BEAMing	Beads, emulsions, amplification and magnetics-based cloning
bPCR	Bridge PCR
BSA	Bulked segregant analysis
CAPS	Cleaved amplified polymorphic sequence
CBD	CAAT box-derived marker
CCD	Charge-coupled device
CDDP	Conserved DNA-derived polymorphism
cDNA	Complementary deoxyribonucleic acid
CE	Capillary electrophoresis
CFLP	Cleavase fragment length polymorphism
CMOS	Complementary metal-oxide-semiconductor
CNV	Copy numbers variation
CoRAP	Conserved region amplification polymorphism
COS	Conserved orthologous set
cPAL	Combinatorial probe-anchor ligation
CRISPR/Cas9	Clustered regularly interspaced short palindromic repeats/ CRISPR associated protein 9

CRoPS	Complexity reduction of polymorphic sequences
D/TGGE	Denaturing/temperature gradient gel electrophoresis
DAF	DNA amplification fingerprinting
DAMD	Directed amplification of minisatellite DNA
DAMD-PCR	Directed amplification of minisatellite DNA-PCR
DArT	Diversity array technology
dATP	Deoxyadenosine triphosphate
dATPαS	Deoxyadenosine α-thiotriphosphate
dCAPS	Derived cleaved amplified polymorphic sequence
dCTP	Deoxycytidine triphosphate
ddNTPs	Dideoxynucleotide triphosphates
ddRAD	Double-digest restriction site-associated DNA
ddRAD-seq	Double digestion restriction site associated DNA sequencing
DFDMs	Direct functional DNA markers
DGGE	Denaturing gradient gel electrophoresis
dGTP	Deoxyguanosine triphosphate
dHPLC	Denaturing high-performance liquid chromatography
DMs	DNA markers
DNA	Deoxyribonucleic acid
DNBS	DNA nanoball sequencing
dNTPs	Deoxyribonucleotide triphosphates
dRAMPs	Digested random amplified microsatellite polymorphism markers
dsDNA	Double-stranded DNA
dTTP	Deoxythymidine triphosphate
dXTP	Deoxynucleoside triphosphate
EDTA	Ethylenediaminetetraacetic acid
ELIDA	Enzymatic luminometric inorganic pyrophosphate detection assay
emPCR	Emulsion PCR
eSSRs	Expressed simple sequence repeats
EST	Expressed sequence tag
EST-SSRs	Expressed sequence tag based simple sequence repeat markers
ezRAD	Standard Illumina TruSeq library preparation kits restriction site associated DNA sequencing
FDMs	Functional DNA markers
FEN	Flap endonuclease
FGS	First generation sequencing
FRET	Fluorescence resonance energy transfer
GAB	Genomics-assisted breeding
GBS	Genotyping by sequencing
GC	Gas chromatography
GS	Genome sequencer
GT-seq	Genotyping-in-thousands by sequencing

HGP	Human genome project
HPLC	High performance liquid chromatography
HRM	High resolution melting analysis
IBM	International business machines
IDH	Isocitrate dehydrogenase
IFDMs	Indirect functional DNA markers
IMP	Inter-MITE polymorphism
InDels	Insertions/deletions
IRAP	Inter-retrotransposon amplified polymorphism
iRRL	Improved reduced-representation librarys
ISA	Inter-simple sequence repeats amplification
ISSR	Inter simple sequence repeats
ISTR	Inverse sequence tagged repeat
ITP	Intron-targeting polymorphism
ITS	Internal transcribed spacer
KASP	Kompetitive allele specific PCR
KASPar	Competitive allele specific PCR
LTRs	Long terminal repeats
MAAP	Multiple arbitrary amplicon profiling
MALDI-TOF MS	Matrix assisted laser desorption ionization time of flight mass spectrometry
MAS	Marker-assisted selection
MDH	Malate dehydrogenase
MEGA-AFLP	Multiplex-endonuclease genotyping approach amplified fragment length polymorphism
MIPs	Molecular inversion probes
MITE-AFLP	Miniature inverted repeat transposable elements-amplified fragment length polymorphism
MITEs	Miniature inverted repeat transposable elements
MLP	Major latex-like protein
MNR	Menadione reductase
MP-PCR	Microsatellite-primed PCR
MPS	Massively parallel sequencing
MPSS	Massively parallel signature sequencing
mRNA	Messenger RNA
MS	Mass spectrometry
MSG	Multiplexed shotgun sequencing
NGS	Next generation sequencing
NIR	Near infrared spectroscopy
NMR	Nuclear magnetic resonance
nsSNP	Non-synonymous single nucleotide polymorphism
OH	Hydroxyl
OLA	Oligonucleotide ligation assay
ONT	Oxford nanopore technologies

ORF	Open reading frame
PAGE	Polyacrylamide gel electrophoresis
paired-end RPLs	paired-end reduced representation libraries
PAV	Presence/absence variations
PCR	Polymerase chain reaction
PCR-RFLP	Polymerase chain reaction-restriction fragment length polymorphism
pERPLs	Paired-end reduced representation libraries
Pfu	DNA polymerase from *Pyrococcus furiosus*
PGI	Phosphoglucoisomerase
PGM	Phosphoglucomutase
PPi	Pyrophosphate
PTP	Picotiter plate
qPCR	Quantitative PCR
QTL	Quantitative trait loci
RAD	Restriction site associated DNA
RAD-seq	Restriction site-associated DNA sequencing
RAMPs	Random amplified microsatellite polymorphisms
RAPD	Random amplified polymorphic DNA
RBIP	Retrotransposon-based insertion polymorphism
rDNA	Ribosomal DNA
REMAP	Retrotransposon microsatellite amplified polymorphism
RESTseq	Restriction fragment sequencing
RFEL	Restriction fragment end labeling
RFLP	Restriction fragment length polymorphism
RLGS	Restriction landmark genome scanning
RNA	Ribonucleic acid
RNA-seq	RNA sequencing
RRL	Reduced-representation library
RRS	Reduced-representation sequencing
RTEs	Retrotransposons
RT-PCR	Reverse transcriptase PCR
SAMPL	Selective amplification of microsatellite polymorphic loci
SBS	Sequencing-by-synthesis
SCAR	Sequence characterized amplified regions
SCoT	Start codon targeted
SFLA	Selective fragment length amplification
SFPs	Single feature polymorphisms
sGBS	Spiked genotyping-by-sequencing
SGS	Second generation sequencing
SKDH	Shikimate dehydrogenase
SkimGBS	Skim genotyping-by-sequencing
SLAF-Seq	Specific length amplified fragment sequencing
SMRT	Single molecule real-time

SMS	Single molecule sequencing
SNP	Single nucleotide polymorphism
SOLiD	Sequencing by oligonucleotide ligation and detection
SPARs	Single primer amplification reactions
SRAP	Sequence-related amplified polymorphism
SRFA	Selective restriction fragment amplification
S-SAP	Sequence-specific amplification polymorphism
SSCP	Single-strand conformation polymorphism
ssDNA	Single-stranded DNA
SSRs	Simple sequence repeats
STMS	Sequence-tagged microsatellite site
STRs	Short tandem repeats
STS	Sequence tagged sites
synSNP	Synonymous single nucleotide polymorphism
Taq	DNA polymerase from *Thermus aquaticus*
TD	Transposon display
TDFs	Transcript-derived fragments
TE	Transposable element
TE-AFLP	Three-endonuclease amplified fragment length polymorphism
tGBS-seq	Tunable genotyping by sequencing
TGGE	Thermal gradient gel electrophoresis
TGS	Third generation sequencing
TILLING	Targeting induced local lesions in genomes
Tm	Melting temperature
TRAP	Target region amplification polymorphism
UGMs	Unigene-derived microsatellites
Vent1	DNA polymerase from *Thermococcus litoralis*
VNTRs	Variable number tandem repeats
WGR	Whole genome resequencing
WGS	Whole genome sequencing
ZMW	Zero-mode waveguide

1 Introduction

Salvia L. is one of the largest genera in the family Lamiaceae, also known as Labiatae, the mint or dead nettle family. The name *Salvia* originates from the Latin word "salvare", which means, "to heal" [1]. The genus *Salvia* naturally occurs in five regions of the world, southern Africa (ca 60 spp.), Central and South America (ca 500 spp.), Europe (ca 36 spp.), western and eastern Asia (ca 100 spp.), and central Asia/Mediterranean (ca 250 spp.). The Mediterranean region is accepted gene center of the genus *Salvia*. This genus consists of about 1000 species and many of which are used for food preservation, flavoring and ornamental plants in parks and gardens. Several species of the genus are also used in medicine and fragrance industries [2–7].

Salvia plants are woody in the lower part of the stem, but annual branches are herbaceous, and perennial, rarely biennial, or annual. Plants are usually 30–150 cm tall and produce attractive flowers in various colors. Due to the invaluable importance of their bioactive components to humans, the demands of global markets for some *Salvia* species are increasing. Because the demands of the global markets cannot be satisfied by commercial production, the natural populations have been over exploited especially in developing countries [3, 4, 8–12].

In the first quarter of the 20th century, scientists discovered that genes are organized on chromosomes. DNA markers are those identifiable polymorphic DNA sequence found at specific locations of the genome and may associate with the inheritance of a trait or linked gene [13–15]. Markers must exist in different forms (polymorphism) so that the chromosome carrying the mutant gene can be distinguished from the chromosome with normal gene by form of the marker it carries [16]. In order to differentiate a polymorphism from a mutation a cut-off value of 1% could be used. If a mutation is successful enough to persist for a while in the population, its percentage increased and it becomes a polymorphism.

Morphological/anatomical markers, phytochemical features (secondary metabolites)/allozyme/protein markers and DNA markers could be used as tools to preserve gene pools and study genetic diversity and do research on the complexities of plant-environment interaction and facilitate future breeding efforts. There exist some advantages and disadvantages of these genetic markers. For instance, some marker types are very convenient to use but have low power of differentiation, some other are complicated but have low power of differentiation [17–30]. Current research uses next generation sequencing based markers such as genotyping by sequencing to identify and map genes and characterization studies [24–28].

The main goal of this chapter is to briefly describe the basic properties of visible, biochemical markers and next generation based DNA markers. Their utilization in *Salvia* L. is provided. Applications of visible, biochemical genetic markers and traditional DNA markers include but not limited: (i) identification of temporal and spatial patterns of genetic variation within and between populations; (ii) describing mating systems, levels of inbreeding and cross-pollinating species; (iii) inferring taxonomic and phylogenetic relationships among and within species; (iv) evaluating the impacts of domestication practices, including management, preservation and improvement; (v) fingerprinting and germplasm identification in breeding and propagation populations; (vi) constructing genetic linkage maps; (vii) map-based cloning of agronomic important genes; (viii) marker assisted breeding or selection; (ix) dissecting polygenic traits into their Mendelian components or quantitative trait loci (QTL).

Next generation sequencing (NGS) based markers such as genotyping by sequencing (GBS) and genome wide resequencing (GWR) technologies produce very large number of genetic markers that could be used in genome selection. Furthermore, GBS opens a new concept next generating breeding that is discussed in this chapter [31, 32].

An ideal DNA maker system is expected to have: (i) high levels of polymorphism, which is the simultaneous occurrence of allelic variability for a trait at the same population; (ii) co-dominant inheritance to allow discrimination of homozygote and heterozygote states; (iii) frequent occurrence in the genome to assay more genomic regions or genes; (iv) selective neutral to environmental conditions, developmental stage of the organism or management practices; (v) easy access (availability), fast assay and highly reproducibility for convenient use and reliability; (vi) high throughput and easy exchange of data between laboratories; (vii) no epistatic interactions so that identification of marker alleles at one locus does not interfere with that at other marker loci; (viii) they have no pleiotropic effects and (ix) multiallelic nature to permit classification of individuals into more than two groups [33–45].

2 Morphological/Anatomical Markers (Visible Markers)

Visible morphological characteristics are the oldest genetic markers used in scientific studies of plants and animals. Commonly used morphological and anatomical markers used in the genus *Salvia* include leaf shape, width and length, stem diameter, plant height, internode length, bract width, length and number, peduncle length and diameter, flowering stem length, number of flowers per inflorescence, inflorescence length, calyx color and length, number of calyx nerves and calyx hairs status, corolla color and length, the number of leaf colored glands, the color of the upper leaf surface, the stem hairs status, the color of shoot flowering, the color of last year shoot, pollen color and shape, and leaf and stem anatomical features. Although these markers represent the actual phenotypes, simple, rapid, and inexpensive, their status is influenced with pleiotropic effect and epistatic interactions along with the stage of plant development. Status of these characters usually varies with environments and evaluation of many morphological traits requires growing the plants to full maturity prior to identification. Efficient markers should be ideally neutral to the effect of environmental and management practices [30, 46–54]. In Table 9.1, several advantages and disadvantages of morphological and anatomical markers are presented.

Table 9.1 Some advantages and disadvantages of morphological and anatomical markers

Advantages	Disadvantages
Simple	Limited number and low polymorphism
Rapid	Most require whole plant
Inexpensive	Expression change depending on the tissue type and development stage
Mendelian inheritance	Dominant type inheritance is observed
Reproducible	Influenced with pleiotropic effect and epistatic interactions
Represent actual genes	Most require suitable genetic stocks

3 Secondary Metabolites/Isozymes (Allozyme)/Protein Markers

3.1 Secondary Metabolites

Plants produce a large number of organic compounds most of which have no direct function in growth and development. These compounds are so called secondary metabolites or phytochemicals [22]. Secondary metabolites of plants are usually specific to certain families and species. Many of these compounds are important in medicinal, cosmetic and food industries. In addition, these plants have important ecological functions such as serving as attractants for pollinators, resistance to pathogens, predators, and abiotic stresses [4, 9, 22, 55].

Secondary metabolites could be divided into three main groups that are based on their structures and the metabolic pathways responsible for their biosynthesis: phenolic compounds; terpenoids; and nitrogen-containing compounds, i.e., alkaloids and glucosinolates. Gas chromatography (GC) interfaced with mass spectrometry (MS), high performance liquid chromatography (HPLC) interfaced with mass spectrometry (MS), capillary electrophoresis, near infrared (NIR) spectroscopy, and nuclear magnetic resonance (NMR) are used to identify and quantify secondary metabolites [7, 8, 22, 56].

Many secondary metabolites include compounds with health-promoting effects. Health-promoting properties of certain secondary metabolites make them attractive genetic markers in selection of naturally grown crops and development of new cultivars. With the use of secondary metabolites as genetic markers, not only individual species but also cells, tissues or organs can be identified and sorted. These features can be also used in marker-assisted selection (MAS). Furthermore, combination of biochemical markers with the omics (metabolomics) enhances our knowledge on gene expressions and regulations [57]. However, the use of secondary metabolites as genetic markers requires specialized and expensive equipments. In addition, the amount and types of secondary metabolites are influenced with environmental effects or management practices [58–61]. Some advantages and disadvantages of the use of secondary metabolites to be used in discrimination and characterization studies are given in Table 9.2.

In the genus *Salvia* monoterpene hydrocarbons (such as pinene, camphene and myrcene), oxygenated monoterpenes (such as cineole, camphor, borneol, verbenone and pinocarvyl acetate), sesquiterpene hydrocarbons (such as bourbonene, caryophyllene, longifolene, aromadendrene, aromadendrene, germacrene, bisabolene and bicyclogermacrene), oxygenated sesquiterpenes (such as isoaromadendrene epoxide, caryophyllene oxide, humulene epoxide, spathulenol, isolongifolol, santalol, longipinocarvone and sclareol oxide), polyphenols (such as rosmarinic acid (labiatenic acid), carnosic acid, salvianolic acid and its derivatives carnosol, rosmanol, epirosmanol, rosmadial and methyl carbonate), phenolic compounds (such as thymol, carvacrol), flavonoids (such as flavones, flavanones, flavonols, isoflavones and chalcones), tanning and other compounds (such as tetraneurin D, isochiapin B,

Table 9.2 Some advantages and disadvantages of secondary metabolites-based genetic markers

Advantages	Disadvantages
Cells, tissues or organs can be differentiated and sorted	Usually require specialized and expensive equipments
Powerful markers in drug discovery	Limited number of available markers
Functional markers	Usually influenced with environmental effects or management practices
Could be effectively used in metabolomic research	Low polymorphisms and reproducibility
Markers themselves have economical values	Usually dominant types of markers are produced

squalene, stigmasterol) have been used to a limited extent to estimate genetic patterns of geographic variation within species. In addition, many studies used these secondary metabolites for species identification, taxonomy and diversity studies as well [6, 9, 11, 60–62].

3.2 *Isozymes (Allozyme)/Protein Markers*

Among the biochemical markers, proteins including isozymes and allozymes have been used for approximately 50 years for various research purposes in plant sciences but these markers have been almost completely replaced by DNA-based markers in recent years. In the past, isozymes and protein markers had been used in identification of phylogenetic relationships, estimation of genetic variability and taxonomy, analysis of population genetics and developmental biology, characterization of genetic resources and management [3, 63–65].

Isozymes can be defined as structurally different forms of an enzyme with the same catalytic function. They are closely related variants of an enzyme encoded by different genes, which may have arisen by gene duplication or polyploidization [66, 67]. On the other hand, allozymes are allelic variants of enzymes encoded by different alleles of gene. Allozymes behave as alleles of a marker locus and useful in linkage analyses. In contrast, strict isozymes are inherited as separate loci and show independent segregation [3]. Generally, the terms isozymes and allozymes are used as synonyms, the term allozyme is seldom used. Isozymes and seed storage proteins have been widely used as genetic markers [68].

Polymorphisms between different forms of isozymes and seed storage proteins are considered result of changes in the amino acid sequences. Changes in the amino acid sequences will affect the electric charge and conformation of the protein. Allelic variations can be detected by starch or polyacrylamide gel electrophoresis and subsequent enzyme-specific stains that contain substrate for the enzyme, cofactors and an oxidized salt. The banding pattern of isozymes called zymograms can be readily interpreted in terms of loci and alleles [3, 63–65, 68, 69].

Protein-based genetic markers utilize two-dimensional polyacrylamide gel electrophoresis (2-D PAGE). In the 2-D PAGE technique, all enzymes and other proteins present in the sample preparation are simultaneously detected as presence (scored as 1) or absence (scored as 0) of spots. However, 2-D PAGE assays are more difficult than that of the allozyme analyses, and the markers are often dominant in their expression [3, 70, 71].

In isozyme and protein (2D-PAGE) analyses, crude protein extracts are isolated from tissue of interest and separated on gels by applying an electrical current (i.e. electrophoresis). Isozymes or proteins in the protein extract migrate to different positions on the gel depending on the electrical charge and size. Isozymes or proteins with different amino acid compositions generally have different charge and/or size, so it is these genetic differences that are revealed as mobility differences on the gel [63, 68–71].

There are some advantages and disadvantages of isozymes and protein markers (Table 9.3). They reflect differences in gene sequences more directly than morphological and secondary metabolite markers, only a small amount of tissue is needed for their analysis. They can be detected at seedling stage or even from seeds and analysis of one marker usually does not interfere with that for other protein-based markers. In addition, compared to morphological and secondary metabolite markers the level of polymorphisms and the number of markers are very high in isozymes and proteins. However, they vary with the tissue, the developmental stage, and the environment. Compared to DNA markers amount and quality of tissue samples required are higher in isozyme and protein markers. Changes in DNA sequence variations are masked at the protein level, reducing the level of detectable variation. In addition, some changes in nucleotide sequence do not change the encoded polypeptide and some polypeptide changes do not alter the mobility of the protein in electrophoresis. These are called null (enzymatically inactive) alleles and synonymous substitutions, respectively [3, 63–69, 71].

In addition, isozymes are usually co-dominant, their analysis is relatively easy, and data interpretation is facilitated by numerous reference data [63, 68, 69, 72]. Sometimes, however, banding patterns present complex profiles arising from

Table 9.3 Some advantages and disadvantages of isozyme and protein markers

Advantages	Disadvantages
Low cost	Requirement for fresh or frozen tissues
Almost universally applicable method	Limited number of available markers (in comparison to DNA markers)
Co-dominant markers	Potentially direct target of selection
Abundance (in comparison to morphological markers)	Low polymorphisms (in comparison to DNA markers)
Ease of use	Not amenable to high-throughput screening
Usually reproducible	Presence of null enzymes
Usually they are tissue and development specific markers	Polymorphisms may not be detected due to silent substitutions

polyploidy or duplicated genes and the formation of intergenic heterodimers. The main disadvantage of allozymes is their relatively low abundance and low level of polymorphisms. In addition, markers with identical electrophoretic mobility (co-migration) may not be homologous for distantly related germplasms. Proteins markers are in fact phenotypic markers, and as such, they may be affected by environmental conditions. The banding profile obtained for a particular allozyme marker may change depending on the type of tissue used for the analysis (e.g. root vs. leaf) [68, 69, 71–73].

Isozymes used in *Salvia* taxonomic studies includes aconitase (ACO), alcohol dehydrogenase (ADH), phosphoglucoisomerase (PGI), phosphoglucomutase (PGM), menadione reductase (MNR), isocitrate dehydrogenase (IDH), 6-phosphogluconate dehydrogenase (6-PGD), malate dehydrogenase (MDH), aspartate aminotransferase (AAT), formate dehydrogenase (FDH), glutamate-oxalacetate transaminase (GOT), menadione reductase, triosephosphate isomerase (TPI) and shikimate dehydrogenase (SKDH) [3, 73]. Most of isozyme allelic frequenced deviate significantly from the expected ratios suggesting presence of null alleles. Some isozymes show intergenic bands with dimeric, trimeric and tetrameric structures. Furthermore, many isozyme markers of the genus *Salvia* have staining intensities varying from a very faint to densely stained bands of ACO, ADH, FDH, GOT, IDH, MNR, 6PGD, SKDH, and TPI. Isoenzyme marker are no longer used in *Salvia* because DNA based markers offer better resolutions along with many other advantages [3, 68, 69, 71–73].

4 DNA Markers (DMs)

A DNA marker can be defined as a polymorphic DNA sequence that can be used to identify genes, chromosomes, individuals or species. DNA markers can be as small as a single base or they can be as long as several hundred or more bases such as a DNA fragment. A marker must exist in different forms (polymorphic) so that variety or genotype carrying a particular form can be distinguished from other variety or genotype with the other forms. For example, a single base polymorphism can differentiate organisms into four groups (A, G, C or T) while two bases can differentiate organisms into 16 groups. DNA marker polymorphisms can result from any substitutions, insertions or deletions (INDELs), or variations in repeats (SSRs) or copy numbers (CNV). Longer DNA sequences produce higher resolution for polymorphisms. This means that longer DNA fragments may have more differentiation polymorphic nucleotide sites. However, the longer the DNA fragments, the ratio of obtaining null alleles and homoplasy may increase due to the detection systems used.

After the first DNA marker technology developed in the 1980s, new types were developed including the polymerase chain reaction (PCR) and these markers have acted as versatile tools in fingerprinting of varieties, mapping of genes and quantitative trait loci (QTLs), marker assisted breeding, positional cloning of genes, identification of chromosomes or/and chromosome segments, inferring and establishing phylogenetic relationships among species, gene pyramiding, and maintenance and utilization of genetic resources [30, 74–78].

DM systems could be classified based on several different criteria, such as first-, second-, third-, and next-generation markers, or hybridization-based markers, polymerase chain reaction (PCR)-based markers or combination of both. In addition, DMs are classified based on; primer types used (such as arbitrary, sequence specific, combination of specific sequence and arbitrary sequence primers). Based on restriction endonuclease digestion before or after PCR and based on detection (single nucleotide polymorphism generated by variation in DNA sequence) or fragment length polymorphisms produced by variation in sequence length) are also used. The location and the functional significance of markers (random, gene-based or functional markers those markers located in or close proximity to a gene) and based on the throughput criteria (low-, medium-, or high-throughput markers) are also used. Furthermore, classification of DNA markers are based on their mode of transmission (biparental or uniparental), and the mode of gene action (dominant or co-dominant markers) [4, 5, 10, 12, 23, 74–85]. The following section summarizes various traditional DNA marker systems along with their utilization in the genus *Salvia*.

4.1 Restriction Fragment Length Polymorphism (RFLP), Derivatives and Synonyms

PCR-RFLP is synonyms of the cleaved amplified polymorphic sequence (CAPS), which is derived (or modified) form of RFLP. In addition, a variation of the CAPS method is called derived cleaved amplified polymorphic sequence (dCAPS) or mismatch PCR-RFLP. Another modification of the RFLP marker system is restriction fragment end labeling (RFEL). The RFLP is the first DNA marker system in which samples of any organisms with probes can be differentiated by analysis of patterns obtained from cleavage of their DNA. For RFLP marker system, genomic DNA is digested with particular restriction endonuclease enzymes and when the two organisms differ genetically, digestion reactions produce DNA fragments varied in their lengths. Polymorphisms within the restriction enzyme recognition sites originate from the DNA rearrangements that occur due to evolutionary processes, recombination and mutations (the gain, loss, or relocation of a restriction site). Length difference polymorphisms are considered to originate from insertion/ deletion, translocation, inversion, and unequal crossing over [4, 53, 86–94].

A summary of the general procedure for RFLP detection is as follows: (i) high molecular weight genomic DNA is digested with a selected restriction enzyme; (ii) the digested DNA fragments are then separated by agarose gel electrophoresis. In most RFLP analyses, the six cutter restriction endonucleases are used because they produce fragments in the size range of 200–20,000 bp fragments, which can be separated on agarose gels. The choice between agarose and polyacrylamide gels is based on the restriction enzyme type used; (iii) the DNA fragments are denatured to single strands and transferred from the gel onto a solid support in such a way that the relative positions of the fragments in the gel are preserved on membranes or filters using a technique referred to as Southern blotting or hybridization [95]; (iv) the DNA fragment is then allowed to hybridize to labeled probe DNA. Probes are DNA or RNA sequences, usually 200–3000 bp in length, obtained from DNA or complementary DNA (cDNA) libraries. A DNA library contains all the genetic information while a cDNA library contains genomic sequences representing the genes expressed in a cell or tissue, from which the mRNA was isolated; (v) in the washing steps, the probe molecules not bound in hybridization are removed, and (vi) in the detection steps, the fragments involved in hybridization with the probe are detected by autoradiography or development ("cold" probes that is based on fluorescent labeling methods) [86–94].

Some advantages and disadvantages of the RFLP marker system are listed in Table 9.4. Moderate level of polymorphisms is detected in the RFLP marker system. RFLP markers show co-dominant inheritance and have higher levels of reproducibility. The RFLP marker system produces fragments of different lengths that are easily separated and detected. With RFLP markers, homozygote individuals can be easily differentiated from heterozygote ones. However, this system requires a large quantities (1–10 μg) of purified, high molecular weight DNA, digestion and hybridization studies. In addition, this marker system requires considerable skill and effort for the development of RFLPs, including the construction of genomic/cDNA libraries for the identification of suitable probes. Furthermore, this marker system is not amenable to automation and high-throughput analysis [86–100].

The RFLP marker system is considered as the first-generation marker type which is based on DNA hybridization. It uses restriction endonuclease digestion, polymorphisms produced by variation in sequence length, the location and the functional significance of markers are random (some of which could be functional markers). It is low-throughput type marker system and can be used to detect biparental and uniparental genetic transmission. However, the RFLP marker system has not been applied in *Salvia* L.

Table 9.4 Some advantages and disadvantages of the RFLP marker system

Advantages	Disadvantages
High reproducibility	Requirement of high quantity and quality of DNA
Co-dominant inheritance	Requirement of specific probe libraries
Transferability between laboratories	Requirement of radioactive or fluorescent probes
Easy to score and analyze	Low level of polymorphism
Provide locus specific markers	Low number of loci analyzed per assay
Probes could be used in related species	Requirement of considerable skill and effort
RFLP markers show Mendelian fashion	Laborious and not amenable for automation

Polymerase chain reaction (PCR) is a molecular biology technique that had been first described in 1968 by Kleppe and co-workers and later on invented by Kary Mullis in 1983 for amplifying small quantities of DNA without using a living organism. PCR involves in one or more oligonucleotide primers that flank the DNA sequences of interest and amplification is achieved by a series of repeated cycles of heat denaturation of the DNA, annealing of the primers to their complementary sequences, and extension of the annealed primers with a thermophilic DNA polymerase. Since the amplified products themselves are also complementary to primers, successive cycles of amplification essentially double the amount of the target amplified DNA in the previous cycle and the result is an exponential accumulation of the specific target fragment or fragments [96].

PCR technique produces microgram quantities of DNA copies from a few DNA segments in a few hours. The DNA produced by PCR is often referred to as amplicons or amplified products or bands. *Taq* DNA polymerase from *Thermus aquaticus, Pfu* (from *Pyrococcus furiosus*) and *Vent*1 (from *Thermococcus litoralis*) polymerases are main enzymes used in amplification reactions [30].

PCR-based marker systems could be categorized as second-generation markers. They require smaller quantity and lower quality of template DNA. PCR-based markers are much more user friendly and amenable to automation. PCR-based marker can be grouped as: (i) markers based on arbitrary sequence primers (may use a single primer); (ii) those markers based on specific sequence primers; (iii) an intermediate group of techniques has been developed that uses either a combination of specific sequence and arbitrary sequence primers or primers composed of both fixed and arbitrary sequences; (iv) some techniques combine restriction digestion of DNA with PCR amplification, and they together may be regarded as a separate group [97–99].

The products of amplification reactions can be separated by electrophoresis using either an agarose or a polyacrylamide gel and are visualized by staining the gel with either ethidium bromide or silver staining, autoradiography, or fluorescence detection. Polyacrylamide gels contain a uniform pore size than agarose gels and allow separation of DNA fragments with a higher resolution. A gel containing 6% acrylamide has a fine network formed by polyacrylamide and can separate DNA fragments differing in length by one or two bases. However, the maximum fragment length that can be separated using this gel is up to 500 bases. In addition, the first generation automatic DNA sequencers used capillary gel electrophoresis because it afforded automation of filling the capillaries with the polymers as well as loading of the samples [97–99].

Cleaved (cleavage) amplified polymorphic sequence (CAPS)/PCR-restriction fragment length polymorphism (PCR-RFLP) and derived-CAPS (dCAPS) marker systems are modifications or derivatives of the RFLP and PCR marker systems, and detect length polymorphism generated by restriction digestion of specifically amplified PCR products from different genotypes [98]. The CAPS system is synonym of PCR-restriction fragment length polymorphism (PCR-RFLP) [99]. The name PCR-RFLP is more appropriate if markers are conversion of RFLP markers into PCR-based markers. In the PCR-RFLP once a useful RFLP marker has been identified, it can be converted into a more convenient and user-friendly PCR-based marker. However, when PCR amplified products other than RFLP markers are used, for example sequence tagged sites (STS), the name cleaved amplified polymorphic sequence (CAPS) is considered be more appropriate [10, 21, 88, 93].

CAPS markers are DNA fragments obtained by PCR using specific 18–30 bp primers, followed by digestion of the fragments with one or more restriction enzymes [21]. The length polymorphisms caused by variation in the occurrence of restriction sites are identified by gel electrophoresis [33, 100]. Critical steps in the CAPS marker system are the DNA extraction, PCR amplification, and the separation of the digested amplification products. A summary of the general procedure for CAPS marker system is as follows: (i) genomic or cytoplasmic DNA extracted is PCR amplified using CAPS specific primer pair. If desired more than one prime pairs could be used in the amplification reactions; (ii) after the amplification reactions, 5–25 μL PCR products are digested using one or more restriction endonucleases; (iii) products are separated using electrophoresis. The choice between agarose and polyacrylamide is based on the fragments length differences [10, 21, 93].

In addition, a variation of the CAPS marker system is derived cleaved amplified polymorphic sequence (dCAPS) or also called mismatch PCR-RFLP. In the dCAPS marker system, one of the primer pair used in the PCR amplification reactions contains a recognition site for a restriction enzyme and one or more mismatches with the target template DNA sequence. These mismatches generate restriction endonuclease recognition site in the PCR product of one allele, but not in that of the other allele [88, 100].

Some advantages and disadvantages of the CAPS/dCAPS marker systems are presented in Table 9.5. It is much easier and less time-consuming than conventional RFLP marker system since it does not require DNA hybridization studies. Expressed sequence tags (ESTs) based-CAPS markers are more useful as genetic markers for comparative mapping studies than those markers derived from non-functional sequences. CAPS markers are inherited usually in a co-dominant type and have high level of reproducibility. It does not require higher molecular weight template DNA and very little amount of DNA such as 10–100 ng per reaction is sufficient for the analysis. There are some disadvantages of the CAPS

Table 9.5 Some advantages and disadvantages of the CAPS/PCR-RFLP and dCAPS marker systems

Advantages	Disadvantages
High reproducibility and very low amount and quality of DNA required	Sequence information required for synthesis of the primers
Easy to score and analyze	Unsuitable for automation
Usually co-dominant markers are obtained	Unsuitable for high-throughput analysis
Transferability between laboratories is high	Produces low level of polymorphism due to limited size of the amplified fragments (300–2000 bp)

marker system. One of the main disadvantages of the CAPS marker system is the failed amplifications or monomorphic fragments of the PCR due to the mutations in the primer binding sequences of the template DNA. However, this has been amended by the use of dCAPS. Both CAPS and dCAPS markers have low level of polymorphisms, not as high as polymorphisms detected by simple sequence repeats (SSRs) and amplified fragment length polymorphisms (AFLPs) [88, 92, 93, 100].

Restriction fragment end labeling (RFEL) is another modification method of the RFLP marker system. This marker system is not widely used in plant species. RFEL marker is produced as follows: (i) total genomic DNA is digested with a restriction endonuclease, (ii) restricted fragments are end labeled with ^{32}P dATP utilizing the Klenow fragment of DNA polymerase, (iii) labeled fragments are separated using 6% polyacrylamide/8 M urea gels coupled with electrophoresis [86].

4.2 *Minisatellites and Directed Amplification of Minisatellite DNA (DAMD-PCR)*

Variable number tandem repeats (VNTRs) are short tandem repeated DNA clustered together and oriented in the same direction in a location of a genome. A minisatellite is a tract of repetitive DNA motifs repeated many times, ranging in length from 10 to 60 base pairs. They are usually 0.2–20 kb long and are usually prominent in the heterochromatin regions such as centromeres and telomeres of chromosomes. Sequences of the terminal regions of minisatellites are segments of non-repetitive sequence, allowing the repeats to be targeted with restriction enzymes and analyzed by DNA hybridization (minisatellite) [75], or amplified by the polymerase chain reaction (DAMD-PCR) [101].

A summary of the general procedure for minisatellite detection is as follows: (i) high-molecular-weight genomic DNA is digested with a selected restriction enzyme that cleaves the DNA outside the regions of VNTR repeats on both sides; (ii) the digested DNA fragments are separated by agarose gel electrophoresis; (iii) the DNA fragments are denatured to single strands and transferred from the gel onto a solid support; (iv) the DNA fragment is then allowed to hybridize to labeled minisatellite probe DNA; (v) the probe molecules not bound in hybridization are removed by repeated washing, and (vi) the fragments involved in hybridization with the probe are detected by autoradiography or by color development [75, 102]. In Table 9.6, some advantages and disadvantages of minisatellite markers are presented.

Polymorphisms between minisatellite DNA sequences are obtained due to recombination and replication errors. Minisatellite markers are particularly useful in those studies involving in the determination of genetic identity, parentage, clonal growth and structure, and identification of varieties and cultivars. Polymorphism rate of the minisatellite markers is enhanced with the use of a multi-locus minisatellite probe. The main advantages of minisatellites are their high level of polymorphisms and high reproducibility (Table 9.7). Disadvantages of minisatellite

Table 9.6 Some advantages and disadvantages of the minisatellite marker system

Advantages	Disadvantages
High reproducibility	Requirement of high quantity and quality of DNA
Co-dominant inheritance	Requirement of Southern hybridization
High level of polymorphism among species	Laborious and not amenable for automation
Easy to score and analyze	Low level of polymorphism within species
Provides locus specific markers	Low number of loci analyzed per assay

Table 9.7 Some advantages and disadvantages of the DAMD-PCR marker system

Advantages	Disadvantages
No sequence data for primer construction are needed	Dominant and moderately complex markers
Low quantities of template DNA are required	Unsuitable for high-throughput analysis and automation
High level of polymorphism and more reproducibility in comparison to RAPD markers	More applicable to inter-specific level studies
Lower reproducibility problems	Nonhomology of similar sized fragments (homoplasy)

markers are similar to RFLP marker system due to the high similarity in methodological procedures.

The minisatellite marker system is considered as the first-generation marker system, it is methodologically very similar to the RFLP marker system. Minisatellite marker system is based on the hybridization and uses restriction endonuclease digestion. Polymorphisms are produced by variations in the fragment length, the location and the functional significance of markers are random and they are low-throughput markers. Minisatellites can be used to detect biparental and uniparental genetic transmission. However, the minisatellite marker system, like RFLP system has not been applied in *Salvia* L.

A modification of minisatellite marker system is direct amplification of minisatellite DNA (DAMD-PCR). This marker system uses plant and animal minisatellite core sequences as primers under relatively high-stringency PCR conditions [101, 103]. Amplification of DAMD-PCR markers is obtained when flanking regions of minisatellites have inverted DNA sequences to which minisatellite core primers bind during PCRs. The use of sequence-specific primers that are obtained from DNA sequences rich in repetitive sequences of plant genomes along with high annealing temperatures used in the PCR amplification enhance the reproducibility of DAMD-PCR markers in comparison to random amplified polymorphic DNA (RAPD) markers [14]. The level of polymorphisms of DAMD-PCR within the species of plant lower than between the species due to the genetic similarities and differences [4, 15, 78, 81].

A summary of the general procedure for DAMD-PCR detection is as follows: (i) DNA is extracted; (ii) minisatellites are amplified using core minisatellite primers that flank the repetitive DNA; (iii) amplified fragments are separated by agarose gel electrophoresis [4, 15, 78, 81].

4.3 Random Amplified Polymorphic DNA (RAPD) with Synonyms and Derivatives

Arbitrarily primed PCR (AP-PCR) is very similar to random amplified polymorphic DNA (RAPD). DNA amplification fingerprinting (DAF) is a modification of the RAPD marker system [104]. These techniques were first use to amplify any species DNA fragments without prior sequences information and they have been collectively termed multiple arbitrary amplicon profiling (MAAP) [96, 105]. The MAAP markers could be converted into other marker systems including sequence characterized amplified regions (SCAR), CAPS, dCAPS, simple sequence repeats (SSRs) [106–108]. Development of SCAR marker is not restricted to RAPD marker system; it can be obtained from other PCR-based markers such as AFLP and other marker systems as well.

The MAAP marker systems produce markedly different amplification profiles, varying from quite simple (RAPD, AP-PCR) to highly complex (DAF) patterns. The key innovation of RAPD, AP-PCR and DAF is the use of a single arbitrary oligonucleotide primer to amplify template DNA without prior knowledge of the target sequence. The differences among these techniques include modifications in the amplification profiles by changing primer length, sequence and annealing, the number of PCR cycles for example two-temperature cycles instead of three-temperature cycles in DAF marker system and alternative methods of fragment separation and staining for example PAGE and $AgNO_3$ stain in AP-PCR and DAF systems [17, 109–113].

AP-PCR, DAF and RAPD marker systems, the presence or absence of amplicons is the result of variations in DNA sequence differences between individuals; therefore, they can be simply detected as DNA fragments that are amplified from one individual but not from another. AP-PCR, DAF and RAPD marker systems produce amplicons from homozygotes with dominant alleles or heterozygotes, but they do not produce amplicons from homozygotes with recessive alleles. In these recessive homozygote individuals, a mismatch of even a single nucleotide can prevent the primer from annealing, hence leads to a loss of allele (band).

The AP-PCR marker system first introduced in Welsh and McClelland [110] uses arbitrary sequence primers of 18–32 bp for PCR amplification studies. In the AP-PCR, initial PCR amplifications are carried out at low stringency and during the subsequent PCR cycles, annealing temperature is increased to a higher stringency. Amplified products are usually electrophoresed using polyacrylamide gel electrophoresis (PAGE) method combined with silver staining or autoradiography.

Although AP-PCR markers are considered essentially the same as RAPD, they differ in terms of primer length, annealing conditions, number of amplified fragments produced, and the types of gels used in electrophoresis [97].

DNA amplification fingerprinting (DAF) uses single 4–8 bp oligonucleotide primers in less stringent two-step PCR amplification conditions. DAF marker system produces a range of up to 100 short amplified products of different lengths mainly depending on the primer and template combination. DAF markers can be resolved and visualized in PAGE combined with silver staining [109].

Among the MAAP marker systems, RAPD is the one of the most used DNA marker systems in the last three decades [111]. A summary of the general procedure for RAPD detection is as follows: (i) isolation of genomic DNA; (ii) amplification of the templates using DNA polymerase chain reaction (PCR) at constant low annealing temperature (generally 34–37 °C) with usually 10 bp arbitrary primers. RAPD system uses single primers serving as both forward and reverse primer annealing complementary strands of DNA template; (iii) the resulting PCR products (usually within the 0.5–5 kb size range) are generally resolved on 1.5–2.0% agarose gels and stained with ethidium bromide or polyacrylamide gels in combination with either $AgNO_3$ staining [114–116], or less frequently fluorescently labeled primers [117, 118] or radioactively labeled primers [13]. Polymorphisms are due to variation in the primer annealing sites and length differences in the amplified sequence between primer annealing sites. These markers contain two short segments in inverted orientation, on opposite strands that are complementary to the primer used.

Some advantages and disadvantages of RAPD marker system are listed in Table 9.8. The main advantage of RAPD markers is that they are quick and easy to assay with low cost. The RAPD marker system requires 20–100 ng template DNAs per reaction. These markers have high genomic abundance and random distribution throughout the genome. However, RAPD markers have low reproducibility [14] and they are not locus-specific markers. Band profiles of RAPD markers cannot be interpreted in terms of loci and alleles. In RAPD marker systems similar sized fragments may not be homologous.

Table 9.8 Some advantages and disadvantages of the RAPD, DAF and MAAP marker systems

Advantages	Disadvantages
No sequence data for primer construction are needed	Dominant and moderately complex markers
Low quantity genomic DNAs are required	Nonhomology of similar sized fragments
Quick and easy to assay with low cost	Reproducibility problems
No pleiotropic and epistatic interactions	Limitations in genomic mapping studies
Low level of reproducibility	Limitations in obeying Mendelian laws
High level of genomic abundance	Difficulty in terms of allele or locus assignment
Could be converted into SCAR markers	Limitations in scoring higher molecular weights
Randomly distributed throughout the genome	Requirements of polyacrylamide gels for DAF and MAAP systems

MAAP markers of interest can be converted into SCAR marker system, which has several advantages over RAPD system. SCAR marker system was first introduced in Michelmore et al. [119] and Martin et al. [120]. Development of SCAR markers includes the following main steps. In the first step, RAPD markers present in one individual but absent in the other individual (for example present in disease resistant lines but absent in susceptible lines) is isolated from an agarose gel, cloned, and the nucleotide sequences are determined. A pair of primers (usually, 20–24 nucleotide bases), one forward and one reverse primer, specific for the two terminal sequences of the RAPD fragment is designed. DNA samples from individuals are amplified particularly at elevated annealing temperatures and the resulting amplicon (usually within the 0.5–2 kb size range) is resolved on a 1.5–2.0% agarose gel stained with ethidium bromide. SCAR polymorphisms are detected as the presence or absence of the band; however, in some cases some of the SCAR markers detect length polymorphism either directly or after digestion of the amplified fragment with a suitable restriction enzyme [106–108, 121].

SCAR markers can be used in physical and genetic mapping, comparative mapping, and phylogenetic relationship studies [106–108]. In Table 9.9, some advantages and disadvantages of the SCAR marker system are presented. SCAR markers are potentially be converted into functional co-dominant DNA markers. In the SCAR system the use of longer and sequence specific PCR primers enhances the reproducible of the obtained markers. Majority of the SCAR markers are co-dominant markers, however, in some situations, dominant markers are also produced when one or both primers partially overlap the site of sequence variation within the amplified regions. High level of reproducibility makes SCAR system attractive to AP-PCR, DAF and RAPD marker systems. The main disadvantage is that its requirements of sequence data to design the PCR primers. In addition, it has lower number of marker produced per reaction in comparison to AP-PCR, DAF and RAPD marker systems [108, 121].

Table 9.9 Some advantages and disadvantages of the SCAR marker system

Advantages	Disadvantages
Have high level of reproducibility	Sequence data for primer construction are needed
Low quantities of template DNA are required	Usually dominant markers are produced
Quick and easy to assay with low cost	Low throughput
No pleiotropic and epistatic interactions	Have limitations in their use in mapping studies
Produces locus specific amplification	Most markers do follow Mendelian genetic laws
Less sensitive to reaction conditions	Requires sequence data for primer construction

4.4 Amplified Fragment Length Polymorphism (AFLP) with Synonyms and Derivatives

Selective fragment length amplification (SFLA) and selective restriction fragment amplification (SRFA) are synonyms to amplified fragment length polymorphism (AFLPs). Several other marker techniques such as, sequence-specific amplification polymorphism (S-SAP), three-endonuclease AFLP (TE-AFLP), multiplex-endonuclease genotyping approach AFLP (MEGA-AFLP) and miniature inverted repeat transposable elements-AFLP (MITE-AFLP) are modifications of AFLP technique. In addition, AFLP markers can be converted to SCAR, CAPS, dCAPS, or sequence tagged site (STS) markers [53, 87, 88, 91, 102, 106–108, 122, 123].

The AFLP marker system is similar to RFLP marker system in term of restriction enzyme digestion and RAPD marker system in term of PCR requirement. Genomic abundance, reproducibility, many informative bands per reaction generated and no requirement of sequence data for primer construction make AFLP marker system powerful genetic marker. Markers generated by AFLP system can be analyzed on automatic sequencers [123]. AFLP polymorphisms may result from; (i) mutations in the recognition sequences of the enzymes used for digestion of the genomic DNA, (ii) mutations in the sequences complementary to AFLP primers; (iii) insertions or deletions within the amplified restriction fragments. AFLP marker system can be applied to any organism without the need for primer sequence development. AFLP marker system usually produces 50–100 fragments whose size varies 80–500 bp per reaction [106–108].

A summary of the general procedure for AFLP marker system is as follows: (i) the first step of the analysis involves DNA extraction; (ii) 300–500 ng genomic DNA is digested with *EcoR*I or *Pst*I (rare cutter) and *Mse*I or *Taq*I (frequent cutter) restriction enzymes; (iii) double-stranded oligonucleotide adaptors are ligated to ends of the fragments that were generated by the restriction enzymes used; (iv) the pre-amplification PCR is performed with primer combinations containing a single nucleotide extension; (v) the selective re-amplification of PCR is performed using primer pairs with 1–3 bases extension; (vi) 50–100 fragments per reaction are obtained and visualized on polyacrylamide gel electrophoresis (PAGE) coupled with silver staining or autoradiography. The use of fluorescence-tagged primers permits the analysis of AFLP fragments by an automated DNA sequencing machine coupled with capillary electrophoresis (CE), which also enables automated data collection and analysis. These are two main separation techniques used in AFLP analysis. These techniques provide maximum resolution of AFLP banding patterns to the level of single-nucleotide length differences [123]. Agarose gel electrophoresis system is not suitable for AFLP analysis because fragment length differences of less than ten nucleotides are difficult to score on agarose gels [76, 124–126].

In Table 9.10, some advantages and disadvantages of AFLP markers systems are presented. It produces highly reproducible and reliable markers and does not require any DNA sequence information for primer development. AFLP system analyzes a

Table 9.10 Some advantages and disadvantages of the AFLP marker system

Advantages	Disadvantages
No requirements of sequence data for primer construction	Laborious, technically demanding, and expensive
High level of polymorphism with effective multiplex ratio	May not be totally randomly distributed around the genome
No pleiotropic and epistatic interactions	More applicable to intra-specific not to interspecific studies
Highly reproducible	Dominant and moderately complex markers
Suitable for automatic sequencers	Nonhomology of similar sized fragments

large number of polymorphic loci per reaction [127]. The main disadvantage of AFLP marker system is that it produces dominant markers. AFLP markers are not randomly distributed around the genomes [128–130]. Other disadvantages include the need for pure DNA templates free of enzyme inhibitors and other contaminants, and the possible non-homology of co-migrating fragments belonging to different loci.

The AFLP marker system has been modified in various ways to achieve specific objectives. Diversity array technology (DArT) and complexity reduction of polymorphic sequences for single nucleotide polymorphisms (CRoPS) are also modification of AFLP marker system and they will be revisited later in this chapter. Other modifications include sequence tagged site (STS), sequence-specific amplification polymorphism (S-SAP), three-endonuclease amplified fragment length polymorphism (TE-AFLP), multiplex-endonuclease genotyping approach amplified fragment length polymorphism (MEGA-AFLP) and miniature inverted repeat transposable elements amplified fragment length polymorphism (MITE-AFLP) [131, 132].

An AFLP marker of interest can be converted into a SCAR marker. In short, DNA sequence from the AFLP fragment is isolated and reamplified using the same AFLP primers that were used in the amplification step. Amplicons are purified, cloned and sequenced. A pair of specific PCR primer is designed and used as SCAR primer pair for SCAR marker system. The same strategy can also be used to generate CAPS, dCAPS, or sequence tagged site (STS) markers. A STS is a sequence of DNA located in a genome and its base composition is known [133, 134]. STS was first developed for the physical mapping of the human genome and later adopted in plants [133]. In plants, STSs could be obtained from AFLP fragments and fragments of RFLP probe representing mapped low copy number sequences [134]. Co-dominant SCAR markers obtained from RAPD marker system could be referred to as STSs [135]. Those STS loci containing genetic polymorphisms are valuable genetic markers. They are co-dominant, highly reproducible, suitable for high throughput and automation and useful for constructing genetic and physical maps from sequence data reported from many different laboratories [136].

Summary of developing a STS marker as follow: (i) an RAPD, RFLP or AFLP fragment extracted, cloned and sequenced; (ii) based on this sequence information, a pair of PCR primers (20–24 bp) is designed for reproducible-specific amplification of the target segment which contains 200–400 bp; (iii) primers are tested for high stringent PCR amplification condition using the genomic DNA as template; (iv) the resulting amplicons are resolved on a 1.5–2.0% agarose gel stained with ethidium bromide [134–136].

4.4.1 Sequence-Specific Amplification Polymorphism (S-SAP)

In the sequence-specific amplification polymorphism (S-SAP) system, the restriction fragments are generated and ligated to the specific adapters as in AFLP marker system. In the amplification step, only one AFLP primer is used, and the other primer is based on long terminal repeats (LTRs) of a transposable element (TE). S-SAP produces dominant markers but this marker system has higher levels of polymorphisms in comparison to AFLP [137]. S-SAP marker system is very similar to the transposon display (TD) procedure of van der Broeck et al. [138]. In the TD procedure, hexa-cutter restriction enzyme is used while S-SAP usually utilizes tetra-cutter or tetra- and hexa-cutter together [138].

4.4.2 Three-Endonuclease AFLP (TE-AFLP)

Three-endonuclease AFLP (TE-AFLP) is another modification of the AFLP marker system. In the TE-AFLP system, three restriction enzymes are used to digest the sample DNA but two sets of adapters are used for amplification of the fragments to increase the discriminatory power of the technique. In the TE-AFLP marker system, one-step amplification procedure is used for fingerprinting studies. Multiplex-endonuclease genotyping approach amplified fragment length polymorphism (MEGA-AFLP) uses four or more endonucleases for digestion of the sample DNA but employs only a single pair of adapters for PCR amplification [139].

4.4.3 Miniature Inverted Repeat Transposable Elements Amplified Fragment Length Polymorphism (MITE-AFLP)

Miniature inverted repeat transposable elements amplified fragment length polymorphism (MITE-AFLP) is another AFLP based method and it is similar to S-SAP. MITE-AFLP uses one AFLP primer and one primer based on a MITE element for the amplification step of the AFLP procedure. This method is also called MITE-transposon display and has been used in grass species including maize and rice. There is no application of MITE-AFLP in the genus *Salvia*.

4.5 Microsatellites or Simple Sequence Repeats (SSRs) with Synonyms and Derivatives

Microsatellites were first described in Litt and Lutty [140] and they are also widely referred to as simple sequence repeats (SSRs). There are several modifications or derivatives of microsatellite marker systems such as single primer amplification reactions (SPARs) or microsatellite-primed PCR (MP-PCR), random (randomly) amplified microsatellite polymorphisms (RAMPs), inter simple sequence repeats (ISSRs), anchored simple sequence repeats (ASSRs), anchored microsatellite-primed PCR (AMP-PCR), or inter-SSR amplification (ISA), selective amplification of microsatellite polymorphic loci (SAMPL), retroposon-microsatellite amplified polymorphism (REMAP) and CAPS-microsatellites [141].

Microsatellite markers are a special version of STS markers, in which a microsatellite locus is amplified using a specific primer pair derived from the unique sequences flanking the SSR locus. Synonyms include sequence-tagged microsatellite site (STMS), short tandem repeats (STRs), simple sequence repeat polymorphism (SSR) [76, 142, 143]. There are several modifications of microsatellites including CAPS-microsatellites and inter-simple sequence repeats (ISSRs) discussed in the next section of this chapter.

SSR is the marker of choice and was widely used in plant molecular research during the last decade of the past century and the first decade of the present century. Microsatellites are highly polymorphic even between closely related lines, reproducible and co-dominant markers exhibiting multi-allelic nature and simple Mendelian inheritance. Development of microsatellite markers requires low amount of DNA, and developed markers can be easily automated for high throughput screening, can be exchanged between laboratories, and are highly transferable between populations [10, 35, 53, 144].

Microsatellites consist of tandemly repeating mono-, di-, tri-, tetra- or penta- and hexa-nucleotide motif units distributed throughout the genomes of most eukaryotic organisms [10, 53, 87]. The differences in the tandemly repeating motifs are caused by DNA polymerase slippage during replication, or slipped strand miss-pairing [145]. It was estimated that many microsatellites have high-mutation rates between 10^{-2} and 10^{-6} mutations per locus per generation, and on average 5×10^{-4} [10, 21, 53, 76].

Development of microsatellite primer pairs include: (i) microsatellite library construction from the target species; (ii) identification of unique microsatellite loci screening with single or mixed simple sequence oligonucleotide probes; (iii) identifying a suitable DNA sequences flanking the microsatellite regions for primer design by sequencing studies; (iv) identification of PCR products representing desired microsatellite alleles; (v) evaluation and interpretation of microsatellite patterns, removing redundant and/or chimeric sequences and (vi) assessing PCR products for polymorphism [10, 53, 146].

A summary of the general procedure for SSR marker validation is as follows: (i) the first step of the analysis involves DNA extraction, only small amounts of

DNA are needed for amplification as the polymerase chain reaction creates an exponential increase in the replicated segments; (ii) PCR is performed using forward and reverse SSR primer pair (generally 20–25 bp). Primer sequences flanking SSR regions are conserved within and closely related species but the SSR sequences mutate frequently by slippage and proofreading errors during DNA replication that primarily change the number of repeats and thus the length of the repeat string; (iii) amplified products of SSR loci are visualized on agarose or polyacrylamide gel electrophoresis coupled with silver staining or capillary electrophoresis (CE) coupled with florescence labeling [76, 147].

Although microsatellite analysis is a single-locus technique, it may be multiplexed during PCR or gel electrophoresis (multi-mixed) when the size ranges of the alleles of different loci does not overlap [148]. The cost and labor of the microsatellite can be reduced by: (i) pooling the PCR products from two or more separate single primer pair-based reactions and running them in a single gel lane or capillary; (ii) using a single PCR reaction tube for simultaneous amplification of two or more SSR loci; or (iii) combining the above two approaches [149, 150]. When PCR primers with different fluorescent labels are used in above approaches, the cost and labor of the microsatellite are further reduced [97, 146, 151]. However, fluorescent labeling of primers is expensive and increases the cost of assays. Also the use of an array tape in place of microtiter plate drastically reduces the amounts of reagents and consumables. However, it is more practical in high throughput analyses. The use of a universal fluorescence-labeled primer like M13 in combination with the normal specific forward and reverse primers could be utilized. In Table 9.11, some advantages and disadvantages are presented.

One of the main disadvantages of SSRs is that development of new SSRs is technically complicated, labor intensive, and costly. The use of expressed sequence tag (EST) databases has been an alternative source of SSRs development [10, 21, 92, 93, 152–157]. The use of ESTs and other DNA sequence has become a fast, efficient, and relatively inexpensive method for development of EST-based SSR markers [10, 158, 159]. However, the use of ESTs and other DNA sequence databases for EST-SSRs is limited to species for which this type of database does not exist. The unique sequences flanking the SSR primers seem to be conserved

Table 9.11 Some advantages and disadvantages of the SSR marker system

Advantages	Disadvantages
Highly reproducible and co-dominant markers	Development of microsatellite is technically complicated, labor intensive, and costly
High level of polymorphism	Unsuitable for studies involving higher taxonomic levels
No pleiotropic and epistatic interactions	Potential presence of null alleles
Suitable for multiplex PCR	Presence of stutter bands
Suitable for automatic sequencers	Homoplasy
Easy for scoring	Low throughput

within species and even across species within a given genus, but rarely across genera. Cross species transferability means that the primers for SSR markers developed for one plant species can be successfully used in related plant species Cross-species transferability of SSRs obtained from ESTs and other DNA sequence has been found to be high but they have lower rate of polymorphisms compared to the SSR markers derived from genomic libraries [10, 21, 92, 93, 152–157]. The SSR markers derived from genome sequences are sometimes termed as genomic SSRs, while those developed from ESTs are referred to as expressed SSRs (eSSRs) or genic SSRs [160]. In addition, SSR markers derived from unigene are called unigene-derived microsatellites (UGMs) [10, 21, 53].

Most SSR polymorphisms cannot be resolved on agarose gels and high resolutions can be achieved through the use of polyacrylamide gels in combination with $AgNO_3$ staining or capillary electrophoresis in combination with fluorescence labeling for automation. The cost of automation for SSRs is high, and due to different genotyping platforms, difficulties are encountered in sharing SSR marker data between laboratories [10, 21, 53, 152].

Two other common problems in the SSR marker analysis arises during the PCR amplification studies; the presence of stutter bands and the null alleles. Stutter bands are artifacts in the SSR marker analysis that occur by DNA slippage during PCR amplification. Stutter bands can complicate the interpretation of the band profiles especially in automation. Occurrence of null alleles is believed to be generated by mutation in the binding site for one or both of the SSR primers, leading to a failure of amplification [151]. The potential occurrence of null alleles increases with the use of SSR primers obtained from unrelated species. The use of null alleles may result in a biased estimate of the allelic and genotypic frequencies [10, 21, 53, 154, 157].

4.5.1 Inter-simple Sequence Repeats (ISSR)

Inter-simple sequence repeat (ISSR) marker system is quick, simple and cost effective marker system [90, 161]. The ISSR marker system overcomes the problems like high cost of AFLP and low reproducibility of RAPD [162]. The ISSR marker system uses microsatellites as primers in a single primer PCR reaction targeting multiple genomic loci to amplify inter simple sequence repeats of different sizes. This technique utilizes 15–30 mer primers with high annealing temperature (45–65 °C). The ISSR marker system produces amplified products usually 200–2000 bp long and amenable to detection by both agarose and polyacrylamide gel electrophoresis. The level of polymorphism is dependent on the detection method used. PAGE in combination with radioactivity was found to be most sensitive, followed by PAGE with $AgNO_3$ staining and then agarose gel with ethidium bromide system of detection [18, 163].

An ISSR fragment contains two identical microsatellite repeat regions oriented in opposite direction at the terminal ends. The ISSR primers can be made of

Table 9.12 Some advantages and disadvantages of the ISSR marker system

Advantages	Disadvantages
No sequence data for primer construction are needed	Dominant markers are produced
High level of polymorphism	Presence of possible non-homology of similar sized fragments
Randomly distributed throughout the genome	Reproducibility problems
Low quantities of template DNA are required	Complex data are produced

di-nucleotide, tri-nucleotide, tetra-nucleotide or penta-nucleotide repeats. An ISSR primer may consist solely of a microsatellite sequence. This kind of ISSR primer is called non-anchored ISSR primer. The other type of ISSR primer is called anchored primer that is made of a microsatellite sequence consisting of two nucleotides at either the 3′ or the 5′ end of the primer [163–165].

The ISSR marker system is easy to use, cheap, has throughput and yield multiple polymorphic loci. It does not require sequence information for primers but has dominant type markers. A major disadvantage of the ISSR marker system, it has poor reproducibility. A summary of the general procedure for ISSR marker system is as follows: (i) the first step of the analysis involves DNA extraction; (ii) PCR is performed using single ISSR primer; (iii) amplified products of ISSR loci are visualized on agarose gel electrophoresis coupled with ethidium bromide staining or polyacrylamide gel electrophoresis coupled with silver staining [163–165].

ISSR marker system, synonyms and modification mentioned above, has several advantages and disadvantages listed in Table 9.12. The main disadvantage of this marker system is the low level of reproducibility and the presence of possible nonhomology of similar sized fragments. Advantages include high level of polymorphism, not requirement of sequence data for primer construction, randomly distributed throughout the genome and low quantities of template DNA are required [163–165].

4.5.2 Single Primer Amplification Reactions (SPARs) or Microsatellite Primed PCR (MP-PCR)

Single primer amplification reactions (SPARs) or also called microsatellite-primed PCR (MP-PCR) use non-anchored primers, which consist of tri-, tetra-, and penta-nucleotide repeats differing from ISSR in which anchored primers are utilized. Polymorphisms using the SPARS or MP-PCR marker systems are detected as variations in the size of the genomic region between the two adjacent microsatellite sequences [166].

4.5.3 Random Amplified Microsatellite Polymorphism (RAMP), Inter-SSR PCR, ASSR, AMP-PCR

Random (randomly) amplified microsatellite polymorphisms (RAMPs), inter-SSR PCR, anchored simple sequence repeats (ASSRs), anchored microsatellite-primed PCR (AMP-PCR), or inter-SSR amplification (ISA) use anchored primers. When 5′-anchored primers are used, the amplified fragment consists of full lengths of two microsatellite sequences as well as the inter-SSR region. When 3′-anchored primers are used, only the region between of two SSRs, including the primer, will be amplified.

The RAMP marker system uses a 5′ anchored SSR primer and a RAPD primer to generate markers termed as randomly amplified microsatellite polymorphisms. Random primer (RAPD) binding site serves as an arbitrary endpoint for the anchored SSR primer-based amplification product. The RAMP products could be digested with a restriction enzyme to yield digested RAMPs (dRAMPs) markers [167].

In the selective amplification of microsatellite polymorphic loci (SAMPL) marker system one AFLP primer with three selective nucleotides and one microsatellite primer are used in combination for the amplification step after the pre-amplification step in the AFLP marker system. SAMPL marker system usually generates dominant markers, but some of the markers may show co-dominant inheritance [168].

4.5.4 Cleavage Amplified Polymorphic Sequence (CAPS)-Microsatellites

CAPS-microsatellite marker system is the combination of the cleavage amplified polymorphic sequence (CAPS) system and the microsatellite marker system. This system is particularly important when EST-based primer pairs are used. The CAPS-microsatellite markers do not only convert monomorphic amplicons of the EST-based primer pairs into polymorphic markers but also produce a high level of polymorphism between closely related genotypes. The polymorphism detected using CAPS-microsatellite technique is based on the sequence differences in the flanking regions outside the microsatellite motifs and within imperfect microsatellite motif variation. Since polymorphisms in CAPS-microsatellites are not due to the sequence length differences but due to insertion/deletion and base substitution events, these markers allow more precise identification of genotypes [10, 21, 93, 154].

4.6 CAAT Box-Derived, Conserved DNA-Derived and Start Codon Targeted (SCoT) Marker Systems

4.6.1 CAAT-Box-Derived Markers (CBD)

The CAAT box-derived marker (CBD) system is a PCR-based marker and its target is CAAT boxes of the promoters of plant genes. The CBD marker system uses a single primer that has five-nucleotide CCAAT core flanked by 10–11 filler nucleotides anchored to the 5′ and 2–3 arbitrary nucleotides anchored to the 3′ end that targets the CAAT box of the promoter sequences of plant genes. Because CBD technique uses single primers in amplification reaction, amplification is obtained when the CAAT boxes of promoters of the two genes occur in a distance suitable for PCR amplification [169].

4.6.2 Conserved DNA-Derived Polymorphism (CDDP)

The conserved DNA-derived polymorphism (CDDP) method uses single 15- to 19-mer primer pairs for PCR amplification using high annealing temperature. Amplification of CDDP marker is obtained when the CAAT boxes of promoters of the two genes occur in a distance suitable for PCR amplification [169]. Sequences of primer pairs are based on conserved DNA regions of a selected set of well-characterized plant genes such as WRKY, MYB, ERF, KNOX, MADS, and ABP1. These genes involve in abiotic/biotic stress responses or developmental processes [169, 170]. Because primers are designed using highly conserved regions of genes, the CDDP markers are also generated in other plant species.

4.6.3 Start Codon Targeted (SCoT) Markers

The start codon targeted (SCoT) marker system is also a PCR-based marker and its target is the short conserved sequence surrounding the translation initiation codon or start codon, ATG, of plant genes [170]. The SCoT primer, which is a single 18 nucleotides long primer, has the following fixed nucleotides: the A, T, G of the start codon (positions +1, +2, +3), G at +4, A at +7, C at +8, and C at +9 and variable number of arbitrary nucleotides on the 5′ side of the ATG nucleotides [169, 171]. In terms of the use of a single primer, lack of sequence information requirement, dominant and two to six amplification products in each PCR, the SCoT marker system is similar to RAPD and ISSR marker systems. On the other hands, the SCoT markers are based on genic regions as compared to the random genomic regions in the cases of RAPD and ISSR marker systems. The cDNA-SCoT technique was developed to be used for a simplified gene expression analysis [169, 172].

4.7 Sequence Related Amplified Polymorphism (SRAP) Marker

Sequence-related amplified polymorphism (SRAP) is another gene-targeted markers based on PCR amplification. It is simple and reliable and targets coding sequences [173]. The SRAP marker system may generate co-dominant markers when the polymorphisms are resulted from the insertions and deletions. On the other hand, it may produce dominant markers when the polymorphisms are due to nucleotide variations affecting primer binding [174, 175].

Target region of the SRAP marker system is open reading frame (ORF) amplification. This system uses two primers of 17 or 18 mers each. Primers have three selective nucleotides in their 3′ ends, followed by a core sequence of four nucleotides (5′ CCGG 3′ in the forward primer and 5′ AATT 3′ in the reverse primer) and a 10 or 11 mer long arbitrary sequence (sometime they are filler sequences) at the 5′ end [176]. The filler sequences are used for the forward and reverse primers contain different nucleotides [53, 174, 177, 178].

The CCGG core sequence in the forward primer is targeted at exons since exons are more frequent in GC-rich regions. The AATT core in the reverse primer is targeted at promoters and introns since they are in AT-rich regions. In the amplification of the SRAP markers, the annealing temperature during the initial five PCR cycles is kept at 35 °C; it is set at 50 °C during the next 35 PCR cycles. The amplified products are separated on denaturing acrylamide gel electrophoresis coupled with silver staining or capillary electrophoresis coupled with florescent labeling [173].

4.8 Target Region Amplification Polymorphism (TRAP)

The target region amplification polymorphism (TRAP) is another PCR-based marker system. Technically the TRAP marker system is similar to SRAP. As in the SRAP, the TRAP system also utilizes two primers of 18-mers each; one of primers is obtained from selected EST and is called fixed primer, while the other primer is an arbitrary primer. The fixed primer is obtained using *in silico* analysis of the EST databases to detect polymorphism around the desired candidate genes. The arbitrary primer of the TRAP marker system has an AT-rich core region such as 5′ AATT 3′, would anneal to an intron or a GC-rich core region such as 5′ CCGG 3′, would anneal to an exon. The amplification of the TRAP markers could be performed similar to that of SRAP but it is better to use a gradient thermal cycler to determine the best amplification parameters. Depending on the species the TRAP marker system can generate 20–40 storable markers ranging from 50 to 900 bp in a single PCR reaction. The TRAP markers are resolved by electrophoresis on polyacrylamide gels or capillary electrophoresis coupled with fluorescent labels [179].

4.9 Restriction Landmark Genome Scanning (RLGS)

This marker system uses direct end labeling of the genomic DNA digested with one or more rare-cutting restriction enzymes and high-resolution two-dimensional electrophoresis. It has a potential of genome scans that equals the work of thousands of PCR. The RLGS technology is a method for high-speed survey of the presence or absence of restriction landmarks throughout a genome and the measurement of their copy number based on the new concept that restriction enzyme sites are used as landmarks. RLGS could be used to genome mapping, the estimation of evolutional distance, and the study of many biological problems, such as mutations and genetic diseases. It may be useful for genome mapping or mapping of the specific mutant locus by use of interspecies crosses. RLGS markers are produced in about seven steps including: (i) reduction of background by using enzyme reactions as a pretreatment to interfere with the incorporation of labeling reagents into nonspecific-damaged sites (blocking); (ii) restriction enzyme digestion (landmark cleavage); (iii) end radio-labeling of the genomic DNA digested with a restriction enzyme; (iv) fragmentation of labeled DNA with another restriction enzyme; (v) first size fractionation; (vi) fragmentation of labeled DNA with restriction enzyme; (vii) second size fractionation, and (viii) autoradiography. Some advantages and disadvantages of this method are presented in Table 9.13 [180].

4.10 Conserved Region Amplification Polymorphism (CoRAP)

The conserved region amplification polymorphism (CoRAP) marker system is similar the TRAP and SRAP marker systems. Differences between these marker systems lie in their core regions of primer sequences. The core primer sequences of the CoRAP marker system is usually found in the introns of plant genes and targets are introns. The CoRAP marker system requires sequence information of the concerned plant species for the design of the fixed primers. When the targeted plant

Table 9.13 Some advantages and disadvantages of the RLGS marker system

Advantages	Disadvantages
No DNA probes are needed	It requires at least a few micrograms of DNA
Ploidy level of the organism can be discriminated by this method	It has limited library coverage in comparison to next generation technologies
Thousands of restriction landmarks can be scanned simultaneously	It needs specially designed instruments, and elaborate gel electrophoresis set-up, a powerful image analysis system
It has high-efficiency scanning capacity, scanning extensibility by using alternate restriction enzyme combinations	Its use is restricted to clinical applications and requirement of autoradiography

species have complementary sequences for CoRAP primers within a suitable distance from each other, PCR products are produced and products will be polymorphic if the intervening sequences had insertions or deletions, which result in length polymorphism among different individuals/strains [181].

The CoRAP marker system utilizes one fixed primer and one arbitrary primer for PCR amplification. The fixed primers is obtained EST databases and the arbitrary primer has the core sequence of CACGC at the 5′ end, followed by 11 arbitrary nucleotides that serve as fillers, and three bases at the 3′ end to serve as selection nucleotides. Amplification of the CoRAP markers are performed relatively higher annealing temperature such as 52–55 °C. Each PCR reaction generates 30–50 fragments of 50–1000 bp. The CoRAP markers are co-dominant and highly reproducible [181].

4.11 Intron-Targeting Polymorphism (ITP) and Conserved Orthologous Set of Markers (COS)

Intron-targeting polymorphism (ITP) marker system uses two primers that are designed based on the sequences of the conserved regions of exons flanking an intron using ESTs available in the database. PCR amplification produces an intronic fragment flanked with the ITP primers at the terminal regions [182]. Because the introns are much less conserved than exons, a high proportion of the amplified fragments, which are designed to in the range 200–1200 bp, are expected to show length polymorphism due to insertions and deletions [23].

Advantages of ITP markers include co-dominant markers and primers are transferable across the species. ITP markers are generated from ESTs, and some of them might give rise to functional markers. Disadvantages are the development of ITP markers require sequence information and the number of marker generated in single reaction is low [23].

Genes with conservation of sequence as well copy number during the evolution of plant species are defined conserved orthologous set (COS). In another word, orthologous genes are homologous sequences descended from the same ancestral sequence for which function of the gene or sequence has been conserved across evolutionary time. Thus, each gene of the orthologous set has an orthologue in all the species of the group and often even in other distantly related species. The genes included in the orthologous set are single-copy genes, but low-copy number genes may also be included. Computational analysis of genomic and EST sequences from a group of related species along with a well-characterized reference species like *Arabidopsis thaliana* (usually, for dicots) or rice (usually, for monocots) is used to identify the conserved set of genes [183]. By screening the consensus EST sequences or the unigene sequences from many plant species, it is feasible to develop COS markers.

4.12 Transposon- and Retrotransposon-Based Marker Systems

Transposable elements (TEs) are DNA sequences that move within a genome. TEs are divided into Group I transposons that are also called retrotransposons (RTEs). Group II transposons are called transposons. RTEs transpose via RNA intermediates to move within a genome while transposons move as DNA molecules. Some RTEs contain long terminal repeats (LTRs) at their terminal elements while others lack LTRs. Among transposons there exist small transposons, which are distributed widely and are plentiful in a number of plant genomes. These types are called miniature inverted repeat transposable elements (MITEs) which are often located in the terminal regions of genes. Because RTEs and transposons are present in plants usually in high copy numbers and are dispersed throughout the genome their variations in the number of copies and the sites of insertion in the genomes of related species could be used as DNA markers [184, 185]. Other techniques include inverse sequence tagged repeat (ISTR) [186]; retrotransposon-based insertion polymorphism (RBIP) [187]; inter-retrotransposon amplified polymorphism (IRAP) and retrotransposon microsatellite amplified polymorphism (REMAP) [188]. ISTR markers are universally applicable by using identical primers derived from coconut copia-like sequences in the fingerprinting of eukaryotic genomes [186].

4.12.1 Inter Retrotransposon Amplified Polymorphism (IRAP) and Inter-MITE (IMP)

The inter-retrotransposon amplified polymorphism (IRAP) marker system uses primers based on LTRs of RTEs to amplify the region between two neighboring insertions of the element. Inter-MITE polymorphism (IMP) marker system is very similar to the IRAP marker system. The IMP marker system uses primer based on MITE-like transposable elements in the place of those derived from retrotransposons in the IRAP marker system [189]. The IRAP marker system produces dominant markers that examine variation in retrotransposon insertion sites. The IRAP marker system can produce 20–30 fragments, which can be separated by high-resolution agarose gel electrophoresis [188].

4.12.2 Retrotransposon Based Insertion Polymorphism (RBIP)

The retrotransposon-based insertion polymorphism (RBIP) marker system uses two set of primers; in one set a pair of primers is designed from the retrotransposon itself and in the second set a pair of primers is designed from flanking DNA of the retrotransposon at the given insertion site to examine insertional polymorphisms for individual retrotransposons [189]. After PCR and agarose gel (or a dot blot assay) electrophoresis, presence or absence of insertion is investigated. When the primer pair derived from the flanking sequences is used for amplification, a product would be obtained if there is no retrotransposon insertion in the region flanked by the primers, but when the primer based on the retrotransposon is used with a primer specific to one of the flanking regions, a PCR product is generated only when the concerned region contains the retrotransposon [190].

Polymorphisms are detected using agarose gel electrophoresis coupled with ethidium bromide staining or by a dot hybridization assay using a reference PCR fragment for hybridization. Although it is a co-dominant and amenable to high-throughput automation marker system, a drawback of the method is that sequence data of the flanking regions is required for primer design [188].

4.12.3 Retroposon Microsatellite Amplified Polymorphism (REMAP)

Retroposon-microsatellite amplified polymorphism (REMAP) is a SSR based marker system. In this marker system two primers are used one is microsatellite primer consisting of repeats and other primer consists long terminal repeat. Thus REMAP may amplify DNA fragments: (i) the segments flanked by an LTR at one end and a microsatellite on the other end, (ii) sequences having a microsatellite locus at both their ends, and (iii) fragments consisting of retrotransposon, which could be called inter-retrotransposon amplified polymorphism (IRAP). IRAP and REMAP markers may produce up to 30 bands per individual. These are highly polymorphic within species [189].

4.13 Cleavase Fragment Length Polymorphism (CFLP)

By analogy to RFLP and single-strand conformation polymorphism analyses, it is called cleavase fragment length polymorphism (CFLP). It is a very powerful and polymorphic marker system and relies on the formation of individual secondary structures that result when DNA is allowed to cool following a brief heat denaturation. These secondary structures serve as substrates for the structure-specific Cleavase I enzyme. By using this enzyme any given DNA with secondary structures can be cleaved into various fragments. Since formation of these secondary structures is remarkably dependent on nucleotide sequence, cleavage results in the generation of a unique collection of fragments for each sequence analyzed. The similarities and differences between DNA sequences are represented in the CFLP patterns generated. CFLP patterns can be formed due to loss or gain of one or more bands, shifts in the position of bands and a change in the relative intensities of bands. The structural fingerprints that result from CFLP analysis provide a ready means of identifying the extent to which sequences are related to each other [191].

The method is accomplished as follows: (i) labeled fragments of DNA are heated to separate the complementary strands; (ii) when the samples are cooled, the single strands of DNA assume folded hairpin like structures; subtle differences in the sequences of the fragments can cause formation of different structures; (iii) enzymatic digestion is performed under conditions of partial cleavage. Cleavase I enzyme cleaves at the 5′ side of these structures, at the junction between duplexed and single-stranded regions before DNA samples are disrupted by reannealing of the complementary strands; and (iv) the fragments are separated by rapid electrophoresis on a denaturing acrylamide gel and visualized by means of the labeled end. Mutational changes are visualized as the appearance or disappearance of bands, band mobility shifts or changes in band signal intensity [191].

5 Single Nucleotide Polymorphism (SNP)

Single nucleotide polymorphism (SNP) is a type of DNA sequence variation that occurs when a single nucleotide (A, T, G or C) differs between the members of a species. The nucleotide polymorphism at a locus is considered as SNP only when frequency of the polymorphism is 1% or more. DNA nucleotide variation below 1% is considered as mutation [66, 192].

Each SNP locus has a specific position in a genome and each is defined by the sequence flanking the polymorphic nucleotide. Densities of SNPs vary within and between genomes of species and SNPs are more frequent in noncoding regions than in the coding regions [66, 192]. A SNP locus can potentially have four alleles, each allele being represented by one of the four DNA nucleotides. However, many SNP loci have three or even two alleles probably due to the fact that variations of SNPs are generated by either transition (C to T (or T to C) and A to G (or G to A) or transversion (A or G to C or T and C or T to A or G)). It is observed that transitions seem to be more frequent than transversions because deamination of 5-methylcytosine results in C to T (in the complementary strand G to A) transitions in genomic regions rich in CpG dinucleotide sequence. Phenotypic effects of SNP loci are caused by altered amino acid sequence, changed splicing pattern of the RNA transcripts or altered the regulatory regions such as promoters [66, 193].

SNPs are the most abundant DNA variations due to their availability in the genomes of different populations. SNP markers are now widely used for the analysis of population structure, association mapping, selection of QTL, and/or evolutionary studies. Due to technological advances, thousands to millions of SNP markers are now available for each plant species. They have the highest resolutions among the DNA marker systems. The fact that in many organisms most polymorphisms observed in the last two decades result from changes in a single nucleotide position. SNP is now considered as marker of choice. Allele specific PCR primers or oligonucleotide probes are required for detection of SNPs. SNPs and their flanking sequences can be obtained through genomic or cDNA library construction and sequencing studies. They can also be found in nucleotide databases using *in silico* approaches and single-stranded pyrosequencing [194–197]. Some advantages and disadvantages of SNP markers are listed in Table 9.14.

Table 9.14 Some advantages and disadvantages of the array based SNP marker system

Advantages	Disadvantages
Many platforms have high level of polymorphism and high level of reproducibility	Sequence information of target genomes is required
Less subject to the problem of homoplasy	Some platforms require sequence information for the species under study
Efficient, fast, and high-throughput analysis	Require dedicated skilled manpower and software programs
Many platforms are cost effective because many data points are produced	The chief limitations of many platforms are the high equipment cost

SNPs are extremely abundant in plant genomes being present one SNP every 100–300 bp and they are relatively easy to detect. They are usually co-dominant markers, efficiently discriminating between homozygous and heterozygous alleles. Using SNP markers a large number of loci can be assessed simultaneously. SNPs are less subject to the problem of homoplasy and amenable to high throughput automation, allowing rapid and efficient genotyping of large numbers of samples. However, development of SNPs involves resequencing of even whole genomes and many SNP detection platforms require dedicated skilled work forces and software programs [193].

SNPs are classified in a variety of ways based on different criteria, including genomic location, the effect on phenotype, etc. SNPs located in the noncoding regions of genome are called noncoding SNPs, while SNPs located in introns are called as intronic SNPs, SNPs located in exons are called exonic SNPs or coding SNPs. An exonic SNP that does not lead to a change in the amino acid sequence of the concerned protein is called a synonymous SNP (synSNP), while a non-synonymous SNP (nsSNP) alters the amino acid sequence [193].

Some of the genic SNPs affect the function of the concerned gene and would give rise to phenotypic effects; these SNPs are termed as functional SNPs or candidate SNPs. However, anonymous SNPs do not affect the function of a gene and do not produce a phenotypic effect; most SNPs belong to this category. A reference SNP is a SNP that serves as a reference point for defining neighboring SNPs. SNPs discovered by mining ESTs or genomic databases are generally called *in silico* SNPs or electronic SNPs [66, 193]. There are different methods of detecting SNPs and following section provides some example of these SNP based DNA marker systems. The use of nucleotide sequence databases such as EST, genomic DNA, mRNA and cRNA sequences for a species is the simplest, most convenient, and highly efficient method for identification of SNPs [198]. There are large numbers of SNP detection platforms dominantly used in clinical application such as BeadXpress, GoldenGate and many others. Here several SNP types utilized in plant species are discussed.

5.1 Single-Strand Conformation Polymorphism (SSCP) and Related Systems

The single-strand conformation polymorphism system (SSCP) is similar and related to denaturing gradient gel electrophoresis (DGGE), thermal gradient gel electrophoresis (TGGE) and denaturing/temperature gradient gel electrophoresis (D/TGGE) techniques. SSCP uses non-denaturing gel or capillary electrophoresis for detection of polymorphism. The DGGE and (D/TGGE) use an increasingly denaturing physical environment/temperature during gel electrophoresis, and TGGE uses temperature gradients to denature double stranded DNA during electrophoresis [199].

SSCP markers are about 200–400 bp PCR products flanked with two primers of 20–25 bp. Shorter DNA molecules provide better resolution. For instance, it is estimated that 100% of the differences in base sequence are revealed by SSCP in DNA molecules up to 200 bp. The SSCP marker system utilizes the principle that the electrophoretic mobility of single strand DNA depends on its secondary structure or conformation. Non-denaturing gel or capillary electrophoresis of single-strand DNA can be used to detect nucleotide sequence variation among the PCR amplified products. The use of SSCP method provides a way to detect nucleotide variation among DNA molecules without having to perform sequencing reactions [200]. SSCP markers are co-dominant and low quantities of template DNAs are required. Limitations of SSCP markers include the need for sequence data to design PCR primers, intensive labor requirements and costly marker development. In addition, SSCP markers may not detect some mutations and high reproducible results are obtained using highly standardized electrophoresis conditions.

A summary of the general procedure for SSCP marker system includes: (i) the first step of the analysis involves DNA extraction; (ii) 20–100 ng genomic DNA is amplified using a pair of primers to obtain identical genomic regions from different individuals of a species; (iii) amplified DNA is first denatured by heating the solution of a double-stranded DNA molecule to 95 °C so that the two strands of the DNA molecules become separated; (iv) DNA solution is then cooled very rapidly so that the complementary strands do not get sufficient time to reform their natural confirmation. Nevertheless, the single strands fold onto themselves, and internal base pairing in short regions leads to the formation of characteristic secondary structures; (v) the secondary structures of the single strands result from folding and internal complementary base pairing in short regions. Because the internal base pairing depends on the base sequence, the differences in conformations of the single-stranded molecules reflect the differences in their base sequences. Two bands will be observed in homozygotes, and the heterozygotes would exhibit four bands. Polymorphisms in secondary structures of the single strands are detected on non-denaturing gel or capillary electrophoresis [200].

5.2 Denaturing High-Performance Liquid Chromatography

In the denaturing high-performance liquid chromatography (dHPLC) method, ion-pair reversed-phase high-performance liquid chromatography is used to differentiate perfectly matched DNA homo-duplexes from hetero-duplexes [97]. dHPLC provides a useful platform for high-throughput mutation detection and SNP discovery. This method is very suitable for high-throughput single nucleotide validation and discovery [201]. Genomic DNA of samples is amplified using two primers and amplified products are mixed with the PCR product of a reference individual that has a known allele at the single nucleotide locus. PCR products along with the reference PCR products are denatured by heading and immediately cooled to permit renaturation. When the reference PCR products and the PCR products are the same size a single peak is obtained. On the other hands, when they are not the same, renaturation produces the two homo-duplexes (corresponding to the two PCR products) as well as two hetero-duplexes formed by pairing between the strands of the two PCR products.

In the dHPLC, two peaks of elution are seen, one for the two homo-duplexes and the other for the two hetero-duplexes. In another words this method exploits the differential retention of homo-duplex and hetero-duplex DNA species under conditions of partial thermal denaturation [201]. dHPLC is an efficient method for SNP scanning with a high level of automation. Yu et al. [201] reported that single-base substitutions and insertions or deletions of up to 1.5 kb in PCR amplified DNA fragments could be detected [201].

5.3 Amplicon Based Sequencing (ABS) for SNP

In the amplicon based sequencing (ABS) method, a pair of specific primers is used for PCR amplification of the desired genomic region, and the amplicon is sequenced for identification of SNPs and InDels. The limitations of the ABS method include: (i) size limitation (10–20 kb); (ii) sequence information for primer pairs; (iii) limitations on the detection of rare SNP alleles and InDels for paralogous and orthologous genes. Paralogous genes are those genes present in the same genome produced by duplication and polyploidization. Orthologous genes that are those genes of different species that perform the same function [202].

When Sanger–Coulson sequencing method is used, this method needs specific primer pair to flank the SNP locus, cloning and sequencing experiments. However, the use of next generation sequencing (NGS) technologies reduces the workload, permits the pooling of tissues (multiplexing), genomic DNAs/cDNAs, or amplicons from different individuals and minimizes false negatives, and ensures detection of genuine SNPs by discrimination of rare alleles from sequencing errors [202].

5.4 Primer Extension and Allele-Specific PCR (AS-PCR)

Primer extension uses specially designed primers to the target PCR amplicons or genomic regions, extending the primer by DNA polymerase using one of the four dideoxynucleotide triphosphates (ddNTPs) labeled with different fluorophores, each giving a different color. The primer is designed that the base at its 3′ end is complementary SNP locus present in the PCR product. In a given reaction only one of the four ddNTPs will be added to the primer, the fluorescence color of the product will permit easy identification of the added ddNTP and the deduction of the allele present at the SNP locus [203]. Reaction products of primer extension are analyzed by either electrophoresis or by matrix assisted laser desorption ionization time of flight mass spectrometry (MALDI-TOF MS). Phosphodiesterase II digestion is used to trim the 5′ ends of the products of primer extension, and the molecular weights of the shortened products are determined by MALDI-TOF MS. This permits an accurate identification of the ddNTP added to the extended primer and deduction of the SNP allele. The MALDI-TOF MS system is very fast, but the equipment is very expensive, and it requires high expertise [204].

Allele-specific PCR uses a pair of primers designed to amplify only one of the alleles at a SNP locus [205]. One of the primers of AS-PCR marker system is based on a conserved sequence present in all the alleles. The other primer is specific to the genomic region having the SNP locus. When the 3′ terminal base of the second primer has perfect match complementary to the SNP allele, a PCR product is obtained and the SNP allele is detected by gel electrophoresis. On the other hand, a mismatch at the 3′ end of the primer prevents amplification. The AS-PCR marker system generates dominant SNPs. The AS-PCR marker system is a very simple and user-friendly method of SNP analysis [205].

5.5 Molecular Beacons

A molecular beacon oligonucleotide consists of 4 parts: (i) a loop which is the 18-30 base pair region of the molecular beacon complementary to the target sequence; (ii) a stem which is the beacon stem typically 5–7 bp long sequence lies on both ends of the loop and the stems of the both ends are complementary to each other; (iii) 5′ fluorophore is attached at the 5′ end of the molecular beacon. It fluoresces in presence of a complementary target; (iv) 3′ quencher which is non-fluorescent. The quencher is covalently attached to the 3′ end of the molecular beacon and when the beacon is in the closed loop shape, it prevents the fluorophore from emitting light [206, 207].

During the assay molecular beacons, primers and PCR product are reacted in a thermal cycler and molecular beacons can report the presence of specific nucleic acids. In the presence of a complementary target, the stem portion of the beacon separates out resulting in the probe hybridizing to the target. In the absence of a complimentary target sequence, the beacon remains closed and there is no appreciable fluorescence. When the beacon unfolds in the presence of the complementary target sequence, the fluorophore is no longer quenched, and the molecular beacon fluoresces. The fluorescence is easily detected in a thermal cycler. For quantitative PCR, molecular beacons bind to the amplified target following each cycle of amplification and the resulting signal is proportional to the amount of template. This information is then used during PCR or reverse transcriptase PCR (RT-PCR) experiments to quantify initial copy number [206–208].

5.6 5′-Nuclease Assay (TaqMan Assay)

This technique uses the 5′-nuclease activity of *Taq* DNA polymerase that cleaves 5′ end of double strand DNA. This assay utilizes two PCR primers for amplification of the genomic region (target region) containing the SNP locus, a specifically designed probe (TaqMan Probe) complementary to that region of the target sequence that has the SNP locus. TaqMan probe has a fluorescent dye (fluorophore) attached to its 5′ end and a quenching dye (quencher) attached to its 3′ end. As long as the fluorophore is located near the quencher, there will be no fluorescence due to the quenching action of the quencher. The base at the 5′ end of the probe has complementary base to the SNP locus. When the 5′ end of the probe is paired properly with the SNP locus in the target sequence, the 5′-nuclease activity of *Taq* polymerase cleaves the whole probe beginning at its 5′ end and fluorophore diffuses away from the quencher; as a result, it generates fluorescence signals. On the other hand, if the base at the 5′ end of the probe is not complementary to the SNP locus, *Taq* DNA polymerase will not be able to cleave the probe at its 5′ end. In this case, fluorescence signal is not generated. TaqMan assay is quick, simple and, the throughput is high. However, the procedure is based on a costly real-time PCR machine, and the costs of labeled probes and other consumables are high [209–211].

5.7 High Resolution Melting Analysis (HRM)

High resolution melting (melt) analysis (HRM) is a real time PCR method based on analysis of the melting curve, which is formed when a dsDNA separates into ssDNAs during increased temperature [212]. The fluorescent dyes that are used for HRM are known as intercalating dyes and have a unique property that they bind specifically to double-stranded DNA. When intercalating dyes are bound, they fluoresce brightly. The fluorescent dye binds to dsDNA during the PCR amplification resulting in an increase in fluorescence level. In the absence of double stranded DNA they have nothing to bind to and they only fluoresce at a low level [213, 214]. After PCR amplification, HRM is started through an incremental increase of the temperature, resulting in a separation of the dsDNA to ssDNA. When the dsDNA separates the fluorescent dye releases and the fluorescence level decreases, this is visualized in a characteristic melt curve. The melting temperature (Tm) of the dsDNA is dependent in the length of the sequence, and the nucleotide distribution in the sequence. GC base pairs have higher melt energy than AT base pairs due to differences in the number of hydrogen bonds. A long sequence with high concentration of GC will have a higher Tm than a shorter sequence with a high concentration of AT. Differences in the DNA sequences result in different shapes of the melt curve. Samples with a homozygote change will have the same shape of the curve but they will melt at different temperatures. Heterozygote will affect the shape and slope of the curve. A differential plot visualizes the curve shape in an easier way, by comparing one selected sample against the other samples. Different nucleotides affect the Tm differently when the dsDNAs separate, it is easier to see a C/T and G/A change than an A/T change in the melt curve shift respectively [212–214].

5.8 *Oligonucleotide Ligation Assay and Allele Specific Oligonucleotide Hybridization*

In the oligonucleotide ligation assay (OLA), two types of oligonucleotides are utilized. One of the oligonucleotides is called as capture oligonucleotide and has complementary bases to the SNP locus itself and the sequence on the upstream of the SNP locus. The other oligonucleotide, called reporter oligonucleotide labeled with a fluorophore, has complementary bases to the sequence on the downstream side of the SNP locus and does not include the SNP locus itself. When the two oligonucleotides are hybridized perfectly with the PCR products representing the target genomic region of an individual, DNA ligase will ligate the two oligonucleotides to generate a product having the combined lengths of the two oligonucleotides. On the other hand, when the base at the 3′ end of the capture oligonucleotide has no complementary base to the SNP allele, the two oligonucleotides stay separate in the solution. Some DNA chip or bead-based techniques are used with the OLA procedure [215, 216].

5.9 *Invasive Cleavage of Oligonucleotide Probes (Invader Assay)*

This method uses a structure-specific flap endonuclease (FEN) enzyme to cut a three-dimensional complex formed by hybridization of allele-specific overlapping oligonucleotides to target DNA containing a SNP locus. Flap endonucleases remove the redundant portions from the 5′ end of a downstream DNA fragment overlapping an upstream DNA fragment called invader. An invader oligonucleotide is designed, with its 3′ ending on the polymorphism to be tested. Two oligonucleotide signal probes are designed, overlapping the polymorphic site and each corresponding to one of the alleles. After displacement of the signal probes by the invader probe, FEN-mediated cleavage occurs only for the perfectly matched allele-specific signal probe. Generation of the cleaved fragment is monitored by using it in a second reaction as an invader probe to cleave a fluorescence resonance energy transfer (FRET) probe. This assay does not require PCR amplification of the locus to be tested and scoring is done using a simple fluorescence plate reader [217].

5.10 Competitive Allele Specific PCR (KASPar/KASP)

Competitive allele specific PCR (KASPar, patented by KBioscience) and Kompetitive Allele Specific PCR (KASP) are used in plant research [218, 219]. KASP/KASPar uses some aspects of the AS-PCR, 5′-nuclease assay and molecular beacons techniques. KASP/KASPar is a competitive allele specific PCR for SNP genotyping assay based on dual Fluorescent Resonance Energy Transfer (FRET). It determines single nucleotide polymorphisms (SNPs), and insertions and deletions (InDels) at specific loci.

The KASP marker system is not massively high-throughput system but genotyping is suitable for a small number of targeted SNPs in a large number of samples. The KASPar utilizes three specific primers consisting of two allele specific forward primers and one common reverse primer, two oligonucleotides with fluorescence resonant energy transfer (FRET) cassette; one labeled with FAM and ROX, the other labeled with HEX and ROX, *Taq* DNA polymerase, dNTPs, $MgCl_2$ and buffer solution.

During the first round of thermal cycling, one of the allele-specific primers and the common reverse primer bind to the template and elongate the tail sequence to the newly synthesized strand. In the second round of thermal cycling, reverse primer and the allele-specific primer bind the templates generated in the first round and elongate the new strands along with the tail at the 5′ terminal complementary to FRET cassettes. In the third round of thermal cycling, one or both of the FRET cassettes bind the template and FRET cassette or cassettes are no longer quenched and emit fluorescence. If the genotype at a given SNP is homozygous, only one of the two possible fluorescent signals will be generated. If the genotype is heterozygous, a mixed fluorescent signal will be generated [218, 219].

Competitive allele-specific hydrolysis probes (TaqMan) PCR is a modification of the TaqMan based allele specific PCR which works by combining a mutant allele-specific primer (ASP) with a wild type allele-specific blocker (ASB) in the same PCR reaction. The use of a molecular blocker completely suppresses the amplification of the wild type allele in order to not interfere with the amplification of the mutant allele and improve the specificity as well as the sensitivity of the assay.

6 Beads/Array-Based Technologies for SNP Markers

The micro-beads-based technique has been used for genotyping based on allele-specific hybridization, allele-specific primer extension, single-base extension, and oligonucleotide ligation assays. Beads-based techniques are similar to microarray and chips-based methods. Instead of flat surfaces used in microarray methods, the bead-based techniques use oligonucleotides attached to micro-beads of 3–5 μm diameter for hybridization. Microarrays or DNA chips are collection of microscopic DNA spots attached to a solid surface, such as glass, plastic, metal or silicon chips containing one end of a large number of different oligonucleotides covalently linked and arranged in spots. Each spot has about 25 nucleotides long single DNA molecule including the nucleotide targets such as SNP around its middle position. The probes are directly synthesized using a parallel procedure involving masks and photolithography on the DNA chips [5, 220, 221].

6.1 Diversity Arrays Technology (DArT)

DArT is as an open source technology, not covered by exclusive patent rights and it is a high throughput, low-cost genotyping system. DArT is essentially similar to procedure of the AFLP marker system except for the use of microarray-based nucleic acid hybridization in the place of gel electrophoresis for the detection of polymorphism. DArT has a great potential for genetic diversity and mapping studies in a number of plant species [29, 222, 223].

DArT system requires a discovery array (panel array) and genotyping array (diversity array). Discovery/panel array preparation procedure involves in the following steps: (i) isolation and purification of the genomic DNAs from individuals of a population or species under the study; (ii) pooling of about 5 ng DNA from each of these individuals and digesting the pooled DNA with the selected restriction enzymes such as *Pst*I and *Taq*I; (iii) ligation of appropriate adapters to the restriction fragments; (iv) PCR amplification of fragments using primers with complementary sequences to the adapter and selective overhangs of the fragments to reduce the complexity. For example, one overhang reduces the number of fragments amplified to one-fourth of the total number. Also another approach for complexity reduction is to digest the genomic DNA with a combination of one rare cutter enzyme like *Pst*I and one frequent cutter enzyme, such as *Taq*I or *BstN*I restriction enzymes; (v) the fragments are cloned, and clones are amplified using vector-specific primers, purified and arrayed onto a solid support [29, 131, 222].

Genotyping/diversity array procedure involves in the following steps: (i) genomic DNA of individuals extracted for the preparations of the DArT panel array is digested individually using the same restriction enzymes such as *Pst*I and *Taq*I; (ii) the restriction fragments are ligated to appropriate adapters; (iii) ligated fragments are PCR amplified using primers complementary to the sequences of the adapter.

Amplification uses selective overhangs of the fragments to reduce the complexity; (iv) amplified products are purified using DNA purification columns or isopropanol precipitation; (v) fragments from one sample are labeled using fluorescent dye such as Cy3 (green) and the fragments from the other sample are labeled using with Cy5 (red); (vi) labeled fragments are mixed and denatured; (vii) mixtures are spotted on the arrays and images are analyzed. Signals are detected and analyzed using specialized software programs such as DArTsoft, which converts signals into scores of 1 or 0, i.e., present or absent. These scores provide the fingerprint of the individual, and these scores are used for statistical analyses in the same way as the scores for RAPD and SSR markers [29, 131, 222, 223].

The DArT marker system detects SNPs that are produced with the restriction sites and the selective PCR primers. This marker system also detects insertions and deletions, structural rearrangements, and copy number variations in the region between the two neighboring restriction sites [29, 131, 222]. Some advantages and disadvantages of the DArT marker system are presented in Table 9.15.

Most of the DArT markers are located in the genetically active regions of the genome. It is a high throughput; highly reproducible marker system that uses automated platform genotypes for thousands of loci in a single assay. It uses very low amount of DNA and requires low costs of development and application. DArT is as effective in detecting polymorphism in a diploid and polyploid species. It does not need sequence information for the species to be studied and it is applicable to all species regardless of how much DNA sequence information is available for that species [224, 225]. However, it is a microarray-based technique that requires demand an extensive investment both in laboratory facility and in skilled work force. It is labor intensive involves many steps, including preparation of genomic representation for the target species, cloning, amplification, fluorescent labeling, data management and analysis which requires dedicated software programs. Furthermore, DArT marker system produces dominant markers limiting its value in some applications [226].

The DArT marker system has been used in the characterization, diversity, parental selections for hybrid breeding, seed purity/product integrity testing and QTL

Table 9.15 Some advantages and disadvantages of the DArT marker system

Advantages	Disadvantages
It is not covered by exclusive patent rights	It requires the use of restriction enzymes, which need high quality of genomic DNA
It has high level of polymorphism and high reproducibility	It produces primarily dominant markers or differences in marker intensity
It does not need prior sequence information for the species	It requires extensive investment both in laboratory facility and skilled manpower
It is a high throughput system	It requires dedicated software programs
It is cost effective, with an estimated cost per data point	It is a microarray-based technique that involves in labor intensive works

mapping studied in a variety of plant species. The DArT marker system has also been used in DNA methylation, association mapping, MAS, and genomic selection [29, 131, 222, 226].

6.2 Single Feature Polymorphisms

Single feature polymorphisms (SFPs) are microarray-based molecular markers that are detected by hybridization of DNA or cRNA to oligonucleotide probes. SFPs are also known as single position polymorphisms. SFP marker system identifies allelic variation in pairs of lines of a species [227, 228]. SFPs detect sequence polymorphisms as SNPs and InDels in or near the sequence represented by the features. Insertions, deletions and substitutions of one or more nucleotides can affect hybridization of DNA to a probe on an array. A target sequence perfectly matching the sequence of a probe will bind with much greater affinity than one with a mismatch sequence. The resulting difference in hybridization intensity between two genotypes for an individual probe is called a single feature polymorphism (SFP), where a feature refers to a probe in the array [228].

The SFP markers occur at high density and cover the whole genome, and their physical locations in the genome are known [229]. All the SFP markers are analyzed simultaneously and rapidly. SFP is highly sensitive, and a new set of markers can be easily identified for any pair of lines. The SFP procedure is relatively simple and can be applied to any species, for which an expression array and, preferably, a physical genome map are available. SFP markers, segregate in a Mendelian fashion, are generally biallelic, and permit rapid mapping of genes/QTLs. The main limitations of SFP technology are high false discovery rates in complex genomes, relatively high cost, and the need for construction of microarrays. Furthermore, SFPs are subject to ascertainment bias due to the use of a reference genome/strain for their discovery.

SFP analysis may use a ready-made gene expression microarray like Affymetrix GeneChips or Nimblegen arrays. Alternatively, an array may be custom-made using expression sequence tags (ESTs), mRNA sequences, known/predicted open reading frames (ORFs) of genomic sequences, unigenes listed in the NCBI database or conserved orthologous sequences. SFP analysis has been used in mapping and gene expression studies of several plant species, including barley, maize, soybean, tomato, wheat, rice, and pigeon pea but not applied to *Salvia*.

6.3 Restriction Site Associated DNA (RAD)

Restriction site associated DNA (RAD) markers are microarray-based DNA markers closely related to RAD-seq, which will be discussed in the next section of this chapter [230]. RADs are detected by hybridization of DNA or cRNA to oligonucleotide probes. RADs are short fragments of DNA adjacent to each instance of a particular restriction enzyme recognition site in a target genome. Each flanking sequence of the restriction enzyme cut site is referred to as a RAD locus or RAD-tag. Usually two RAD tags would be recovered for each restriction site, and each of them is a potential RAD marker. RAD markers (tags/loci) represent polymorphisms in the recognition sites for the restriction enzyme used for the preparation of the assay sample. The RAD markers are anonymous, dominant, and scored as presence/absence [230–233].

Polymorphic RAD markers are identified by using RAD-tag samples prepared from two species for competitive hybridization with a suitable microarray. A polymorphic RAD marker is detected when a microarray probe hybridizes with the RAD-tag from only one of the two samples. The microarray used for hybridization may be a genomic tiling array, a cDNA array, or an oligonucleotide microarray. A genomic tiling array is a high-density microarray made up of oligonucleotide probes, which together span the entire genome of an organism. cDNA and oligonucleotide microarrays would identify a much smaller number of RAD markers in comparison to the genomic tiling arrays [234–236].

The development of RAD tags involves the following steps: (i) digestion of genomic DNA with a specific restriction enzyme such as *Eco*RI and reaction is cleaned; (ii) linkers (5′-biotinylated) are ligated to fragments and reaction is cleaned; (iii) ligated DNA fragments are randomly sheared by sonication into fragments smaller than the average distance between restriction sites, leaving small fragments with restriction sites attached to the biotinylated linkers; (iv) DNA fragments are immobilized on streptavidin-coated micro-beads; and (v) RAD-tags from the beads are removed by digestion at the original restriction sites; (vi) RAD-tag samples are labeled with different fluorophores and cloned before they are spotted onto a solid support for preparing the microarray; (vii) RAD-tags from each of a number of samples, when hybridized on to a microarray, allows high-throughput identification and/or typing of differential hybridization patterns [231, 234–236].

7 Next Generation Sequencing (NGS) Technologies

Two sequencing methods were developed and published in 1977. One of these methods was the Sanger method developed by Frederick Sanger and Alan Coulson. It is a sequencing-by-synthesis (SBS) method. The Sanger method relies on a combination of deoxy- and dideoxy-labeled chain terminator nucleotides. The other method was Maxam–Gilbert sequencing method developed by Allan Maxam and Walter Gilbert in 1976–1977. Maxam–Gilbert sequencing method uses chemical modification followed by cleavage at specific sites [237, 238].

Frederick Sanger and his colleagues developed chain terminator sequencing (Sanger sequencing) in 1977. The development of this technique won Sanger the Nobel Prize in Chemistry in 1980. From the 80s to the mid-2000,

Sanger sequencing dominated the DNA sequencing platform, bringing successful completion of the Human Genome Project (HGP) in 2003. Although this technique has been replaced by next generation sequencing methods, it is still used today for smaller-scale projects [239, 240].

Two important findings enhanced to make DNA sequencing possible. The first one was the incorporation of radioactive nucleotide by DNA polymerase and the second one was a single nucleotide separation by polynucleotide length via electrophoresis through polyacrylamide gels. The plus and minus DNA sequencing was the first method utilized these findings. However, the first DNA sequencing considered the real birth of 'first-generation' DNA sequencing was the Maxam and Gilbert technique, which was based on DNA polymerase to generate fragments that contained 5′-radiolabeled DNA terminals. In this sequencing method, DNA fragment is treated with chemicals that break the fragment at specific bases; after running on a polyacrylamide gel the position of specific nucleotides can be determined and therefore sequence inferred. The next DNA sequencing method was Sanger's chain-termination or dideoxy technique. It uses chemical analogues of the deoxyribonucleotide triphosphates (dNTPs) and dideoxynucleotide triphosphates (ddNTPs). ddNTPs lack the 3′ hydroxyl (OH) and cannot form a bond with the 5′ phosphate of the next dNTP. Chain terminators procedure enables sequencing of fragments of up to 700–800 bases. In Sanger's chain-termination method, the two complementary strands of a DNA fragment are sequenced in separate reactions for an enhanced reliability of the sequence data [239, 240].

The dideoxy technique was improved by replacing phospho-ortritrium-radiolabeling with fluorometric detection allowing the reaction to occur in one reaction tube instead of four and detection through capillary-based electrophoresis. The first-generation automated DNA sequencers developed by Applied Biosystem Instruments (ABI) used the Sanger method. In Table 9.16, some advantages and disadvantages of first generation sequencing methods are presented.

First generation sequencing uses Sanger method with fluorescent dye-terminator reagents for single-reaction sequencing rather than the usual four separate reactions. These sequencers were later improved by including computers to collect, store, and analyze the sequencing data. These new methods became known as next-generation sequencing (NGS) because they were designed to employ massively parallel strategies to produce large amounts of sequences from multiple samples at very

Table 9.16 Some advantages and disadvantages of first generation sequencing methods

Advantages	Disadvantages
Not as toxic and less radioactivity than Maxam and Gilbert method	Poor quality in the first 15–40 bases of the sequence
Easier to automate and get started	Requires DNA fragments cloned before sequencing and expensive due to relatively large volumes of chemicals
Highly accurate long sequence reads of about 700 base pairs	700–1000 nucleotides long DNA fragments in a single reaction

high-throughput and at a high degree of sequence coverage to allow for the loss of accuracy of individual reads when compared to Sanger sequencing [240–242].

The most advanced version of automated Sanger sequencers is capable of sequencing up to 1 kb for 96 individual specimens at a time. In the last few years, a series of high-throughput sequencing devices have been commercially introduced based on different chemistries, detection techniques and platforms. These NGS technologies can potentially generate several hundred thousand to tens of millions of sequencing reads in parallel. This massively parallel throughput sequencing capacity can generate sequence reads from fragmented libraries of a specific genome (i.e. genome sequencing); from a pool of cDNA library fragments generated through reverse transcription of RNA molecules (i.e. RNA-seq or transcriptome sequencing); or from a pool of PCR amplified molecules (i.e. amplicon sequencing). In all cases, sequences are generated without the need of a conventional, vector-based cloning procedure that is typically used to amplify and separate DNA templates [241–243].

NGS is a high-throughput DNA sequencing methodology that makes use of mass parallelization to process million sequences concurrently. The process of running thousands of analytes at a time is known as a multiplexing. De novo genome sequencing and genome resequencing are used for the analysis of SNPs, InDels, SSRs and copy numbers variation (CNV), structural variation and transcript profiling; etc. The following sections provide information on DNA sequencing methods and platforms summarizing the past, present and future generations [243].

8 Second Generation Sequencing (SGS)

NGS technologies can be classified into two main categories. The first group, Second Generation Sequencing (SGS) approaches are PCR-based technologies, which include four commercially available platforms: Roche 454 Genome Sequencer (Roche Diagnostics Corp., Branford, CT, USA), HiSeq 2000 (Illumina Inc., San Diego, CA, USA), AB SOLiD System (Life Technologies Corp., Carlsbad, CA, USA) and Ion Personal Genome Machine (Life Technologies, San Francisco, CA, USA). The second generation sequencing technologies differ from the Sanger method in terms of three major improvements: (i) instead of requiring bacterial cloning of DNA fragments, NGS technologies rely on the preparation of sequencing libraries in a cell free system; (ii) instead of hundreds, thousands-to-many-millions of sequencing reactions are produced in parallel; (iii) the sequencing output is directly detected without the need for electrophoresis [244–250].

The second generation sequencing technologies involves the following eight general steps: (i) extraction of DNA from the sample, (ii) fragmentation of the DNA and size selection of DNA fragments, (iii) ligation of adapters, (iv) amplification of fragments using primers targeted to known adapter sequences, (v) immobilization of amplified templates by arraying them spatially, (vi) sequencing the reactions in

Table 9.17 Main differences between FGS and SGS

First generation sequencing (FGS)	Second generation sequencing (SGS)
Much suitable for de novo genome sequencing research	Much suitable for genome resequencing research
Regular PCRs or amplification through bacterial species are used	Emulsion PCR (emPCR), bridge PCR (bPCR) or rolling cycle amplification methods are used
Sanger sequencing (chain-termination method) is used	Four different sequencing methods are used Reversible chain-termination Pyrosequencing Sequencing by ligation (SOLiD) Semiconductor sequencing (Ion Proton)
Plasmid based library preparation methods are used	Adapter-barcode based libraries are used
Single reaction per target	Massive parallelization and multiplexing are used
All SGS techniques replicate DNA before sequencing is done	All SGS techniques replicate DNA before sequencing is done

large numbers in parallel, (vii) data generation and checking for accuracy of sequence generated, (viii) aligning the sequences generated to a reference genome [242, 245, 251–258]. In Table 9.17, some differences between first and second generation sequencing technologies are presented.

The basic characteristics of second-generation sequencing technology include: (i) fragmented genomic DNA or cDNA or in some new approaches direct RNA sequences are ligated to linker and/or adapter sequences for construction of template libraries without the need for cloning via a foreign host cell; (ii) the DNA fragments are clonally amplified by emulsion PCR with micro-beads or by solid-phase PCR using primers attached to a solid surface (bridge amplification) or rolling circle amplification, or isothermal amplification in order to generate sufficient single-stranded DNA molecules and detectable signal for producing sufficiently reliable sequencing data [252–258].

Polonies (polymerase colonies) are discrete clonal amplifications of a single DNA molecule, grown on a solid-phase surface. This approach greatly improves the signal-to-noise ratio. Polonies can be generated using several techniques that include solid-phase PCR in polyacrylamide gels, bridge PCR, rolling-circle amplification, BEAMing (beads, emulsions, amplification and magnetics)-based cloning on beads and massively parallel signature sequencing (MPSS) to generate clonal bead arrays. Polony sequencing technology is an open-source platform and the software and protocols are free and do not require licensing or a fee for use. Polony sequencing uses paired-tag library emulsion PCR to amplify the target DNA, and sequencing by ligation to detect DNA bases [259, 260].

All second generation sequencing methods have the following shared attributes: (i) a sequencing library created using fragmented DNA that have linkers (adapters) at their terminal ends; (ii) amplification of DNA fragments using one of the three most used method (emulsion PCR, bridge PCR or rolling cycle amplification); (iii) direct or indirect detection of each nucleotides incorporated (synthesized or ligated) during the sequencing reaction. Most of SGS methods are capable of paired-end reads [242, 254–258].

The template for sequencing is single-stranded DNA (ssDNA), which can be prepared from genomic DNA, BAC clones, PCR products, and cDNAs. Sequencing experiments may need to sequence some specific regions identified by linkage studies. In such cases, methods like "enrichment," "genome partitioning," or "genome capture" can be used for template sample preparation. Genomic DNA and in some cases RNA molecules are usually fragmented into a suitable size of 300–700 bp using sonication (high sound energy) or nebulization (forces DNA through a small hole) or enzymatic digestion by DNase I or fragmentase (i.e., shredase). Nextera tagmentation technology can be implemented using a transposase enzyme to simultaneously fragment and insert adapter sequences into the dsDNA and thereby reduce sample handling and preparation time. The fragmented genomic DNAs are then end-repaired by polymerase and exonuclease activity. The 3′ ends are phosphorylated, while 5′ ends are adenylated [242, 253–258, 261].

Genomic DNA sample is fragmented in a modal size of 100–200 bp with random ends by using sonication. DNA fragments generated by sonication are polished to produce a library of DNA fragments with blunt, 5′-phosphorylated ends that are ready for ligation. The end polishing is accomplished by using the T4 DNA polymerase, which can fill in 5′ overhangs via its polymerase activity and recess 3′ overhangs via its 3′-5′ exonuclease activity. The phosphorylation of 5′ ends is accomplished by T4 polynucleotide kinase. Ligation of double-stranded DNA adaptors to the DNA fragments is accomplished by use of T4 DNA ligase. Since the double-stranded adaptors do not have 5′ phosphates and contain a 5′ overhang on one end, this prevents ligation in the incorrect orientation. The lack of 5′ phosphate leads to generation of a nick on one strand of the ligated DNA. This will be nick translated and removed during an upcoming amplification step [143, 257, 261, 262].

Adapters may contain elements for the immobilization of library molecules on a solid surface beads or the surface of a glass slide, bar-coding sequences for amplification. Either one or both adapters may contain sequencing priming sites. In some library types, such as for 'dual index' sequencing, both adapters contain an index to improve the multiplexing capacity up to 96 unique barcodes. In addition, specific adapters are added to both the ends of the much larger fragments to facilitate their circularization.

In some studies, researchers also use size selection of the fragmented DNA before adapter ligation step. However size selection of library is usually performed after ligation steps and in this step, unligated adaptors are removed and ~200 to 250 bp adaptor-ligated library are selected by polyacrylamide gel electrophoresis or other gel free methods [143, 257, 261, 262].

The size-selected fragment library (~200 to 250 bp) is amplified to a small extent to prepare for downstream emulsion PCR. The initial phase of amplification involves a step for nick translation of the nick remaining at the 5′ ligation junction of the adaptors. This nick will be translated to the end of the adaptor to remove the nick completely. The next steps use standard PCR with primers specific to the P1 and P2 adaptors to amplify the library. The polymerase used for amplification can also carry out the nick translation step, and therefore only one enzyme is needed. After adapter ligation, the size selection occurs through gel electrophoresis and PCR selection [242, 257].

All SGS platforms use one of several different methods including emulsion PCR (e.g. BEAMing digital PCR, 454 sequencing), bridge amplification (e.g. Illumina sequencing), rolling circle amplification (e.g. Complete Genomics) and amplification by template walking on the solid surface (SOLiD). Nanoballs are constructed as follows: sheared DNA molecules are ligated with specific adapters called "half adaptors," one to each end of the sheared fragments. The molecules are then circularized through intramolecular ligation to form a complete adaptor. A series of proprietary methods shorten and linearize the DNA circles. A second set of half adaptors is added and the linear molecules are recircularized. The resulting library is composed of DNA circles, which contain ~70 base pairs (bp) from each end of the original genomic DNA fragment, separated by the two adaptors. The genomic distance between the DNA fragments (referred to as "mate-pair distance") is 250–500 bp. The tight distribution of the mate-pair distance has important implications for the quality of variant calls across the genome [263–265].

8.1 Emulsion PCR

Prior to emulsion PCR (emPCR) DNA libraries with appropriate 5′ and 3′ ends are prepared by fragmentation and adapter ligation. In summary, the genomic DNA is fragmented by either high sound energy (sonication) or forces DNA through a small hole (nebulization). Adapters are ligated to the DNA fragments so that the fragments bind to the emulsion micro-beads. The double stranded DNAs with adapters are denatured by heating the DNA. DNA fragments with adapters are made single stranded and immobilized onto individual suitably oligonucleotide-tagged micro-beads. Bead-DNA conjugates are emulsified using aqueous amplification reagents in oil, ideally producing emulsion droplets containing only one micro-bead per one fragment. Each micro-bead is emulsified in a water-in-oil droplet with PCR reagents. emPCR is conducted within oil droplets and involves the steps denaturation, annealing, synthesis. Firstly, the strand is elongated with DNA polymerase and dNTPs. Then the double-strand is denatured, allowing the strand to ligate to another site on the surface of the micro-bead. Eventually, 1 million copies of the target are amplified on the surface of each bead. After completion of the final amplification, the emulsion is broken using isopropanol and detergent buffer. For micro-bead enrichment, the solution is vortexed, centrifuged, and magnetically separated. The resulting solution is a suspension of empty, clonal and non-clonal micro-beads, which will be filtered in the next step. A fiber-optic chip filled with a field of micro-wells, known as a PicoTiterPlate, is then washed with the emulsion, allowing a single bead to drop into each well [259, 266].

The library fragments are amplified through emulsion PCR thermal cycling into individual water: oil micro-reactors that contain PCR ingredients. Each library fragment is amplified on the surface of one bead in a single micro-reactor, generating billions of copies of the same fragment covering the surface of the DNA bead. The amplified beads are recovered from emulsion oil followed by an enrichment step to keep only the amplified beads and discard the failed ones [259, 266–268].

8.2 *Bridge Amplification*

In the bridge PCR (bPCR) amplification method, isothermal PCRs produce clusters of clonal DNA populations in a planar solid-phase. bPCR widely used in Illumina sequencing platforms. For bPCR amplification target DNA is fragmented using one of the method previously mentioned and adapters are attached to fragmented DNA. The DNA is denatured into single-stranded molecules. These fragments are floated onto a flow cell. Flow cells have corresponding sequences to the adapters. Adapter attached the DNA fragments are ligated to the sister adapters attached onto the flow cell. Lowering the temperature induces bending of the single-stranded DNA to the two lawn-oligos (adapters on the flow cell) creating base pairing. Isothermal PCR replicates the DNA. Bridge amplification occurs with regular PCR components added onto the slide. Chemical denaturation and repeating generate clusters until clusters densities of up to 2000 molecules in each channel of flow cell. One of the main disadvantages of bPCR is that each of the multiple amplification cycles requires the addition of fresh mixture containing polymerase, followed by a chemical denaturation step, making it cost much more than regular PCR. bPCR also requires a specialized pumping utilities and systems [247, 263, 269–271].

DNA fragment to oligo hybridization on the flow cell occurs by active heating and cooling steps. This is followed by a subsequent incubation with the amplification reactants and an isothermal polymerase that generates millions of clusters of the library fragments. In the sequencing step, each cluster is supplied with polymerase and four differentially labeled fluorescent nucleotides that have their 3-OH chemically inactivated. This blocking modification ensures that only a single base will be incorporated per flow cycle. After each nucleotide is incorporated, an excitation followed by an imaging step takes place to identify the incorporated nucleotide in each cluster. A chemical deblocking treatment removes the fluorescent group and allows the incorporation of the following nucleotide during the next flow cycle. The sequence of each cluster is computed and subjected to quality filtering to eliminate low-quality reads [239, 263–265, 269–271].

9 Next Generation Sequencing Chemistries and Platforms

The main disadvantages of second generation sequencing (SGS) technologies are their relatively labor intensive procedures and production of short reads that make genome assembly more difficult and requiring the development of novel alignment algorithms (Table 9.18).

Commercially available SGS technologies include (i) enzymatic luminometric inorganic pyrophosphate detection assay (ELIDA), pyrosequencing chemistry of Roche (454), (ii) fluorescent reversible terminators chemistry of Illumina, (iii) sequencing by ligation (Applied Biosytems) and (iv) semiconductor sequencing (Ion Torrent). Sequencing in these chemistries is first captured on micro-beads in a way so that one fragment is attached per micro-bead.

Each DNA fragment on each bead is amplified in a water-in-oil emulsion containing PCR reagents (emPCR). After amplification steps, micro-beads with clonally amplified products are distributed into the wells of a fiber optic slide (454 and Ion Torrent: one bead per well) or on a glass slide (SOLiD) [242, 244, 245, 251, 253–258, 272–275].

Table 9.18 Comparison of some sequencing chemistries and platforms

Platform	Read length (bp)	Run (h)	Error rate (%)	Sequencing chemistry
First generation sequencing				
Sanger/Life Tech/85 kb	800	2	0.3	Dideoxy terminator
Second generation sequencing				
454 GS FLX +/Roche/0.7 Gb	700	24/48	1	Pyrosequencing
GS Junior/Roche/70 Mb	500	18		Pyrosequencing
HiSeq/Illumina/1500 Gb	2 × 150	27/ 240	0.8	Reversible terminators
MiSeq/Illumina/15 Gb	2 × 300	27	0.8	Reversible terminators
SOLiD/Life Technologies/ 120 Gb	50	336	0.01	Ligation
Retrovolocity/BGI/3000 Gb	50	336	0.01	Nanoball/ligation
Ion Proton/Life Technologies/ 100 Gb	200	2–5	1.7	Proton detection
Ion PGM/Life Technologies/ 2 Gb	200	2–5	1.7	Proton detection
Third generation sequencing				
SMRT/Pac Bio/1 Gb	>10,000	1–2	12.9	Real-time SMS
Heliscope/Helicos/25 Gb	35	192	0.2	Real-time SMS
Oxford Nanopore Technologies/1 Gb	>5000	48/72	34	Real-time SMS
Electron microscopy/ZS	7200	14		Real-time SMS

9.1 Pyrosequencing Chemistry and Platforms

Pyrosequencing is the first of the second-generation sequencing (SGS) technologies introduced in 2005. Pyrosequencing is also commonly referred to as 454 sequencing and is a sequencing-by-synthesis (SBS) method. The pyrosequencing technique relies on the detection of pyrophosphate molecules, which are detected by an enzymatic luminometric inorganic pyrophosphate (PPi) detection assay (ELIDA) through the generation of a light signal following the conversion of PPi into ATP. The amount of generated light signal is directly proportional to the number of nucleotides incorporated. The method can generate up to 700 Mb in a 23-h run with an average read length of around 750 nucleotides [276–279].

After library preparation and emPCR amplification, the capture beads along with the attached DNA fragments are removed from the emulsion and loaded into the wells of a PicoTiter Plate (PTP). The size of wells is such that only a single bead can be loaded in each well. Each of the wells can hold only one micro-bead with amplified DNA. The reaction mixture contains the template DNA, sequencing primer, adenosine-5′-phosphosulfate (APS), luciferin, the Klenow fragment, ATP sulfurylase, luciferase, and apyrase. The nucleotides dCTP, dGTP, and dTTP, and deoxyadenosine α-thiotriphosphate (dATPαS) are added to this reaction mixture sequentially one after the other. dATPαS is used in the place of dATP because it can be used by luciferase for light generation only after it has been used for DNA synthesis. In contrast, dATP will be used for producing light even when it is not used for DNA synthesis [264, 265, 270–282].

Four layers of engineered beads are deposited into the PTP. From bottom to top, diluted pyrosequencing enzyme micro-beads, DNA amplified micro-beads, pyrosequencing enzyme micro-beads and, finally, PPiase micro-beads. All micro-bead layers are deposited by centrifugation and one of the four dNTPs is added sequentially one after the other. If the next sequence of the growing strand is complementary to the dXTP (X is A, T, C or G), then PPi is released. PPi and ASP are reacted by ATP sulfurylase and ATP and sulphate are produced. ATP and luciferin along with oxygen are reacted by luciferase enzyme and AMP, PPi, oxyluciferin, carbon dioxide and light are obtained. Any remaining deoxynucleoside triphosphate (dXTP) and ATP are degraded by apyrase and washed away. Chemiluminescence signal is sensed by a sensitive charge-coupled device (CCD) camera. The computer software uses the chemiluminescence data to deduce the base sequence of the template DNA segment attached to every bead [277–283].

The last two commercially available pyrosequencing platforms were Genome Sequencer (GS) FLX System and GS FLX Titanium series. The current GS FLX + system provides 200 nucleotide flow cycles to generate up to 800-bp sequencing reads. The generated raw signals are processed by a 454-pyrosequencing analysis software and then screened by various quality filters to remove poor-quality sequences. However, Roche company announced that they ceased supporting these sequencing machines, reagents and chemicals after 2016. In Table 9.19, some advantages and disadvantages of pyrosequencing method are presented.

Table 9.19 Some advantages and disadvantages of 454-pyrosequencing method

Advantages	Disadvantages
Produces long sequencing reads (500–700 bp) that make easier to map to a reference genome	It has relatively low throughput in comparison to other SGS methods
It is useful for *de novo* genome assemblies	It has high reagent cost and has high error rates in the homopolymer repeats
Run times are relatively fast, it could be observed in real time	It is no longer supported by the service provider in 2016
It could be performed using natural nucleotides and the overall error rate is approximately 1%	It may produce artifacts of single base pair deletions or insertions by signal-to-noise threshold problems

9.2 *Fluorescent Reversible Terminators Chemistry and Platforms*

It is next second-generation (SGS) platform that utilizes a reversible chain terminators approach coupled with bridge amplification on the surface of a flow cell [244, 284]. Flow cells are silica or teflon slides of eight lengthwise lengths. They also contain a dense lawn of primers to be used in the next step of solid phase PCR. The flow cells are subjected to isothermal bridge amplification, created clusters densities of up to 2000 molecules. Reverse strands are cleaved and washed away and fluorescent reversible terminators sequencing is started. This sequencing is a modification of the Sanger sequencing method. It uses a cyclic method that consists of three main steps such as nucleotide incorporation, fluorescence imaging and cleavage [244, 253, 272, 275]. In Table 9.20, some advantages and disadvantages of this sequencing method are presented.

The fluorescent reversible chain terminators approach uses modified ddNTP nucleotide with a cleavable fluorescent dye attached to the base and reversible blocking group on the 3′ carbon of the sugar moiety. The sequencing method can be outlined as follows: (i) sequencing preparation begins with lengths of DNA that have specific adapters on either end; (ii) the second step is to anneal primers to the target DNAs; (iii) adapter containing DNA fragments are washed over a flow cell filled with specific oligonucleotides that hybridize to the ends of the fragments; (iv) each fragment is then replicated to make a cluster of identical fragments; (vi) the next step is to initiate and block DNA synthesis temporarily with a blocking group on the 3′ carbon of the sugar moiety. Upon each cycle, just one dNTP binds to the elongating strand and emits a fluorescent dye color. Depending on the color emitted, nucleotide is recorded, the blocking group and fluorescent dye are removed, and a 3′ hydroxyl is generated so that the growing strand can now elongate. These steps are repeated to end of the fragment. In each cycle, the terminators are reversed so that the process can repeat and nucleotides can continue to be added in subsequent cycles. Unlike pyrosequencing, in the fluorescent reversible terminators sequencing all four bases are present for the polymerization step and

Table 9.20 Some advantages and disadvantages of fluorescent reversible terminators chemistry and platforms

Advantages	Disadvantages
High-throughput and less cost	Sample loading is technically challenging
Produce paired-end reads that are about 250 bases in length each. It could produce 4 billion fragments in a paired-end fashion with 125 bases for each read in a single run	Requirement for sequence complexity
The use of fluorophore chain terminators linked to the dNTPs eliminates the error in base sequence determination when the same base is present at two or more consecutive positions in the template strand	It produces relative short-read length because of optical signal decay and dephasing limiting the application of these technologies in situations where no reference sequence is available to align, assign and annotate the short sequences
Homopolymer regions can be accurately sequenced. The chemical deblocking step is performed prior to the next nucleotide incorporation	The fluorescence properties of the four dyes used in this method tend to produce substitutions of A for C, G for T, and vice versa in the sequence data

only a single molecule is incorporated per cycle. This platform provides paired-end reads that would be helpful for de novo assemblies, detection of insertions/deletions and other genomic mutations. After the forward strand is sequenced, dsDNA clusters are regenerated by bridge amplification and DNA is denatured. The forward base strand is cleaved and the newly synthesized strands are sequenced. This is usually accomplished using index 1 and index 2 primers [244, 253, 272, 275, 276, 284].

9.3 Sequencing by Oligonucleotide Ligation and Detection (SOLiD) and Platforms

Sequencing by oligonucleotide ligation and detection (SOLiD) was first introduced in 2007 by Life Technologies and released in 2008 by Applied Biosystems Instruments [245]. It is the third second generation sequencing method after the reversible chain terminators technology. The library preparation in SOLiD is very similar to Roche/454. The limitation of this system is the short sequence read length (50 bases); however, accuracy is very high (Table 9.21).

SOLiD sequencing is performed by using a set of 16 oligonucleotides for hybridization with the template DNA and ligation to the 5′ end of the anchor primer, which provides a free 5′ phosphate group for ligation. Each oligonucleotide is 8–9 bases long, labeled with fluorophore at the 5′ end, and each member of a set of 16 oligos has a unique combination of two nucleotides at its 3′ end. At a given time, four specific oligonucleotides of the set, each labeled with a different fluorophore, are pumped in and allowed to pair at their 3′ ends with the template DNA.

Table 9.21 Some advantages and disadvantages of SOLiD method

Advantages	Disadvantages
It is suitable for small- and large-scale projects involving whole genomes, exomes and transcriptomes	The major disadvantages are the short read lengths (50–75 bp), the very long run times of 7–14 days, and the need for state-of-the-art computational infrastructure and expert computing personnel for analysis of the raw data
It is claimed to have lower error rates, 99.94% accuracy with generating up to 1410 million reads	Less-well-suited for de novo genome assembly problems sequencing palindromic strands
The positives to this method is that it is easy to implement with off-the-shelf reagents	A downside to this method is its limitation to short reads, and the time it takes to ligate oligonucleotides on and off

The 3′ ends of the oligonucleotides paired with the template DNA are ligated to the 5′ ends of the anchor primer molecules, the color of fluorescence is recorded. After removing of the unpaired 5′ ends of the oligonucleotides, a new set of four oligonucleotides is added and the steps of the first cycle are repeated [245].

When the oligonucleotide matches the template sequence, it is ligated onto the primer and the unincorporated nucleotides are washed away. A CCD camera captures the different colors attached to the primer. Each fluorescence wavelength corresponds to a particular dinucleotide combination. After image capture, the fluorescent tag is removed and new set of oligonucleotides are injected into the flow cell to begin the next round of DNA ligation.

The SOLiD system typically uses five rounds of ten SOLiD cycles for each sequence template. In each cycle, sequencing primers are hybridized to DNA templates amplified by emulsion PCR followed by hybridization of fluorescently labeled probes to complementary sequences. The labeled probes are then ligated to the primers by DNA ligase and the fluorescent signal on the ligated probes is detected [245, 285].

There are five main steps in sequencing by ligation as follow: (i) a known sequence is flanked onto the target DNA strand; (ii) an oligonucleotide is ligated to flank the target DNA strand; (iii) the pool of oligonucleotides with partially degenerate sequences tagged at the 3′ ends with a fluorescent dye from a pool are mixed in with the target DNA and allowed to hybridize with target DNA sequence. DNA ligase joins the molecule in the 3′-5′ direction to the anchor when its bases match the unknown DNA sequence and color light is emitted [245].

The colors vary depending on the known query position; (iv) fluorescent labels are cleaved away, regenerating at 5′-phosphate group on the ends of the ligated probes. This will allow the next oligonucleotides to be ligated onto the rest of the unknown sequences; (v) another round of ligation. This process in steps 4 and 5 are repeated until the nanomers have reached the end of the unknown DNA sequence. After this, the anchor sequence is reduced by one nucleotide, and the process is repeated. After five cycles of oligonucleotide hybridization and ligation steps, the DNA is melted and the newly synthesized DNA strands are removed. A new anchor primer one base shorter than the adaptor is ligated at their 3′ ends with the template DNA and the color of fluorescence is recorded. Again, five cycles of hybridization and ligation are carried out, and fluorescence from each cycle is recorded. The data from the two repeats of ligation reactions are compared and analyzed to obtain the base sequence of the template strand. The repeat hybridization run using one base shorter primer allows each base to be examined twice and to fill any gaps that may remain after the first run [242, 285].

9.4 *Ion Semiconductor Sequencing Chemistry and Platforms*

It is the forth second-generation technology introduced in 2010 by Ion Torrent (acquired by Life Technologies). This sequencing chemistry uses complementary metal-oxide-semiconductor (CMOS) chip technology to capture chemistry changes instead of light and translate these changes into digital data [252, 280]. It is called post-light sequencing technology since it does not use fluorescence or luminescence. This sequencing chemistry could generate up to 10 Gb of sequence data per run with 100 bp or 200 bases read-length protocols and sample multiplexing in a typical run that lasts just 2 h (Table 9.22) [252, 286].

The sequencing system of ion semiconductor sequencing chemistry consists of an electronic detection system that is interfaced with the chip, a microprocessor to process the signals, and a fluidics system for regulating the reagent flow over the chip. The ion chip has 3.5-μm-diameter wells, each of which is located directly over each sensor. The well depth and diameter of wells ensure that only a single micro-bead is loaded in each well. During the sequencing, bPCR products attached to 2-μm-diameter beads are transferred into the loading port of the ion chip; the four dNTPs are added one at a time. When the DNA polymerase adds a dNTP to a

Table 9.22 Some advantages and disadvantages of semi-conductor (the ion proton system) technology of Ion Torrent method

Advantages	Disadvantages
No requirement for optical scanning and fluorescent nucleotides	100 bp or 200 bases read-length sequencing
Fast run times; a typical run takes only a few hours	it has difficulties in detecting the sequences of homopolymer regions
Broad range of applications	Rate limited by dNTP flow
Real-time sequencing and no need for expensive optical equipment	Short read lengths compared to Sanger sequencing and pyrosequencing
No need for modified nucleotides or special enzymes	Assembly limitations in homeologs and higher ploidy levels
Low upfront and operating costs	Cost not as low as Illumina

primer/growing chain, there is net release of one proton (H^+), which produces a change in the pH of the surrounding solution. The signals from H^+ ions are used for direct non-optical identification of the bases present in the DNA template. The sensor located at the bottom of each well detects the change in the pH. The change in pH is proportional to the number of nucleotides incorporated (0.02 pH units for each dNTP molecule added). The signal generation and detection takes ~4 s. Any homopolymers (multiple of the same base) result in a strong signal. The unused nucleotides are removed by washing before the new dNTP is added [242, 252, 253, 256, 285, 286].

9.5 DNA Nanoball Sequencing Chemistry and Platforms

DNA nanoball sequencing (DNBS) of Complete Genomics and combinatorial probe-anchor ligation (cPAL) of Revolocity System are hybridization and ligation type sequencing. In the library preparation of DNBS, small fragments (440–500 bp) of genomic DNA or cDNA are amplified into DNA nanoballs by rolling-circle replication that requires the construction of complete circular templates before the generation of nanoballs. In the library construction of the revolocity system, DNA fragments are ligated with two synthetic DNA sequences called half adaptors, one to each end. Adapters-ligated fragments are then circularized through intramolecular ligation to form a complete adaptor. A series of proprietary methods shorten and linearize the DNA circles. A second set of half adaptors is added and the linear molecules are recircularized. The library is composed of DNA circles, which contain about 70 bp from each end of the original genomic DNA fragment, separated by the two adaptors [287].

The DNA nanoballs are deposited onto an arrayed flow cell, with one nanoball per well and sequenced at high density. Up to 10 bases of the template are read in 5′ and 3′ direction from each adapter. Since only short sequences, adjacent to adapters, are read, this sequencing format resembles a multiplexed form of mate-pair sequencing similar to using Exact Call Chemistry in SOLiD sequencing. Ligated sequencing probes are removed, and a new pool of probes is added, specific for different interrogated positions. The cycle of annealing, ligation, washing, and image recording is repeated for all 10 positions adjacent to one terminus of one adapter. This process is repeated for all seven remaining adapter termini. The major disadvantage of DNBS is the short length of reads and the length of time for the sequencing projects [288–290].

The ligation-based combinatorial probe-anchor ligation (cPAL) chemistry consists of repeating cycles of anchor hybridization, probe ligation, imaging, and removal of the anchor-probe complex. Sequencing initiates with the hybridization of an anchor sequence to one of the synthetic adaptors. To sequence a given position, a pool of probes is introduced with fluorescent labels associated with bases at that position. When the probe with the appropriate sequence hybridizes adjacent to the anchor, the anchor and the probe ligate to each other, thus stabilizing the interaction between the probe and the template DNA molecule. Unligated probes are washed away and the flow cell is imaged by simultaneous high-speed four-color detection. After a base is read, the entire anchor-probe complex is removed and the substrate is ready for another cycle of sequencing, using a different combination of probe [242, 285, 287–290].

10 The Third Generation Sequencing (TGS) Technologies

Third generation sequencing (TGS) technologies are considered those next generation technologies that capable of sequencing single molecules, without the requirement for DNA amplification (library preparation) shared in all the second generation sequencing technologies [291–293]. The TGS methods do not use PCR amplification for template preparation unlike second-generation sequencing (SGS) platforms because TGS platforms sequence single DNA molecules. Other advantages offered by TGS technologies are lower cost, high throughput, faster turnaround, and longer reads [291–295]. These advantages of TGS methods enhanced their utilization in determination of insertions, deletions, translocations and other structural changes. TGS methods often produce more uniform coverage as they are not as sensitive to GC content as SGS technologies [291–295].

The two main limitations of TGS technologies are (i) high error rate, which could be effectively, removed using error correction algorithms and (ii) the cost relative to second-generation approaches. De novo genomes assembled using SGS technologies are often of lower quality compared with those genomes sequenced using older methods primarily due to short-reads of the SGS [296]. The third generation sequencing is based on single molecule sequencing (SMS) technologies, which are non-PCR based and do not include an amplification step prior to sequencing. Two single-molecule sequencing systems exist: HeliScope (Helicos Bio-Sciences Corp., Cambridge, MA, USA) and PacBio RS SMRT system (Pacific Biosciences, Menlo Park, CA, USA) [275, 297].

The technologies being developed for TGS are quite diverse and can broadly be classified into six different categories: (i) virtual terminators sequencing; (ii) captured DNA polymerase sequencing; (iii) nanopore sequencing; (iv) electronic detection, (v) fluorescence energy transfer, and (v) direct imaging of individual molecules [291–295].

10.1 Virtual Terminators Sequencing

Helicos BioSciences (filed for bankruptcy early in 2012) commercialized the first single molecule sequencing (SMS) technology in 2008 [280–283, 298]. This sequencing system is a combination of sequencing by hybridization and sequencing by synthesis using a DNA polymerase [299]. Sample preparation does not require ligation or PCR amplification and, therefore, largely avoids the GC content and size biases observed in other technologies. Its sequencing chemistry was dependent on virtual terminators' ddNTPs [280–283, 298, 299].

In summary library preparation of TGS starts with random fragmentation of genomic DNA ranging 100–200 bp in length and series of adenosine bases are added to the 3′ end of the DNA fragments to create poly-A tails. Poly-A tailed DNA fragments are hybridized to poly-T anchor oligonucleotides covalently linked to the surface of the flow cell and adenosines at the terminal of the DNA fragments are fluorescently labeled so that the instrument can identify the position of each template molecule on the flow cells prior to sequencing. The fluorescent nucleotides are modified (virtual terminators' ddNTPs) to stop the polymerase extension until the incorporated nucleotide's fluorescence is captured and the images are recorded with a highly sensitive charge-coupled device (CCD) camera connected to a fluorescent microscope like instrument. A washing step then takes place to wash off the unincorporated nucleotides as well as the by-products of the previous cycle. After washing, fluorescent labels on the extended strands are chemically cleaved and removed. Another cycle of single-base extension, label cleaving and imaging follow. The HeliScope sequencing read lengths range from 25 to over 60 bases could generate up to 28 Gb of sequence data per run at about 99% accuracy. It generates short reads that may cause ambiguous read mapping due to the presence of paralogous sequences and repetitive elements in plant genomes [253, 280–283, 298, 299].

10.2 Nanopore Sequencing and Platforms

Nanopore is considered a third-generation technology because it enables the sequencing of single molecules in real time [300]. The idea behind sequencing using nanopore is that the conductivity of ion currents in the pore changes when the strand of DNA passes through it. The flow of ion current depends on the shape of the molecule translocating through the pore. Since nucleotides have different shapes, each nucleotide is recognized by its effect on the change of the ionic current. The main advantage of nanopore approach is that sample preparation is minimal compared to SGS methods, and long read lengths compared to the earlier TGS [299]. In addition, there are no amplification or ligation steps required before sequencing. However, the main problem with this technology is the requirement to optimize the speed of DNA translocation through the nanopore to ensure the high error rates of base calling [301–304].

Oxford Nanopore Technologies (ONT), the first company offering nanopore sequencers uses electrical current in a small, mobile phone sized USB device, which was first released to end users in an early access trial in 2014. There are other platforms utilizing nanopore technologies such as Roche's Nanopore and NanoTag sequencing approaches. Nanopore sequencing uses pores formed from proteins, such as α-hemolysin, a biological protein channel system in *Staphylococcus aureus* [301–304].

The NanoTag sequencing approach uses a DNA replication enzyme to sequence a template strand with single base precision as base-specific engineered NanoTags are captured by the nanopore. Electrochemically, each of the four NanoTags interact with the nanopore recognition site differently, partially blocking the ion current by a characteristic amount which results in a tag-specific electronic signature. Each base sequentially binds to the cyclodextrin located on the inside of the nanopore. This binding creates a disturbance in the electric current passing through the nanopore, which generates characteristic signal for each DNA base. This signal is sensed by an electronic device and is converted into base sequence data [301–303].

In another approach, double stranded DNA is denatured by a processive enzyme, which denatures one of the strands through a biological nanopore embedded in a synthetic membrane, across which a voltage is applied. As the ssDNA passes through the nanopore, the different bases prevent ionic flow in a distinctive manner, allowing the sequence of the molecule to be inferred by monitoring the current at each channel [287, 301–303, 305].

10.3 Single-Molecule Real-Time (SMRT) Sequencing (Captured Polymerase Sequencing)

Single molecule real-time (SMRT) sequencing chemistry developed by Pacific Biosciences and released in 2010 and 2015. DNA template molecules and DNA polymerase are immobilized at the bottom of an extremely small well called a zero-mode waveguide (ZMW) that effectively reduce the detection volume [253, 306]. The ZMW focuses the input energy from an excitation laser precisely to the zone containing the immobilized DNA polymerase. This technology uses the natural capacity of DNA polymerase to incorporate ten or more nucleotides per second in several thousand parallel ZMWs. All four nucleotides linked to different fluorophores through the terminal phosphate (dNTP analogs) are then added simultaneously and measured in real time. Unincorporated nucleotides pass rapidly in and out of the detection volume, too quickly for a measurable fluorescent signal to be recorded. However, when a nucleotide binds the active site of the DNA polymerase, its motion is sufficiently slowed for the fluorescent signal to be detected. As the nucleotide is incorporated, the fluorophore is cleaved off as the phosphodiester bond is formed. The free fluorophore then rapidly diffuses out of the detection volume, terminating the fluorescent signal for that particular nucleotide incorporation event. CCD cameras continuously monitor the extremely high number of ZMWs as a series of observed pulses that are converted into single molecular traces representing the template sequences [291–295, 307].

PacBio machines are also capable of producing very long reads, up to and exceeding 10 kb in length, which are useful for de novo genome assemblies. However, an improved technology allows sequencing of up to 20 kb fragments, and efforts are being made to increase it to 40 kb. It requires no amplification step for sample preparation as it involves a single molecule sequencing-by-synthesis approach. During the sequencing workflow, the complimentary DNA strand is synthesized from the single-stranded template by the action of DNA polymerase captured at the bottom of each waveguide. Each DNA polymerase molecule will use a single DNA fragment as template to add the fluorophore-labeled dNTPs to the primer/growing chain. Since the detection system is focused onto the DNA polymerase molecule, the liberated fluorophore molecules do not interfere with the detection process. The DNA polymerase can sequence the DNA fragment more than once, producing multiple coverage of the same molecule [275, 308, 309].

10.4 Other Third-Generation Sequencing Technologies

International Business Machines (IBM) is developing a DNA transistor that would electronically identify individual bases in a single DNA molecule. NABsys is trying to develop the existing solid-state technologies for whole-genome sequencing based on electronic detection of bases. The Starlight technology uses fluorescence resonance energy transfer (FRET) for real-time sequencing of single DNA molecules. Another technology uses a specialized technique with a high-resolution (sub-angstrom) transmission electron microscope for identification of the DNA bases by direct imaging of the base sequence [273, 274, 310].

The sequence of long, intact DNA molecules can be visualized and identified by using electron microscopy. The first report on the successful application of electron microscopy for NGS was genomes of 3.3 and 7.2 kb length that were sequenced by enzymatically incorporating modified bases that contained atoms of increased atomic weight and allowed for the direct visualization and identification of individually labeled bases [273, 274].

In this sequencing process, the double strands of the DNA sample are separated into single strands using common enzymes and reactions. Then, the single-stranded DNA is labeled by PCR using dNTPs attached to heavy atom metal labels that can be separated into identifiable electron microscope-generated images showing large black dots, small black dots, and large gray dots along the DNA molecule linearized by ZSG threading [273–275].

Standard image-based technologies perform the reads and analysis of the labeled DNA using image analysis software that provides sequence data in real time. The sequenced molecules are reads in the range of 5–50 kb in length that are useful for de novo genome assembly and for analysis of full haplotypes and copy number variants. The company ZS Genetics offers a service to provide accurate, long read, single-molecule DNA sequences using the NGS electron microscopy platform [273–275]. It is obvious that new technologies will challenge the sequencing *status quo*. It seems that single molecular sequencing platforms will challenge each other with their new platforms. Together, these long-read technologies have an opportunity to challenge to the limitations of short-read technologies [273–275].

Drop-seq utilizes a micro fluidic partitioning method not unlike the methods described earlier for partitioning DNA into subhaploid amounts for phasing studies. Rather than partitioning DNA, this method partitions individual cells into droplets and associates a unique barcode with the RNA from that individual cell [311–314].

11 Genotyping by Sequencing Method (GBS)

Genotyping by sequencing (GBS) is a generic name for a range of methods, which use NGS data for the genotyping of individuals and populations. GBS methods use enzyme-based complexity reduction (using restriction endonucleases to target only a small portion of the genome) or mechanical shearing coupled with DNA bar-coded adapters to produce multiplex libraries of samples ready for NGS sequencing. Library production requires genome partitioning. The most common method to divide a genome into partitions is the shotgun-sequencing approach in which genomic DNA is fragmented to pieces that are between 50 and 300 kb in size by partial digestion with a restriction endonuclease or by mechanical shearing. These fragments are size fractionated by pulsed-field gel electrophoresis or some other gel free systems. Purified fragments are subsequently cloned into one or more libraries of bacterial artificial chromosomes for amplification and sequencing [314–316].

Multiplexed shotgun sequencing (MSG) approach is an advantageous genome partition method. This approach is widely used in the next generation sequencing (NGS) technologies. Briefly MSG performed as follows: (i) genomic DNA is extracted and fragmented with a restriction enzyme usually producing sticky ends; (ii) barcode adaptors specific to a sample are ligated to DNA fragments; (iii) different samples with specific barcodes are pooled; (iv) pooled products from different samples are size selected; (v) selected products are PCR-amplified and then (vi) PCR-products are sequenced on NGS platforms such as Illumina Genome Analyzer [247, 315–321].

Although the term genotyping by sequencing (GBS) method was first introduced to plant science by Elshire et al. [314] it had been already available since the earliest form of GBS methods such as complexity reduction of polymorphic sequences (CRoPS), restriction site-associated DNA sequencing (RAD-seq) and reduced-representation library (RRL). Whole genome sequencing and resequencing (WGS and WGR) along with GBS methods usually utilize various RRS methods [315–321].

GBS methods potentially produce polymorphisms of SNPs, insertions/deletions (InDels), microsatellites (SSRs) and copy number variation [248, 322–324]. All these types of markers could be called as GBS markers. There are two different source of SNP polymorphism in GBS. Polymorphisms could present at the restriction enzyme recognition sites and sequences either between the two restriction enzyme recognition sites that could be between a restriction enzyme recognition site and randomly sheared site or between the two restriction enzyme cut sites [230, 234, 325].

Polymorphisms outside of the restriction enzymes recognition sites are usually co-dominant markers while polymorphisms detected at the enzyme recognition sites are dominant markers. Co-dominant GBS markers could differentiate homozygous individuals from that of heterozygous while dominant type GBS markers could not differentiate hybrids such as F_1 from one of its parents. Production of co-dominant and dominant GBS markers may relate with the restriction enzymes utilized. Thus, selection of restriction enzymes in library preparation of GBS methods is very important in several aspects. It is always advisable to use enzymes with recognition sequences that are rarely mutated in the genome and that are insensitive/resistant to DNA modifying enzymes and should generate a unique set of target sites with dense distribution along chromosomes and should have even coverage per site across all individuals. Mutations in the restriction enzyme recognition sites can jeopardize genotype calling accuracy in individuals because they are dominant markers, rather than the co-dominant markers. The use of dominant markers generated with the GBS methods decreases their use efficiency in marker-assisted selection [45, 324, 326–330].

Preparation of GBS libraries using fragment size selective amplification and amplification of libraries for sequencing could be amplified using specifically designed effective primer pairs. These primers could be designed with different binding positions controlling the number of sites amplified. Fragment size selection based on gel-free size selection with micro-beads could reduce uneven coverage of reads per site [331].

All GBS methods use adapters that may contain barcode sequences along with sequences of selective amplification and final sequencing. Well-designed adapters with proper length, type (double or single stranded) and shape improve the quality of library by reduces the fabrication of artifacts among the sequenced reads. In addition, properly obtained libraries specific to individuals with distinct and effective barcodes would decrease the inconsistency in the number of reads per sample and inconsistency in the number of sites sequenced per sample [330, 331].

GBS methods have been widely adopted as efficient approaches to genotyping using NGS technologies. GBS methods are useful because: (i) they provide effective tools and approaches for simultaneous marker discoveries (SNPs, SSRs, CNV) and effective genotyping; (ii) GBS methods provide opportunities to study predetermined areas of genomes; (iii) they effectively reduce ascertainment bias compared to arrays and conventional DNA markers; (iv) GBSs can identify not only SNPs but small insertions, deletions, microsatellites and copy number differences; (v) these methods provide a low cost per sample due to highly multiplexed system for constructing libraries for NGS; (vi) GBS methods are suitable for population and germplasm characterization, plant genetics, and genetic mapping of complex traits through genome-wide association studies (GWAS), and estimation of breeding values in genomic selection (GS) [323–325, 328–341].

GBS methods can be performed either through a reduced-representation sequencing (RRS) or a whole-genome resequencing (WGR) approaches. The presence of repetitive elements in plants can represent a significant challenge for de novo assembly, alignment to a reference sequence and sequence comparison for variant discovery. The choice of whether to sequence the entire genome or a reduced portion of it is generally dictated by several factors, including repetitive content, ploidy, and presence or absence of homeologs. The use of reduced-representation resequencing is generally preferred [323–325, 331, 335–347].

GBS methods could be mainly divided in two approaches; reduced representation sequencing (RRS) and whole-genome resequencing (WGR) approaches, listed in Table 9.23. Although both RRS and WGR methods profit from prior genomic information, reference sequence is a prerequisite only for WGR methods. This relative independence from prior genomic information means that RRS shows particular promise for characterizing the genomes of non-model species and previously neglected crops like the genus *Salvia*. The sequencing read depth can be affected by some biological factors of a target species, including: genome size, genome complexity, ploidy, and expected heterozygosity. Read depth differs between RRS and WGR. Low read depth in WGR methods is typically less than 1x and this low read depth can cause problems when genotyping heterozygotes. On the other hands, read depth in GBS methods is grater but varies from 1x to 15x depending on the type of GBS methods used. Read depth in GBS methods can be increased by reduced numbers of genotypes per library, use of rare cutting restriction enzymes, double digestion, and multiple sequencing runs for a library [331, 337–347].

Sequencing a large population of a species with a large genome is best suited to RRS methods. However, when a reference genome is available, WGR methods will be more efficient because they provide higher marker density and thus more precise anchoring. In large or highly repetitive genomes, however, RRS approaches will substantially lower the costs while achieving adequate results [315–327, 331–347].

11.1 Complexity Reduction of Polymorphic Sequences (CRoPS)

Complexity reduction of polymorphic sequences (CRoPS) is the first RRS method [333] that is based on AFLP method and next-generation sequencing technologies. It is also called scalable genotyping by sequencing (GBS) [4]. CRoPS can be used in any organisms, irrespective of genome complexity and size for medium to large-scale genotyping. It enables selection of SNPs, which are located in low-, or single-copy genome sequences and therefore has a high conversion rate to genotyping assays.

In the CRoPS method, tagged complexity-reduced libraries of two or more genetically diverse samples are prepared by AFLP method and sequenced using NGS technologies such as Genome Sequencer (GS) 20/GS FLX (Roche) platform. In CRoPS method (i) total genomic DNA is digested using two restriction enzymes such as *Hpa*II and *Mse*I; (ii) adapters specific to the restriction enzymes used are ligated to the digested products; (iii) product mixture is diluted and is used as a template in a selective pre-amplification step; (iv) pre-amplification reaction is

Table 9.23 Some genotyping-by-sequencing (GBS) methods currently available, divided into reduced-representation sequencing (RRS) and whole genome resequencing (WGR) methods

Type	Methods	References
RRS	Reduced representation shotgun sequencing (RRS)	Altshuler et al. [332]
RRS	Complexity reduction of polymorphic sequences (CRoPS)	Van Orsouw et al. [333]
RRS	Restriction site-associated DNA sequencing (RAD-Seq)	Baird et al. [230]
RRS	Reduced-representation library (RRL)	van Tassell et al. [192]
WGS	Sliding window WGR	Huang et al. [334]
RRS	Paired-end reduced representation libraries (pERPLs)	Kerstens et al. [335]
WGS	Parental inference WGR	Xie et al. [336]
RRS	Multiplexed shotgun genotyping (MSG)	Andolfatto et al. [323]
RRS	Simple genotyping-by-sequencing (GBS)	Elshire et al. [314]
RRS	Two-enzyme genotyping-by-sequencing (GBS)	Poland et al. [325]
RRS	Double-digest RAD sequencing (ddRAD)	Peterson et al. [235]
RRS	Sequence-based genotyping (SBG)	Truong et al. [39]
RRS	Paired-end reduced representation libraries	Deschamps et al. [337]
RRS	Type IIB endonucleases restriction-site associated DNA (2b-RAD)	Wang et al. [234]
RRS	ezRAD	Toonen et al. [236]
RRS	Restriction fragment sequencing (RESTseq)	Stolle and Moritz [338]
RRS	Specific length amplified fragment sequencing (SLAF-Seq)	Sun et al. [339]
RSS	Scalable Genotyping by Sequencing (GBS)	Sonah et al. [40]
RSS	Genotyping by genome reducing and sequencing	Chen et al. [340]
RSS	GBS with one enzyme digest	Beissinger et al. [342]
RSS	Ion Torrent Genotyping by Sequencing	Mascher et al. [248]
RSS	Flexible and scalable GBS	Heffelfinger et al. [343]
RSS	GBS with two enzyme digest	Gardner et al. [344]
RSS	Improved RRLs (iRRL)	Greminger et al. [345]
WGS	Parental inference WGR with individualized model	Rowan et al. [346]
RRS	Genotyping-in-Thousands by sequencing (GT-seq)	Campbell et al. [202]
RRS	Spiked genotyping-by-sequencing (sGBS),	Rife et al. [347]
WGS	Skim genotyping-by-sequencing (SkimGBS)	Bayer et al. [41]
RRS	RAD Capture (Rapture)	Ali et al. [331]
WGS	Whole-genome shotgun (WGS) SMRT sequencing	Du et al. [341]
RRS	Tunable Genotyping-By-Sequencing (tGBS)	Ott et al. [324]

diluted usually 20-fold and used for the second selective amplification step; (v) PCR products of each sample are purified and equal amounts of the two or more samples are pooled and further treated as one fragment library sample; (vi) emulsion PCR and bead enrichment are carried out according to the standard GS 20 protocol (Roche Applied Science); (vii) sequencing is performed according to the manufacturer's instructions (Roche Applied Science); and (viii) resulting sequences are clustered and aligned, and the alignments are mined for SNPs using appropriate bio-informatics tools [333].

11.2 Reduced-Representation Sequencing (RRS)

GBS methods are derivatives or improvement of approaches that have mainly evolved from reduced representation (library) sequencing (RRS). The use of RRLs for SNP discovery was first described using Sanger sequencing [332]. In this method, pools of DNA from multiple individuals are reduced in complexity by the type II DNA restriction enzyme digestion and fragments produced by complete digestion are size selected. The use of restriction enzyme digestion has the advantages of reducing the fraction of the genome present in the RRL by one to two orders of magnitude and ensuring that independently constructed libraries contain nearly identical fragment populations [230, 333].

Other strategies for genome reduction such as multiplexed amplification of target sequences, molecular inversion probes (MIPs) or the use of probes to capture DNA fragments by direct hybridization prior to sequencing are available but in comparison to the use of restriction enzyme they can be labor intensive and rely heavily on existing sequence information, thus potentially limiting their value in large and highly divergent populations or species [331].

van Tassell et al. [192] described an efficient, single step method for SNP discovery, validation and characterization method, RRS that uses deep sequencing of reduced representation libraries from specified target populations. RRS approach is suitable for simultaneous de novo discovery of high-quality SNPs and population characterization of allele frequencies of any species with at least a partially sequenced genome [192].

In RRS method (i) various samples from the related organisms are pooled and pooled genomic DNA samples are digested with a type II DNA restriction enzymes; (ii) digested DNA samples are size selected; (iii) selected DNA fragments are ligated with adapters required for sequencing on the NGS platform; (iv) ligated fragments are again size selected and purified; (v) purified DNA fragments are amplified and the PCR products are sequenced using an Illumina platform [192, 335, 337].

One of the main limitations of this method is that it requires reference sequence of the species under study. A reference genome sequence is used to order SNPs within the sequence assembly. However, this challenge may be overcome by genotyping linkage mapping populations or by using comparative genomic information to infer likely or relative genome position [314].

11.3 Restriction Site Associated DNA Sequencing (RAD-Seq)

There are different types of RAD-seq methods including the original RAD discussed in the previous section, RAD-Seq, 2b-RAD-Seq, ddRAD-Seq, ezRAD-Seq, RAD Capture (Rapture). Microarray-based RAD techniques can only assay a fraction of segregating polymorphisms. As an alternative, RAD analyses can be incorporated into high-throughput next-generation sequencing, RAD-seq. In the RAD-seq method after the ligation of adapters and random shearing, the sheared ends are repaired and second adapters are ligated. Second adapters are Y shaped adapters that have divergent ends. Because the second adaptors have reverse amplification primer sites, the reverse amplification primer is unable to bind to the second adapter unless the complementary sequence is filled in during the first round of forward elongation originating from the forward amplification primer. The structure of this adapter ensures that only first adapter ligated RAD tags will be amplified during the final PCR amplification step [230, 235, 236, 331]. Although this approach reduce the nonspecific amplicon sequences, size selection steps are required for low error rated NGS results [324].

In order to different DNA samples can be prepared with different barcodes to allow for sample tracking when multiple samples are sequenced in the same reaction, the first adapter contains a short DNA sequence barcode. In addition, the first and second adapters may contain paired end sequencing primer sites. High level of multiplexing of RAD-seq produces higher genetic marker density than microarray methods [314, 331]. However, the use of Y shaped adapters may produce high percentage of non-amplifiable DNA fragments [331].

In brief RAD-seq is performed as follows: (i) genomic DNA samples are individually digested with a restriction enzyme of choice; (ii) adaptors with nucleotide barcodes for unique identification of each sample are ligated; (iii) fragments with 300–700 bp are size selected; (iv) different DNA samples are pooled and DNA fragments with adapters are randomly sheared by sonication; (v) repeated ends are ligated with a second type adapters; (vi) purified fragments are PCR amplified and sequenced using a NGS such as reversible dideoxy based Illumina sequencing which uses either sequencing one (one read, single end) or both (two reads, paired end) ends of each fragment and currently gives reads of up to 300 bp in length [39, 41, 235, 314, 324, 325, 339].

11.4 Double Digestion Restriction Site Associated DNA Sequencing (ddRAD-Seq)

Double digestion restriction site associated DNA sequencing (ddRAD-seq) method is a modification of the RAD-seq method [230]. ddRAD-seq uses a double digestion of genomic DNA with two restriction enzymes and eliminates the shearing step of the RAD-seq [230]. Compared to the RAD-Seq protocol, the workflow of preparation of ddRAD libraries is simpler, quicker and substantially cheaper [230, 235]. Due to the typical use of a rare and a common cutting enzyme as in the AFLP protocols, ddRAD results in fewer sequenced sites than RAD-Seq, facilitating higher sequence coverage and/or more individuals multiplexed within a single sequencing lane. In the ddRAD-seq method, higher multiplexing is also possible due to combinational multiplex indexing, whereby a first barcode is introduced in the ligation step and a second during the PCR. Therefore, a larger number of samples can potentially be sequenced in a single lane than with the other RAD techniques [235, 324, 330, 331].

11.5 Type IIB Restriction Enzyme Digestion Restriction Site Associated DNA Sequencing (2b-RAD-Seq)

The main difference between RAD-seq and 2b-RAD is the use of type IIB restriction enzymes [234]. Type II restriction enzymes digest the genomic DNA at both sides of the recognition site at a fixed distance, resulting in protruding blunt ends in 2b-RAD-seq. The resultant products are short genomic DNA fragments with identical size at each IIB restriction enzyme site in the genome. Library preparation in the 2b-RAD-seq protocol is simpler than RAD-seq. Specific adaptors are ligated to the 2b-restriction enzyme digested fragments, and specific barcodes are added to each sample through PCR amplification using degenerated linkers. Amplified products are pooled and sequenced typically using Illumina technology [230, 234, 235].

11.6 ezRAD

This method is based on the use of proprietary kits of standard Illumina TruSeq library preparation kits [236]. ezRAD differs from RAD-seq in its flexibility to use any restriction enzyme (or combination of enzymes). It is argued that the simplicity of ezRAD makes it particularly useful for the discovery of single nucleotide polymorphisms and targeted amplicon sequencing in natural populations of non-model organisms [236]. However, the main limitation of this method is that it requires the proprietary kits of Illumina.

11.7 RAD Capture (Rapture)

Rapture protocol employs physical enrichment of RAD tags rather than PCR-based enrichment. This approach is based on RAD and sequence capture methods and thus it combines the benefits of both RAD methods, i.e., very inexpensive and rapid library preparation for many individuals as well as high specificity in the number and location of genomic loci analyzed [331]. Rapture method uses in-solution capture approach to reduce genome complexity and thereby allows an increased number of individuals to be analyzed with massively parallel sequencing (MPS). Because capturing only genomic regions of interest prior to MPS is more economical than sequencing the entire genome for many studies [331].

Briefly this method involves in following steps: (i) genomic DNA is digested with a restriction enzyme such as *Sbf*I-HF; (ii) SbfI/PstI biotinylated barcode containing RAD adapters are ligated to DNA fragments; (iii) DNA fragments of different samples are pooled and precipitated onto micro-beads; (iv) DNA fragments are sheared using sonication and cleaned; (v) DNA fragments are liberated, targets are captured and sequenced using Illumina platforms. Capturing can generate high sequence depth at targeted loci [331].

Main advantages of RAD-seq include: (i) it does not require any prior genomic knowledge; (ii) it allows generation of population-specific genotype data; (iii) a suitable number of markers for an application can be selected by choice of restriction enzyme, and the number of markers can be increased almost indefinitely by using additional enzymes; (iv) RAD tags create a reduced representation of the genome, allowing over sequencing of the nucleotides next to restriction sites and detection of SNPs; (v) it is amenable to genotyping pooled populations for bulk segregant analysis and also multiplexed genotyping of individuals for fine-scale mapping [230, 235, 236].

One of the main drawbacks of the original RAD-seq is that shearing by sonication is random and variable, potentially hindering the efficiency and the reproducibility of RAD-seq. However, this random shearing step can also be a benefit, as the variable size of the genomic fragments anchored at the restriction enzyme cut site facilitates the assembly of a contigs based on the paired-end reads. In addition, RAD-seq may contain a very high percentage of non-amplifiable DNA fragments due to the use of Y adapters [230, 235, 314].

In comparison to RAD-seq, 2b-RAD-seq allows runs of shorter read length of 33–36 bp due to the smaller size of the fragments. Short sequencing reads are less amenable for alignment to reference genome assemblies, and hinders the design of primer pair for validation of individual SNP assays due to lack of SNP flanking sequence. The use of type IIB restriction enzymes in 2b-RAD-seq theoretically facilitates the sampling and sequencing of identical sites across individuals, circumventing the potential bias of RAD-seq caused by the random shearing step [230]. 2b-RAD-seq also avoids the time-consuming and potentially error-prone size selection step, which characterizes the majority of other RAD methods [330]. 2b-RAD allows removal of loci exhibiting strand bias. The possibility to produce individually bar-coded library allows targeted adjustment before pooling to obtain more equal representation of individual samples [230, 234].

The ddRAD protocol is more flexible than RAD-seq or 2b-RAD in terms of targeted marker density; the number of fragments and SNPs can be readily tailored by combining different restriction enzyme pairs. However, the workflow of the ddRAD-seq is still more complex than the 2b-RAD protocol and requires a size-selection step [230, 235, 330].

Rapture method has several drawbacks including a relatively high library preparation costs prior to capture and low multiplexing capacity during capture. In comparison to RAD-seq, rapture outperforms increased numbers of mapped fragments and better coverage per locus and requires fewer sequence data to achieve the same coverage [330, 331].

The four main weaknesses of RAD-seq and other RSS methods include: (i) low coverage, (ii) allele dropout, (iii) PCR duplicates, and (iv) variation in coverage. The first weakness or the limitations of most forms of genotyping is that only restricted SNPs are assayed across a population. While this is efficient for the identification of major crossover recombination used for genetic mapping, the resolution is fixed by the restriction site density. In allele dropout, mutations in the restriction enzyme recognition site prevent cutting and can thus lead to erroneous genotyping. This is potentially more of a problem for ddRAD than other RAD methods [230, 331].

The concept of PCR duplicates is due to preferential amplification of certain clonal DNA fragments derived from the original genomic DNA fragments. PCR duplicates can give rise to the situation where one allele is overrepresented in the resulting sequence data and causes problems with differentiating homozygous and heterozygous individuals at that locus. However, this does not affect the PCR-free systems of the third generation sequencing platforms. Finally, variance in coverage between loci can be caused by an amplification bias towards fragments of shorter length and with higher GC content. Beyond these common errors, the frequent use of methylation-sensitive enzymes in RAD-seq introduces an ascertainment bias [328, 330].

11.8 Elshire Genotyping by Sequencing (GBS)

Most GBS methods use one or more restriction enzymes to facilitate genome complexity reduction amongst individuals or populations, and provide fixed starting points for sequencing. Since target is reduced to a subset of a genome with the restriction enzymes, the resulting sequence dataset can provide higher read coverage per locus while allowing higher level of multiplexing with uniquely bar-coded adapters for different samples [314].

GBS developed in Elshire et al. [314] combines genomic DNA digestion and ligation steps of adapters. Briefly (i) sample DNA, adapters and a restriction enzyme are combined in a single reaction to ligate adapters to generated sticky ends; (ii) sets of 48 or 96 DNA samples, each with a different barcode adapter ligated in the previous reaction are pooled and purified using a commercial kit; (iii) DNA fragments of the pooled library are PCR amplified using primers contained complementary sequences for amplifying restriction fragments with ligated adapters; (iv) single stranded DNA fragments are ligated onto the Illumina sequencing flow cell; and (v) sequencing reactions are performed using single end or pair end approaches of the Illumina platforms [247, 314, 324].

11.9 Two-Enzyme Genotyping-by-Sequencing

In the two enzyme GBS method [325] (i) genomic DNA is digested using two restriction enzymes such as *Pst*I and *Msp*I and reactions are inactivated; (ii) barcode containing adapter and common Y-adapter are ligated and enzyme is inactivated prior to pooling the samples; (iii) ligated samples are pooled and PCR-amplified in a single tube, producing a single library from 48 samples; (iv) pooled samples are sequenced on a single lane of Illumina platform.

11.10 Sequence-Based Genotyping (SBG)

Sequence based genotyping (SBG) is another type of GBS methods for simultaneous marker discovery and co-dominant scoring, using next-generation sequencing [39]. SBG offers users several advantages including a generic sample preparation method, a highly robust genome complexity reduction strategy to facilitate de novo marker discovery across entire genomes, and a uniform bioinformatics workflow strategy to achieve genotyping goals tailored to individual species, regardless of the availability of a reference sequence. The most distinguishing features of SBG technology are the ability to genotype any population structure, regardless whether parental data is included, and the ability to co-dominantly score SNP markers segregating in populations.

SBG incorporates the high-throughput capacity of NGS platforms, and the proven, reproducible and robust genome complexity reduction capabilities of AFLP, to score random SNP markers across an entire genome. Using SBG, genome-wide SNP discovery and genotyping of large populations can be attained in a single experiment, without the need for a prior knowledge of a reference genome sequence. Depending upon the user's needs, this method allows for the customization of the type of complexity reduction required, the optimal number of samples to be analyzed, as well as the desired number of SNPs.

In SBG method (i) genomic DNA samples of individuals are digested with two restriction enzymes (*Eco*RI and *Mse*I, rare and frequent cutting enzymes, respectively); (ii) adapters specific to the restriction enzyme cut sites containing sample identification tags are ligated to the restriction fragments to construct SBG libraries; (iii) SBG libraries are PCR amplified; (iv) different samples containing different bar-coded adapters are pooled; (v) pooled products are sequenced using Illumina sequencing platforms; (vi) SNPs are mined between the samples and simultaneously genotyped using the SBG bioinformatics analysis workflow [39].

11.11 Restriction Fragment Sequencing (RESTseq)

Restriction fragment sequencing (RESTseq) method [338] is another type of GBS based on Ion Torrent platforms. However RESTseq can be modified to be using in any other NGS platforms. Typical RESTseq starts with the (i) digestion of genomic DNA with a frequent cutting restriction endonuclease to generate a high number of fragments; (ii), DNA fragments are purified and ligated with NGS platform specific adapter pairs (Ion Torrent PGM (Life Technologies), MiSeq (Illumina) and 454 GS Junior (Roche); (iii) adapters ligated DNA fragments are digested with another frequent cutting restriction endonuclease to further reducing the library in complexity; (iv) adapters contain barcode and primer sequences for pooling, amplification and sequencing purposes; (v) the remaining uncleaved fragments are recovered in a size selection step, removing cleaved fragments and adapter dimers from the sample; (vi) different samples containing different brocaded adapters are pooled and if necessary, an additional PCR can be used to amplify the library using the suitable primers matching adapter sequences on either end and (vii) the retrieved library can be sequenced according to the manufacturer's instructions such as using Ion Torrent PGM, MiSeq and 454 GS Junior. One of the obligate constraint of this GBS method is that restriction endonucleases may not have a restriction site within the ligated sequencing adapters [338].

The RESTseq method is applicable for non-model organisms without a sequenced reference genome. Although RESTseq is similar to main GBS methods including the double digest RAD, it has the advantage of a direct control over the complexity of the library due to the digestion after adapter ligation, size selection as well as the more unbiased fragment distribution [330, 331].

11.12 Tunable Genotyping by Sequencing (tGBS-seq) Method

Tunable genotyping by sequencing (tGBS-seq) method [324] uses single stranded adapter (oligonucleotide), size selection and selective PCR amplification steps. During tGBS, genomic DNA is subjected to double-digestion with two restriction enzymes in the same reaction, producing DNA fragments with a 5′ overhangs on one end and a 3′ overhangs on the other. In contrast to other methods, which employ double-stranded adapters [235, 344], a single-strand oligonucleotide (oligo) is ligated to each overhang. One of the oligos is unique to an individual sample and contains a DNA barcode (called barcode oligo) while the other oligo is common to all samples and contains a universal sequence (called universal oligo) for subsequent library construction [324].

All digestion-ligation products are pooled and the pooled products are purified before using a selective PCR. Selective PCR uses two primers, one of which is selective and one of the other is non-selective primer. The selective primer is designed to be the reverse complement of the universal ligation oligo; however, it extends an additional 1–3 nucleotides (selective bases) at its 3′ end which can only perfectly anneal to a subset of the genomic fragments created by restriction enzyme digestion and oligo ligation, thus reducing the number of targeted sites to be amplified [324].

As a result, genomic fragments that include the complement of the selective bases and the universal oligo will be preferentially amplified. The non-selective primer used in selective PCR matches the 5′ end of the barcode oligo. Because this primer will anneal and amplify the sequence preceding the barcode, the primer itself does not need to be designed match the barcode, reducing primer complexity and cost. After selective PCR, products are purified and used as template for the second PCR (final PCR). Two primers compatible with the appropriate sequencing platform are used to create the sequencing library [324].

11.13 Specific Locus Amplified Fragment Sequencing (SLAF-Seq)

Specific-locus amplified fragment sequencing (SLAF-seq) is an efficient method of genotyping by sequencing for large-scale genotyping [339]. SLAF-seq does not require reference genome sequences and polymorphism information. Large populations with large numbers of loci can be genotyped simultaneously with barcode-multiplexed sequencing of SLAF-seq. The use of appropriate restriction enzyme(s) in the pre-designed reduction schemes (enzymes and sizes of restriction fragments could be evaluated using training data), repetitive sequences can be avoided, and the selected fragment number can be decided for personalized research purposes to maintain the balance between marker density and population size [329].

SLAF-seq method produces several different types of polymorphism in the same set of analysis. Markers of SLAF-seq are mostly SNPs (SNPs-only), insertion–deletions (InDels-only), and both types (SNPs & InDels). All these marker types could be used in mapping and other application of SLAF-seq. In the analysis of SLAF-seq the number of SLAFs must be suitable for the specific needs of the research undertaken, the SLAF-must be evenly distributed through the sequences to be studied and repeated SLAF markers must be avoided [329]. In order to ensure these requirement, optimization studies called pre-designed reduced representation scheme are performed before SLAF-seq studies [329, 339].

Briefly, the SLAF method is performed as follows; (i) genomic DNA is digested with a selected restriction enzyme such *Rsa*I or *Mse*I or *Hae*III varying depending on the genome; (ii) after repairing the ends of the DNA fragments, dual-index adapters are ligated. After the ligation, additional restriction enzyme digestions could be performed if desired; (iii) ligated products are selectively amplified using adapter specific primer pair with barcode 1; (iv) different samples amplified using different barcodes are pooled; (v) fragments are size selected (450–500 bp)

Table 9.24 Some advantages and disadvantages of the SLAF marker system

Advantages	Disadvantages
It produces deep sequencing to ensure genotyping accuracy	It requires the use of restriction enzymes, which need high quality of genomic DNA
Reduced representation strategy to reduce sequencing costs	Markers are primarily dominant (present or absent) or differences in intensity
Its pre-designed reduced representation scheme optimizes marker efficiency	Produces short reads that require reference sequences for alignment studies
Its double barcode system increases population size to be studied	It requires pre-designed reduced representation scheme which increases the costs and labor

using gel or gel free systems; (vi) selected fragments are amplified using primer pair with barcode 2; (vii) size selected and prepared ssDNAs are subjected to bridge PCR amplification and sequenced using an Illumina high-throughput pair-end sequencing platform. Table 9.24. summarizes some advantages and disadvantages of the SLAF marker system [339].

11.14 Comparison of Library Preparation Approaches for GBS Method

There are several genomic complexity reduction strategies not based on restriction enzymes, including long-range PCR amplification of specific genomic regions, use of molecular inversion probes, and various RNA, DNA hybridization/sequence capture methods. These enrichment strategies are time-consuming, technologically challenging, and cost-prohibitive for assaying large numbers of samples. Reducing genome complexity with restriction enzymes, however, is easy, quick, extremely specific, highly reproducible, and may reach important regions of the genome that are inaccessible to sequence capture approaches. By choosing appropriate restriction enzymes, repetitive regions of genomes can be avoided and lower copy regions can be targeted much higher efficiency, which tremendously simplifies computationally challenging alignment problems in species with high levels of genetic diversity.

Genotyping by Sequencing (GBS) libraries are generally constructed by digesting DNA with one or two restriction enzymes and analyzing the subsequent library with NGS platforms such as Illumina or Ion Torrent. However, literature presents many different approaches after the restriction enzyme digestion steps. These approaches can be classified into four main categories [330].

First is GBS library preparation with selective fragment amplification approaches including the CRoPS [333], scalable GBS [40] and tunable genotyping by sequencing (tGBS-seq) methods [324].

Second is GBS without fragment size selection approaches including simple genotyping-by-sequencing (GBS) [314], two-enzyme genotyping-by-sequencing [325], genotyping by genome reducing and sequencing [340].

Third is GBS with fragment semi-size selection approaches including restriction-site associated DNA sequencing (RAD-seq) [230], paired-end reduced representation libraries (paired-end RPLs) [337], double-digest RAD-seq (ddRAD-seq) [235], specific length amplified fragment sequencing (SLAF-Seq) [339] and flexible and scalable GBS [343].

Fourth is GBS with fragment size selection approaches including reduced representation shotgun sequencing (RRS) [332], reduced representation libraries (RRLs) [192], paired-end reduced representation libraries (paired-end RPLs) [335], type IIB endonucleases restriction-site associated DNA (2b-RAD) [234], GBS with one enzyme digest [342], restriction fragment sequencing (RESTseq) [338], ezRAD [236], GBS with two enzyme digest [344] and improved RRLs (iRRL) [345].

12 Transcriptome/Exome/RNA Sequencing

Transcriptome/exome/RNA sequencing analysis (RNA-seq) is able to reveal genes that are being actively expressed in specific tissue and species of interest, and facilitate the discovery of potential molecular marker of SNPs, SSRs or InDels, some of which could be functional DNA markers (FDMs). This type of analysis is useful in non-model organisms where the full genome data is still not available for comparison. Sequences that are targets for transcriptome/exome/RNA sequencing analysis do not contain repetitive genomic regions and rich in regulatory sequences 5′-UTR, 3′-UTR, miRNA and gene bodies. Furthermore these regulatory sequences and genes are present in only those genes that are transcribed in a particular tissue/organ during the given developmental stage and under the environmental conditions [76, 348–350].

Transcriptome analysis is an attractive alternative to whole genome analysis. It provides abundant information on gene expression, gene regulation and amino acid content of proteins. However, it is limited to only those genes that are transcribed in the concerned tissue/organ during the given developmental stage and under the environmental conditions prevailing at the time of sample collection. Therefore, a fair number of organs/tissues, developmental stages, and environments should be sampled to ensure the representation of most, if not all, of the genes present in the genome of the concerned species [348–350].

For RNA sequencing of mRNA, RNA with polyA tails is isolated from total RNA and reverse transcribed to cDNA with reverse transcriptase and polyT or polyU primers before being treated much the same way as the fragmented genomic DNA. To isolate noncoding RNAs (micro, small, and long) from total RNA, these sequences are selectively ligated to 3′ and 5′ adapters and reverse transcribed to cDNA. The cDNA libraries are labeled with barcode sample tags to enable the libraries to be pooled and therefore maximize the sequence output as a multiplex amplicon-sequencing step for each sequencing run. After library construction, the DNA fragments are clonally amplified by emulsion PCR with micro beads or by bridge PCR using primers attached to a solid surface in order to generate sufficient single-stranded DNA molecules and detectable signal for producing sufficiently reliable sequencing data [348, 349, 351–354].

An example of RNA-sequencing (mRNA-seq, transcriptome-seq) technologies using Illumina sequencing platforms is performed as follows: (i) the poly-(A) mRNA is purified; (ii) mRNA samples are fragmented before the cDNA synthesis to avoid the priming bias; (iii) RNA fragments are reversely transcribed into first strand cDNA using reverse transcriptase; (iv) second strand cDNA synthesis is carried out; (v) the double-stranded cDNA is end repaired using T4 DNA polymerase, Klenow fragment and T4 polynucleotide kinase; (vi) after addition of single adenine bases to the 3′-ends adapters with a single 'T' base overhang at their 3′ ends are ligated; (vii) ligation products are purified and size selected, separating on an agarose gel through electrophoresis; (viii) to selectively enrich and amplify the cDNA fragments, PCR is performed using two primers and the amplified PCR products are purified and a paired-end-read protocol is applied on the Illumina platform. Data analysis and base calling are performed by suitable software programs [350, 353, 354].

13 Whole-Genome Sequencing (WGS)

Sequencing of whole genomes involves considerable time, labor, and financial and other resources. Therefore, this approach for DNA marker discovery should remain the method of choice wherever resources and other considerations do not preclude this option. In order to reduce time, labor and cost for discovery of SNPs using whole genome sequencing, several alternative approaches and resequencing methods have been developed [334, 346, 347].

The shotgun sequencing (also known as shotgun cloning), is a method involves randomly breaking up genomic DNA sequences into lots of small pieces and cloned for library construction. Random clones are picked from this library and used for sequencing. The sequenced fragments are then assembled together by *in silico* analyses that find where fragments overlapped for the sequencing of long DNA strands. In the sequencing studies using this approach or other methods the sequencing depth and coverage should be large enough (for example 20x) not only to reduce the errors but also to ensure that sequence of a given genomic region is available from enough number of individuals/lines to allow SNP discovery. The sequencing depth for the whole genome is the average number of times each base of the genome (the entire genome) of an individual has been sequenced. The general formula for coverage or depth (C) is C = LN/G, where L is the read length, N is the total number of reads, and G is the length (bp) of the haploid genome of the concerned species. Read is a raw sequence that comes from a sequencing machine and it is usually 100–800 bp long [332, 355–357].

The reduced representation approaches sample a fraction of the whole genome for sequencing to reduce the cost and labor for genotyping. For instance, hypomethylated regions of a genome can be obtained for sequencing. The genomic DNA of the target individual is digested with a 5-methylcytosine-sensitive restriction enzyme and the digest is subjected to electrophoresis; fragments of 100–600 bp are separated and used for sequencing by an NGS technology. Alternatively, exome or transcriptome sequences could be used for genotyping studies. The sequence capture approaches can be applied when the genomic region of interest is known and a closely related reference genome sequence is available. For instance, the sequence capture approach of NimbleGen SeqCap uses oligonucleotide probes synthesized on microarray slides, and the lengths of the probes are adjusted to obtain a uniform melting temperature. The genomic DNA fragments are hybridized with the microarray and the captured fragments are used for sequencing. On the other hand, Agilent SureSelect method, implements in-solution target sequence capture using biotinylated RNA probes of 120 nucleotides. The genomic DNA fragments already ligated to sequencing adapters are hybridized with the probes, and the hybridized fragments are separated by exploiting the high affinity of biotin for streptavidin. The RainDance Targeted Sequencing System, in contrast, uses two rounds of PCR to specifically amplify fragments from the targeted genomic region. The target-specific primers are used for the first round of PCR amplification. In the second round of PCR, universal primers with the partial NGS adaptor sequences at their 3′ ends are used. The PCR products generated from the second round of PCR are directly used for sequencing. In addition, molecular inversion probes could also be used for whole genome sequencing studies [220, 350, 358].

14 Whole Genome Resequencing (WGR)

Whole genome resequencing (WGR) methods based on NGS technologies and platforms are theoretically capable of identification all genetic variants among individuals. Although WGR provides complete resolution of any genome, it is cost-prohibitive and may be unnecessary for many studies involving a large number of individuals. The parental genomes with high-quality sequences and a reference sequence are often required for WGR. It differs from RRS, in the lack of complexity reduction steps before sequencing. WGR is well suited to genotyping biparental cross populations with complex, small- to moderate-sized genomes. It provides the lowest cost per marker data point. Compared to WGR methods RRS approaches differ in their suitability for various tasks, but demonstrate similar costs per marker data point. However RRS approaches are generally better suited for de novo applications and more cost-effective when genotyping populations with large genomes or high heterozygosity. On the other hands, WGR offers the greatest cost-efficiency per marker data point, and is particularly useful when recombination is high and many markers are needed for a well-resolved genetic map in a species with a small- to moderate-sized genome. WGR has the added benefit of increasing the chances of finding causative SNPs, InDels or genes, which allow development of 'perfect' markers. In the light of the decreasing costs of sequencing, the use of WGR to increase the resolution of mapping studies is likely to become more common in the future [334, 346, 347].

WGR allows the discovery of a huge number of DNA markers such as SNPs, InDels, copy number variations (CNV), and presence/absence variations (PAV) in crops and provides deep insight into genome evolution. Moreover, the combination of WGR with bulked segregant analysis (BSA) allows rapid identification of genes and causal mutations in crops. The application of NGS technologies in targeting induced local lesions in genomes (TILLING) provides a robust approach to discover rare mutations in populations. In the past few years, a number of WGR-based approaches have been emerged to exploit the full potential of DNA sequencing [41, 334, 336, 341, 346].

A typical WGR method is performed as follows: genomic DNA is fragmented to about 500 bp by sonication and the fragments are end repaired, and dATPs are added to generate a protruding 3′ A for ligating with the adaptor carrying a three-base index. Three-bp indexes are linked to adapters and the indexed DNA samples are run on 2% agarose gels, and fragments of 150–180 bp are recovered and purified. Each sample is amplified by PCR for about 18 cycles. DNA samples of individuals with different indexes are mixed in an equal molar concentration and loaded into one lane of the Illumina GA for 36-cycle sequencing, with the Illumina PhiX sample used as control. Image analysis and base calling were performed using Illumina GA processing pipeline [336, 341, 346].

In another type of WGR method, genomic DNA samples are digested using dsDNA shearase. The enzyme reactions are stopped by adding a final 50 mM ethylenediaminetetraacetic acid (EDTA) and purified while DNA fragments are bound on the beads. Samples are A-tailed and purified. Adapters with barcodes are ligated to the fragments. Usually in current technology eight samples are pooled and concentrated eluting in lower volumes. Samples are PCR amplified and cleaned and libraries are sequenced using Illumina NGS platform. Parental genotypes are prepared separately while F2 individuals are pooled as mentioned above [41, 334, 336, 341, 346].

Skim genotyping by sequencing [41] is another WGS method that uses low-coverage whole genome sequencing for high-resolution genotyping. Skim-GBS is a two-stage method requires a reference genome sequence, genomic reads from parental individuals and sequences of individuals obtained from the parents. In the Skim-GBS method, the DNA sequences of parents are mapped to the reference genome and SNPs are identifies using a suitable software such as SGSautoSNP. Sequences of individuals are mapped to the same reference and comparison with the parental SNP file enables the calling of the parental genotype.

Most of second generation based GBS technologies use DNA enrichment methods that rely heavily on fragmentation of genomic DNA prior to amplification, resulting in relatively short sequencing templates. Characterization of complex genome requires larger size fragments that are required to successfully span confounding sequence elements, such as extended repeats or homeologus within the target. Third generation sequencing (TGS) platforms can overcome some inherent disadvantages of NGS and are expected to promote the applications of WGR-based approaches and revolutionize plant breeding, genomics and genetics research. Current TGS platforms are capable of producing significantly larger read lengths and sequencing through traditionally difficult sequence templates with high GC content [341, 359].

15 Functional DNA Markers (FDMs)

Traditional and NGS-based DNA markers (DMs) are single (such as SNPs) or larger nucleotide sequences (fragments) that are located within or between gene bodies (promoters, 5′-UTRs, exons, introns and 3′UTRs). DM polymorphisms result from substitutions, insertions or deletions (INDELs), variation in repeats (SSRs) and copy numbers (CNV). Any DMs associated with phenotypic/physiologic trait variations are referred as functional DNA markers (FDMs) also called as gene based markers or perfect markers. Functional DNA markers could be divided into two main groups. Those FMs that closely associated with the phenotypic trait variations are called direct (D) functional markers (DFDMs) whereas those FMs that less or not directly associated with the phenotypic traits due to recombination and genetic interaction are called indirect (I) functional markers (IFDMs) [360, 361].

FDMs have several advantages over DMs. FDMs are not lost due to the recombination between marker and gene of interest because they represent polymorphism in the actual gene sequences. This makes FDMs much more meaningful in plant breeding, marker assisted selection and genomic selection because the number of DNA markers required for foreground selection is reduced to the number of genes to be selected. FDMs are more useful in determination of population dynamics, germplasm collections, and monitoring evolutionary changes because they are not affected by non-functional allelic variations. Large number of DFDMs generated using GBS technologies could be used in genomic-assisted breeding (GAB) and genomic selection (GS). GBS and GS may allow researchers to conduct selection on a novel germplasm or species without needing any prior molecular tools, or conservation biologists to determine population structure without prior knowledge of the genome or diversity in the species [176, 360, 362].

Some of limitations of DFDMs include: once direct functional molecular markers have been developed they need to be evaluated in different genetic backgrounds in order to obtain more precise estimates of the phenotypic effects of different markers and this is the reason for only a small fraction of the genes of different crop species have been functionally characterized. Development of DFDMs requires complete or nearly complete genome sequence information or nearly complete genome sequence information, thus, it is application is very limited for little or no genomic information is available. Main disadvantages of DFDMs are their requirement of the use of forward genetic approaches such as map-based gene isolation, expression profiling, and sequence homology based characterization of genes from other species. In addition, genome editing which is an alternative to standard breeding processes, and genetic transformation are required. Reverse genetic approaches such as insertion mutagenesis using transposes or T-DNA and Targeting Induced Local Lesions IN Genomes (TILLING) need to be employed [336, 341, 346]. There are some traditional DNA markers assigned as functional markers in *Salvia* discussed in the next sections.

16 Traditional DNA Markers Used in *Salvia*

Various traditional (conventional) DNA marker systems and technologies have been used in *Salvia* aiming for different purposes [2, 9, 38, 52, 80, 363–368]. One of the most pursued aims was to study molecular taxonomy and genetic relationship of *Salvia* L. Studies of Walker and Sytsma [2, 363] and Will and Clasen-Bockhoff [31, 83] along with several others [364, 365] revealed that *Salvia* is polyphyletic with four distinct evolutionary lineages. Xu et al. [366] used rDNA internal transcribed spacer (ITS) regions to identify *S. miltiorrhiza* and related *Salvia* species. The nucleotide sequences of *Salvia* samples showed obvious diversity, and each *Salvia* species and *S. miltiorrhiza* population were found to have a unique sequence in their ITS regions. This indicates that ITS regions of *Salvia* species is useful in taxonomic and genetic studies.

Danilo-Stojanovicv et al. [367] studied current genetic patterns and genealogical history of 83 individuals from eight natural populations of *S. officinalis* based on sequences of two chloroplast intergenic spacers, *rps16–trnK* and *rpl32–trnL*. The authors presented relationship between the genetic profile and chemical composition of the indigenous populations. Their study indicated that wild populations contain a great genetic diversity and they can be utilized to improve cultivated populations by introducing genetic diversity through sexual reproduction [367]. De Mattia et al. [368] studied five chloroplast genes using six different genera (*Mentha, Ocimum, Origanum, Salvia*, *Thymus* and *Rosmarinus*) from the family Lamiaceae to identify bar-coding regions suitable for plant identification studies. Bar-coding regions consisted of *rpoB*, *rbcL*, *matK* and *trnH-psbA* genes with universal primers. Results suggested that the non-coding *trnH-psbA* intergenic spacer was the most suitable marker for molecular identification followed by *matK*.

Among conventional DNA markers; random amplified polymorphic DNA (RAPD) [9, 165, 369–373], amplified fragment length polymorphism (AFLP) [38, 79, 80, 125, 132, 370, 374–376] and inter simple sequence repeat (ISSR) [52, 79, 85, 163–165, 372, 377, 378] markers have been widely used while the DNA marker systems simple sequence repeats (SSRs), also called microsatellite [10, 351, 379–383] has limited number of utilization in *Salvia*. On the other hands, markers from sequence-related amplified polymorphism (SRAP) [52, 82, 177, 178] are applied in some extend in *Salvia* research. In addition, there are applications of some novel markers systems such as polymerase chain reaction-restriction fragment length polymorphism (PCR-RFLP) [4], conserved region amplification polymorphism (CoRAP) [181] and directed amplification of minisatellite DNA region (DAMD-PCR) [4, 81] in *Salvia*.

Boszormenyi et al. [9] reported a strong relationship between the chemical profile and the genetic variability of RAPD markers in *S. officinalis*. Cahill [371] used RAPD markers to assess genetic diversity among 38 wild and domesticated accessions of *S. hispanica* collected throughout Mesoamerica. The study used 18 primers, 13 of which produced polymorphic markers across samples. Genetic diversity was higher among wild varieties than all domesticated varieties and modern commercial domesticated varieties. It was found that there was a slight loss of genetic diversity accompanying domestication and a near lack of genetic diversity in modern commercial varieties. Javan et al. [372] studied eight species of *Salvia* using RAPD and ISSR markers. Fifteen RAPD primers produced 220 markers all of which were polymorphic while the same number of ISSR primers produced 245 amplified bands, 241 of which were polymorphic.

Zhang et al. [125] characterized more than 200 germplasms of *S. miltiorrhiza* by ISSR. The authors stated that integrated investigation of the phenotypic, genetic, and phytochemical diversity could identify germplasms with desired traits such as good yield root biomass and high content of active constituents. Erbano et al. [85] using ISSR marker system characterized 72 *Salvia* individuals representing three populations in Brazil. Populations were clustered into 4 main groups and largely recovered the geographical origin of each sample [85].

Safaei et al. [164] studied morphological and genetic variability and species relationship among 39 samples of *Salvia* species naturally occurring at Alborz mountain region in Iran based on ISSR markers. Their results indicated that there was no correlation between genetic distance and geographical distance. The authors also found that there was some degree of gene flow among the species studied. Sarrou et al. [163] reported that the correlation between matrices obtained from ISSR data and metabolic profiles of *S. officinalis* grown in Greece was non-significant based on ISSR markers. However, the author stated that they were able to identify parent lines for breeding programs based on the differences in polyphenol contents and desired essential oil compositions.

Yousefiazarkhanian et al. [165] used ISSR and RAPD molecular techniques to evaluate the genetic relationships among twenty-one ecotypes of eight *Salvia* species growing in different geographical regions in Iran. RAPD used 15 primers while ISSR used 8 primers; altogether they produced a total of 280 bands, most of which were polymorphic. The authors stated that there were no clear differences between the two marker systems, both had similar efficiency in detecting genetic polymorphisms with remarkable ability to differentiate the closely related ecotypes of *Salvia*. In another research, Yousefiazarkhanian et al. [378] studied genetic variation between and within *S. aethiopsis*, *S. macrosiphon* and *S. virgata* ecotypes by ISSR. The authors stated that the diversity within the species was lower (18%) than the between species (82%).

Shu et al. [38] reported that an AFLP marker was tightly linked to the dominant male sterile gene with at a distance of 6.89 cM in *S. miltiorrhiza*. The authors suggested that this marker could be converted to PCR-based assay for large-scale selection of fertile plants in marker-assisted selection (MAS) at the seedling stage. Zhang et al. [125] identified two AFLP markers; one tightly linked to the drought stress gene at a distance of 7.02 cM and one closely associated with the drought stress gene at a distance of 4.66 cM [376]. All terpenes levels were significantly correlated with the AFLP genetic clusters, suggesting close correspondence between genetic groups and chemical profiles. Yang et al. [132] obtained 2300 transcript-derived fragments (TDFs) in *S. miltiorrhiza* and *S. castanea* using cDNA-AFLP method. The quantitative PCR (qPCR) analysis indicated that the expression of 9 TDFs was positively correlated with accumulation of phenolic compounds and tanshinones. Al-Gharaibeh et al. [80] characterized 12 *S. spinosa* and 13 *S. syriaca* populations from three phytogeographical regions (Mediterranean, Irano-Turanian and Saharo-Arabian) based on AFLP markers. The author concluded that genetic distance was significantly correlated to environmental distance, but not to geographic distance.

Wang et al. [181] used conserved region amplification polymorphism (CoRAP) method to investigate the applicability of CoRAP in *Salvia* by studying samples from different geographical origins. In their study CoRAP used two primers: fixed and arbitrary primers. The fixed primers were derived from target EST sequences deposited in GenBank; and the core sequence (CACGC) of the arbitrary primers is a conserved region found in most introns. CoRAP markers were in the range of 50–1000 bp and successfully genotyped *Salvia* samples.

Karaca et al. [4] stated DAMD-PCR and PCR-RFLP markers for selection of *Salvia* with higher essential oil yield. The authors reported that there were correlation between the genetic diversity obtained with DAMD-PCR marker and the essential oil contents of some *Salvia* species. Ince and Karaca [81] reported a total of 70 species-specific DAMD-PCR markers that could differentiate 5 *Salvia* species. The authors stated that these markers could be used as diagnostic markers for identification of selected *Salvia* species.

Aghaei et al. [178] studied 54 *Salvia* samples belonging to *S. virgata*, *S. nemorosa*, *S. officinalis*, *S. sclarea* and *S. cereal* species collected from different regions in Iran based on SRAP markers. A total of 265 markers were produced using 14 primer combinations. SRAP marker system targets open reading frames as functional regions in genomes, thus are capable of being functional markers. Saebnazar and Rahmani [82] studied genetic diversity of 11 *Salvia* species using SRAP technique. A total of 205 DNA amplicons were produced with 25 selective primer combinations. SRAP markers were able to detect high polymorphism among *Salvia* genotypes. The authors pointed out that SRAP technique is a reliable tool for differentiation of *Salvia* species at inter and intra-species level.

Radosavljevi et al. [380] developed 9 microsatellite markers isolated from *S. officinalis*. This study is among those that have shown that rare species have less genetic variation than wide spread species. Xu et al. [381] downloaded 10,494 ESTs from the dbEST database for eSSRs development. In their study eSSRs were considered for primer design that fitted the following criteria: a minimum length of 10 bp; at least five repeat units for di-, tri-, tetra-, penta- and hexa-nucleotide eSSRs. The authors reported 14 EST-SSRs with high level of polymorphism and transferability among forty-one samples of *S. miltiorrhiza* from eight regions in China. Ince and Karaca [10] reported 22 eESTs based microsatellite markers called E-microsatellites. The authors concluded that ESTs of related species deposited in GenBank were reliable sources for the development of e-microsatellite markers. Their results indicated that cross-transferability of e-microsatellite markers decreased as the evolutionary distance among the tested *Salvia* species increased.

Peng et al. [52] utilized ISSR and SRAP markers for genetic diversity and characterization studies of 53 *S. miltiorrhiza*. Out of the 100 ISSR primers and 100 SRAP primer combinations screened, 13 ISSRs and 7 SRAPs were selected to evaluate the level of polymorphism and discriminating capacity. Their results showed that the 13 ISSRs generated 190 repeatable amplified bands, 177 of which were polymorphic (93.2%), with an average of 13.6 polymorphic fragments per primer. The seven SRAPs produced 286 repeatable amplified bands, 266 of which were polymorphic (93.4%), with an average of 38.1 polymorphic fragments per primer. Cluster analysis readily separated different morphological accessions, wild and cultivated controls based on morphological traits. Song et al. [177] applied ISSR and SRAP markers to assess the level and pattern of genetic diversity in a total of 62 individuals from five cultivated populations of *S. miltiorrhiza*. A total of 11 primers consisting of 5 ISSR and 6 SRAP were utilized and amplified a total of 244 markers, 120 of which were from ISSR and 122 of which were from SRAP. In both marker systems, high levels of genetic diversity at the species level were detected.

Wang et al. [23] developed and utilized intron polymorphism using primer pairs designed according to the exon sequences flanking the target intron of major latex-like protein (MLP) gene. Intron polymorphism included single nucleotide polymorphisms (SNPs) and InDels. The authors suggested that MLP intron is a valuable DNA barcode for intra-specific study of *S. miltiorrhiza* populations, and the developed markers could serve as a useful tool for molecular identification of specific cultivar and geographically distinct populations of *S. miltiorrhiza*.

There are also some DNA marker technologies that lie between the low throughput and high throughput technologies. These medium throughput technologies include DArT, KASpar and some DNA microarrays SNP assays [384]. Microarrays can capture large numbers of sequence variations between different samples. However, restrictions in microarray design and the number of probes on the microarrays limit the applications of this technology [220]. Olarte et al. [5] reported the microarray for *Salvia* species enriched with polymorphic and divergent DNA sequences. Array consisted of a total of 285-subtracted genomic DNA fragments obtained for fifteen *Salvia* genotypes. Developed microarray was capable of differentiating *S. officinalis* and *S. miltiorrhiza* from their closely related species and was also able to reveal genetic relationships consistent with geographical origins.

17 Next Generation Sequencing Based DNA Markers Used in *Salvia*

NGS-based DNA markers use sequence-based polymorphisms not as fragment polymorphisms as used in some traditional (conventional) DNA marker systems. As previously mentioned currently there exist approximately 20 NGS-based DNA marker systems, all may be referred as genotyping by sequencing (GBS). GBS technologies not only capable of high throughput massively paralleled genotyping of breeding population and natural population, but also genotype mutagenized population using TILLING, genome editing technologies such as Clustered Regularly Interspaced Short Palindromic Repeats (CRISPR/Cas9) [385] or identification of suitable mutations within populations. High throughput technologies including GBS and several RNA-seq, de novo transcriptome sequencing technologies have been applied to *Salvia* [12, 28, 348, 351–354, 386–388].

Using high-throughput sequencing Yan et al. [351] produced 12,959 cDNA clones and sequenced as single-pass sequencing from their 5′ ends in *Salvia miltiorrhiza*. A total of 10,288 ESTs (with sizes equal or greater than 100 bp) were selected and assembled into 1288 contigs, leaving 2937 singletons, for a total of 4225 unigenes. Based on the metabolic enzyme assignment, expression patterns of 14 secondary metabolic enzyme genes in different organs and under different treatments were verified using real-time PCR analysis. In addition, a total of 122 microsatellites were identified from the ESTs, with 89 having sufficient flanking sequences for primer design. This set of ESTs represents a significant proportion of the *S. miltiorrhiza* transcriptome and gives preliminary insights into the gene complement of *S. miltiorrhiza*.

Liu et al. [28] used the specific length amplified fragment sequencing (SLAF-seq), a NGS-based-GBS method in *S. miltiorrhiza* for large-scale single nucleotide polymorphisms discovery. The authors constructed a linkage map using two parents and their 96 F_1 individuals. Their final linkage map consisted of 5164 SLAFs in 8 linkage groups (LGs) and spanned 1516.43 cM, with an average distance of 0.29 cM between adjacent markers. The results of SLAF provided not only a platform for mapping quantitative trait loci but also offered a critical new tool for *S. miltiorrhiza* biotechnology and comparative genomics. The number of SNP markers obtained from this single GBS study is much larger than the previous DNA markers.

Several kinds of transcriptome sequencing platforms such as first generation Sanger sequencing and second generation 454 pyrosequencing and Illumina sequencing technologies have been used for discovering related genes involved in secondary metabolism pathways in *S. miltiorrhiza* [351, 353, 386, 389]. Two *S. miltiorrhiza* genotypes (BH18 and ZH23) with different phenolic acid concentrations were used for de novo RNA sequencing (RNA-seq) [12]. A total of 170,787 transcripts and 56,216 unigenes were obtained. In the same study, Song et al. [12] a total of 670 differentially expressed genes were identified between the two genotypes including genes involved in the phenylpropanoid biosynthesis pathway and the genes involved in the lignin biosynthesis pathway [12].

A transcriptome deep sequencing analysis of *S. miltiorrhiza* has been studied and 56,774 unigenes (average length of 467 bases) have been obtained. Among the transcriptome sequences, 34,340 unigenes were annotated and 2545 unigenes were assigned to specific pathways. A total of 1539 unigenes were identified as part of five major, secondary-metabolite pathways, covering almost all nodes in the phenylpropanoid and terpenoid pathways [353].

Ge et al. [382] performed de novo transcriptome sequencing on two different branching type plants using the Illumina paired-end sequencing technology. This study was clamed as the first *S. splendens* transcriptomic data generated from massively parallel sequencing through de novo assembly since majority of the NGS applied to *Salvia* was for *S. miltiorrhiza*. Ge et al. [382] also obtained SSRs using NGS data that might contribute to marker-assisted selection. The authors determined 2453 SSRs from transcriptome sequencing for *S. splendens* using the Illumina GA IIx sequencing platform. The number of cross-transferable SSR markers is very low in *Salvia*, NGS based detection of SSRs could enhance the development of cross-genera transferable SSR markers in *Salvia* [10].

18 Conclusions and Future Prospects

Morphological markers are available in *Salvia*, however, the number of such markers is limited [46, 49, 50]. A few types of proteins and isozymes (allozymes) exist and they have been used in characterization studies of several species of *Salvia* [3, 63, 73]. On the other hands, although a large numbers of secondary metabolites that could be used as biochemical markers exist in *Salvia* their dominant type and tissue specific expressions limit their utility as genetic markers [12, 22, 390]. The number of DNA marker technologies is well below in *Salvia* compared to other plant species [377, 391–393]. Low throughput DNA marker methods have been used for species identification in *Salvia* including chloroplast and mitochondrial DNA regions, and the internal transcribed spacer region of nuclear ribosomal DNA [4, 349, 392]. There is a still need additional and much powerful genetic markers in *Salvia* that could be very useful not only for the accurate identification of closely related species but for that could also be linked to desirable agricultural traits [394–396]. The number of species in the genus *Salvia* is very large and its taxonomy still needs revising studies [4, 31, 363]. Furthermore, application of DNA markers in *Salvia* breeding has not reached the desired level. Traditional molecular markers have limitations in throughput in comparison to array technologies. A few array technologies have been applied to *Salvia*, however, the requirements of array design and the number of probes on the arrays along with high cost limit their application [5, 220, 390].

Last decade witnessed the development of a large number of high throughput GBS genotyping methods, many of which vary in their costs per marker data point, in the types of data produced and in errors and potential biases. Researchers need to consider genome size and complexity, population size and structure before selecting suitable GBS systems, methods and software tools [397–400]. Currently two types of high throughput technologies, GBS and WGR, exist in genotyping studies. GBS is a cost-effective unique tool for genomics-assisted breeding in a range of plant species and it is suitable for *Salvia*. GBS combines marker discovery and genotyping of large populations in the absence of a reference genome sequence. Genotyping cost of GBS per individual is lowest in comparison to array-based and other NGS-based markers. However, GBS has limited genome coverage and some GBS technologies produce dominant type of markers [28, 314, 324]. Low coverage limitation of GBS could be improved with the use of third next generation sequencing (TGS) based technologies, however, currently many of the TGS technologies have higher error rate [197, 394, 395].

Due to its complex genomes with varying ploidy levels along with huge number of species distributed in several continents, the genus *Salvia* L. requires further detailed genetic studies. Functional and general genomic DNA markers based on NGS based GBS and WGR are required in conservation and maintenance studies of wild accessions, identification of agronomical important traits, and the transfer of these traits into elite sage varieties. In addition, more-efficient bioinformatic approaches to data storage and analysis are needed to keep up with the increasing speed of data production and analysis. In the future, it is obvious that agricultural genomics will switch from simple marker assisted selection to genomic selection based on next generation functional marker technologies.

Acknowledgements The authors thank Ph.D. candidates Emine Uygur Gocer and Adnan Aydin for their helps in reference preparation. Finally, we thank our teachers and researchers who contributed and touched our lives.

References

1. Topcu G (2006) Bioactive triterpenoids from *Salvia* species. J Nat Prod 69:482–487
2. Walker JB, Sytsma KJ, Treutlein J, Wink M (2004) *Salvia* (Lamiaceae) is not monophyletic: implications for the systematics, radiation, and ecological specializations of *Salvia* and tribe *Mentheae*. Am J Bot 91:1115–1125
3. Sudarmono S, Okado H (2007) Speciation process of *Salvia isensis* (Lamiaceae), a species endemic to serpentine areas in the Ise-Tokai district, Japan, from the viewpoint of the contradictory phylogenetic trees generated from chloroplast and nuclear DNA. J Plant Res 120:483–490
4. Karaca M, Ince AG, Ay ST, Turgut K, Onus AN (2008) PCR-RFLP and DAMD-PCR genotyping for *Salvia* species. J Sci Food Agric 88:2508–2516
5. Olarte A, Mantri N, Nugent G, Wohlmuth H, Li CG, Xue C, Pang E (2013) A gDNA microarray for genotyping *Salvia* species. Mol Biotech 54:770–783
6. Lopresti A (2016) *Salvia* (Sage): a review of its potential cognitive-enhancing and protective effects. Drugs R&D. https://doi.org/10.1007/s40268-016-0157-5
7. Hatipoglu SD, Zorlu N, Dirmenci T, Goren AC, Ozturk T, Topcu G (2016) Determination of volatile organic compounds in fourty five *Salvia* species by thermal desorption-GC-MS technique. Rec Natl Prod 10:659–700
8. Bernotiene G, Nivinskiene O, Butkiene R, Mockute D (2007) Essential oil composition variability in sage (*Salvia officinalis* L.). Chemija 18:38–43

9. Boszormenyi A, Hethelyi E, Farkas A, Horvath G, Papp N, Lemberkovics E, Szoke E (2009) Chemical and genetic relationships among sage (*Salvia officinalis* L.) cultivars and Judean sage (*Salvia judaica* Boiss.). J Agric Food Chem 57:4663–4667
10. Ince AG, Karaca M (2015) E-microsatellite markers for some naturally occurring *Salvia* species in the Mediterranean region. Turk J Biol 39:69–77
11. Cvetkovic I, Stefkov G, Karapandzova M, Kulevanova S, Satovic Z (2015) Essential oils and chemical diversity of Southeast European populations of *Salvia officinalis* L. Chem Biodivers 12:1025–1039
12. Song Z, Guo L, Liu T, Lin C, Wang J, Li X (2017) Comparative RNA-sequence transcriptome analysis of phenolic acid metabolism in *Salvia miltiorrhiza*, a traditional Chinese medicine model plant. Int J Genom 2017:9364594
13. Pammi S, Schertz K, Xu G, Hart G, Mullet JE (1994) Random amplified-polymorphic DNA markers in sorghum. Theor Appl Genet 89:80–88
14. Karaca M, Ince AG (2008) Minisatellites as DNA markers to classify bermudagrasses (*Cynodon* spp.): confirmation of minisatellite in amplified products. J Genet 87:83–86
15. Gocer EU, Karaca M (2016) Genetic characterization of some commercial cotton varieties using Td-DAMD-PCR markers. J Sci Eng Res 3:487–494
16. Utomo HS, Wenefrida I, Linscombe SD (2012) Progression of DNA marker and the next generation of crop development. In: Goyal A (ed) Crop Plant. https://doi.org/10.5772/29674
17. Jain A, Bhatia S, Banga S, Prakash S (1994) Potential use of random amplified polymorphic DNA (RAPD) technique to study the genetic diversity in Indian mustard (*Brassica juncea*) and its relationship to heterosis. Theor Appl Genet 88:116–122
18. Joshi SP, Gupta VS, Aggarwal RK, Ranjekar PK, Brar DS (2000) Genetic diversity and phylogenetic relationship as revealed by inter simple sequence repeat (ISSR) polymorphism in the genus *Oryza*. Theor Appl Genet 100:1311–1320
19. de Vicente MC, Fulton T (2003) Using molecular marker technology in studies on plant genetic diversity. Online. IPGRI, Rome, and Institute for Genetic Diversity, Ithaca
20. Ince AG, Karaca M, Aydin A (2011) Comparison of multiple DNA alignment algorithms for Labiatae molecular phylogeny inferences. Planta Med 77:1280
21. Karaca M, Ince AG, Aydin A, Ay ST (2013) Cross-genera transferable e-microsatellite markers for 12 genera of the Lamiaceae family. J Sci Food Agric 93:1869–1879
22. Borrelli GM, Trono D (2016) Molecular approaches to genetically improve the accumulation of health-promoting secondary metabolites in staple crops-a case study: the lipoxygenase-B1 genes and regulation of the carotenoid content in pasta products. Int J Mol Sci 17:1177–1210
23. Wang H, Hao N, Chen L, Li G (2016) Development of intron polymorphism markers in major latex-like protein gene for locality-level and cultivar identification of *Salvia miltiorrhiza*. SpringerPlus 5:1919
24. Anderson MW, Schrijver I (2010) Next generation DNA sequencing and the future of genomic medicine. Genes 1:38–69
25. Barabaschia D, Tondellia A, Desiderioa F, Volanteb A, Vaccinoc P, Valeb G, Cattivellia L (2016) Next generation breeding. Plant Sci 242:3–13
26. Bhat JA, Ali S, Salgotra RK, Mir ZA, Dutta S, Jadon V, Tyagi A et al (2016) Genomic selection in the era of next generation sequencing for complex traits in plant breeding. Front Genet 7:221
27. Robin JD, Ludlow AT, LaRanger R, Wright WE, Shay WJ (2016) Comparison of DNA quantification methods for next generation sequencing. Sci Rep 6:24067
28. Liu T, Guo Y, Pan Y, Zhao Q, Wang J, Song Z (2016) Construction of the first high density genetic linkage map of *Salvia miltiorrhiza* using specific length amplified fragment (SLAF) sequencing. Sci Rep 6:24070
29. Wenzl P, Carling J, Kudrna D, Jaccoud D, Huttner E, Kleinhofs A, Kilian A (2004) Diversity arrays technology (DArT) for whole-genome profiling of barley. Proc Nat Acad Sci U S A 101:9915–9920
30. Aydin A, Karaca M (2016) An alternative way to assign the chromosome location of genes/markers in cotton. FEBS J 283:335

31. Will M, Clasen-Bockhoff R (2017) Time to split *Salvia* (Lamiaceae)-new insights from old world *Salvia* phylogeny. Mol Phylogenet Evol 109:33–58
32. Hamona P, Grover CE, Davis AP, Rakotomalala JJ, Raharimalala NE, Albert VA et al (2017) Genotyping-by-sequencing provides the first well-resolved phylogeny for coffee (*Coffea*) and insights into the evolution of caffeine content in its species GBS coffee phylogeny and the evolution of caffeine content. Mol Phylogenet Evol 109:351–361
33. Konieczny A, Ausubel FM (1993) A procedure for mapping *Arabidopsis* mutations using co-dominant ecotype-specific PCR-based markers. Plant J 4:403–410
34. Yu J, Gu WK, Provvidenti R, Weeden NF (1995) Identifying and mapping two DNA markers linked to the gene conferring resistance to pea enation mosaic virus. J Am Soc Hort Sci 120:730–733
35. Liu ZJ, Cordes JF (2004) DNA marker technologies and their applications in aquaculture genetics. Aquaculture 238:1–37
36. Gupta PK, Rustgi S, Mir RR (2008) Array-based high-throughput DNA markers for crop improvement. Heredity 101:5–18
37. Chutimanitsakun Y, Nipper RW, Cuesta-Marcos A, Cistue L, Corey A, Filichkina T, Johnson EA, Hayes PM (2011) Construction and application for QTL analysis of a restriction site associated DNA (RAD) linkage map in barley. BMC Genomics 12:4
38. Shu Z, Wang Z, Mu X, Liang Z, Guo H (2012) A dominant gene for male sterility in *Salvia miltiorrhiza* Bunge. PLoS ONE 7:e50903
39. Truong HT, Ramos AM, Yalcin F, de Ruiter M, van der Poel HJA et al (2012) Sequence-based genotyping for marker discovery and co-dominant scoring in germplasm and populations. PLoS ONE 7(5): e37565. https://doi.org/10.1371/journal.pone.0037565
40. Sonah H, Bastien M, Iquira E, Tardivel A, Legare G, Boyle B et al (2013) An improved genotyping by sequencing (GBS) approach offering increased versatility and efficiency of SNP discovery and genotyping. PLoS ONE 8:e54603
41. Bayer PE, Ruperao P, Mason AS, Stiller J, Chan CK, Hayashi S et al (2015) High-resolution skim genotyping by sequencing reveals the distribution of crossovers and gene conversions in *Cicer arietinum* and *Brassica napus*. Theor Appl Genet 128:1039–1047
42. Chen J, Wang N, Fang LC, Liang ZC, Li SH, Wu BH (2015) Construction of a high-density genetic map and QTLs mapping for sugars and acids in grape berries. BMC Plant Biol 15:28
43. Lateef DD (2015) DNA marker technologies in plants and applications for crop improvements. J Biosci Med 3:7–18
44. Kim C, Guo H, Kong W, Chandnani R, Shuang LS, Paterson AH (2016) Application of genotyping by sequencing technology to a variety of crop breeding programs. Plant Sci 242:14–22
45. Robledo D, Palaiokostas C, Bargelloni L, Martinez P, Houston R (2017) Applications of genotyping by sequencing in aquaculture breeding and genetics. Rev Aquac. https://doi.org/10.1111/raq.12193
46. Corsi G, Bottega S (1999) Glandular hairs of *Salvia officinalis*: new data on morphology, localization and histochemistry in relation to function. Ann Bot 84:657–664
47. Kaya A, Demirci B, Başer KHC (2003) Glandular trichomes and essential oil of *Salvia glutinosa* L. South Afr J Bot 69:422–427
48. Avato P, Fortunato IM, Ruta C, D'Elia R (2005) Glandular hairs and essential oils in micropropageted plants of *Salvia officinalis* L. Plant Sci 169:29–36
49. Baran P, Ozdemir C (2006) The morphological and anatomical characters of *Salvia napifolia* Jacq. in Turkey. Bangladesh J Bot 35:77–84
50. Kaya A, Goger F, Başer KHC (2007) Morphological, anatomical and palynological characteristics of *Salvia halophia* endemic to Turkey. Nord J Bot 25:351–358
51. Dajic-Stevanovic Z, Sostaric I, Marin PD, Stojanovic D, Ristic M (2008) Population variability in *Thymus glabrescens* Willd. from Serbia: morphology, anatomy and essential oil composition. Arch Biol Sci 60:475–483
52. Peng L, Rub M, Wang B, Wang Y, Lia B, Yu J, Liang L (2014) Genetic diversity assessment of a germplasm collection of *Salvia miltiorrhiza* Bunge. based on morphology, ISSR and SRAP markers. Biochem Syst Ecol 55:84–92

53. Ince AG, Karaca M, Elmasulu SY (2014) New microsatellite and CAPS-microsatellite markers for clarifying taxonomic and phylogenetic relationships within *Origanum* L. Mol Breed 34:643–654
54. Karaca M, Ince AG, Aydin A, Elmasulu SY, Turgut K (2015) Microsatellites for genetic and taxonomic research in thyme (*Thymus* L.). Turk J Biol 39:1406–1420
55. Karaca M, Ince AG, Elmasulu SY, Onus AN, Turgut K (2005) Coisolation of genomic and organelle DNAs from 15 genera and 31 species of plants. Anal Biochem 343:353–355
56. Rzepa J, Wojtal T, Staszek D, Grygierczyk G, Labe K, Hajnos M, Kowalska T, Waksmundzka-Hajnos M (2009) Fingerprint of selected *Salvia* species by HS–GC–MS analysis of their volatile fraction. J Chromatogr Sci 47:575–580
57. Elansary HO, Mahmoud EA (2015) Basil cultivar identification using chemotyping still favored over genotyping using core barcodes and possible resources of antioxidants. J Essent Oil Res 27:82–87
58. Maksimovic M, Vidic D, Milos M, Solic ME, Abadzic S, Siljak-Yakovlev S (2007) Effect of the environmental conditions on essential oil profile in two dinaric *Salvia* species: *S. brachyodon* Vandas and *S. officinalis* L. Biochem Syst Ecol 35:473–478
59. Lakusic B, Ristic M, Slavkovska V, Stojanovic D, Lakusic D (2013) Variations in essential oil yields and compositions of *Salvia officinalis* (Lamiaceae) at different developmental stages. Bot Serbica 37:127–139
60. Hanlidou E, Karousoua R, Lazari D (2014) Essential oil diversity of *Salvia tomentosa* Mill. in Greece. Chem Biodivers 11:1205–1215
61. Kharazian N (2014) Chemotaxonomy and flavonoid diversity of *Salvia* L. (Lamiaceae) in Iran. Acta Bot Brasilica 28:281–292
62. Liu Z, Liu Y, Liu C, Song Z, Li Q, Zha Q, Lu C, Wang C, Ning Z, Zhang Y, Tian C, Lu A (2013) The chemotaxonomic classification of *Rhodiola* plants and its correlation with morphological characteristics and genetic taxonomy. Chem Cent J 7:118
63. Hadacova V, Ondrej M (1972) Isoenzymy. Biologicke Listy 37:1–25
64. Bretting PK, Widrlechner MP (1995) Genetic markers and horticultural germplasm management. HortScience 30:1349–1356
65. Staub JE, Serquen FC, Gupta M (1996) Genetic markers, map construction, and their application in plant breed. HortScience 31:729–741
66. Clevenger J, Chavarro C, Pearl SA, Ozias-Akins P, Jackson SA (2015) Single nucleotide polymorphism identification in polyploids: a review, example, and recommendations. Mol Plant 8:831–846
67. Fua D, Masonb AS, Xiaoa M, Yanc H (2016) Effects of genome structure variation, homeologous genes and repetitive DNA on polyploid crop research in the age of genomics. Plant Sci 242:37–46
68. Tanksley SD, Orton TJ (1983) Isozymes in plant genetic and breeding. (Developments in plant genetics and breeding, 1). Elsevier, Amsterdam
69. Hamrick JL, Godt JW (1990) Allozyme diversity in plant species. In: Brown AHD, Clegg MT, Kahler AL, Weir BS (eds.) Plant population genetics, breeding and genetic resources, Sinauer, Sunderland, pp 43–63
70. Kephart SR (1990) Starch gel electrophoresis of plant isozymes: a comparative analysis of techniques. Am J Bot 77:693–712
71. Rabilloud T (2002) Two-dimensional gel electrophoresis in proteomics: old, old fashioned, but it still climbs up the mountains. Proteomics 2:3–10
72. Muller-Starck G (1998) Isozymes. In: Karp A, Isaac PG, Ingram D (eds) Molecular tools for screening biodiversity: plants and animals. Chapmann Hall, London, pp 75–78
73. Sudarmono S, Okada H (2008) Genetic differentiations among the populations of *Salvia japonica* (Lamiaceae) and its related species. Hayati J Biosci 15:18–26
74. Bostein D, White RL, Skolnick M, Davis RW (1980) Construction of a genetic linkage map in man using restriction fragment length polymorphisms. Am J Hum Genet 32:314–333
75. Jeffreys AJ, Wilson V, Thein SL (1985) Hypervariable “minisatellite” regions in human DNA. Nature 314:67–73

76. Karaca M (2001) Characterization of *Cynodon* spp. and *Gossypium* spp. genomes using molecular and cytological techniques (PhD dissertation). Mississippi State University, Mississippi State
77. Jonah PM, Bello LL, Lucky O, Midau A, Moruppa SM (2011) Review: the importance of molecular markers in plant breeding programs. Glob J Sci Front Res 11:0975–5896
78. Ince AG, Karaca M (2011) Genetic variation in common bean landraces efficiently revealed by Td-DAMD-PCR markers. Plant Omics J 4:220–227
79. Zhang Y, Deng K, Xie L, Li X, Wang B, Chen L (2009) Genetic diversity analysis and conservation of the Chinese herb *Salvia miltiorrhiza* collected from different geographic origins in China. Afr J Biotechnol 8:4849–4855
80. Al-Gharaibeh MM, Hamasha HR, Rosche C, Lachmuth S, Wesche K, Hensen I (2011) Environmental gradients shape the genetic structure of two medicinal *Salvia* species in Jordan. Plant Biosyst 145:274–277
81. Ince AG, Karaca M (2012) Species-specific touch-down DAMD PCR markers for *Salvia* species. J Med Plants Res 6:1590–1595
82. Saebnazar A, Rahmani F (2013) Genetic variation among *Salvia* species based on sequence-related amplified polymorphism (SRAP) marker. J Plant Physiol Breed 3:71–78
83. Will M, Clasen-Bockhoff R (2014) Why Africa matters: evolution of old world *Salvia* L. (Lamiaceae) in Africa. Ann Bot Lond 114:61–83
84. Will M, Schmalz N, Classen Bockhoff R (2015) Towards a new classification of *Salvia* s.l.: (re)establishing the genus Pleudia Raf. Turk J Bot 39:693–707
85. Erbano M, Schuhli GSE, Santos EPD (2015) Genetic variability and population structure of *Salvia lachnostachys*: implications for breeding and conservation programs. Int J Mol Sci 16:7839–7850
86. van Steenbergen TJM, Collomstt SD, Hermans PWM, De Graaff J, Plasterk RHA (1995) Genomic DNA fingerprinting by restriction fragment end labeling. Proc Natl Acad Sci U S A 92:5572–5576
87. Powell W, Morgante M, Andre C, Hanafey M, Vogel J, Tingey S, Rafalski A (1996) The comparison of RFLP, RAPD, AFLP and SSR (microsatellite) markers for germplasm analysis. Mol Breed 2:225–238
88. Neff MM, Neff JD, Chory J, Pepper AE (1998) dCAPS, a simple technique for the genetic analysis of single nucleotide polymorphisms: experimental applications in *Arabidopsis thaliana* genetics. Plant J 14:387–392
89. Gocer EU, Karaca M (2016) Cleaved minisatellite locus (CML) markers for fingerprinting of cotton cultivars grown in Turkey. FEBS J 283:334
90. Kojima T, Nagaoka T, Ogihara Y (1998) Genetic linkage map of ISSR and RAPD markers in Einkorn wheat in relation to that of the RFLP markers. Theor Appl Genet 96:37–45
91. Hruba M (2007) dCAPS method: advantages, troubles and solution. Plant Soil Environ 53:417–420
92. Ince AG, Karaca M, Onus AN (2010) CAPS-microsatellites: use of CAPS method to convert non-polymorphic microsatellites into useful markers. Mol Breed 25:491–499
93. Karaca M, Ince AG (2011) New non-redundant microsatellite and CAPS-microsatellite markers for cotton (*Gossypium* L.). Turk J Field Crops 16:172–178
94. Papa S, Bacu A (2016) RFLP analysis of cpDNA of *Salvia officinalis* L. of Northern Albania can serve to elucidate genetic diversity among close natural populations. Eur J Biotechnol Genet Eng 3:65–74
95. Southern EM (1975) Detection of specific sequences among DNA fragments separated by gel electrophoresis. J Mol Biol 98:503–517
96. Saiki RK, Gelfand DH, Stoffel S, Scharf SJ, Higuchi R, Horn GT, Mullis KB, Erlich HA (1988) Primer directed enzymatic amplification of DNA with a thermostable DNA polymerase. Science 239:487–491
97. de Vienne D, Santoni S, Falque M (2003) Principal sources of molecular markers. In: Vienne DD (eds) Molecular markers in plant genetics and biotechnology. Science Publishers Inc., Plymouth, pp 3–41

98. Williams MNV, Pande N, Nair M, Mohan M, Bennet J (1991) Restriction fragment length polymorphism analysis of polymerase chain reaction products amplified from mapped loci of rice (*Oryza sativa* L.) genomic DNA. Theor Appl Genet 82:489–498
99. Maeda M, Uryu N, Murayama N, Ishii H, Ota M, Tsuji K, Inoko H (1990) A simple and rapid method for HLA-DP genotying by digestion of PCR-amplified DNA with allele specific restriction endonucleases. Hum Immunol 27:111–121
100. Ince AG, Karaca M (2011) Early determination of sex in jojoba plant by CAPS assay J Agric Sci 149:327–336
101. Heath DD, Iwama GK, Devlin RH (1993) PCR primed with VNTR core sequence yields species specific patterns and hypervariable probes. Nucl Acids Res 21:5782–5785
102. Jones CJ, Edwards KJ, Castaglione S, Winfield MO, Sala F et al (1997) Reproducibility testing of RAPD, AFLP and SSR markers in plants by a network of European laboratories. Mol Breed 3:381–390
103. Ince AG, Karaca M (2015) Td-DAMD-PCR assays for fingerprinting of commercial carnations. Turk J Bot 39:147–159
104. Williams JGK, Kubelik AR, Livak KJ, Rafalski JA, Tingey SV (1990) DNA polymorphisms amplified by arbitrary primers are useful as genetic markers. Nucl Acids Res 18:6531–6535
105. Caetano-Anolles G (1994) MAAP—a versatile and universal tool for genome analysis. Plant Mol Biol 25:1011–1026
106. Li G, Park Y-J (2012) SCAR Markers for discriminating species of two genera of medicinal plants, *Liriope* and *Ophiopogon*. Genet Mol Res 11:2987–2995
107. Jogaiah S, Sharathchandra RG, Raj N, Vedamurthy AB, Shetty HS (2014) Development of SCAR marker associated with downy mildew disease resistance in pearl millet (*Pennisetum glaucum* L.). Mol Biol Rep 41:7815–7824
108. Bhagyawant SS (2016) RAPD-SCAR markers: an interface tool for authentication of traits. J Biosci Med 4:1–9
109. Caetano-Anolles G, Bassam BJ, Gresshoff PM (1991) High resolution DNA amplification fingerprinting using very short arbitrary oligonucleotide primers. Nat Biotechnol 9:553–557
110. Welsh J, McClelland M (1990) Fingerprinting genomes using PCR with arbitrary primers. Nucl Acids Res 18:7213–7218
111. Ince AG, Karaca M, Onus AN (2010) A reliable gender diagnostic PCR assay for jojoba (*Simmondsia chinensis* (Link) Schneider). Genet Res Crop Evol 57:773–779
112. Micheli MR, Bova R, Calissano P, Dambrosio E (1993) Randomly amplified polymorphic DNA fingerprinting using combinations of oligonucleotide primers. Biotechniques 15: 388–390
113. Caetano-Anolles G, Bassam BJ, Gresshoff PM (1993) Enhanced detection of polymorphic DNA by multiple arbitrary amplicons profiling of endonuclease digested DNA: identification of markers linked to the super nodulation locus in soybean. Mol Gen Genet 241:57–64
114. Huff DR, Peakall R, Smouse PE (1993) RAPD variation within and among natural populations of outcrossing buffalograss (*Buchloe dactyloides* (Nutt.) Engelm. Theor Appl Genet 86:927–934
115. Vejl P (1997) Identification of genotypes in hop (*Humulus lupulus* L.) by RAPD analysis using program Gel Manager for Windows. Rostlinna Vyroba 43:325–331
116. Hollingsworth WO, Christie CB, Nichols MA, Neilson HF (1998) Detection of variation among and within asparagus hybrids using random amplified DNA (RAPD) markers. N Z J Crop Hort Sci 26:1–9
117. Corley Smith GE, Lim CJ, Kalmar GB, Brandhorst BP (1997) Efficient detection of DNA polymorphisms by fluorescent RAPD analysis. Biotechniques 22:690–692
118. Weller JW, Reddy A (1997) Fluorescent detection and analysis of RAPD amplicons using the ABI PRISM DNA sequencers. In: Micheli MR, Bova R (eds) Fingerprinting methods based on arbitrarily primed PCR. Springer Lab Manual, Berlin, pp 81–92
119. Michelmore RW, Paran I, Kesseli RV (1991) Identification of markers linked to disease resistance genes by bulked segregant analysis: a rapid method to detect markers in specific genomic regions using segregating populations. Proc Nat Acad Sci U S A 88:9828–9832

120. Martin GB, Williams JGK, Tanksley SD (1991) Rapid identification of markers linked to a *Pseudomonas* resistance gene in tomato by using random primers and near-isogenic lines. Proc Nat Acad Sci U S A 88:2336–2340
121. Paran I, Michelmore RW (1993) Development of reliable PCR based markers linked to downy mildew resistance genes in lettuce. Theor Appl Genet 85:985–993
122. Alonso C, Perez R, Bazaga P, Medrano M, Herrera CM (2016) MSAP markers and global cytosine methylation in plants: a literature survey and comparative analysis for a wild-growing species. Mol Ecol Resour 16:80–90
123. Vos P, Hogers R, Bleeker M, Reijans M, van de Lee T, Hornes M et al (1995) AFLP: a new concept for DNA fingerprinting. Nucl Acids Res 21:4407–4414
124. Li A, Hu BQ, Xue ZY, Chen L, Wang WX, Song WQ, Chen CB, Wa CG (2011) DNA methylation in genomes of several annual herbaceous and woody perennial plants of varying ploidy as detected by MSAP. Plant Mol Biol Rep 29:784–793
125. Zhang Y, Guo L, Shu Z, Sun Y, Chen Y, Liang Z, Guo H (2013) Identification of amplified fragment length polymorphism (AFLP) markers tightly associated with drought stress gene in male sterile and fertile *Salvia miltiorrhiza* Bunge. Int J Mol Sci 14:6518–6528
126. Velasco-Ramirez AP, Torres-Moran MI, Molina-Moret S, Sanchez-Gonzalez JJ, Santacruz-Ruvalcaba F (2014) Efficiency of RAPD, ISSR, AFLP and ISTR markers for the detection of polymorphisms and genetic relationships in camote de cerro (*Dioscorea* spp.). Elec J Biotech 17:65–71
127. Russell RJ, Fuller JD, Macaulay M, Hats BG, Jahoor A, Powell W, Waugh R (1997) Direct comparison of levels of genetic variation among barley accessions detected by RFLPs, AFLPs, SSRs and RAPDs. Theor Appl Genet 95:714–722
128. Alonso-Blanco C, Peeters AJM, Koornneef M, Lister C, Dean C, van den Bosch N, Pot J, Kuiper MTR (1998) Development of an AFLP based linkage map of Ler, Col and Cvi *Arabidopsis thaliana* ecotypes and construction of a Ler/Cvi recombinant inbred line population. Plant J Cell Mol Biol 14:259–271
129. Young WP, Schupp JM, Keim P (1999) DNA methylation and AFLP marker distribution in the soybean genome. Theor Appl Genet 99:785–790
130. Saal B, Wricke G (2002) Clustering of amplified fragment length polymorphism markers in a linkage map of rye. Plant Breed 121:117–123
131. Akbari M, Wenzl P, Caig V, Carling J, Xia L, Yang S, Uszynski G et al (2006) Diversity arrays technology (DArT) for high- throughput profiling of the hexaploid wheat genome. Theor Appl Genet 113:1409–1420
132. Yang D, Ma P, Liang X, Liang Z, Zhang M, Shen S, Liu H, Liu Y (2012) Metabolic profiles and cDNA-AFLP analysis of *Salvia miltiorrhiza* and *Salvia castanea* Diel f. *tomentosa* Stib. PLoS ONE 7:e29678
133. Olsen M, Hood L, Cantor C, Botstein D (1989) A common language for physical mapping of human genome. Science 245:1434–1435
134. Blake TK, Kadyrzhanova D, Shepherd KW, Islam AKMR, Langridge PL et al (1996) STSPCR markers appropriate for wheat-barley introgression. Theor Appl Genet 82:715–721
135. Naik S, Gill KS, Prakasa Rao VS, Gupta VS, Tamhankar SA, Pujar S, Gill BS, Ranjekar PK (1998) Identification of a STS marker linked to the *Aegilops speltoides* derived leaf rust resistance gene *Lr28* in wheat. Theor Appl Genet 97:535–540
136. Reamon-Buttner SM, Jung C (2000) AFLP derived STS markers for the identification of sex in *Asparagus officinalis* L. Theor Appl Genet 100:432–438
137. Waugh R, Bonar N, Baird E, Thomas B, Graner A, Hayes P, Powell W (1997) Homology of AFLP products in three mapping populations of barley. Mol Gen Genet 255:311321
138. van der Broeck D, Maes T, Sauer M, Zethoff J, de Keukeleire P, Hauw MD, van Montagu M, Gerats T (1998) Transposon display identifies individual transposable elements in high copy number lines. Plant J 13:121–129
139. Van der Wurff AWG, Chan YL, van Straalen NM, Schouten J (2000) TE-AFLP: combining rapidity and robustness in DNA fingerprinting. Nucl Acids Res 28:e105

140. Litt M, Luty JA (1989) A hypervariable microsatellite revealed by in vitro amplification of a dinucleotide repeat within the cardiac muscle actin gene. Am J Hum Genet 44:397–401
141. Vartia S, Villanueva-Canas JL, Finarelli J, Farrell ED, Collins PC et al (2016) A novel method of microsatellite genotyping-by-sequencing using individual combinatorial barcoding. Royal Soc Open Sci 3:150565
142. Beckmann JS, Soller M (1990) Toward a unified approach to genetic mapping of eukaryotes based on sequence tagged microsatellite sites. Biotech 8:930–932
143. Senan S, Kizhakayil D, Sasikumar B, Sheeja TE (2014) Methods for development of microsatellite markers. Not Sci Biol 6:1
144. Gupta PK, Varshney RK, Sharma PC, Ramesh B (1999) Molecular markers and their applications in wheat breeding. Plant Breed 118:369–390
145. Matsuoka Y, Vigouroux Y, Goodman MM, Sanchez GJ, Buckler E, Doebley J (2002) A single domestication for maize shown by multilocus microsatellite genotyping. Proc Nat Acad Sci U S A 99:6080–6084
146. Squirrell J, Hollingsworth PM, Woodhead M, Russell J, Lowe AJ, Gibby M, Powell W (2003) How much effort is required to isolate nuclear microsatellites from plants? Mol Ecol 12:1339–1348
147. Eisen JA (1999) Mechanistic basis for microsatellilte instability. In: Goldstein DB, Schlotterer C (eds) Microsatellites: evolution and applications. Oxford University Press, Oxford, pp 34–48
148. Ghislain M, Spooner DM, Rodriguez F, Villamon F, Nunez J, Vasquez C, Waugh R, Bonierbale M (2004) Selection of highly informative and user-friendly microsatellites (SSRs) for genotyping of cultivated potato. Theor Appl Genet 108:881–890
149. Mitchell SE, Kresovich S, Jester CA, Hernandez CJ, Szewc-McFadden AK (1997) Application of multiplex-PCR and fluorescence-based, semi-automated allele sizing technology for genotyping plant genetic resources. Crop Sci 37:617–624
150. Ribaut JM, Jiang C, Gonzalez-de-Leon D, Edmeades GO, Hoisington DA (1997) Identification of quantitative trait loci under drought conditions in tropical maize. II. Yield components and marker assisted selection strategies. Theor Appl Genet 94:887–896
151. Gupta PK, Varshney RK (2000) The development and use of microsatellite markers for genetic analysis and plant breeding with emphasis on bread wheat. Euphytica 113:163–185
152. Ince AG, Karaca M, Bilgen M, Onus AN (2008) Digital differential display tools for mining microsatellite containing organism, organ and tissue. Plant Cell Tiss Organ Cult 94:281–290
153. Ince AG, Karaca M, Onus AN (2010) Polymorphic microsatellite markers transferable across *Capsicum* species. Plant Mol Biol Rep 28:285–291
154. Ince AG, Karaca M, Onus AN (2010) Differential expression patterns of genes containing microsatellites in *Capsicum annuum* L. Mol Breed 25:645–658
155. Polat E, Ince AG, Karaca M, Onus AN (2010) Mining and utilization of mushroom ESTs for microsatellites. Conserv Genet 11:1123–1126
156. Ince AG, Karaca M, Onus AN (2011) Exact microsatellite density differences among *Capsicum* tissues and development stages. J Agric Sci 17:291–299
157. Ince AG (2012) A contig-based microsatellite marker approach and its application in *Cichorium* ESTs. Rom Biotech Lett 17:7177–7186
158. Kota R, Varshney RK, Thiel T, Dehmer KJ, Graner A (2001) Generation and comparison of EST-derived SSRs and SNPs in barley (*Hordeum vulgare* L.). Hereditas 135:145–151
159. Kantety RV, La Rota M, Matthews DE, Sorrells ME (2002) Data mining for simple sequence repeats in expressed sequence tags from barley, maize, rice, sorghum and wheat. Plant Mol Biol 48:501–510
160. Varshney RK, Graner A, Sorrells ME (2005) Genic microsatellite markers in plants: features and applications. Trends Biotech 23:48–55
161. Zietkiewicz E, Rafalski A, Labuda D (1994) Genome fingerprinting by simple sequence repeat (SSR)-anchored polymerase chain reaction amplification. Genomics 20:176–183

162. Bornet B, Branchard M (2001) Nonanchored inter simple sequence repeat (ISSR) markers: reproducible and polymorphic tools for genome fingerprinting. Plant Mol Biol Rep 19: 209–215
163. Sarrou E, Ganopoulos I, Xanthopoulou A, Masuero D, Martens S, Madesis P, Mavromatis A, Chatzopoulou P (2017) Genetic diversity and metabolic profile of *Salvia officinalis* populations: implications for advanced breeding strategies. Planta. https://doi.org/10.1007/s00425-017-2666-z
164. Safaei M, Sheidai M, Alijanpoor B, Noormohammadi Z (2016) Species delimitation and genetic diversity analysis in *Salvia* with the use of ISSR molecular markers. Acta Bot Croat 75:45–52
165. Yousefiazarkhanian M, Asghari A, Ahmadi J, Asghari B, Jafari AA (2016) Genetic diversity of *Salvia* species assessed by ISSR and RAPD markers. Notulae Botanicae Horti Agrobotanici Cluj-Napoca 44:431–436
166. Caldeira RL, Carvalho OS, Lage RCG, Cardoso PCM, Oliveira GC (2002) Sequencing of simple sequence repeat anchored polymerase chain reaction amplification products of *Biomphalaria glabrata*. Mem Inst Oswaldo Cruz 97:23–26
167. Becker J, Heun M (1995) Barley microsatellites: alleles variation and mapping. Plant Mol Biol 27:835–845
168. Witsenboer H, Vogel J, Michelmore RW (1997) Identification, genetic localization, and allelic diversity of selectively amplified microsatellite polymorphic loci in lettuce and wild relatives (*Lactuca* spp.). Genome 40:923–936
169. Heikrujam M, Kumar J, Agrawal V (2015) Genetic diversity analysis among male and female Jojoba genotypes employing gene targeted molecular markers, start codon targeted (SCoT) polymorphism and CAAT box-derived polymorphism (CBDP) markers. Meta Gene 16:90–97
170. Collard BCY, Mackill DJ (2009) Conserved DNA-derived polymorphism (CDDP): a simple and novel method for generating DNA markers in plants. Plant Mol Biol Report 27:558–562
171. Singh AK, Rana MK, Singh S, Kumar S, Kumar R, Singh R (2014) CAAT box-derived polymorphism (CBDP): a novel promoter-targeted molecular marker for plants. J Plant Biochem Biotech 23:175–183
172. Wu J, Wang Z, Shi Z, Zhang S, Ming R, Zhu S, Khan MA et al (2013) The genome of pear (*Pyrus bretschneideri* Rehd.). Genome Res 23:396–408
173. Li G, Quiros CF (2001) Sequence-related amplified polymorphism (SRAP), a new marker system based on a simple PCR reaction: its application to mapping and gene tagging in *Brassica*. Theor Appl Genet 103:455–461
174. Kumar LS (1999) DNA markers in plant improvement: an overview. Biotechnol Adv 17:143–182
175. Bargish TA, Rahmani F (2016) SRAP markers based genetic analysis of *Silene* species. J Trop Biol Conserv 13:57–70
176. Poczai P, Varga I, Laos M, Cseh A, Bell N, Valkonen JPT, Hyvönen J (2013) Advances in plant gene-targeted and functional markers: a review. Plant Methods 9:6
177. Song Z, Li X, Wang H, Wang H (2010) Genetic diversity and population structure of *Salvia miltiorrhiza* Bge in China revealed by ISSR and SRAP. Genetica 138:241–249
178. Aghaei Z, Talebi M, Rahimmalek M (2017) Assessment of genetic diversity within and among sage (*Salvia*) species using SRAP markers. Plant Genet Res 15:279–282
179. Hu J, Vick BA (2003) Target region amplification polymorphism: a novel marker technique for plant genotyping. Plant Mol Biol Reporter 21:289–294
180. Hatada I, Hayashizaki Y, Hirotsune S, Komatsubara H, Mukai T (1999) A genomic scanning method for higher organisms using restriction sites as landmarks. Proc Nat Acad Sci 88:9523–9527
181. Wang Q, Zhang B, Lu Q (2009) Conserved region amplification polymorphism (CoRAP), a novel marker technique for plant genotyping in *Salvia miltiorrhiza*. Plant Mol Biol Rep 27:139–143

182. Choi HK, Kim D, Uhm T, Limpens E, Lim H, Mun JH, Kalo P, Penmetsa RV, Seres A, Kulikova O, Roe BA, Bisseling T, Kiss GB, Cook DR (2004) A sequence-based genetic map of *Medicago trunculata* and comparison of marker colinearity with *M. sativa*. Genetics 166:1463–1502
183. Fulton TM, Van der Hoeven R, Eannetta NT, Tanksley SD (2002) Identification, analysis, and utilization of conserved ortholog set markers for comparative genomics in higher plants. Plant Cell 14:1457–1467
184. Minghetti PP, Dugaiczyk A (1993) The emergence of new DNA repeats and the divergence of primates. Proc Nat Acad Sci U S A 90:1872–1876
185. Shimamura M, Yasue H, Ohshima K, Abe H, Kato K, Kishiro T, Goto M, Munechika I, Okada N (1997) Molecular evidence from retroposons that whales form a clade within even-toed ungulates. Nature 388:666–670
186. Rohde W (1996) Inverse sequence-tagged repeat (ISTR) analysis: a novel and universal PCR-based technique for genome analysis in the plant and animal kingdom. J Genet Breed 50:249–261
187. Flavell AJ, Knox MR, Pearce SR, Ellis TH (1998) Retrotransposon-based insertion polymorphisms (RBIP) for high throughput marker analysis. Plant J 16:643–650
188. Kalendar R, Grob T, Regina M, Suoniemi A, Schulman AH (1999) IRAP and REMAP: two new retrotransposon-based DNA fingerprinting techniques. Theor Appl Genet 98:704–711
189. Agarwal M, Shrivastava N, Padh H (2008) Advances in molecular marker techniques and their applications in plant sciences. Plant Cell Rep 27:617–631
190. Kalendar R, Flavell AJ, Ellis THN, Sjakste T, Moisy C, Schulman AH (2011) Analysis of plant diversity with retrotransposon-based molecular markers. Heredity 106:520–530
191. Oldenburg MC, Siebert M (2000) New cleavase fragment length polymorphism method improves the mutation detection assay. Biotechniques 28:351–357
192. van Tassell CP, Smith TP, Matukumalli LK, Taylor JF, Schnabel RD, Lawley CT et al (2008) SNP discovery and allele frequency estimation by deep sequencing of reduced representation libraries. Nat Methods 5:247–252
193. Boutet G, Carvalho SA, Falque M, Peterlongo P, Lhuillier E, Bouchez O, Lavaud C, Pilet-Nayel ML, Rivière N, Baranger A (2016) SNP discovery and genetic mapping using genotyping by sequencing of whole genome genomic DNA from a pea RIL population. BMC Genomics 17:121
194. Miller S, Bartlett W, Chandrasekaran S, Simpson S, Edwards M, Booth IR (2003) Domain organization of the MscS mechanosensitive channel of *Escherichia coli*. EMBO J 22:36–46
195. Coles ND, Coleman CE, Christensen SA, Jellen EN, Stevens MR, Bonifacio A, Rojas-Beltran JA, Fairbanks DJ, Maughan PJ (2005) Development and use of an expressed sequenced tag library in quinoa (*Chenopodium quinoa* Willd.) for the discovery of single nucleotide polymorphisms. Plant Sci 168:439–447
196. Foster JT, Allan GJ, Chan AP, Rabinowicz PD et al (2010) Single nucleotide polymorphisms for assessing genetic diversity in castor bean (*Ricinus communis*). BMC Plant Biol 10:13
197. Zou XL, Shi C, Austin RS, Merico D, Munholland S, Marsolais F et al (2014) Genome-wide single nucleotide polymorphism and insertion-deletion discovery through next-generation sequencing of reduced representation libraries in common bean. Mol Breed 33:769–778
198. Jalili V, Matteucci M, Masseroli M, Ceri S (2016) Indexing next-generation sequencing data. Inf Sci 20:1–20
199. Hayashi K (1992) PCR-SSCP: a method for detection of mutations. Genet Anal Tech App 9:73–79
200. Orita M, Iwahana H, Hayashi K, Sekiya T (1989) Detection of polymorphisms of human DNA by gel electrophoresis as single-strand conformation polymorphisms. Proc Nat Acad Sci U S A 86:2766–2770
201. Yu B, Sawyer NA, Chiu C, Oefner PJ, Underhill PA (2006) DNA mutation detection using denaturing high-performance liquid chromatography (DHPLC). Curr Protoc Hum Genet. https://doi.org/10.1002/0471142905.hg0710s48

202. Campbell NR, Stephanie AH, Shawn RN (2015) Genotyping-in-thousands by sequencing (GT-seq): a costeffective SNP genotyping method based on custom amplicon sequencing. Mol Ecol Resour 15:855–867
203. Podini D, Vallone PM (2009) SNP genotyping using multiplex single base primer extension assays. Methods Mol Biol 578:379–391
204. Singhal N, Kumar M, Kanaujia PK, Virdi JS (2015) MALDI-TOF mass spectrometry: an emerging technology for microbial identification and diagnosis. Front Microbiol 6:791
205. Okayama H, Curiel DT, Brantly ML, Holmes MD, Crystal RG (1989) Rapid, nonradioactive detection of mutations in the human genome by allele-specific amplification. J Lab Clin Med 114:105–113
206. Tyagi S, Marras SA, Kramer FR (2000) Wavelength-shifting molecular beacons. Nat Biotechnol 18:1191–1196
207. Vet JA, van der Rijt BJ, Blom HJ (2002) Molecular beacons: colorful analysis of nucleic acids. Expert Rev Mol Diagn 2:77–86
208. Schlachter S, Chan K, Marras SAE, Parveen N (2017) Detection and differentiation of lyme spirochetes and other tick-borne pathogens from blood using real-time PCR with molecular beacons. Methods Mol Biol 1616:155–170
209. Holland PM, Abramson RD, Watson R, Gelfand DH (1991) Detection of specific polymerase chain reaction product by utilizing the 5′-3′ exonuclease activity of *Thermus aquaticus* DNA polymerase. Biochemistry 88:7276–7280
210. Suzuki N, Yoshida A, Nakano Y (2005) Quantitative analysis of multi species oral biofilms by TaqMan Real-Time PCR. Clin Med Res 3:176–185
211. Gibson NJ (2006) The use of real-time PCR methods in DNA sequence variation analysis. Clin Chim Acta 363:32–47
212. Bosmali I, Ganopoulos I, Madesis P, Tsaftaris A (2012) Microsatellite and DNA-barcode regions typing combined with High Resolution Melting (HRM) analysis for food forensic uses: a case study on lentils (*Lens culinaris*). Food Res Int 46:141–147
213. Tong YR, Jiang C, Huang LQ, Cui ZH, Yuan Y (2014) Molecular identification of Radix Notoginseng powder by DNA melt curve analysis. Chin J Pharm Anal 34:1384–1390
214. Osathanunkul M, Suwannapoom C, Osathanunkul K, Madesis P, de Boer H (2016) Evaluation of DNA barcoding coupled high resolution melting for discrimination of closely related species in phytopharmaceuticals. Phytomedicine 23:156–165
215. Bui MH, Stone GG, Nilius AM, Almer L, Flamm RK (2003) PCR-oligonucleotide ligation assay for detection of point mutations associated with quinolone resistance in *Streptococcus* pneumoniae. Antimicrob Agents Chemother 47:1456–1459
216. Deng JY, Zhang XE, Mang Y, Zhang ZP, Zhou YF, Liu Q, Lu HB, Fu ZJ (2004) Oligonucleotide ligation assay-based DNA chip for multiplex detection of single nucleotide polymorphism. Biosens Bioelectron 19:1277–1283
217. Olivier M (2005) The Invader assay for SNP genotyping. Mutat Res 3:103–110
218. Neelam K, Brown-Guedira G, Huang L (2013) Development and validation of a breeder friendly KASPar marker for wheat leaf rust resistance locus Lr21. Mol Breed 31:233–237
219. Zhao S, Li A, Li C, Xia H, Zhao C, Zhang Y, Hou L, Wang X (2017) Development and application of KASP marker for high throughput detection of AhFAD2 mutation in peanut. Electron J Biotechnol 25:9–12
220. Cui G, Huang L, Tang X, Zhao J (2011) Candidate genes involved in tanshinone biosynthesis in hairy roots of *Salvia miltiorrhiza* revealed by cDNA microarray. Mol Biol Rep 38:2471–2478
221. Srivastava A, Sawant SV, Jena SN (2015) Microarray-based large scale detection of single feature polymorphism in *Gossypium hirsutum* L. J Genet 94:669–676
222. Jaccoud D, Peng K, Feinstein D, Kilian A (2001) Diversity arrays: a solid state technology for sequence information independent genotyping. Nuc Acids Res 29:e25
223. Xia L, Peng K, Yang S, Wenzl P, Carmen M, De Vicente MC, Fregene M, Kilian A (2005) DArT for high-throughput genotyping of cassava (*Manihot esculenta*) and its wild relatives. Theor Appl Genet 110:1092–1098

224. Lu X, Xiao B, Li Y, Gui Y, Wang Y, Fan L (2013) Diversity arrays technology (DArT) for studying the genetic polymorphism of flue-cured tobacco (*Nicotiana tabacum*). J Zhejiang Univ Sci B 14:570–577
225. Sanchez-Sevilla JF, Horvath A, Botella MA, Gaston A, Folta K, Kilian A, Denoyes B, Amaya I (2015) Diversity arrays technology (DArT) marker platforms for diversity analysis and linkage mapping in a complex crop, the octoploid cultivated strawberry (*Fragaria* x *ananassa*). PLoS ONE 10(12): e0144960. https://doi.org/10.1371/journal.pone.0144960
226. Lezar S, Myburg AA, Berger DK, Wingfield MJ, Wingfield BD (2004) Development and assessment of microarray-based DNA fingerprinting in *Eucalyptus grandis*. Theor Appl Genet 109:1329–1336
227. Winzeler EA, Richards DR, Conway AR, Goldstein AL, Kalman S, McCullough MJ, McCusker JH, Stevens DA, Wodicka L, Lockhart DJ, Davis RW (1998) Direct allelic variation scanning of the yeast genome. Science 21:1194–1197
228. Saxena RK, Cui X, Thakur V, Walter B, Close TJ, Varshney RK (2011) Single feature polymorphisms (SFPs) for drought tolerance in pigeonpea (*Cajanus* spp.). Funct Integr Genomics 11:651–657
229. Hill TA, Ashrai H, Reyes-Chin-Wo S, Yao J, Stoffel K et al (2013) Characterization of *Capsicum annuum* genetic S. diversity and population structure based on parallel polymorphism discovery with a 30 k unigene pepper GeneChip. PLoS ONE 8(2): e56200. https://doi.org/10.1371/journal.pone.0056200
230. Baird NA, Etter PD, Atwood TS, Currey MC, Shiver AL, Lewis ZA, Selker EU, Cresko WA, Johnson EA (2008) Rapid SNP discovery and genetic mapping using sequenced RAD markers. PLoS ONE 3(10): e3376. https://doi.org/10.1371/journal.pone.0003376
231. Miller MR, Dunham JP, Amores A, Cresko WA, Johnson EA (2007) Rapid and cost-effective polymorphism identification and genotyping using restriction site associated DNA (RAD) markers. Genome Res 17:240–248
232. Pfender WF, Saha MC, Johnson EA, Slabaugh MB (2011) Mapping with RAD (restriction-site associated DNA) markers to rapidly identify QTL for stem rust resistance in *Lolium perenne*. Theor Appl Genet 122:1467–1480
233. Davey JW, Blaxter ML (2010) RADSeq: next-generation population genetics. Brief Funct Genomics 9:416–423
234. Wang S, Meyer E, McKay JK, Matz MV (2012) 2b-RAD: a simple and flexible method for genome-wide genotyping. Nat Methods 9:808–810
235. Peterson BK, Weber JN, Kay EH, Fisher HS, Hoekstra HE (2012) Double digest RADseq: an inexpensive method for de novo SNP discovery and genotyping in model and non-model species. PLoS ONE 7(5): e37135. https://doi.org/10.1371/journal.pone.0037135
236. Toonen RJ, Puritz JB, Forsman ZH, Whitney JL, Fernandez-Silva I, Andrews KR, Bird CE (2013) ezRAD: a simplified method for genomic genotyping in non-model organisms. PeerJ 1:e203
237. Sanger F, Nicklen S, Coulson AR (1977) DNA sequencing with chain-terminating inhibitors. Proc Natl Acad Sci U S A 74:5463–5467
238. Maxam AM, Gilbert W (1977) A new method for sequencing DNA. Proc Natl Acad Sci U S A 74:560–564
239. Shendure J, Ji H (2008) Next-generation DNA sequencing. Nat Biotechnol 26:1135–1145
240. Sims D, Sudbery I, Ilott NE, Heger A, Ponting CP (2014) Sequencing depth and coverage: key considerations in genomic analyses. Nat Rev Genet 15:121–132
241. Franca LTC, Carrilho E, Kist TBL (2002) A review of DNA sequencing techniques. Q Rev Biophys 35:169–200
242. Buermans HPJ, den Dunnen JT (2014) Next generation sequencing technology: advances and applications. Biochim Biophys Acta 1842:1932–1941
243. Van Dijk EL, Auger H, Jaszczyszyn Y, Thermes C (2014) Ten years of next-generation sequencing technology. Trends Genet 30:418–426
244. Bentley DR, Balasubramanian S, Swerdlow HP, Smith GP, Milton J et al (2008) Accurate whole human genome sequencing using reversible terminator chemistry. Nature 456:53–59

245. Valouev A, Ichikawa J, Tonthat T, Stuart J, Ranade S, Peckham H, Zeng K, Malek JA, Costa G, McKernan K, Sidow A, Fire A, Johnson SM (2008) A high-resolution, nucleosome position map of *C. elegans* reveals a lack of universal sequence-dictated positioning. Genome Res 18:1051–1063
246. Wang Z, Gerstein M, Snyder M (2009) RNA-Seq: a revolutionary tool for transcriptomics. Nat Rev Genet 10:57–63
247. Kozarewa I, Kozarewa I, Ning Z, Sanders MJ, Berriman M, Turner DJ (2009) Amplification-free Illumina sequencing-library preparation facilitates improved mapping and assembly of (G^+ C)-biased genomes. Nat Methods 6:291–295
248. Mascher M, Wu S, Amand PS, Stein N, Poland J (2013) Application of genotyping-by-sequencing on semiconductor sequencing platforms: a comparison of genetic and reference-based marker ordering in barley. PLoS ONE 8(10): e76925. https://doi.org/10.1371/journal.pone.0076925
249. Jiang Z, Zhou X, Li R, Michal JJ, Zhang S, Dodson MV, Zhang Z, Harland RM (2015) Whole transcriptome analysis with sequencing: methods, challenges and potential solutions. Cell Mol Life Sci 72:3425–3439
250. Goodwin S, Gurtowski J, Ethe-Sayers S, Deshpande P, Schatz MC, McCombie WR (2015) Oxford Nanopore sequencing, hybrid error correction, and de novo assembly of a eukaryotic genome. Genome Res 25:1750–1756
251. Korlach J, Marks PJ, Cicero RL, Gray JJ, Murphy DL, Roitman DB, Pham TT et al (2008) Selective aluminum passivation for targeted immobilization of single DNA polymerase molecules in zero-mode waveguide nanostructures. Proc Natl Acad Sci U S A 105:1176–1181
252. Rothberg JM, Hinz W, Rearick TM, Schultz J, Mileski W et al (2011) An integrated semiconductor device enabling non-optical genome sequencing. Nature 475:348–352
253. Quail MA, Smith M, Coupland P, Otto TD, Harris SR et al (2012) A tale of three next generation sequencing platforms: comparison of Ion Torrent, Pacific Biosciences and Illumina MiSeq sequencers. BMC Genomics 13:341
254. Loman NJ, Misra RV, Dallman TJ, Constantinidou C, Gharbia SE, Wain J, Pallen MJ (2012) Performance comparison of benchtop high-throughput sequencing platforms. Nat Biotechnol 30:434–439
255. Schneider GF, Dekker C (2012) DNA sequencing with nanopores. Nat Biotechnol 30:326–328
256. Bragg LM, Stone G, Butler MK, Hugenholtz P, Tyson GW (2013) Shining a light on dark sequencing: characterising errors in Ion Torrent PGM Data. PLoS Comput Biol 9:e1003031
257. Balasubramanian S (2015) Solexa sequencing: decoding genomes on a population scale. Clin Chem 61:21–24
258. Karlsson E, Larkeryd Forsman M, Stenberg P (2015) Scaffolding of a bacterial genome using MinION nanopore sequencing. Sci Rep 5:11996
259. Tawfik DS, Griffiths AD (1998) Man-made cell-like compartments for molecular evolution. Nat Biotechnol 16:652–656
260. Shendure J, Porreca GJ, Reppas NB, Lin X, McCutcheon JP, Rosenbaum AM, Wang MD, Zhang K, Mitra RD, Church GM (2005) Accurate multiplex polony sequencing of an evolved bacterial genome. Science 309:1728–1732
261. Yegnasubramanian S (2013) Preparation of fragment libraries for next-generation sequencing on the applied biosystems SOLiD platform. Methods Enzymol 529:185–200
262. Myhrvold C, Baym M, Hanikel N, Ong LL, Gootenberg JS, Yin P (2017) Barcode extension for analysis and reconstruction of structures. Nat Commun 8:14698
263. Shao K, Ding W, Wang F, Li H, Ma D, Wang H (2011) Emulsion PCR: a high efficient way of PCR amplification of random DNA libraries in aptamer selection. PLoS ONE 6(9):e24910
264. Khan SA (2005) Plasmid rolling-circle replication: highlights of two decades of research. Plasmid 53:126–136
265. Ali MM, Li F, Zhang Z, Zhang K, Kang DK, Ankrum JA, Le XC, Zhao W (2014) Rolling circle amplification: a versatile tool for chemical biology, materials science and medicine. Chem Soc Rev 43:3324–3341

266. Nakano M, Komatsu J, Matsuura S, Takashima K, Katsura S, Mizuno A (2003) Single-molecule PCR using water-in-oil emulsion. J Biotechnol 24:117–124
267. Head SR, Komori HK, LaMere SA, Whisenant T, van Nieuwerburgh F, Salomon DR, Ordoukhanian P (2014) Library construction for next-generation sequencing: overviews and challenges. BioTech 56:61–77
268. Williams R, Peisajovich SG, Miller OJ, Magdassi S, Tawfik DS, Griffiths AD (2006) Amplification of complex gene libraries by emulsion PCR. Nat Methods 3:545–550
269. Fedurco M, Romieu A, Williams S, Lawrence I, Turcatti G (2006) BTA, a novel reagent for DNA attachment on glass and efficient generation of solid-phase amplified DNA colonies. Nucl Acids Res 34:e22
270. Voelkerding KV, Dames SA, Durtschi JD (2009) Next-generation sequencing: from basic research to diagnostics. Clin Chem 55:641–658
271. Miles TD, Martin FN, Coffey MD (2015) Development of rapid isothermal amplification assays for detection of *Phytophthora* spp. in plant tissue. Techniques 105:265–278
272. Ju J, Kim DH, Bi L, Meng Q, Bai X, Li Z et al (2006) Four-color DNA sequencing by synthesis using cleavable fluorescent nucleotide reversible terminators. Proc Natl Acad Sci U S A 103:19635–19640
273. Bell DC, Thomas WK, Murtagh KM, Dionne CA, Graham AC, Anderson JE, Glover WR (2012) DNA base identification by electron microscopy. Microsc Microanal 18:1049–1053
274. Mankos M, Shadman K, Persson HH, N'Diaye AT, Schmid AK, Davis RW (2014) A novel low energy electron microscope for DNA sequencing and surface analysis. Ultramicroscopy 145:36–49
275. Wall JD, Tang LF, Zerbe B, Kvale MN, Kwok PY, Schaefer C, Risch N (2014) Estimating genotype error rates from high-coverage next-generation sequence data. Genome Res 24:1734–1739
276. Luo C, Tsementzi D, Kyrpides N, Read T, Konstantinidis KT (2012) Direct comparisons of Illumina vs. Roche 454 sequencing technologies on the same microbial community DNA sample. PLoS ONE 7:e30087
277. Huse SM, Huber JA, Morrison HG, Sogin ML, Welch DM (2007) Accuracy and quality of massively parallel DNA pyrosequencing. Genome Biol 8:R143
278. Gilles A, Meglecz E, Pech N, Ferreira S, Malausa T, Martin JF (2011) Accuracy and quality assessment of 454 GS-FLX Titanium pyrosequencing. BMC Genomics 12:245
279. Pareek CSP, Smoczynski R, Tretyn A (2011) Sequencing technologies and genome sequencing. J Appl Genet 52:413–435
280. Liu L, Li Y, Li S, Hu N, He Y et al (2012) Comparison of next-generation sequencing systems. J Biomed Biotechnol. https://doi.org/10.1155/2012/251364
281. Guzvic M (2013) The history of DNA sequencing. J Med Biochem 32:301–312
282. Soon WW, Hariharan M, Snyder MP (2013) High-throughput sequencing for biology and medicine. Mol Syst Biol 9:640
283. Kulski JK, Suzuki S, Ozaki Y, Mitsunaga S, Inoko H, Shiina T (2014) Phase HLA genotyping by next generation sequencing: a comparison between two massively parallel sequencing bench-top systems, the Roche GS Junior and Ion Torrent PGM. In: Xi Y (ed) HLA and associated important diseases. Croatia Intech 141–181
284. Turcatti G, Romieu A, Fedurco M, Tairi AP (2008) A new class of cleavable fluorescent nucleotides: synthesis and optimization as reversible terminators for DNA sequencing by synthesis. Nucleic Acids Res 36:e25
285. Frey KG, Herrera-Galeano JE, Redden CL, Luu TV, Servetas SL, Mateczun AJ, Mokashi VP, Bishop-Lilly KA (2014) Comparison of three next-generation sequencing platforms for metagenomic sequencing and identification of pathogens in blood. BMC Genomics 15:1–14
286. Malapelle U, Vigliar E, Sgariglia R, Bellevicine C, Colarossi L, Vitale D et al (2015) Ion Torrent next-generation sequencing for routine identification of clinically relevant mutations in colorectal cancer patients. J Clin Pathol 68:64–68

287. Goodwin S, McPherson JD, McCombie WR (2016) Coming of age: ten years of next-generation sequencing technologies. Nat Rev Genetics 17:333–351
288. Anderson JP, Reynolds BL, Baum K, Williams JG (2010) Fluorescent structural DNA nanoballs functionalized with phosphate-linked nucleotide triphosphates. Nano Lett 10: 788–792
289. Porreca G (2010) Genome sequencing on nanoballs. Nat Biotechnol 28:43–44
290. Morey M, Fernandez-Marmiesse A, Castineiras D, Fraga JM, Couce MA, Cocho JA (2013) A glimpse into past, present, and future DNA sequencing. Mol Genet Metabol 110:3–24
291. Pushkarev D, Neff NF, Quake SR (2009) Single-molecule sequencing of an individual human genome. Nat Biotechnol 27:847–852
292. Hart C, Lipson D, Ozsolak F, Raz T, Steinmann K, Thompson J, Milos PM (2010) Single-molecule sequencing: sequence methods to enable accurate quantitation. Methods Enzymol 472:407–430
293. Thompson JF, Milos PM (2011) The properties and applications of single-molecule DNA sequencing. Genome Biol 12:217
294. Koren S, Harhay GP, Smith TP, Bono JL, Harhay DM, Mcvey SD, Radune D, Bergman NH, Phillippy AM (2013) Reducing assembly complexity of microbial genomes with single-molecule sequencing. Genome Biol 14:R101
295. Ritz A, Bashir A, Sindi S, Hsu D, Hajirasouliha I, Raphael B (2014) Characterization of structural variants with single molecule and hybrid sequencing approaches. Bioinformatics 30:3458–3466
296. Eid J, Fehr A, Gray J, Luong K, Lyle J et al (2009) Real-time DNA sequencing from single polymerase molecules. Science 323:133–138
297. Larsen PA, Heilman AM, Yoder AD (2014) The utility of PacBio circular consensus sequencing for characterizing complex gene families in non-model organisms. BMC Genomics 15:720
298. Mardis ER (2011) A decade's perspective on DNA sequencing technology. Nature 470: 198–203
299. Harris TD, Buzby PR, Babcock H, Beer E, Bowers J et al (2008) Single-molecule DNA sequencing of a viral genome. Science 320:106–109
300. Wang Y, Yang Q, Wang Z (2014) The evolution of nanopore sequencing. Front Genet 5:449
301. Wescoe ZL, Schreiber J, Akeson M (2014) Nanopores discriminate among five C5-cytosine variants in DNA. J Am Chem Soc 136:16582–16587
302. Laver T, Harrison J, O'Neill PA, Moore K, Farbos A, Paszkiewicz K, Studholme DJ (2015) Assessing the performance of the Oxford nanopore technologies MinION. Biomol Detect Quant 3:1–8
303. Jain M, Fiddes IT, Miga KH, Olsen HE, Paten B, Akeson M (2015) Improved data analysis for the MinION nanopore sequencer. Nat Meth 12:351–356
304. Ashton PM, Nair S, Dallman T, Rubino S, Rabsch W et al (2015) MinION nanopore sequencing identifies the position and structure of a bacterial antibiotic resistance island. Nat Biotechnol 33:296–300
305. Stoddart D, Heron A, Mikhailova E, Maglia G, Bayley H (2009) Single nucleotide discrimination in immobilized DNA oligonucleotides with a biological nanopore. Proc Natl Acad Sci U S A 106:7702–7707
306. Levene MJ, Korlach J, Turner SW, Foquet M, Craighead HG, Webb WW (2003) Zero-mode waveguides for single-molecule analysis at high concentrations. Science 299:682–686
307. Carneiro MO, Russ C, Ross MG, Gabriel SB, Nusbaum C, DePristo MA (2012) Pacific biosciences sequencing technology for genotyping and variation discovery in human data. BMC Genomics 13:375
308. English AC, Richards S, Han Y, Wang M, Vee V et al (2012) Mind the gap: upgrading genomes with Pacific Biosciences RS long-read sequencing technology. PLoS ONE 7(11): e47768. https://doi.org/10.1371/journal.pone.0047768
309. Cao MD, Ganesamoorthy D, Cooper MA, Coin LJM (2016) Realtime analysis and visualization of MinION sequencing data with npReader. Bioinformatics 32:764–766

310. Edwards SV (2013) Next-generation QTL mapping: crowdsourcing SNPs, without pedigrees. Mol Ecol 22:3885–3887
311. Hindson BJ, Ness KD, Masquelier DA, Belgrader P, Heredia NJ, Makarewicz AJ, Bright IJ (2011) High-throughput droplet digital PCR system for absolute quantitation of DNA copy number. Anal Chem 83:8604–8610
312. Hodkinson BP, Elizabeth A (2015) Grice next-generation sequencing: a review of technologies and tools for wound microbiome research. Adv Wound Care 4:50–58
313. Zhao Z, Gu H, Sheng X, Yu H, Wang J, Huang L, Wang D (2016) Genome-wide single-nucleotide polymorphisms discovery and high-density genetic map construction in cauliflower using specific-locus amplified fragment sequencing. Front Plant Sci 7:334
314. Elshire RJ, Glaubitz JC, Sun Q, Poland JA, Kawamoto K, Buckler ES, Mitchell SE (2011) A robust, simple genotyping-by-sequencing (GBS) approach for high diversity species. PLoS ONE 6(5): e19379. https://doi.org/10.1371/journal.pone.0019379
315. He J, Zhao X, Laroche A, Lu ZX, Liu HK, Li Z (2014) Genotyping-by-sequencing (GBS), an ultimate marker-assisted selection (MAS) tool to accelerate plant breeding. Front Plant Genet Genom 5:1–8
316. Yang H, Li C, Lam HM, Clements J, Yan G, Zhao S (2015) Sequencing consolidates molecular markers with plant breeding practice. Theor Appl Genet 128:779–795
317. Voss-Fels K, Rod J (2016) Snowdon understanding and utilizing crop genome diversity via high-resolution genotyping. Plant Biotechnol J 14:1086–1094
318. Zhu F, Cui QQ, Hou ZC (2016) SNP discovery and genotyping using genotyping-by-sequencing in Pekin ducks. Sci Rep 6:36223
319. Scheben A, Batley J, Edwards D (2017) Genotyping-by-sequencing approaches to characterize crop genomes: choosing the right tool for the right application. Plant Biotechnol J 15:149–161
320. Furuta T, Ashikari M, Jena KK, Doi K, Reuscher S (2017) Adapting genotyping-by-sequencing for rice F_2 populations. Gene Genom Genet 7:881–893
321. Stetter MG, Schmid KJ (2017) Analysis of phylogenetic relationships and genome size evolution of the *Amaranthus* genus using GBS indicates the ancestors of an ancient crop. Mol Phylogenet Evol 109:80–92
322. Tsai CC, Shih HC, Wang HV, Lin YS, Chang CH, Chiang YC, Chou CH (2015) RNA-Seq SSRs of moth orchid and screening for molecular markers across genus *Phalaenopsis* (*Orchidaceae*). PLoS ONE 10(11): e0141761. https://doi.org/10.1371/journal.pone.0141761
323. Andolfatto P, Davison D, Erezyilmaz D, Hu TT, Mast J, Sunayama-Morita T, Stern DL (2011) Multiplexed shotgun genotyping for rapid and efficient genetic mapping. Genome Res 21:610–617
324. Ott A, Liu S, Schnable JC, Yeh CT, Wang C, Schnable PS (2017) Tunable genotyping-by-sequencing (tGBS®) enables reliable genotyping of heterozygous loci. BioRxiv. https://doi.org/10.1101/100461
325. Poland JA, Rife TW (2012) Genotyping-by-sequencing for plant breeding and genetics. Plant Genome 5:92–102
326. Kang YJ, Lee T, Lee J, Shim S, Jeong H, Satyawan D, Kim MY, Lee SH (2016) Translational genomics for plant breeding with the genome sequence explosion. Plant Biotechnol J 14:1057–1069
327. Pavan S, Marcotrigiano AR, Ciani E, Mazzeo R, Zonno V, Ruggieri V, Lotti C, Ricciardi L (2017) Genotyping-by-sequencing of a melon (*Cucumis melo* L.) germplasm collection from a secondary center of diversity highlights patterns of genetic variation and genomic features of different gene pools. BMC Genomics. https://doi.org/10.1186/s12864-016-3429-0
328. Kurdyukov S, Bullock M (2016) DNA methylation analysis: choosing the right method. Biology. https://doi.org/10.3390/biology5010003
329. Guo Y, Shi G, Liu Z, Zhao Y, Yang X, Zhu J, Li K, Guo X (2015) Using specific length amplified fragment sequencing to construct the high-density genetic map for *vitis* (*Vitis vinifera* L. × *Vitis amurensis* Rupr.). Front. Plant Sci 6:393

330. Jiang Z, Wang H, Michal JJ, Zhou X, Liu B, Woods LCS, Fuchs RT (2016) Genome wide sampling sequencing for SNP genotyping: methods, challenges and future development. Int J Biol Sci 12:100–108
331. Ali OA, O'Rourke SM, Amish SJ, Meek MH, Luikart G, Jeffres C, Miller MR (2016) RAD capture (Rapture): flexible and efficient sequence-based genotyping. genetics. Genetics 202:389–400
332. Altshuler D, Pollara VJ, Cowles CR, Van Etten WJ, Baldwin J, Linton L, Lander ES (2000) An SNP map of the human genome generated by reduced representation shotgun sequencing. Nature 407:513–516
333. Van Orsouw NJ, Hogers RCJ, Janssen A, Yalcin F, Snoeijers S, Verstege E et al (2007) Complexity reduction of polymorphic sequences (CRoPS™): a novel approach for large-ccale polymorphism discovery in complex genomes. PLoS ONE 2(11): e1172. https://doi.org/10.1371/journal.pone.0001172
334. Huang X, Feng Q, Qian Q, Zhao Q, Wang L, Wang A, Guan J, Fan D, Weng Q, Huang T, Dong G, Sang T, Han B (2009) High-throughput genotyping by whole-genome resequencing. Genome Res 19:1068–1076
335. Kerstens HDH, Crooijmans RPMA, Dibbits BW, Vereijken A, Okimoto R, Groenen MAM (2011) Structural variation in the chicken genome identified by paired-end next-generation DNA sequencing of reduced representation libraries. BMC Genomics. https://doi.org/10.1186/1471-2164-12-94
336. Xie W, Feng Q, Yu H, Huang X, Zhao Q, Xing Y, Yu S, Han B, Zhang Q (2010) Parent-independent genotyping for constructing an ultrahigh-density linkage map based on population sequencing. Proc Nat Acad Sci U S A 107:10578–10583
337. Deschamps S, Llaca V, May GD (2012) Genotyping-by-sequencing in plants. Biology 1:460–483
338. Stolle E, Moritz RFA (2013) RESTseq efficient benchtop population genomics with RESTriction fragment SEQuencing. PLoS ONE 8(5): e63960. https://doi.org/10.1371/journal.pone.0063960
339. Sun X, Liu D, Zhang X, Li W, Liu H, Hong W, Jiang C, Guan N, Ma C et al (2013) SLAF-seq: an efficient method of large-scale de novo SNP discovery and genotyping using high-throughput sequencing. PLoS ONE 8(3): e58700. https://doi.org/10.1371/journal.pone.0058700
340. Chen Q, Ma Y, Yang Y, Chen Z, Liao R, Xie X, Wang Z, He P et al (2013) Genotyping by genome reducing and sequencing for outbred animals. PLoS ONE 8(7): e67500. https://doi.org/10.1371/journal.pone.0067500
341. Du H, Yu Y, Ma Y, Gao Q, Cao Y, Zhuo Chen Z et al (2017) Sequencing and de novo assembly of a near complete indica rice genome. Nat Commun 8:15324
342. Beissinger TM, Hirsch CN, Sekhon RS, Foerster JM, Johnson JM et al (2013) Marker density and read depth for genotyping populations using genotyping-by-sequencing. Genetics 193:1073–1081
343. Heffelfinger C, Fragoso CA, Moreno MA, Overton JD, Mottinger JP, Zhao H, Tohme J, Dellaporta SL (2014) Flexible and scalable genotyping-by-sequencing strategies for population studies. BMC Genomics 15:979
344. Gardner KM, Brown P, Cooke TF, Cann S, Costa F, Bustamante C, Velasco R, Troggio M, Myles S (2014) Fast and cost-effective genetic mapping in apple using next-generation sequencing. Genes Genomes Genet 4:1681–1687
345. Greminger MP, Stolting KN, Nater A, Goossens B, Arora N, Bruggmann R et al (2014) Generation of SNP datasets for orangutan population genomics using improved reduced-representation sequencing and direct comparisons of SNP calling algorithms. BMC Genomics 15:16
346. Rowan BA, Patel V, Weigel D, Schneeberger K (2015) Rapid and inexpensive whole-genome genotyping-by-sequencing for crossover localization and fine-scale genetic mapping. Genes Genomes Genet 5:385–398

347. Rife TW, Wu S, Bowden RL, Poland JA (2015) Spiked GBS: a unified, open platform for single marker genotyping and whole-genome profiling. BMC Genomics 16:248
348. Wei T, Deng K, Zhang Q, Gao Y, Liu Y, Yang M, Zhang L, Zheng X, Wang C, Liu Z, Chen C, Zhang Y (2017) Modulating AtDREB1C expression improves drought tolerance in *Salvia miltiorrhiza*. Front Plant Sci 8:52
349. Batovska J, Cogan NOI, Lynch SE, Blacket MJ (2017) Using next-generation sequencing for DNA barcoding: capturing allelic variation in ITS2. Gene Genomes Genet 7:19–29
350. Waiho K, Fazhan H, Shahreza MS, Moh JHZ, Noorbaiduri S, Wong LL et al (2017) Transcriptome analysis and differential gene expression on the testis of orange mud crab, *Scylla olivacea*, during sexual maturation. PLoS ONE 12(1): e0171095. https://doi.org/10.1371/journal.pone.0171095
351. Yan YP, Wang ZZ, Tian W, Zhong MD, David S (2010) Generation and analysis of expressed sequence tags from the medicinal plant *Salvia miltiorrhiza*. Sci China Life Sci 53:273–285
352. Li Y, Sun C, Luo HM, Li XW, Niu YY, Chen SL (2010) Transcriptome characterization for *Salvia miltiorrhiza* using 454 GS FLX. Acta Pharm Sinica 45:524–529
353. Hua WP, Zhang Y, Song J, Zhao LJ, Wang ZZ (2011) De novo transcriptome sequencing in *Salvia miltiorrhiza* to identify genes involved in the biosynthesis of active ingredients. Genomics 98:272–279
354. Du Q, Li C, Li D, Lu S (2015) Genome-wide analysis, molecular cloning and expression profiling reveal tissue specifically expressed, feedback-regulated, stress-responsive and alternatively spliced novel genes involved in gibberellin metabolism in *Salvia miltiorrhiza*. BMC Genomics 16:1087
355. Davey JW, Hohenlohe PA, Etter PD, Boone JQ, Catchen JM, Blaxter ML (2011) Genome-wide genetic marker discovery and genotyping using next-generation sequencing. Nat Rev Genet 12:499–510
356. Kumar S, You FM, Cloutier S (2012) Genome wide SNP discovery in flax through next generation sequencing of reduced representation libraries. BMC Genomics 13:684
357. Jia Q, Tan C, Wang J, Zhang XQ, Zhu J, Luo H, Yang J, Westcott S, Broughton S, Moody D, Li C (2016) Marker development using SLAF-seq and whole-genome shotgun strategy to fine map the semi-dwarf gene *ari-e* in barley. BMC Genomics 17:911
358. Levy SE, Myers RM (2016) Advancements in next-generation sequencing. Ann Rev Genom Hum Genet 17:95–115
359. Bowers J, Mitchell J, Beer E, Buzby PR, Causey M, Efcavitch JW et al (2009) Virtual terminator nucleotides for next-generation DNA sequencing. Nat Methods 6:593–595
360. Michael TP, van Buren R (2015) Progress, challenges and the future of crop genomes. Curr Opin Plant Biol 24:71–81
361. Kage U, Kumar A, Dhokane D, Karre S, Kushalappa AC (2016) Functional molecular markers for crop improvement. Critic Rev Biotechnol 36:917–930
362. Salgotra RK, Gupta BB, Stewart CN Jr (2014) From genomics to functional markers in the era of next-generation sequencing. Biotechnol Lett 36:417–426
363. Walker JB, Sytsma KJ (2007) Staminal evolution in the genus *Salvia* (Lamiaceae): molecular phylogenetic evidence for multiple origins of the staminal lever. Ann Bot 100:375–391
364. Jenks AA, Walker JB, Kim SC (2011) Evolution and origins of the Mazatec hallucinogenic sage, *Salvia divinorum* (Lamiaceae): a molecular phylogenetic approach. J Plant Res 124:593–600
365. Li QQ, Li MH, Yuan QJ, Cui ZH, Huang LQ, Xiao PG (2013) Phylogenetic relationships of *Salvia* (Lamiaceae) in China: evidence from DNA sequence datasets. J Syst Evol 51:184–195
366. Xu Z, Peters RJ, Weirather J, Luo H, Liao B, Zhang X et al (2015) Full-length transcriptome sequences and splice variants obtained by a combination of sequencing platforms applied to different root tissues of *Salvia miltiorrhiza* and tanshinone biosynthesis. Plant J 82:951–961
367. Danilo Stojanovicv D, Aleksic JA, Jancic I, Jancic R (2015) A Mediterranean medicinal plant in the continental Balkans: a plastid DNA-based phylogeographic survey of *Salvia officinalis* (Lamiaceae) and its conservation implications. Willdenowia 45:103–118

368. De Mattia F, Bruni I, Galimberti A, Cattaneo F, Casiraghi M, Labra M (2011) A comparative study of different DNA barcoding markers for the identification of some members of Lamiacaea. Food Res Int 44:693–702
369. Skoula M, Hilali IE, Makris AM (1999) Evaluation of the genetic diversity of *Salvia fruticosa* Mill. clones using RAPD markers and comparison with the essential oil profiles. Biochem Syst Ecol 27:559–568
370. Guo BL, Feng YX, Zhao YJ (2002) Review of germplasm resources studies on *Salvia miltiorrhiza*. Zhongguo Zhong yao za zhi = Zhongguo zhongyao zazhi = China J Chin Materia Medica 27:492–495
371. Cahill PJ (2004) Genetic diversity among varieties of Chia (*Salvia hispanica* L.). Genet Resour Crop Evol 51:773–781
372. Javan ZS, Rahmani F, Heidari R (2012) Assessment of genetic variation of genus *Salvia* by RAPD and ISSR markers. Aust J Crop Sci 6:1068–1073
373. Kaya EM, Uysal UD, Ozturk N, Cenkci S, Tuncel M (2015) Determination of DNA in certain *Salvia* species by capillary gel electrophoresis. J Liq Chromatogr Relat Technol 38:1417–1425
374. Wang B, Zhang Y, Chen CB, Li XL, Chen RY, Chen L (2007) Analysis on genetic diversity of different *Salvia miltiorrhiza* geographical populations in China. Chin J Chin Mater Medica 32:1988–1991
375. Braglia L, Casabianca V, De Benedetti L, Pecchioni N, Mercuri A, Cervelli C, Ruffoni B (2011) Amplified fragment length polymorphism markers for DNA fingerprinting in the genus *Salvia*. Plant Biosyst 145:274–277
376. Rapposelli E, Melito S, Barmina GG, Foddai M, Azara E, Scarpa GM (2015) AFLP fingerprinting and essential oil profiling of cultivated and wild populations of Sardinian *Salvia desoleana*. Genet Resour Crop Evol 62:959–970
377. Zhang Y, Li X, Wang Z (2013) Diversity evaluation of *Salvia miltiorrhiza* using ISSR markers. Biochem Genet 51:707–721
378. Yousefiazarkhanian M, Asghari A, Ahmadi J, Asghari B, Jafari AA (2015) Genetic diversity assessment of some *Salvia* sp. ecotypes based on ISSR markers. Biol Forum 7:286–288
379. Deng KJ, Zhang Y, Xiong BQ, Peng JH, Zhang T, Zhao XN, Ren ZL (2009) Identification, characterization and utilization of simple sequence repeat markers derived from *Salvia miltiorrhiza* expressed sequence tags. Acta Pharma Sinica 44:1165–1172
380. Radosavljevic I, Satovic Z, Jakse J, Javornik B, Greguras D, Jug-Dujakovic M, Liber Z (2012) Development of new microsatellite markers for *Salvia officinalis* L. and its potential use in conservation-genetic studies of narrow endemic *Salvia brachyodon* Vandas. Int J Mol Sci 13:12082–12093
381. Xu G, Liu C, Huang L, Wang X et al (2013) Development of new EST-derived SSRs in *Salvia miltiorrhiza* (Labiatae) in China and preliminary analysis of genetic diversity and population structure. Biochem Syst Ecol 51:308–313
382. Ge X, Chen H, Wang H, Shi A, Liu K (2014) De novo assembly and annotation of *Salvia splendens* transcriptome using the Illumina platform. PLoS ONE 9(3): e87693. https://doi.org/10.1371/journal.pone.0087693
383. Resetnik I, Baricevic D, Batir Rusu D, Carovic-Stanko K, Chatzopoulou P, Dajic-Stevanovic Z et al (2016) Genetic diversity and demographic history of wild and cultivated/naturalised plant populations: evidence from dalmatian dage (*Salvia officinalis* L., Lamiaceae). PLoS ONE 11(7): e0159545. https://doi.org/10.1371/journal.pone.0159545
384. Sansaloni CP, Petroli CD, Carling J, Hudson CJ, Steane DA, Myburg AA, Grattapaglia D, Vaillancourt RE, Kilian A (2010) A high-density diversity arrays technology (DArT) microarray for genome-wide genotyping in eucalyptus. Plant Methods. https://doi.org/10.1186/1746-4811-6-16
385. Bortesi L, Fischer R (2015) The CRISPR/Cas9 system for plant genome editing and beyond. Biotechnol Adv 33:41–52

386. Yang L, Ding G, Lin H, Cheng H, Kong Y et al (2013) Transcriptome analysis of medicinal plant *Salvia miltiorrhiza* and identification of genes related to tanshinone biosynthesis. PLoS ONE 8:e80464
387. Wang B, Sun W, Li Q, Li Y, Luo H et al (2014) Genome-wide identification of phenolic acid biosynthetic genes in *Salvia miltiorrhiza*. Planta 241:711–725
388. Fu YB (2015) Understanding crop genetic diversity under modern plant breeding. Theor Appl Genet 128:2131–2142
389. Gao W, Sun HX, Xiao H, Cui G, Hillwig M et al (2014) Combining metabolomics and transcriptomics to characterize tanshinone biosynthesis in *Salvia miltiorrhiza*. BMC Genomics 15:73
390. Grdisa M, Jug-Dujakovic M, Loncaric M, Carovic-Stanko K et al (2015) Dalmatian sage (*Salvia officinalis* L.): a review of biochemical contents, medical properties and genetic diversity. Agriculturae Conspectus Scientificus 80:69–78
391. Xu H, Wang ZT, Cheng KT, Wu T, Gu LH, Hu ZB (2009) Comparison of rDNA ITS sequences and tanshinones between *Salvia miltiorrhiza* populations and *Salvia* species. Bot Stud 50:127–135
392. Liber Z, Zidovec V, Bogdanovic S, Radosavljevic I, Prusa M, Filipovic M et al (2014) Genetic diversity of dalmatian sage (*Salvia officinalis* L.) as assessed by RAPD markers. Agriculturae Conspectus Scientificus 79:77–84
393. Zhang T, Wang J, Wang D, Wang H (2015) Development of an authentication system for genuine radix *Salviae Miltiorrhizae* (*Salvia miltiorrhiza*) using SNP markers. Planta Med Lett 2:e65–e68
394. Mardis ER (2013) Next-generation sequencing platforms. Ann Rev Anal Chem 6:287–303
395. Heather JM, Chain B (2016) The sequence of sequencers: the history of sequencing DNA. Genomics 107:1–8
396. Brozynska M, Furtado A, Henry RJ (2016) Genomics of crop wild relatives: expanding the gene pool for crop improvement. Plant Biotechnol J 14:1070–1085
397. Mader E, Lohwasser U, Borner A, Novak J (2010) Population structures of Genebank accessions of *Salvia officinalis* L. (Lamiaceae) revealed by high resolution melting analysis. Biochem Syst Ecol 38:178–186
398. Ng SB, Buckingham KJ, Lee C, Bigham AW, Tabor HK et al (2010) Exome sequencing identifies the cause of a mendelian disorder. Nat Genet 42:30–35
399. Ma Y, Yuan L, Wu B, Li X, Chen S, Lu S (2012) Genome-wide identification and characterization of novel genes involved in terpenoid biosynthesis in *Salvia miltiorrhiza*. J Exp Bot 63:2809–2823
400. Hao G, Shi R, Tao R, Fang Q, Jiang X, Ji H, Feng L, Huang L (2013) Cloning, molecular characterization and functional analysis of 1-hydroxy-2-methyl-2-(E)-butenyl-4-diphosphate reductase (HDR) gene for diterpenoid tanshinone biosynthesis in *Salvia miltiorrhiza* Bge. f. alba. Plant Physiol Biochem 70:21–32

Genetic Engineering and Manipulation of Metabolite Pathways in *Salvia* Spp.

Vasil Georgiev and Atanas Pavlov

Abstract Plants from the genus *Salvia* have been recognized as medicinal herbs all over the world since earliest times. They are known to accumulate diverse range of bioactive phytochemicals, including polyphenols (rosmarinic and salvianolic acids), triterpenes (ursolic and oleanolic acids), diterpenes (tanshinones, carnosic acid), flavonoids and sterols, etc. Nowadays, the powerful methods of genetic and metabolic engineering, synthetic biology and combinatorial biosynthesis have been widely applied for improvement of commercial crops yields and for increasing their medicinal value by modulating the accumulation of biologically active phytochemicals. Over the past few years, these techniques have been applied in *Salvia* plants and in vitro systems, but the research still remains limited to few species. In this chapter, we summarized the recent achievements in genetic engineering of *Salvia* species, with special attention on metabolite engineering of phenolic biosynthesis and terpenoids biosynthesis pathways. Some aspects of the applications of functional genes, cloned by *Salvia* species, for the needs of synthetic biology and combinatorial biosynthesis are reviewed as well.

Keywords Metabolic engineering · Synthetic biology · Terpenoid biosynthesis Phenolic biosynthesis · Combinational genetic

Abbreviations

GRAS	Generally recognized as safe
FDA	Food and drug administration
EMA	European medicines agency
NCBI	National center for biotechnology information
CDS	Coding DNA sequence

V. Georgiev · A. Pavlov
University of Food Technologies, 26 Maritza Blvd., 4002 Plovdiv, Bulgaria

V. Georgiev (✉) · A. Pavlov
Laboratory of Applied Biotechnologies, Institute of Microbiology, Bulgarian Academy of Sciences, 139 Ruski Blvd., 4000 Plovdiv, Bulgaria
e-mail: vasgeorgiev@gmail.com

V. Georgiev and A. Pavlov (eds.), *Salvia Biotechnology*,
https://doi.org/10.1007/978-3-319-73900-7_10

EST	Expressed sequence tags
DNA	Deoxyribonucleic acid
cDNA	complementary DNA
nrITS	nuclear internal transcribed spacer
CaMV35S	Cauliflower mosaic virus 35S promoter
hptII	hygromycin phosphotransferase gene
ATM	Activation tagging mutagenesis
T-DNA	Transferred DNA
LEA	Late embryogenesis abundant
AtDREB	*Arabidopsis thaliana* dehydration-responsive-element-binding protein
SAMDC	S-adenosylmethionine decarboxylase
AtEDT	*Arabidopsis thaliana*-enhanced drought tolerance
R2R3-MYB	R2R3 type -myeloblastosis protein
WRKY	WRKY transcription factor
C4H	Cinnamic acid 4-hydroxylase
TAT	Tyrosine aminotransferase
PAL	Phenylalanine ammonialyase
4CL	4-Coumaroyl:CoA ligase
HPPR	4-Hydroxyphenylpyruvate reductase
CYP	Cytochrome P450 reductase
RAS	Rosmarinic acid synthase
RNA	Ribonucleic acid
mRNA	messenger RNA
RNAi	RNA interference
AtPAP	*Arabidopsis thaliana* production of anthocyanin pigment
CCR	Cinnamoyl-CoA reductase
COMT	Caffeic acid O-methyltransferase
MEP	Methyl-D-erythritol 4-phosphate pathway
MVA	Mevalonic acid pathway
DXS	1-Deoxy-D-xylulose 5-phosphate synthase
DXP	1-Deoxy-D-xylulose 5-phosphate
MCT	2-C-methyl-D-erythritol 4-phosphate cytidylyltransferase
CMK	4-(cytidine 5′-diphospho)-2-C-methyl-D-erythritol kinase
MDS	2-C-methyl-D-erythritol 2,4-cyclodiphosphate synthase
HDS	4-hydroxy-3-methylbut-2-enyl diphosphate synthase
HDR	4-hydroxy-3-methylbut-2-enyl diphosphate reductase
GPS	Geranyl diphosphate synthase
GGPS	Geranylgeranyl diphosphate synthase
PSY	Phytoene synthase
AACT	Acetyl-CoA C-acetyltransferase
HMGS	Hydroxymethylglutaryl-CoA synthase
HMGR	Hydroxymethylglutaryl-CoA reductase
MK	Mevalonate kinase

PMK	5-Phosphomevalonate kinase
MDK	Mevalonate pyrophosphate decarboxylase
SQS	Squalene synthase
FPS	Farnesyl diphosphate synthase
IDI	Isopentenyl diphosphate isomerase
bHLH	basic helix-loop-helix
GSG	Gly-Ser repeats linker

1 Introduction

Plants from the genus *Salvia* have been important ingredients in traditional medicine all over the world since earliest times. Species as *Salvia officinalis* L. (Garden Sage) and *S. tomentosa* Mill. (Balsamic Sage), used as seasoning in food and for tea preparations, *S. sclarea* L. (Clary Sage) and *S. lavandulifolia* Vahl. (Spanish Sage), being important sources of essential oils for cosmetics, *S. hispanica* L. (Chia), used for food and oil production, *S. miltiorrhiza* Bunge (Danshen), used to treat coronary artery diseases, and *S. divinorum* Epling & Jativa (Diviner's Sage), used as hallucinogenic psychoactive plant, are only few examples of that huge genus (including over 900 species), that have been appreciated and cultivated by humans since centuries [1–8]. Nowadays, some of these species are one of the most economically important cultured crops in the globe. For example, the annual production of *S. officinalis* L. essential oil (Dalmatian Type Sage Oil) and oleoresin exceed 1750 tons per year [4]. Recently, *S. officinalis* L. leaves and essential oil have been listed as "Generally Recognized as Safe" (GRAS) for use as a seasoning, spices or flavoring agent by the United States Food and Drug Administration (FDA), and have been registered as traditional herbal medicinal products in the European Community by the European Medicines Agency (EMA) [9, 10]. Other examples could be given with the growing demand for *S. sclarea* L. essential oil for the needs of cosmetic industry (annual production of more than 100 tons per year), and the expanded market for Chia seeds, which production cover 24% of Argentina agricultural industry and is now is the major contributor to the country economy [4, 11]. Based on *Salvia*'s traditional use as medicinal and culinary herbs, as well as the wide perception of the modern concept for healthy lifestyle supported by the consumption of high quality nutraceuticals, we could easily conclude that the demand for *Salvia* products will continue to grow exponentially in the future. It is obviously, that to supply such demand, new strategies, leading to drastic improvements in yields and medicinal quality of *Salvia* plants should be applied. The advanced methods of genetic and metabolic engineering, synthetic biology and combinatorial biosynthesis could be used as powerful tools to overcome any obstacles limiting the accumulation of biologically active phytochemicals, and thus, to improve significantly the crops yields. Nowadays, many economically important crops have been genetically improved and their yields have been successfully increased [12]. However, despite the increasing number of reports for genetic improvement of industrial crops, the researches on genetic modifications of *Salvia* plants are very limited and focused only on few species. This could be due to the scantly information about genome sequences of plants, belonging to that genus. For example, in one of the world's most popular GenBank, the National Center for Biotechnology Information (NCBI), there are very low number of coding DNA sequence (CDS) and expressed sequence tags (EST) submitted so far for plants, belonging to the genus *Salvia* (Table 1). Moreover, the major number of CDS

Table 1 Numbers of accessions (CDS and EST), associated with *Salvia*, found in the National Center for Biotechnology Information (https://www.ncbi.nlm.nih.gov/) database (as of September, 2017)

Salvia species	Number of nucleotide sequence entries found (mRNA, partial and/or complete CDS)	Number of EST sequence entries found (cDNA Libraries, mRNA sequences)
Salvia miltiorrhiza Bunge	987	10,288
Salvia guaranitica A.St.-Hil. ex Benth.	39	–
Salvia officinalis L.	20	–
Salvia hispanica L.	16	–
Salvia sclarea L.	15	29
Salvia divinorum Epling & Játiva	14	–
Salvia pomifera L.	12	–
Salvia fruticosa Mill.	9	1465
Salvia coccinea Buc'hoz ex Etl.	8	–
Salvia columbariae Benth.	8	–
Salvia sp. BC-2016	5	–
Salvia splendens Sellow ex J.A. Schultes	4	–
Salvia yangii B.T. Drew	4	–
Salvia castanea Diels f. *tomentosa* Stib.	4	–
Salvia macrosiphon Boiss.	3	–

accessions (987 from 1148) and EST accessions (10,288 from 11,782) are related to one species—*Salvia miltiorrhiza* Bunge (Table 1).

The lack of sequence data makes harder the design of primers, and thus, complicates the processes of cloning and sequencing the functional and regulatory genes from closely related *Salvia* species. Moreover, recent study, based on investigation of infrageneric relationships within *Salvia* species by using the chloroplast DNA regions *rbcL* (71 species studied) and *trnL-F* (36 species studied) was found that the genus *Salvia* is not monophyletic and the investigated species were classified to belong to tree clades [13]. More recent research, including 220 *Salvia* species, sequenced by nuclear internal transcribed spacer (*nrITS*) and variable chloroplast marker (*rpl32-trnL*), confirmed that *Salvia* is polyphyletic genus with four distinct evolutionary lineages (four clades) [14]. The high diversity of genus *Salvia* could be an additional complication to genes cloning from different *Salvia* species, since the low interspecies similarity of genes decrease effectiveness in primers design. However, the recent progress in transcriptome sequencing, sequencing of the complete chloroplast genome, and the full sequencing of the *Salvia miltiorrhiza* Bunge genome highlighted the genes, responsible for the biosynthesis of the main bioactive secondary metabolites in *Salvia* spp. (phenylpropanoids, phenolics, terpenoids, tanshinones) and paved the way for the metabolite engineering of *Salvia* plants [15–21].

In this chapter, we review the recent progress in development of transgenic *Salvia* plants, metabolite engineering of phenolic biosynthesis and terpenoids biosynthesis pathways, as well as the advancements in synthetic biology and combinatorial biosynthesis of valuable compound by using genes with *Salvia* origin.

2 Development of Transgenic *Salvia* Plants

With the advancement of plant transformation technology, the genetic modification of crop plants has been used to improve many varieties, but only a few examples of *Salvia* species could be found. Experiments with activation tagging mutagenesis (ATM) showed that transgenic *Salvia miltiorrhiza* Bunge plants, obtained after transformation with pTAG8 binary vector, caring a viral CaMV35S enhancer and *hptII* (hygromycin phosphotransferase) as selectable marker, had increased leaves size with bigger guard cells, stomata and trichomes [22]. These transgenic plants accumulated 3.7 times more tanshinone I and 2 times more tanshinone IIA than the control plants [22]. However, the biggest disadvantage of ATM remains the unpredictable result after transformation, since the *Agrobacterium tumefaciens* introduce its T-DNA randomly into the plant genome. This obstacle could be overcome by cloning and overexpression of entire key gene into the transgenic plant. Wu et al. demonstrated, that cloning of the gene, responsible for biosynthesis of late embryogenesis abundant (LEA) proteins in *Salvia miltiorrhiza* Bunge, and its overexpression under the control of CaMV35S promoter in transgenic *S. miltiorrhiza* plants could significantly increase the plant tolerance to abiotic stress as increased salinity and drought [23]. Salinity and drought are very important factors, that could significantly limited plant growth. In another study, the gene responsible for synthesis of S-adenosylmethionine decarboxylase (SAMDC) a key enzyme of biosynthetic pathway of polyamines, have been isolated from *S. miltiorrhiza* plant and cloned and overexpressed in transgenic *Nicotiana tabacum* L. [24]. The transgenic tobacco plants showed increased drought tolerance and high rates of survival after 18 d drought treatment [24]. The authors concluded, that *S. miltiorrhiza SAMDC* could be applied in other species to improve their tolerance to drought stress. To improve the ability of *S. miltiorrhiza* to survive under drought stress, the *Arabidopsis thaliana* (L.) Heynh. stress-inducible genes, responsible for the synthesis of dehydration-responsive-element-binding proteins 1A, 1B and 1C (*AtDREB1A*, *AtDREB1B* and *AtDREB1C*), have been cloned in transgenic *S. miltiorrhiza* plants under the constitutive control of CaMV35S promoter [25–27]. The resulted transgenic plants showed high tolerance to drought stress, but the transgenic *S. miltiorrhiza* plants overexpressing *AtDREB1C* showed significant dwarfing, whereas the transgenic plants overexpressing *AtDREB1A* showed slight dwarfing, and the transgenic plants overexpressing *AtDREB1B* showed no stunting growth [25–27]. However, overexpression of stress related genes could have effect on secondary metabolite production in transgenic *Salvia* plants. The cloning and overexpression of *Arabidopsis thaliana*-enhanced drought tolerance 1 (*AtEDT1*) gene in *S. miltiorrhiza* plants resulted in increased drought tolerance and elongated root system of transgenic plants, but also caused an significant increase in accumulation of salvianolic acids and decreased the tanshinones production [28]. In addition to functional genes, cloning and overexpression of regulatory genes may have even more drastic effect on secondary metabolites accumulation in transgenic *Salvia* plants. The important transcription factors, such as R2R3-MYB and WRKY have been cloned and characterized in *Salvia miltiorrhiza* Bunge plants [29, 30]. It was demonstrated, that overexpression of *S. miltiorrhiza MYB39* had negative effect on accumulation of phenolic acids in transgenic *S. miltiorrhiza* plants by down-regulating the genes, responsible for synthesis of 4-hydroxylase (C4H) and tyrosine aminotransferase (TAT)—the key enzymes from phenylpropanoid and tyrosine pathways [31]. On the other hand, the cloning and overexpression of *S. miltiorrhiza MYC2* have been found to upregulate the expression of the genes from phenylpropanoid and tyrosine pathways, and thus, to cause an 1.88 times increase in concentration of salvianolic acid B, accumulated by transgenic *S. miltiorrhiza* plants [32]. However, the authors reported that no tanshinones were found into transgenic plants overexpressing *MYC2* [32].

3 Metabolic Engineering of Phenolic Biosynthesis in *Salvia*

Polyphenols, and especially rosmarinic and salvianolic acids are considered to be the main phytochemicals, responsible for the remarkable antioxidant activities of *Salvia* extracts. The biosynthesis pathway, leading to the formation of these phenolic acids have been well established in other species as *Coleus blumei* Benth. [33]. Recently, a specific step have been found to occur in *Salvia miltiorrhiza* Bunge plants (Fig. 1) [21].

Generally, there are two pathways, which joined, leaded to rosmarinic and salvianolic acids synthesis. These are the phenylpropanoid pathway and the tyrosine pathway (Fig. 1). The first enzyme in phenylpropanoid pathway is the phenylalanine ammonia lyase (PAL), followed by cinnamate 4-hydroxylase (C4H) and 4-coumarate-CoA ligase (4CL). All these enzymes catalyzed the conversion of L-phenylalanine into 4-coumaroyl-CoA. The tyrosine pathway starts with the conversion of L-tyrosine into 4-hydroxyphenylpyruvic acid by tyrosine aminotransferase (TAT), and then the conversion of 4-hydroxyphenylpyruvic acid into 4-hydroxyphenyllactic acid by 4-hydroxyphenylpyruvate reductase (HPPR). In *Salvia miltiorrhiza* Bunge plants, the last one is then catalyzed into 3,4-dihydroxyphenllactic acid by an unknown cytochrome P450 reductase (CYP450) (Fig. 1). Rosmarinic acid synthase (RAS) connects the phenylpropanoids and tyrosine pathways by catalyzing the conversion of the 4-coumaroyl-CoA and 4-hydroxyphenyllactic acid (or 3,4-dihydroxyphenllactic acid in *S. miltiorrhiza*) into 4-coumaroyl-3′,4′-dihydroxyphenllactic acid. The last one is converted by cytochrome P450 reductase (CYP98A78) into rosmarinic acid, which then is converted to salvianolic acid probably by the action of unknown laccase [21, 34, 35]. Because rosmarinic and salvianolic acids are secondary metabolites, participating in cell defense, their biosynthesis is normally activated under stress or in response to elicitors [36, 37]. However, to achieve constant increase in their production, an engineering of their biosynthetic pathway is necessary. It was demonstrated, that cloning and overexpression of the genes for transcription factors MYC2a and MYC2b, participating in jasmonates signaling system in plant cells, could significantly increase the production of phenolic acids in transgenic *Salvia miltiorrhiza* Bunge plants [35]. Another effective strategy, based on silencing of chalcone synthase (CHS) gene, responsible for synthesis of the entry enzyme of the side flavonoids biosynthesis branch from the phenylpropanoid pathway, have been applied in transgenic *S. miltiorrhiza* hairy root culture [38]. By using RNAi-silencing, the authors succeed to redirect the metabolite flow to production of phenolic acids rather than to flavonoids [38]. Another effective strategy for increasing phenolic acid content in *Salvia* is the combinational genetic. For example, ectopic expression of *Arabidopsis* MYB transcription factor *AtPAP1*, which stimulates the production of phenolics in transgenic *S. miltiorrhiza*, in combination with the co-suppression of two key enzyme genes of phenylpropanoids side branch of lignans biosynthesis—cinnamoyl-CoA reductase (CCR) and caffeic acid *O*-methyltransferase (COMT), was found to increase the accumulation of salvianolic acid B up to 3 times in transgenic *S. miltiorrhiza* plants [39].

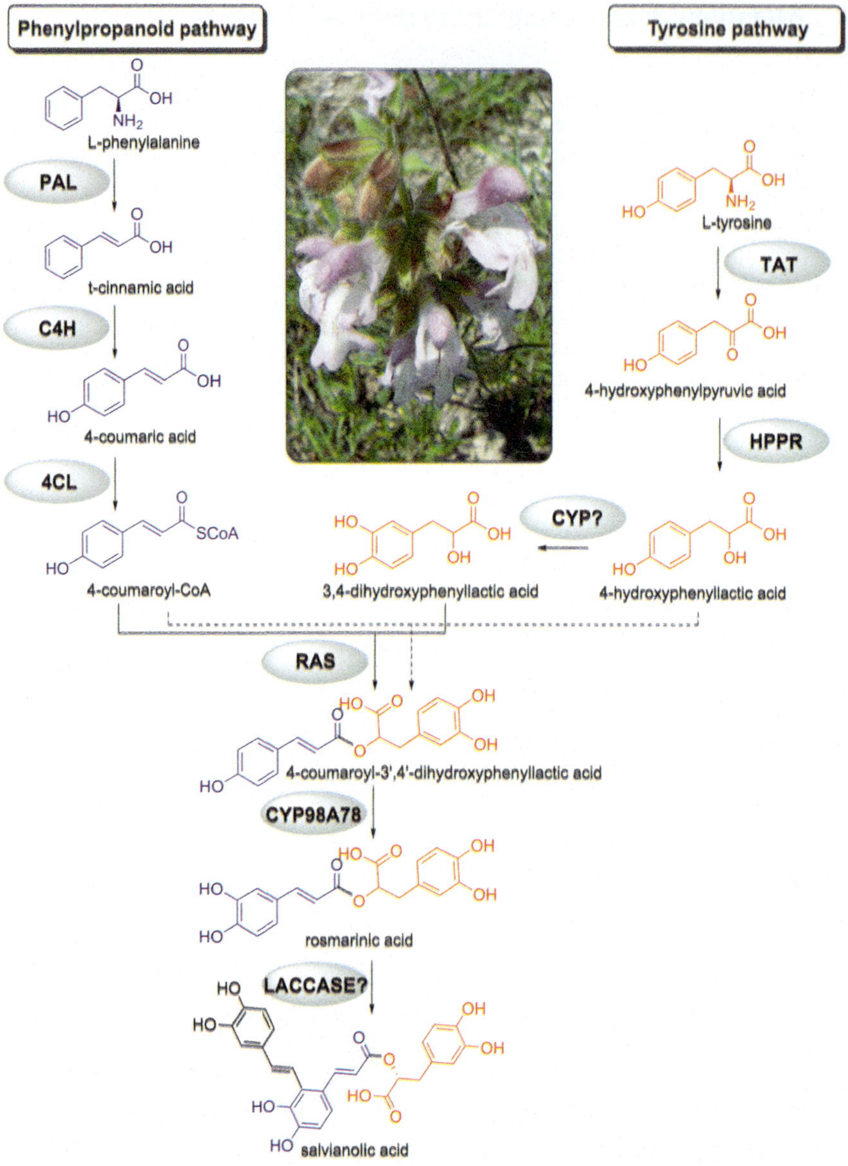

Fig. 1 Biosynthesis of phenolic acids in *Salvia*. The general biosynthetic pathway of rosmarinic and salvianolic acids is showed with solid arrows, whereas, the specific pathway, found in *Salvia miltiorrhiza* is showed with dashed arrows. PAL—phenylalanine ammonialyase; C4H—cinnamic acid 4-hydroxylase; 4CL—4-coumaroyl: CoA ligase; TAT—tyrosine aminotransferase; HPPR—4-hydroxyphenylpyruvate reductase; CYP?—unknown cytochrome P450; RAS—rosmarinic acid synthase; CYP98A78—cytochrome P450 98A78; LACCASE?—unknown laccase

4 Metabolic Engineering of Terpenoid Biosynthesis in *Salvia*

Terpenoids, including monoterpenes, diterpenes (labdane-types and abietane-type), triterpenes, etc. are considered as the main group of pharmacologically active phytochemicals found in *Salvia* plants, which have wide range of activities including anti-inflammatory, cardio- and neuro-protective, antimicrobial, cytotoxic and anticancer etc. However, all these terpenoids are derived by using two building blocks—dimethylallyl pyrophosphate and isopentenyl pyrophosphate (Fig. 2).

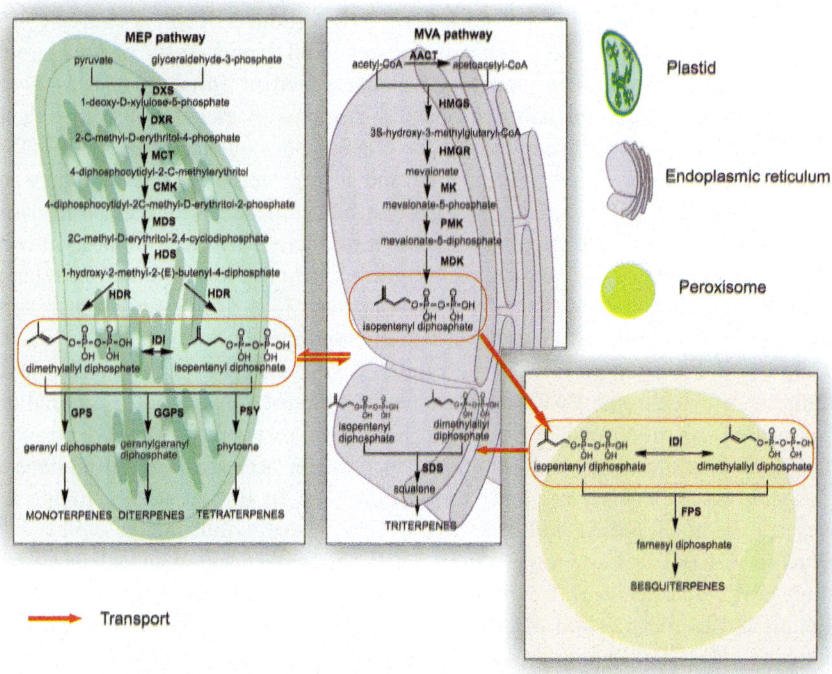

Fig. 2 Biosynthesis of terpenoids in plants. Key enzymes in the individual pathways are presented in bold. Red arrows indicate transport of the terpenoids building blocks—dimethylallyl diphosphate and isopentenyl diphosphate. MEP—2-C-methyl-D-erythritol 4-phosphate pathway; MVA—mevalonic acid pathway; DXS—1-deoxy-D-xylulose 5-phosphate synthase; DXP—1-deoxy-D-xylulose 5-phosphate; MCT—2-C-methyl-D-erythritol 4-phosphate cytidylyltransferase; CMK—4-(cytidine 5′-diphospho)-2-C-methyl-D-erythritol kinase; MDS—2-C-methyl-D-erythritol 2,4-cyclodiphosphate synthase; HDS—4-hydroxy-3-methylbut-2-enyl diphosphate synthase; HDR—4-hydroxy-3-methylbut-2-enyl diphosphate reductase; GPS—geranyl diphosphate synthase; GGPS—geranylgeranyl diphosphate synthase; PSY—phytoene synthase; AACT—acetyl-CoA C-acetyltransferase; HMGS—hydroxymethylglutaryl-CoA synthase; HMGR—hydroxymethylglutaryl-CoA reductase; MK—mevalonate kinase; PMK—5-phosphomevalonate kinase; MDK—mevalonate pyrophosphate decarboxylase; SQS—squalene synthase; FPS—farnesyl diphosphate synthase; IDI—isopentenyl diphosphate isomerase

Dimethylallyl pyrophosphate and isopentenyl pyrophosphate can be produced via two distinct metabolic pathways, the mevalonic acid pathway (MVA) and the 2-C-methyl-D-erythritol 4-phosphate pathway (MEP) (Fig. 2) [40, 41]. It was considered, that MVA pathway has been evolved in archaea, whereas the MEP pathway is evolved in prokaryotic organisms [41]. Moreover, MEP pathway is localized in the plastids, whereas the MVA pathway is localized in cytosol, endoplasmic reticulum and peroxisome (Fig. 2). MEP pathway generates both isopentenyl diphosphate and dimethylallyl diphosphate precursors, whereas the MVA pathway leads to formation of only isopentenyl diphosphate, which is further isomerized by isopentenyl diphosphate isomerase to dimethylallyl diphosphate [40, 41]. MEP is the main pathway leading to the formation of monoterpenes, diterpenes (labdane-types and abietane-type) and tetraterpenes in plants, whereas triterpenes and sesquiterpenes are considered to be products of MVA pathway [41]. Many genes, encoding enzymes from the terpenoid biosynthesis pathway (both MVA and MEP) have been cloned and characterized in *Salvia* [34, 40, 42–45]. Recently, five genes, encoding gibberellin-responsive element binding factor (GRAS) transcription regulators have been cloned and sequenced from EST cDNA library generated by *Salvia miltiorrhiza* and it was demonstrated that they are closely associated with increased tanshinone biosynthesis by *Salvia miltiorrhiza* hairy roots [46]. Another important transcription factor, the basic helix-loop-helix (bHLH), have been cloned and characterized in *Salvia miltiorrhiza*, and its relation to tanshinones biosynthesis have been demonstrated [47]. However, for increasing terpenoids production, key enzymes from terpenoid metabolite pathway have to be manipulated. For example, RNAi-silencing of *CYP76AH1*, the gene responsible for synthesis of cytochrome P450 reductase responsible for conversion of miltiradiene into tanshinones intermediate compound ferruginol, resulted in drastic decrease in ferruginol concentrations and significant increase in accumulated miltiradiene in transgenic *Salvia miltiorrhiza* hairy root culture [48]. In another study, cloning and overexpression of *Salvia miltiorrhiza* 4-hydroxy-3-methylbut-2-enyl diphosphate reductase (HDR), a key enzyme from MEP pathway, resulted in enhanced tanshinones in transgenic *Salvia miltiorrhiza* hairy root culture [43]. Interestingly, cloning and overexpression of another key gene, the hydroxymethylglutaryl-CoA reductase (HMGR), belonging to the MVA pathway, was also found to increase tanshinones production in transgenic *Salvia miltiorrhiza* hairy root culture [44]. Kai et al. reported for the first time a complex metabolic engineering push-pull strategy for enchanting tanshinones biosynthesis by transgenic *Salvia miltiorrhiza* hairy root culture simultaneously overexpressing tree genes (DXS and GGPS from MEP pathway) and (HMGR from MVA pathway) and achieved remarkable increase in tanshinones production (about 4.74 times higher than the control) [49]. All these examples demonstrated the unlimited potential of metabolite engineering as a tool for improvement of secondary metabolites production by *Salvia* plants, but also outline the necessity of deep understanding of the existing relationships and regulatory mechanisms hidden behind the complexity of biosynthetic pathways.

5 Role of *Salvia* Spp. in Development of Synthetic Biology Platforms

Over the past decade there has been a large increase in the amount of research focused on development of synthetic biology platforms and combinatorial biosynthesis for discovery new drug [41, 50]. Whereas the metabolic engineering could be a powerful tool for development of transgenic plants with improved secondary metabolites content, the synthetic biology offers the advantages to transfer the entire biosynthetic pathways and their heterologous expression in fast-growing and easy-to-cultivate microorganisms. Moreover, by applying the methods of combinatorial biosynthesis, enzymes and pathways from different organisms could be combined to produce new-to-nature natural products, which could significantly expand the structural diversity of pharmaceutically active natural products [51]. Many genes, originated from Salvia plants have been used to design synthetic biology platforms for terpenoids and phenolics production [41, 52–55]. Ingea et al. used wild type yeast *Saccharomyces cerevisiae* to develop an improved strains, with increased monoterpenes and sesquiterpenes productivity and sustaining high terpenoid production levels for an prolonged time period [56]. They used the existing genes HMG2, ERG20 and IDI1 of the yeast sterol biosynthetic pathway, expressed together with two terpenoid synthase genes from *Salvia fruticosa* (cineole synthase 1) and *Salvia pomifera* (sesquiterpene synthase) and found that the modified yeast cells produced a range of terpenoids which are also present in the *Salvia* plant's essential oils [56]. Recently, a modular terpenoid production platform aiming to reconstruct the chemical diversity of labdane-type diterpenoids biosynthesis in yeast using module-specific parts has been proposed (Fig. 3a) [57].

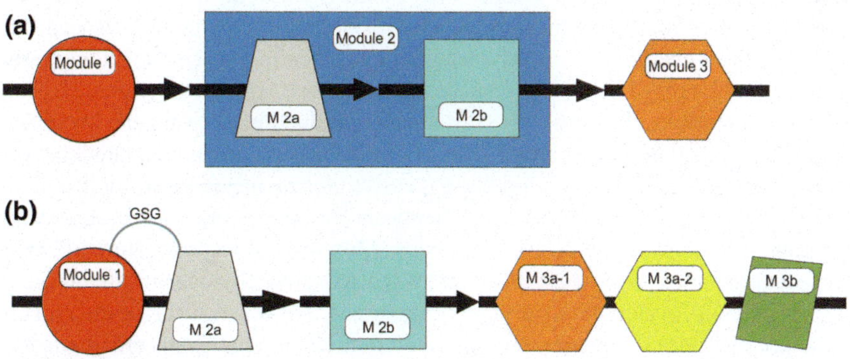

Fig. 3 Schematic representation of the modular design and the platform configuration used for diterpene biosynthesis in yeast: **a** General design of the modular labdane-type diterpenoids production platform; **b** General design of engineered yeast platform for production of carnosic acid-related diterpenoids in three steps with a shorter linker (GSG) between the M 1 and M 2a parts. Adapted from [57, 58]

According to this platform, terpenoid biosynthesis is split into three conceptual modules. Module 1 is assigned to prenyl-diphosphate synthesis. Module 2 consist of two sub-modules, M 2a—responsible for the synthesis of the labdane diphosphate precursor, and M 2b—responsible for the conversion of diphosphate precursor into labdane skeleton. Module 3 is responsible for terpenoid skeleton decoration (Fig. 3a) [57]. Recently, an improved by several modifications platform has been proposed (Fig. 3b) [58]. This platform has included tree modules, Module 1 consisted of a yeast farnesyl diphosphate synthase, Module 2 consist of two sub-modules. Moreover, sub-module M 2a, consist of *S. fruticosa* copalyl diphosphate synthase, and, to facilitate metabolic channeling, it was fused with Module 1 by shorter Gly-Ser repeats linker (GSG) (Fig. 3b) [58]. Sub-module M2b consist of *S. pomifera* miltiradiene synthase and was responsible for terpenoid cyclization to convert copalyl diphosphate to miltiradiene. Module 3 consists of two sub-modules. M 3a consist of various CYP (M 3a-1 and M 3a-2) responsible for terpenoid oxidation events. Sub-module M 3b consists of cytochrome P450 reductase and is responsible for terpenoid skeleton decoration (Fig. 3b) [58].

6 Conclusions and Future Prospects

Recent advances in sequencing and transcriptomic technologies pushed the research on discovery of more and more functional and regulatory genes in a new level. Nevertheless the scantly sequence data for *Salvia* species, found in GenBank databases, the number of cloned and characterized genes continues to grow exponentially. The elucidation of the general biosynthetic pathways, leading to accumulation of pharmaceutically important secondary metabolites, such as terpenoids (monoterpenes, diterpenes, triterpenes) polyphenols (rosmarinic and salvianolic acids, flavonoids), sterols, etc. in *Salvia* species paved the way for development of highly productive plants with improved phytochemicals profiles. The advanced methods, such as metabolite engineering, synthetic biology and combinatorial biosynthesis now can be used to develop highly productive, economically feasible platforms for sustainable production of valuable medicinal phytochemicals with *Salvia* origin. The synthetic biology discovered new horizons in commercialization of traditionally expensive *Salvia* metabolites, by providing possibility to produce complex phytochemicals by using simple and fast growing expression systems, such as microbial cells. Moreover, the combinatorial biosynthesis can be used to produce new, often undiscovered phytochemicals, which is very important for new drugs discovery programs. However, to favor the most of these technologies, sequence information of different functional and regulatory genes from more members *Salvia* genus should be collected. This will help for better understanding the specific biosynthetic and regulatory mechanisms existing in this remarkable genus.

References

1. Scholey A, Camfield D, Owen L, Pipingas A, Stough C (2011) Functional foods and cognition In: Saarela M (ed) Functional Foods, 2nd edn. Woodhead Publishing, pp 277–308. https://doi.org/10.1533/9780857092557.2.277
2. Berdahl DR, McKeague J (2015) Rosemary and sage extracts as antioxidants for food preservation. In: Shahidi F (ed) Handbook of antioxidants for food preservation. Woodhead Publishing, pp 177–217. https://doi.org/10.1016/B978-1-78242-089-7.00008-7
3. Doğan S, Turan P, Doğan M, Arslan O, Alkan M (2007) Variations of peroxidase activity among Salvia species. J Food Eng 79(2):375–382. https://doi.org/10.1016/j.jfoodeng.2006.02.001
4. Panda H (2005) Cultivation and utilization of aromatic plants. Asia Pacific Business Press INC., Delhi
5. Tissier A (2012) Trichome specific expression: promoters and their applications. In: Çiftçi YÖ (ed) Transgenic plants-advances and limitations. InTech, pp 353–378. https://doi.org/10.5772/32101
6. Ullah R, Nadeem M, Khalique A, Imran M, Mehmood S, Javid A, Hussain J (2016) Nutritional and therapeutic perspectives of Chia (*Salvia hispanica* L.): a review. J Food Sci Technol 53(4):1750–1758. https://doi.org/10.1007/s13197-015-1967-0
7. Zhou L, Zuo Z, Chow MSS (2005) Danshen: an overview of its chemistry, pharmacology, pharmacokinetics, and clinical use. J Clin Pharmacol 45(12):1345–1359. https://doi.org/10.1177/0091270005282630
8. Bücheler R, Gleiter CH, Schwoerer P, Gaertner I (2005) Use of nonprohibited hallucinogenic plants: increasing relevance for public health? A case report and literature review on the consumption of *Salvia divinorum*. Pharmacopsychiatry 38(1):1–5. https://doi.org/10.1055/s-2005-837763
9. United States Food and Drug Administration (FDA) (2011) 21 CFR Part 582.20. Substances Generally Recognized as Safe. National Archives and Records Administration, Washington, DC
10. European Directorate for the Quality of Medicines (2010) European Pharmacopoeia, 7th edn (PhEur 7.0). European Pharmacopoeia Commission, Strasbourg, France
11. Mohd Ali N, Yeap SK, Ho WY, Beh BK, Tan SW, Tan SG (2012) The promising future of Chia, *Salvia hispanica* L. J Biomed Biotechnol 2012:9. https://doi.org/10.1155/2012/171956
12. Benkeblia N (2017) Phytonutritional improvement of crops. Wiley-Blackwell, NJ
13. Walker JB, Sytsma KJ, Treutlein J, Wink M (2004) Salvia (Lamiaceae) is not monophyletic: implications for the systematics, radiation, and ecological specializations of Salvia and tribe Mentheae. Am J Bot 91(7):1115–1125. https://doi.org/10.3732/ajb.91.7.1115
14. Will M, Claßen-Bockhoff R (2017) Time to split Salvia s.l. (Lamiaceae)—new insights from old world Salvia phylogeny. Mol Phylogenet Evol 109:33–58. https://doi.org/10.1016/j.ympev.2016.12.041
15. Wenping H, Yuan Z, Jie S, Lijun Z, Zhezhi W (2011) De novo transcriptome sequencing in Salvia miltiorrhiza to identify genes involved in the biosynthesis of active ingredients. Genomics 98(4):272–279. https://doi.org/10.1016/j.ygeno.2011.03.012
16. Yang L, Ding G, Lin H, Cheng H, Kong Y, Wei Y, Fang X, Liu R, Wang L, Chen X, Yang C (2013) Transcriptome analysis of medicinal plant *Salvia miltiorrhiza* and Identification of Genes Related to Tanshinone Biosynthesis. PLoS ONE 8(11):e80464. https://doi.org/10.1371/journal.pone.0080464
17. Gao W, Sun H-X, Xiao H, Cui G, Hillwig ML, Jackson A, Wang X, Shen Y, Zhao N, Zhang L, Wang X-J, Peters RJ, Huang L (2014) Combining metabolomics and transcriptomics to characterize tanshinone biosynthesis in *Salvia miltiorrhiza*. BMC Genom 15(1):73. https://doi.org/10.1186/1471-2164-15-73

18. Song Z, Guo L, Liu T, Lin C, Wang J, Li X (2017) Comparative RNA-sequence transcriptome analysis of phenolic acid metabolism in *Salvia miltiorrhiza*, a traditional Chinese medicine model plant. Int J Genomics 2017:10. https://doi.org/10.1155/2017/9364594
19. Qian J, Song J, Gao H, Zhu Y, Xu J, Pang X, Yao H, Sun C, Xe Li, Li C, Liu J, Xu H, Chen S (2013) The complete chloroplast genome sequence of the medicinal plant *Salvia miltiorrhiza*. PLoS ONE 8(2):e57607. https://doi.org/10.1371/journal.pone.0057607
20. Xu H, Song J, Luo H, Zhang Y, Li Q, Zhu Y, Xu J, Li Y, Song C, Wang B, Sun W, Shen G, Zhang X, Qian J, Ji A, Xu Z, Luo X, He L, Li C, Sun C, Yan H, Cui G, Li X, Xe Li, Wei J, Liu J, Wang Y, Hayward A, Nelson D, Ning Z, Peters Reuben J, Qi X, Chen S (2016) Analysis of the genome sequence of the medicinal plant *Salvia miltiorrhiza*. Mol Plant 9 (6):949–952. https://doi.org/10.1016/j.molp.2016.03.010
21. Xu Z, Luo H, Ji A, Zhang X, Song J, Chen S (2016) Global identification of the full-length transcripts and alternative splicing related to phenolic acid biosynthetic genes in *Salvia miltiorrhiza*. Front Plant Sci 7(100). https://doi.org/10.3389/fpls.2016.00100
22. Ho H-S, Vishwakarma RK, Chen EC-F, Tsay H-S (2013) Activation tagging in *Salvia miltiorrhiza* can cause increased leaf size and accumulation of tanshinone I and IIA in its roots. Botanical Studies 54:37. https://doi.org/10.1186/1999-3110-54-37
23. Wu Y, Liu C, Kuang J, Ge Q, Zhang Y, Wang Z (2014) Overexpression of SmLEA enhances salt and drought tolerance in *Escherichia coli* and *Salvia miltiorrhiza*. Protoplasma 251 (5):1191–1199. https://doi.org/10.1007/s00709-014-0626-z
24. Liu Y, Sun G, Zhong Z, Zheng X, Deng K (2017) Overexpression of SAMDC gene from *Salvia miltiorrhiza* enhances drought tolerance in transgenic tobacco (*Nicotiana tabacum*). J Agric Biotechnol 25(5):729–738
25. Wei T, Deng K, Liu D, Gao Y, Liu Y, Yang M, Zhang L, Zheng X, Wang C, Song W, Chen C, Zhang Y (2016) Ectopic expression of DREB transcription factor, AtDREB1A, confers tolerance to drought in transgenic *Salvia miltiorrhiza*. Plant Cell Physiol 57(8): 1593–1609. https://doi.org/10.1093/pcp/pcw084
26. Wei T, Deng K, Gao Y, Liu Y, Yang M, Zhang L, Zheng X, Wang C, Song W, Chen C, Zhang Y (2016) Arabidopsis DREB1B in transgenic *Salvia miltiorrhiza* increased tolerance to drought stress without stunting growth. Plant Physiol Biochem 104(Supplement C):17–28. https://doi.org/10.1016/j.plaphy.2016.03.003
27. Wei T, Deng K, Zhang Q, Gao Y, Liu Y, Yang M, Zhang L, Zheng X, Wang C, Liu Z, Chen C, Zhang Y (2017) Modulating AtDREB1C expression improves drought tolerance in *Salvia miltiorrhiza*. Front Plant Sci 8(52). https://doi.org/10.3389/fpls.2017.00052
28. Liu Y, Sun G, Zhong Z, Ji L, Zhang Y, Zhou J, Zheng X, Deng K (2017) Overexpression of AtEDT1 promotes root elongation and affects medicinal secondary metabolite biosynthesis in roots of transgenic *Salvia miltiorrhiza*. Protoplasma 254(4):1617–1625. https://doi.org/10.1007/s00709-016-1045-0
29. Li C, Lu S (2014) Genome-wide characterization and comparative analysis of R2R3-MYB transcription factors shows the complexity of MYB-associated regulatory networks in *Salvia miltiorrhiza*. BMC Genom 15(1):277. https://doi.org/10.1186/1471-2164-15-277
30. Li C, Li D, Shao F, Lu S (2015) Molecular cloning and expression analysis of WRKY transcription factor genes in *Salvia miltiorrhiza*. BMC Genom 16(1):200. https://doi.org/10.1186/s12864-015-1411-x
31. Zhang S, Ma P, Yang D, Li W, Liang Z, Liu Y, Liu F (2013) Cloning and characterization of a putative R2R3 MYB transcriptional repressor of the rosmarinic acid biosynthetic pathway from *Salvia miltiorrhiza*. PLoS ONE 8(9):e73259. https://doi.org/10.1371/journal.pone.0073259
32. Yang N, Zhou W, Su J, Wang X, Li L, Wang L, Cao X, Wang Z (2017) Overexpression of smMYC2 increases the production of phenolic acids in *Salvia miltiorrhiza*. Front Plant Sci 8 (1804). https://doi.org/10.3389/fpls.2017.01804
33. Petersen M, Simmonds MSJ (2003) Rosmarinic acid. Phytochemistry 62(2):121–125. https://doi.org/10.1016/S0031-9422(02)00513-7

34. Ma X-H, Ma Y, Tang J-F, He Y-L, Liu Y-C, Ma X-J, Shen Y, Cui G-H, Lin H-X, Rong Q-X, Guo J, Huang L-Q (2015) The biosynthetic pathways of tanshinones and phenolic acids in *Salvia miltiorrhiza*. Molecules 20(9):16235
35. Zhou Y, Sun W, Chen J, Tan H, Xiao Y, Li Q, Ji Q, Gao S, Chen L, Chen S, Zhang L, Chen W (2016) SmMYC2a and SmMYC2b played similar but irreplaceable roles in regulating the biosynthesis of tanshinones and phenolic acids in *Salvia miltiorrhiza.* Sci Rep 6:22852. https://doi.org/10.1038/srep22852
36. Zhang S, Yan Y, Wang B, Liang Z, Liu Y, Liu F, Qi Z (2014) Selective responses of enzymes in the two parallel pathways of rosmarinic acid biosynthetic pathway to elicitors in *Salvia miltiorrhiza* hairy root cultures. J Biosci Bioeng 117(5):645–651. https://doi.org/10.1016/j.jbiosc.2013.10.013
37. Song Z, Li X (2015) Expression profiles of rosmarinic acid biosynthesis genes in two *Salvia miltiorrhiza* lines with differing water-soluble phenolic contents. Ind Crops Prod 71 (Supplement C):24–30. doi:https://doi.org/10.1016/j.indcrop.2015.03.081
38. Zhang S, Li H, Liang X, Yan Y, Xia P, Jia Y, Liang Z (2015) Enhanced production of phenolic acids in *Salvia miltiorrhiza* hairy root cultures by combing the RNAi-mediated silencing of chalcone synthase gene with salicylic acid treatment. Biochem Eng J 103 (Supplement C):185–192. https://doi.org/10.1016/j.bej.2015.07.019
39. Zhang Y, Yan Y-P, Wu Y-C, Hua W-P, Chen C, Ge Q, Wang Z-Z (2014) Pathway engineering for phenolic acid accumulations in *Salvia miltiorrhiza* by combinational genetic manipulation. Metab Eng 21(Supplement C):71–80. https://doi.org/10.1016/j.ymben.2013.10.009
40. Ma Y, Yuan L, Wu B, Xe Li, Chen S, Lu S (2012) Genome-wide identification and characterization of novel genes involved in terpenoid biosynthesis in *Salvia miltiorrhiza*. J Exp Bot 63(7):2809–2823. https://doi.org/10.1093/jxb/err466
41. Arendt P, Pollier J, Callewaert N, Goossens A (2016) Synthetic biology for production of natural and new-to-nature terpenoids in photosynthetic organisms. Plant J 87(1):16–37. https://doi.org/10.1111/tpj.13138
42. Trikka FA, Nikolaidis A, Ignea C, Tsaballa A, Tziveleka L-A, Ioannou E, Roussis V, Stea EA, Božić D, Argiriou A, Kanellis AK, Kampranis SC, Makris AM (2015) Combined metabolome and transcriptome profiling provides new insights into diterpene biosynthesis in *S. pomifera* glandular trichomes. BMC Genom 16:935. https://doi.org/10.1186/s12864-015-2147-3
43. Hao G, Shi R, Tao R, Fang Q, Jiang X, Ji H, Feng L, Huang L (2013) Cloning, molecular characterization and functional analysis of 1-hydroxy-2-methyl-2-(E)-butenyl-4-diphosphate reductase (HDR) gene for diterpenoid tanshinone biosynthesis in *Salvia miltiorrhiza* Bge. f. alba. Plant Physiol Biochem 70(Supplement C):21–32. https://doi.org/10.1016/j.plaphy.2013.05.010
44. Dai Z, Cui G, Zhou S-F, Zhang X, Huang L (2011) Cloning and characterization of a novel 3-hydroxy-3-methylglutaryl coenzyme A reductase gene from *Salvia miltiorrhiza* involved in diterpenoid tanshinone accumulation. J Plant Physiol 168(2):148–157. https://doi.org/10.1016/j.jplph.2010.06.008
45. Božić D, Papaefthimiou D, Brückner K, de Vos RCH, Tsoleridis CA, Katsarou D, Papanikolaou A, Pateraki I, Chatzopoulou FM, Dimitriadou E, Kostas S, Manzano D, Scheler U, Ferrer A, Tissier A, Makris AM, Kampranis SC, Kanellis AK (2015) Towards elucidating carnosic acid biosynthesis in lamiaceae: functional characterization of the three first steps of the pathway in *Salvia fruticosa* and *Rosmarinus officinalis*. PLoS ONE 10(5): e0124106. https://doi.org/10.1371/journal.pone.0124106
46. Bai Z, Xia P, Wang R, Jiao J, Ru M, Liu J, Liang Z (2017) Molecular cloning and characterization of five SmGRAS genes associated with tanshinone biosynthesis in *Salvia miltiorrhiza* hairy roots. PLoS ONE 12(9):e0185322. https://doi.org/10.1371/journal.pone.0185322

47. Zhang X, Luo H, Xu Z, Zhu Y, Ji A, Song J, Chen S (2015) Genome-wide characterisation and analysis of bHLH transcription factors related to tanshinone biosynthesis in *Salvia miltiorrhiza*. Sci Rep 5:11244. https://doi.org/10.1038/srep11244
48. Ma Y, Ma X-H, Meng F-Y, Zhan Z-L, Guo J, Huang L-Q (2016) RNA interference targeting CYP76AH1 in hairy roots of *Salvia miltiorrhiza* reveals its key role in the biosynthetic pathway of tanshinones. Biochem Biophys Res Commun 477(2):155–160. https://doi.org/10.1016/j.bbrc.2016.06.036
49. Kai G, Xu H, Zhou C, Liao P, Xiao J, Luo X, You L, Zhang L (2011) Metabolic engineering tanshinone biosynthetic pathway in *Salvia miltiorrhiza hairy root cultures*. Metab Eng 13 (3):319–327. https://doi.org/10.1016/j.ymben.2011.02.003
50. Yadav VG, De Mey M, Giaw Lim C, Kumaran Ajikumar P, Stephanopoulos G (2012) The future of metabolic engineering and synthetic biology: Towards a systematic practice. Metab Eng 14(3):233–241. https://doi.org/10.1016/j.ymben.2012.02.001
51. Sun H, Liu Z, Zhao H, Ang EL (2015) Recent advances in combinatorial biosynthesis for drug discovery. Drug Des Dev Ther 9:823–833. https://doi.org/10.2147/dddt.s63023
52. Kampranis SC, Makris AM (2012) Developing a yeast cell factory for the production of terpenoids. Comput Struct Biotechnol J 3:e201210006. https://doi.org/10.5936/csbj.201210006
53. Zebec Z, Wilkes J, Jervis AJ, Scrutton NS, Takano E, Breitling R (2016) Towards synthesis of monoterpenes and derivatives using synthetic biology. Curr Opin Chem Biol 34 (Supplement C):37–43. https://doi.org/10.1016/j.cbpa.2016.06.002
54. Zhang H, Liu Q, Cao Y, Feng X, Zheng Y, Zou H, Liu H, Yang J, Xian M (2014) Microbial production of sabinene—a new terpene-based precursor of advanced biofuel. Microb Cell Fact 13(1):20. https://doi.org/10.1186/1475-2859-13-20
55. Ignea C, Pontini M, Maffei ME, Makris AM, Kampranis SC (2014) Engineering monoterpene production in yeast using a synthetic dominant negative geranyl diphosphate synthase. ACS Synth Biol 3(5):298–306. https://doi.org/10.1021/sb400115e
56. Ignea C, Cvetkovic I, Loupassaki S, Kefalas P, Johnson CB, Kampranis SC, Makris AM (2011) Improving yeast strains using recyclable integration cassettes, for the production of plant terpenoids. Microb Cell Fact 10(1):4. https://doi.org/10.1186/1475-2859-10-4
57. Ignea C, Athanasakoglou A, Ioannou E, Georgantea P, Trikka FA, Loupassaki S, Roussis V, Makris AM, Kampranis SC (2016) Carnosic acid biosynthesis elucidated by a synthetic biology platform. Proc Natl Acad Sci 113(13):3681–3686. https://doi.org/10.1073/pnas.1523787113
58. Ignea C, Athanasakoglou A, Andreadelli A, Apostolaki M, Iakovides M, Stephanou EG, Makris AM, Kampranis SC (2017) Overcoming the plasticity of plant specialized metabolism for selective diterpene production in yeast. Sci Rep 7(1):8855. https://doi.org/10.1038/s41598-017-09592-5

Microbial Diversity of Lactic Acid Bacteria in Genus *Salvia*

Tsvetanka Teneva-Angelova and Dora Beshkova

Abstract In this chapter we provide evidences that genus *Salvia* is alternative ecological niche for isolation of lactic acid microbiota with new metabolic activities and characteristics, with subsequent potential possibility for in situ cultivation in milk, in order to produce fermented dairy products with beneficial effects. For the first time we describe *Salvia scabiosifolia* Lam., *Salvia ringens* Sibth. & Sm., *Salvia officinalis* L. and *Salvia blepharophylla* Brandegee ex Epling as potential sources for isolation of various types of plant-derived lactic acid bacteria, belonging to the genus *Lactococcus*, *Enterococcus* and *Streptococcus*, unlike the species *Salvia tomentosa* Mill.

Keywords Genus *Salvia* · Microbial diversity · Lactic acid bacteria Isolation · 16S rDNA

1 Introduction

In recent years it has been observed an increased interest in lactic acid bacteria isolated from alternative non-dairy sources, because they possess a broad metabolic profile and unique flavor-forming activities [1]. We are evidencing a continuous challenge for diversified food demand by using new food ingredients and starter combinations, isolated from the natural habitats, carrying specific health benefits and/or new sensory characteristics [2].

Medicinal plants are various and unstudied ecosystem for isolation of lactic acid bacteria [3, 4]. More than 200 years ago shepherds from the Rhodope Mountains region of Bulgaria have prepared a yogurt-type product by crushing the roots of cornel (*Cornus mas*) and afterwards they have mixed the obtained juice with sheep's milk. Katrandjiev [5] used other plants (*Berberis vulgaris* and *Paliurus*

T. Teneva-Angelova · D. Beshkova (✉)
Laboratory of Applied Biotechnologies, The Stephan Angeloff Institute of Microbiology, Bulgarian Academy of Sciences, 139 Ruski Blvd, 4000 Plovdiv, Bulgaria
e-mail: beshkova@yahoo.com

V. Georgiev and A. Pavlov (eds.), *Salvia Biotechnology*,
https://doi.org/10.1007/978-3-319-73900-7_11

aculeatus) for the same purpose. Girginoff [6] pointed out *Matricaria chamomilla*, *Prunus spinosa*, Alfalfa (*Medicago sativa*) as possible sources for isolation of yogurt-starter bacteria.

Salvia species are one of the most widespread and used medicinal plants by humans. *Salvia* is one of the largest genera of flowering plants, spread almost every where around the world, except within Australia [7]. The species belonging to this genus are spread mainly within three areas—in Europe (along the region of the Mediterranean), in Southeast Asia, as well as in Central and South America [8]. In Bulgaria, they are mainly found along the Black Sea coast, Central and Eastern Balkan Mountains, Strandzha Mountains, the Eastern Rhodopes as well as throughout the Thracian Plain, up to about 700 m altitude [9]. The *Salvia* plant has been known and used for centuries. Its name comes from the Latin "*Salveo*" which means health or healing. Exists extensive information about *Salvia* in regard to its healing properties and prophylactic effect. Over the years, people have attributed almost magical powers to *Salvia*, like healing all kinds of diseases and it has also been considered to be a miraculous herb called "The herb of centenarians". It is a carrier of various bioactive components [10–14], which possess antimicrobial, anti-allergic, antioxidant, antitumor and immunomodulatory activities, as well as anesthetic effect of the central and peripheral nervous system, moreover there is permission to be used as harmless by humans.

The microbial biodiversity of the *Salvia* genus has not yet been studied and the plant is of scientific interest, being an ecosystem for isolation of various microbial populations. Each particular plant species provides a unique environment in terms of competing microorganisms, natural plant antagonists, as well as accessibility, type and concentration of substrate under various physical factors. These conditions allow growth of a characteristic epiphytic flora, from which originates a population and a chain of fermenting microorganisms, when the plant material is harvested and prepared for fermentation. One of the main factors of epiphytic colonization of bacteria on the leaves is the presence of carbon and other nutrients [15]. Glucose, fructose and sucrose are dominant carbon sources on the leaves' or on the stem's surface of the plant [16].

For the first time it is reported a biodiversity of lactic acid bacteria, isolated from the medicinal plant *Salvia*, belonging to the *Lamiaceae* family.

2 Biodiversity of Lactic Acid Bacteria in Genus *Salvia*

The microbiota of 5 species of genus *Salvia* was examined—*S. scabiosifolia* Lam., *S. ringens* Sibth. & Sm., *S. officinalis* L., *S. tomentosa* Mill., *S. blepharophylla* Brandegee ex Epling (Fig. 1). Four of those species (*S. scabiosifolia*, *S. ringens*, *S. officinalis*, *S. tomentosa*) are typical for Bulgaria. It was used the aerial part of the studied plants (flowers, leaves and stems), which were collected from their natural habitats in Bulgaria or from the experimental field of the Institute of Biodiversity and Ecosystem Research (IBER)—Bulgarian Academy of Sciences (BAS), except

Fig. 1 Species of genus *Salvia*: **a** *S. scabiosifolia* Lam., **b** *S. ringens* Sibth. & Sm., **c** *S. tomentosa* Mill., **d** *S. blepharophylla* Brandegee ex Epling, **e** *S. officinalis* L.

S. blepharophylla, which was taken from the Botanical garden at the Technical University of Dresden, Germany.

Salvia scabiosifolia Lam. (Crimena *Salvia*) is an endemic, protected species included in the Red Data Book of the Republic of Bulgaria with endangered, disappearing and extinct biological species living on the territory of the country, *Salvia ringens* Sibth. & Sm. (Macedonian *Salvia*) and *Salvia tomentosa* Mill. (Large-flowered *Salvia*) have a limited spread, while on the other hand *S. officinalis* L. (Common sage) is widespread throughout the territory of Bulgaria. *S. blepharophylla* Brandegee ex Epling is not common in Bulgaria and therefore it was taken from the Botanical garden of the Technical University in Dresden, Germany. *Salvia scabiosifolia* Lam. was represented with only one single habitat in the region of Polsko Kosovo, Byala municipality, Rousse district, Bulgaria.

The plant material was collected from six regions in Bulgaria and one region in Germany as follows: *S. scabiosifolia* Lam.—Rousse region (natural habitat) and also from Sofia region (IBER—BAS); *S. ringens* Sibth. & Sm.—Varna region and Shumen plateau (natural habitats), Sofia region (IBER—BAS); *S. tomentosa* Mill.—Sofia region (IBER—BAS); *S. officinalis* L.—Eastern Rhodopes (natural habitats) and Sofia region (IBER—BAS); *S. blepharophylla* Brandegee ex Epling—Dresden, Germany.

After isolation and conducted screening using morphological characteristics, catalase test, Gram staining, oxidase test, indole reaction, production of CO_2 from glucose, bacterial isolates, referred to the group of lactic acid bacteria were obtained. The steps for isolation of presumptive lactic acid bacteria are presented on Fig. 2. Selected potential lactobacteria had entirely cocci-shape and showed differences in cell morphology (single, in pairs, in various chains; size 0.3–2.0 μm), as well as in the morphology of the colony [shiny, with different colors (white, yellow and light beige), shape (circular—convex or flat, with entire or undulate margins) and size (1.1–3.1 mm)]. They were Gram-positive, catalase-negative, indole-negative, oxidase-negative and homofermentative.

Genotype-based methods, like 16S rDNA sequencing, are a precise addition to phenotypic methods for identification of lactic acid bacteria, isolated from certain sources [17–23]. 29 strains lactic acid bacteria, isolated from genus *Salvia* were identified using 16S rDNA sequencing. The isolated strains belonged to the species *Streptococcus thermophilus*, *Lactococcus lactis*, *Enterococcus faecium*,

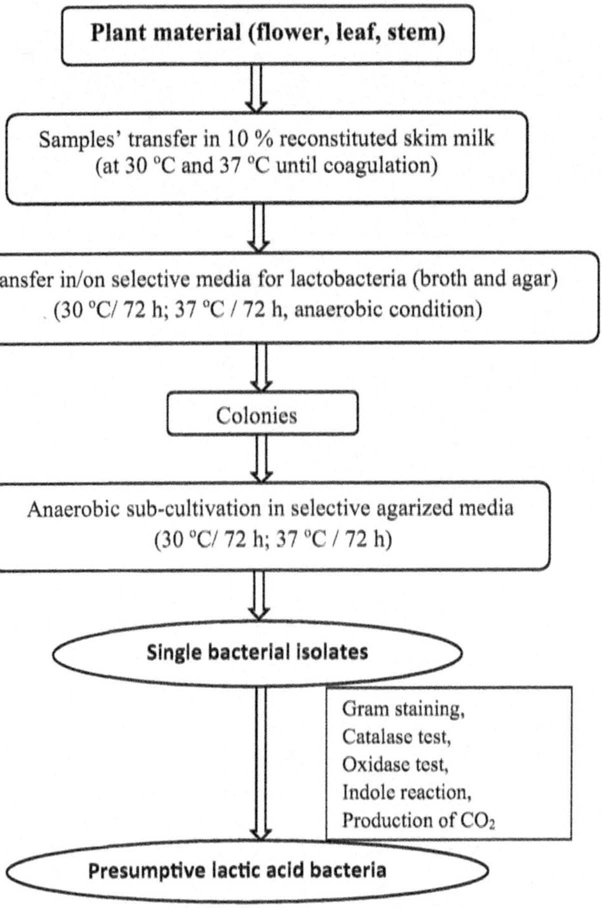

Fig. 2 Scheme for isolation of bacterial isolates (presumptive lactic acid bacteria) from *Salvia* species

Enterococcus mundtii and *Enterococc uscasseliflavus* [24]. In Table 1 are presented isolated strains lactic acid bacteria from the studied *Salvia* species as well as their relevant accession numbers in data base NCBI (National Center for Biotechnology Information).

From the isolated plant-derived lactic acid bacteria 29 are identified as follows: 3 strains *Streptococcus thermophilus* (isolated from *S. blepharophylla* Brandegee ex Epling); 6 strains *Lactococcus lactis* (isolated from *S. scabiosifolia* Lam. and *S. ringens* Sibth. & Sm.); 13 strains *Enterococcus faecium* and 4 strains *Enterococcus mundtii* (isolated from *S. scabiosifolia* Lam., *S. ringens* Sibth. & Sm. and *S. officinalis* L.) and also 3 strains *Enterococcus casseliflavus* (isolated from *S. ringens* Sibth. & Sm.). The flowers of the plant turned out to be the most favorite part for habitation of lactic acid bacteria (18 strains), followed by the leaves

Table 1 Plant-derived lactic acid bacteria isolated from *Salvia* species

Salvia plant species	Plant-derived lactic acid bacteria	Accession numbers[a]
S. scabiosifolia Lam.	*Lc. lactis* SsfRU61[b]	KR265160.1
	Lc. lactis SsfRU62	KR265161.1
	E. faecium SsfIB21	KR265154.1
	E. faecium SsfIB22	KR265155.1
	E. faecium SsfIB32	KR265157.1
	E. faecium SsfIB33	KR265158.1
	E. faecium SsfIB34	KR265159.1
	E. mundtii SsfIB24	KR265156.1
S. ringens Sibth. & Sm.	*Lc. lactis* SrfV181	KR265147.1
	Lc. lactis SrfV183	KR265148.1
	Lc. lactis SrfIB232	KR265145.1
	Lc. lactis SrfIB233	KR265146.1
	E. faecium SrsV161	KR265153.1
	E. mundtii SrsV151	KR265151.1
	E. casseliflavus SrlSH132	KR265149.1
	E. casseliflavus SrlSH141	KR265150.1
	E. casseliflavus SrlSH144	KR265151.1
S. officinalis L.	*E. faecium* SofIB271	KR265136.1
	E. faecium SofIB277	KR265137.1
	E. faecium SofIB279	KR265138.1
	E. faecium SolIB284	KR265139.1
	E. faecium SosIB301	KR265142.1
	E. faecium SosIB312	KR265143.1
	E. faecium SosIB328	KR265144.1
	E. mundtii SolIB285	KR265140.1
	E. mundtii SolIB288	KR265141.1
S. blepharophylla Brandegee ex Epling	*S. thermophilus* SbfGER352	KR265133.1
	S. thermophilus SbfGER373	KR265134.1
	S. thermophilus SbfGER401	KR265135.1
S. tomentosa Mill.	Non isolated	

[a]Accession numbers in NCBI [24]

[b]The abbreviation of the isolated lactic acid strains from the relevant plant samples includes: initials of the relevant plant species; an index indicating the location of the plant species or an initial of a nearby location to the region in question in Bulgaria, or abbreviated spelling of Germany (For example SsfRU61: Ss—*Salvia scabiosifolia*, f—flower, RU—Rousse, 61—number of colony)

(6 strains) and by the stem (5 strains). Lactobacteria have not been isolated only from the species *S. tomentosa* Mill.

In the available scientific literature, there are no data on scientific investigations about *Salvia* as a source for isolation of lactic acid bacteria, and this is the first research team, who studies their biodiversity in the *Salvia* genus. Lactic acid bacteria, isolated from other herbs have been tested by different scientific groups, proving their biodiversity as well as thatevery single plant species provides unique conditions for growing of different lactobacteria types. Michaylova et al. [25] have found out that the following plant species—*Calendula officinalis*, *Capsella bursapastoris*, *Chrysanthemum*, *Cichorium intybus*, *Colchicum*, *Cornus mas*, *Dianthus*, *Hedera*, *Nerium oleander*, *Plantago lanceolata*, *Rosa* and *Tropaeolum*are appropriate sources for *S. thermophilus* isolation, as meanwhile the species *Calendula officinalis*, *Cornus mas*, *Galantus nivalis* and *Prunus spinosa* are appropriate for the *Lb. bulgaricus* isolation. Other genus *Lactobacillus* representatives have also been isolated from the herb *Phyllanthus niruri* [26], *Lb. paraplantarum* from *Camellia sinensis* leaves [27], and *Lb. rhamnosus* from *Panax gingseng* C.A. Meyer [28] and also from the herb *Hottuynia Cordata* Thunb. [29]. *Lb. curvatus*, *P. pentosaceus* and *Lc. lactis* have been isolated from the leaves of a Malaysian herb *Polygonum minus* [21]. Magnusson et al. [30] have isolated the representatives of genera *Lactobacillus* and *Pediococcus*, as well as *E. hirae* from the following medicinal plants: *Hepatica*, *Coltsfoot*, dandelion and clover. It has been isolated *S. thermophilus* and *E. faecium* from different parts of plants *Geranium sanguimeum* L. and *Hypericum perforatum* L., in Bulgaria [28].

The above mentioned results confirm the thesis, that medicinal plants are a suitable alternative source for isolation of lactic acid microflora.

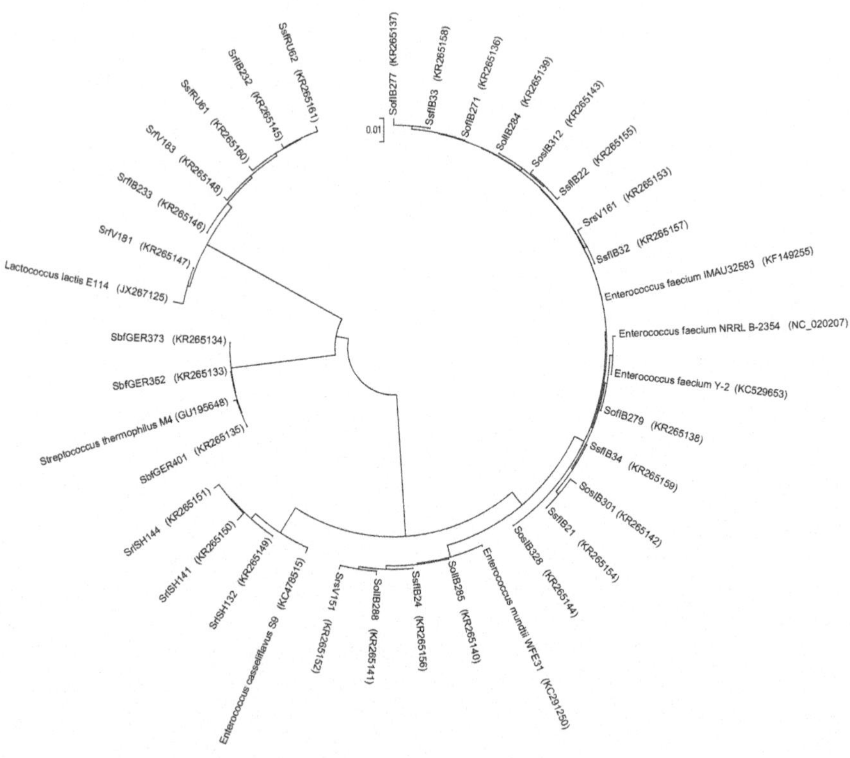

Fig. 3 Phylogenetic tree based on the 16S rDNA gene sequences of identified lactic acid bacteria, isolated from medicinal plants *S. scabiosifolia* Lam., *S. ringens* Sibth. & Sm. and *S. officinalis* L. and *S. blepharophylla* Brandegee ex Epling

Using software Mega 4.1 and basing on data derived from carried out 16S rDNA gene sequence analysis, it was created a phylogenetic tree, on which are shown the relationships among the studied strains (Fig. 3). The phylogenetic tree demonstrates the diversity of the isolated lactic acid bacteria from plant origin, which belongs to the following genera *Enterococcus*, *Streptococcus* and *Lactococcus*, and it also shows the phylogenetic position of the strains, which are divided at the same time in several clusters and sub-clusters.

3 Characterization of Plant-Derived Lactic Acid Bacteria

3.1 Acid-Producing Activity

A significant characteristic of lactic acid bacteria is their acid-producing activity, i.e. production of lactic acid (LA) during their homofermentative or heterofermentative metabolism. The data for isomeric forms of LA produced by plant lactobacteria showed that 82.8% synthesize L(+)-LA, and 17.2% D(−)/L(+)-LA (Fig. 4). Moreover, in the plant-derived lactobacteria, producing both isomeric forms of lactic acid, the presence of L(+)-LA is dominant. There were no representatives producing only D(−)-LA. The highest activity of L(+)-LA synthesis was observed in the strains *S. thermophilus*, isolated from *S. blepharophylla* Brandegee ex Epling (8.34–15.56 g/L). *Lc. lactis*, *E. casseliflavus* and *E. mundtii* strains produced L(+)-LA in range of 4.16–5.71 g/L, while *E. faecium*—4.38–9.22 g/L (Fig. 4). From studied lactic acid bacteria, 3 strains *E. faecium* (SsfIB21, SsfIB22,

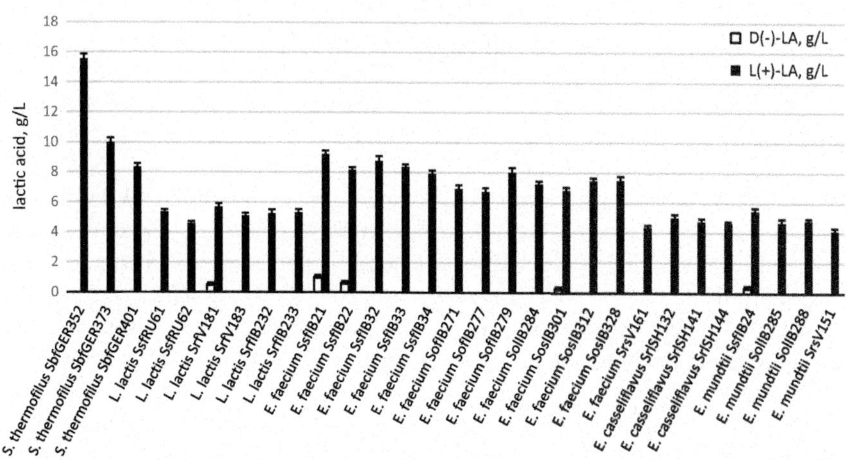

Fig. 4 Production of lactic acid (isomeric forms) by lactic acid bacteria strains, isolated from *S. scabiosifolia* Lam. (Ss), *S. officinalis* L. (So), *S. blepharophylla* Brandegee ex Epling (Sb); *S. ringens* Sibth. & Sm. (Sr): ■—L(+)-lactic acid; □—D(−)-lactic acid

SosIB301), 1 strain *E. mundtii* (SsfIB24) and 1 strain *Lc. lactis* (SrfV181) synthesized also minimal concentrations of D(−)-LA.

The observed concentrations of LA produced by selected plant-derived lactic acid bacteria are comparable to those, produced by lactobacteria, isolated from dairy products—*L. lactis* LL6-5.52 g L(+)-LA/L, *S. thermophilus* ST5-8.79 g L(+)-LA/L, *E. faecium* EF2-6.51 g L(+)-LA/L.

Many researchers were reported about isolated bacterial isolates from various plant matérial, producing only L(+)-LA [17, 31] or producing only D(−)/L(+)-LA [18, 32]. Unlike our data, Tamang et al. [32] and Kostinek et al. [17] have isolated cocci producing only D(−)-LA.

3.2 Physiological Characteristics

Growth parameters (temperature, pH and presence of NaCl) of lactic acid bacteria are essential physiological characteristics of the relevant type, important for their application in various technological fermentation processes. Almost all identified strains, belonging to *Lactococcus lactis*, *Enterococcus faecium*, *Enterococcus mundtii* and *Enterococcu scasseliflavus*, showed very good growth in a wide temperature range (15–45 °C), good and weak growth at 4 °C and absence of growth at 55 °C; except the strains of *Streptococcus thermophilus*, which demonstrate a weak growth at temperature of 4 and 15 °C and a very good growth in the range of 30–45 °C.

The pH range for growth of plant-derived lactic acid bacteria of genus *Enterococcus* was extended, i.e. very good growth was obtained in the range of 7.0–9.6, good growth at pH 5.0, and absence of growth at pH 3.0. Strains of *Streptococcus thermophilus* and *Lactococcus lactis* had very good growth in the range of 7.0–8.0, good growth at pH 5.0 and pH 9.6, and absence of growth at pH 3.0.

The identified strains of *Enterococcus faecium* and *Enterococcus mundtii* showed very good growth in a medium, containing NaCl, at concentrations from 3.0 to 6.5%, good growth at 8.5% NaCl and no growth at 10% NaCl. The *Enterococcus casseliflavus* strains showed good growth in range from 3.0 to 8.5% NaCl and no growth at 10% NaCl. In regard to *Lactococcus lactis* strains has been observed expressed strain dependencein terms of haloolerance, with strain growth ranging from 3.0 to 6.5% NaCl, showing very good, good or weak growth; while at 8.5 and 10% NaCl—no growth. The growth of strains *Streptococcus thermophilus* was inhibited in the presence of NaCl, throughout the whole examined range (3.0–10.0% NaCl).

Analogous temperature growth range (15–45 °C) for lactococci, isolated from fermented tea leaves Tanasupawat et al. [18] and (10–45 °C) for lactococci from the *Polygonum minus* herb Baradaran et al. [21] were reported. The same authors found out that the studied isolates grew well within a pH range of 3.0–7.0 and 4.0–8.5, which corresponds to the results reported by Tamang et al. [32] for lactococci from other plant sources. Tamang et al. [32], Baradaran et al. [21] and Alemayehu et al. [23] determined a high level of halotolerance (10% NaCl and 6.5% NaCl, respectively), but Kimoto et al. [31] and Tanasupawat et al. [18] proved a lower level (4.0% NaCl) for lactococci, isolated from different plant sources.

The studied plant-derived lactic acid bacteria showed better growth characteristics in a broader range of temperature, pH and halotolerance in comparison to those of lactic acid bacteria with dairy origin (very good growth of: *L. lactis* LL6 at 30–37 °C, pH 7.0, 3.0% NaCl; *S. thermophilus* ST5 at 30–45 °C, pH 7.0, no growth in presence of NaCl; *E. faecium* EF2 at 30–45 °C, pH 7.0–9.6, 3.0–4.5% NaCl).

3.3 Resistant Profile of Plant-Derived Lactic Acid Bacteria Against Herb Extracts

For centuries, it has been claimed that extracts from many plant species, which have been and still are used in foods and beverages, had beneficial health effects. It's important to examine the effectiveness of isolated plant-derived lactic acid bacteria for survival in presence of plant extracts of medicinal plants *Salvia officinalis* L., *Geranium sanguineum* L. and *Hypericum perforatum* L., aiming their incorporation into starter communities (in the presence of plant extracts) in order to obtain new fermented dairy products with health benefits.

It has been studied the behavior of plant-derived lactic acid bacteria regarding hydroethanolic extracts (70% ethanol), both from the plant species, from which they were isolated (*Salvia* species) and from other plant extracts (*Geranium sanguineum* L. and *Hypericum perforatum* L.), dissolved in 30% ethanol, in the range 0.05–500 mg/mL. Essential in the selection of the test plants was their applicability in the food industry, as well as the fact that they were carriers of different biological activities. All isolated strains of *Salvia* were tested against an extract only of *Salvia officinalis* L. (Common sage), which has been used in medicine and cooking for centuries and is widespread.

In Fig. 5 are presented the maximum concentrations of plant extracts, which did not inhibit the development of the studied plant-derived lactic acid bacteria. *S. thermophilus* was resistant up to 25–30 mg/mL of *Salvia* extract, but *Lc. lactis* up to 40–50 mg/mL. In *E. faecium* strains was observed a different behavior to the studied extract—resistance from 30 to 120 mg/mL, while *E. mundtii* and *E. casseliflavus* were resistant to concentrations 100 mg/mL. The behavior of the isolated strains of the 3 genera was similar in terms of the extracts from *Geranium sanguineum* L. (10–30 mg/mL). In *Hypericum perforatum* L. extracts, the resistance of strains from the genera *Lactococcus* and *Enterococcus* was up to 0.1–0.5 mg/mL, while *S. thermophilus* strains have been inhibited at lower concentrations (0.05 mg/mL).

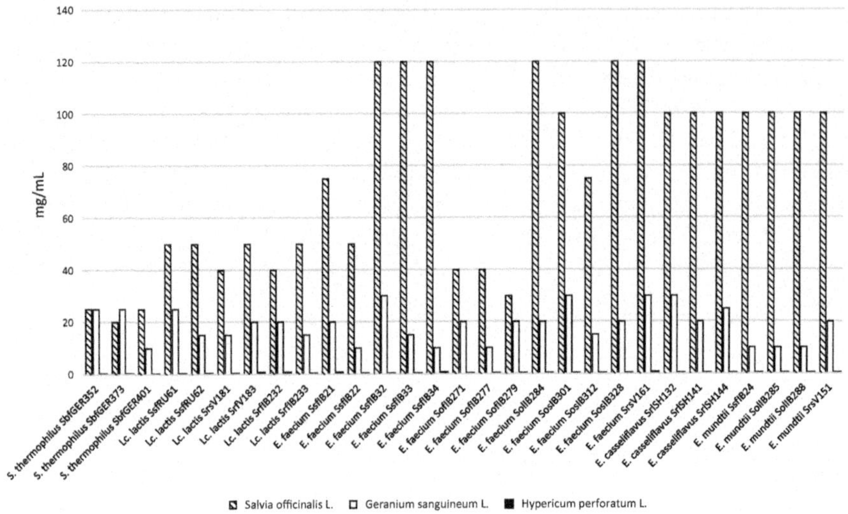

Fig. 5 Resistance of plant-derived lactic acid bacteria, isolated from genus *Salvia*, against hydroethanolic extracts from: ▧—*Salvia officinalis* L., □—*Geranium sanguineum* L. and ■—*Hypericum perforatum* L.

The behavior of plant-derived lactic acid bacteria to hydroethanolic extracts dissolved in water has also been studied, in regard their application during the technological process for production of fermented foods. The results obtained showed that dissolved in water extracts inhibited lactic acid bacteria at higher concentrations. Strains of the *Salvia officinalis* extract were inhibited at the following concentrations: *Streptococcus thermophilus*—150 mg/mL, *Lactococcus lactis* 100–300 mg/mL, while strains of the genus *Enterococcus*—200–500 mg/mL. In extracts of *Hypericum perforatum* L. the resistance of strains of *Streptococcus*, *Lactococcus* and *Enterococcus* was in the range 0.1–10 mg/mL. In *Geranium sanguineum* L. extracts, dissolved in water, was observed inhibition (up to 15–40 mg/mL) comparable to the inhibition of the same extract dissolved in 30% ethanol.

4 Conclusions and Future Prospects

For the first time is presented the biodiversity of lactic acid bacteria from *Salvia* genus. From the studies species (*S. scabiosifolia* Lam., *S. officinalis* L., *S. blepharophylla* Brandegee ex Epling; *S. ringens* Sibth. & Sm. and *S. tomentosa* Mill.) have been isolated representatives from the genera *Lactococcus*, *Streptococcus* and *Enterococcus*. Most representatives of lactic acid bacteria were isolated from *S. ringens* Sibth. & Sm.—*Lactococcus lactis*, *Enterococcus faecium*, *Enterococcus mundtii* and *Enterococcus casseliflavus*. *S. scabiosifolia* Lam. was a source for isolation of *Lactococcus lactis*, *Enterococcus faecium*, and *Enterococcus mundtii*; *S. officinalis* L.—*Enterococcus faecium* and *Enterococcus mundtii*; *S. blepharophylla* Brandegee ex Epling—only a source of *Streptococcus thermophilus*. From *S. tomentosa* Mill. have not been isolated lactic acid bacteria. The flowers of the studied plants appeared to be the most suitable part of the plants for isolation of lactic acid bacteria, in comparison to the rest aerial plant parts (leaves and stems).

The isolated lactic acid bacteria exhibited better physiological and biochemical characteristics (lactic acid production, temperature, pH, and halotolerance) in comparison to the lactic acid bacteria of milk origin, which is an advantage for their future incorporation in starter communities for the production of fermented foods.

The established resistance to plant extracts allows the same to be included during the technological process, in order to obtain functional foods by incorporating bioactive components (flavonoids, phenols and essential oils) during the cultivation of new starter communities, which are carriers of specific health benefits of unique flavour-forming activities.

A direct consequence of the studied microbial biodiversity will be the ability for modeling the microbial communities with desirable biological activities and a specific impact on a milk matrix, in order to be obtained a balanced and healthy dietary-food regime. Studied plant-derived strains, with desirable activities and possibility for growing in presence of plant extracts (carriers of biologically active substances), have perspective for future application in tailor-made health food.

References

1. Van Hylckama Vlieg JE, Rademaker JL, Bachmann H, Molenaar D, Kelly WJ, Siezenm RJ (2006) Natural diversity and adaptive responses of *Lactococcus lactis*. Curr Opin Biotechnol 17:183–190. https://doi.org/10.1016/j.copbio.2006.02.007
2. Teusink B, Smid EJ (2006) Modeling strategies for the industrial exploitation of lactic acid bacteria. Nat Rev Microbiol 4:46–56. https://doi.org/10.1038/nrmicro1319
3. Kelly WJ, Davey GP, Ward LJ (1998) Characterization of lactococci isolated from minimally processed fresh fruit and vegetables. Int J Food Microbiol 45:85–92. https://doi.org/10.1016/S0168-1605(98)00135-4
4. Siezen RJ, Starrenburg MJC, Boekhorst J, Renckens B, Molenaar D, van Hylckama Vlieg JET (2008) Genome-scale genotype-phenotype matching of two *Lactococcus lactis* isolates from plants identifies mechanisms of adaptation to the plant niche. Appl Environ Microbiol 74:424–436. https://doi.org/10.1128/AEM.01850-07
5. Katrandjiev K (1962) The Bulgarian yogurt. Publishing House of Bulgarian Academy of Science, Bulgaria (in Bulgarian)
6. Girginoff T (1959) Microbiology of the animal products. State Publishing House for Agriculture Literature, Sofia, p 52 (in Bulgarian)
7. Ali MR, Aboud AS (2010) Antimicrobial activities of aqueous and methanolic extracts from *Salvia officinalis* and *Salix acmophylla* used in the treatment of wound infection isolates. Ibn Al-Haitham J Pure Appl Sci 23:25–38
8. Ulubelen A (2000) Chemical constituents. Terpenoids in the genus *Salvia*. In: Kintzios SE (ed) Sage. The Genus *Salvia*, vol 14. Harwood Academic Publ., Amsterdam, pp 55–68

9. Markova M (1989) *Labiatea*. In: Jordanov D (ed) Flora na Bulgaria 9:236–474 (in Bulgarian)
10. Habibi Z, Eftekhar F, Samiee K, Rustaiyan A (2000) Structure and antibacterial activity of a new labdane diterpenoid from *Salvia leriaefolia*. J Nat Prod 63:270–271. https://doi.org/10.1021/np990287h
11. Nieto M, García EE, Giordano OS, Tonn CE (2000) Icetexane and abietane diterpenoids from *Salvia gilliessi*. Phytochemistry 53:911–915. https://doi.org/10.1016/S0031-9422(99)00480-X
12. Rauter AP, Branco I, Lopes RG, Justino J, Silva VMF, Noronha JP, Cabrita EJ, Brouard I, Bermejo JA (2007) A new lupine triterpenetriol and anticholinesterase activity of *Salvia sclareoides*. Fitoterapia 78:474–481. https://doi.org/10.1016/j.fitote.2007.02.013
13. Lu YR, Foo LY (2002) Polyphenolics of *Salvia*—a review. Phytochemistry 59:117–140. https://doi.org/10.1016/S0031-9422(01)00415-0
14. Marchev A, Ivanov I, Denev P, Nikolova M, Gochev V, Stoyanova A, Pavlov A, Georgiev V (2015) Acetylcholinesterase inhibitory, antioxidant and antimicrobial activities of *Salvia tomentosa* Mill. essential oil. J BioSci Biotechnol 4:219–229
15. Andrews JH, Harris RF (2000) The ecology and biogeography of microorganisms on plant surfaces. Annu Rev Phytopathol 38:145–180. https://doi.org/10.1146/annurev.phyto.38.1.145
16. Mercier J, Lindow SE (2000) Role of leaf surface sugars in colonization of plants by bacterial epiphytes. Appl Environ Microbiol 66:369–374. https://doi.org/10.1128/AEM.66.1.369-374.2000
17. Kostinek M, Specht I, Edward VA, Pinto C, Egounlety M, Sossa C, Mbugua S, Dortu C, Thonart P, Taljaard L, Mengu M, Franz CMAP, Holzapfel WH (2007) Characterisation and biochemical properties of predominant lactic acid bacteria from fermenting cassava for selection as starter cultures. Int J Food Microbiol 114:342–351. https://doi.org/10.1016/j.ijfoodmicro.2006.09.029
18. Tanasupawat S, Pakdeeto A, Thawai C, Yukphan P, Okada S (2007) Identification of lactic acid bacteria from fermented tea leaves (miang) in Thailand and proposals of *Lactobacillus thailandensis* sp. nov., *Lactobacillus camelliae* sp. nov., and *Pediococcussiamensis* sp. nov. J Gen Appl Microbiol 53:7–15
19. Chen Y-S, Wu H, Yanagida F (2010) Isolation and characteristics of lactic acid bacteria isolated from ripe mulberries in Taiwan. Braz J Microbiol 41:916–921
20. Kpikpi EN, Glover RLK, Dzogbefia VP, Nielsen DS, Jakobsen M (2010) Isolation of lactic acid bacteria from kantong, a condiment produced from the fermentation of kapok (*Ceiba pentandra*) seeds and cassava (*Manihot esculentum*) flour. Rep Opin 2:1–7
21. Baradaran A, Foo HL, Sieo CC, Rahim RA (2012) Isolation, identification and characterization of lactic acid bacteria from *Polygonum minus*. Rom Biotechnol Lett 17:7245–7252
22. Venturi M, Guerrini S, Granchi L, Vincenzini M (2012) Typing of *Lactobacillus sanfranciscensis* isolates from traditional sourdoughs by combining conventional and multiplex RAPD-PCR profiles. Int J Food Microbiol 156:122–126. https://doi.org/10.1016/j.ijfoodmicro.2012.03.011
23. Alemayehu D, Hannon JA, McAuliffe O, Ross RP (2014) Characterization of plant-derived lactococci on the basis of their volatile compounds profile when grown in milk. Int J Food Microbiol 172:57–61. https://doi.org/10.1016/j.ijfoodmicro.2013.11.024
24. Teneva-Angelova T, Beshkova D (2015) Genus *Salvia*—ecosystem for isolation of lactic acid bacteria. J Microbiol Biotechnol Food Sci 5:103–108. https://doi.org/10.15414/jmbfs.2015.5.2.103-108
25. Michaylova M, Minkova S, Kimura K, Sasaki T, Isawa K (2007) Isolation and characterization of *Lactobacillus delbrueckii* ssp. *bulgaricus* and *Streptococcus thermophilus* from plants in Bulgaria. FEMS Microbiol Lett 269:160–169. https://doi.org/10.1111/j.1574-6968.2007.00631.x
26. Venugopalan V, Dinesh MS, Geetha KS (2010) Enhancement of antimicrobial potential of *Phyllanthus niruri* by fermentation. J Herb Med Toxicol 4:167–175

27. Gharaei-Fathabad E, Eslamifar M (2011) Isolation and applications of one strain of *Lactobacillus paraplantarum* from tea leaves (*Camellia sinensis*). Am J Food Technol 6:429–434. https://doi.org/10.3923/ajft.2011.429.434
28. Teneva-Angelova T, Beshkova (2016) Non-traditional sources for isolation of lactic acid bacteria. Ann Microbiol 66:449–459. https://doi.org/10.1007/s13213-015-1127-9
29. Nguyen THK, Doan VTT, Ha LD, Nguyen HN (2013) Molecular cloning, expression of minD gene from *Lactobacillus acidophilus* VTCC-B-871 and analyses to identify *Lactobacillus rhamnosus* PN04 from Vietnam *Hottuynia cordata* Thunb. Indian J Microbiol 53:385–390. https://doi.org/10.1007/s12088-013-0384-1
30. Magnusson J, Ström K, Roos S, Sjögren J, Schnürer J (2003) Broad andcomplex antifungal activity among environmental isolates of lactic acid bacteria. FEMS Microbiol Lett 219:129–135. https://doi.org/10.1016/S0378-1097(02)01207-7.1016/
31. Kimoto H, Nomura M, Kobayashi M, Okamoto T, Ohmomo S (2004) Identification and probiotic characteristics of *Lactococcus* strains from plant materials. JARQ 38:111–117. https://doi.org/10.6090/jarq.38.111
32. Tamang JP, Tamang B, Schillinger U, Franz CM, Gores M, Holzapfel WH (2005) Identification of predominant lactic acid bacteria isolated from traditionally fermented vegetable products of the Eastern Himalayas. Int J Food Microbiol 105:347–356. https://doi.org/10.1016/j.ijfoodmicro.2005.04.024

The manufacturer's authorised representative in the EU is Springer Nature Customer Service Centre GmbH, Europaplatz 3, 69115 Heidelberg, Germany. If you have any concerns regarding our products, please contact ProductSafety@springernature.com

Printed and bound by CPI Group (UK) Ltd, Croydon, CR0 4YY
29/07/2026
02180515-0005